Karl Marx

SELECTED WRITINGS

Karl Marx

SELECTED WRITINGS

EDITED BY
DAVID McLELLAN

OXFORD UNIVERSITY PRESS

Oxford University Press, Walton Street, Oxford OX2 6DP

Oxford New York Toronto
Delhi Bombay Calcutta Madras Karachi
Petaling Jaya Singapore Hong Kong Tokyo
Nairobi Dar es Salaam Cape Town
Melbourne Auckland

and associated companies in
Berlin Ibadan

Oxford is a trade mark of Oxford University Press

Published in the United States
by Oxford University Press, New York

Introduction and Compilation © David McLellan 1977

First printed 1977
Reprinted 1978, 1982, 1984, 1985, 1987, 1988, 1990

British Library Cataloguing in Publication Data
Marx, Karl
Selected writings.
1. Marxian economics
I. McLellan, David
335.4'12   HB97.5
ISBN 0-19-876037-X
ISBN 0-19-876038-8 Pbk

Printed in Hong Kong

Contents

A. From Volume One

Abbreviations

MEGA K. Marx, F. Engels, *Historisch-kritische Gesamtausgabe. Werke, Schriften, Briefe*, 13 vols. Completed, Frankfurt, Berlin, and Moscow, 1927 ff.

MESC K. Marx, F. Engels, *Selected Correspondence*, Moscow, 1935.

MESW K. Marx, F. Engels, *Selected Works*, Moscow, 1935, 2 vols.

MEW K. Marx, F. Engels, *Werke*, 41 vols, Berlin, 1956 ff.

Introduction

The aim of this book is to present as comprehensive and balanced a selection of Marx's writings as possible. I have forgone the opportunity of writing an extended introduction offering either 'potted' biography or an interpretation of Marx's thought. A biography can readily be obtained elsewhere and such contextual details as are necessary for an understanding of each extract are provided with it. An interpretation—to be worthwhile—would have to be fairly lengthy and involve the exclusion of some of Marx's texts. Nevertheless a few words on the principles of selection are necessary.

The most evident difficulty confronting a portrayal of Marx's thought is that he is politically a controversial figure. And there is the additional difficulty that Marx was a prolific writer, in different styles and contexts, and left half of it unpublished so that it only emerged piecemeal during the years after his death. Up until very recently the most accessible large selection of Marx's works was issued by the Russian Communists and their allies who claimed to be the political incarnation of Marx's ideas. Naturally, they saw Marx from their own point of view and their selection had two deficiencies. First, it ignored Marx's early writings. These were published around 1930 and reveal a more philosophical, humanist Marx, that many thought incompatible with the economic, materialist Marx of Stalinist orthodoxy.

Although there was considerable controversy about whether the young or the old Marx was the real Marx and whether there was or was not a continuity in Marxist thought, any selection that ignored these early writings would be seriously deficient. So recent editions of Marx have tended to make selections from the Moscow selections and supplement them with some of the early writings. But the Moscow selections had another drawback: they consisted almost entirely of Marx's political writings together with some simple summaries of his economic doctrines. Over recent years increasing attention has been paid to the three works that Marx produced between 1857 and 1867—the *Grundrisse*, the *Theories of Surplus Value*, and *Capital*. On almost any reading of Marx these constituted his main theoretical contribution: yet, apart from a few pages of *Capital*, they are absent from the Moscow selection which tends to concentrate on Marx's political writings. Believing that it was not enough just to augment the Moscow selection yet again with some extracts from *Capital*, I have tried to have a fresh look at the whole corpus of Marx's writings in the light of recent scholarship. This has involved translating certain passages that have never been published in England before and being prepared to include very short extracts when necessary—although I have tried to avoid being too 'bitty'. The main interest in recent years among interpreters of Marx has focused on his methodology and on his contribution to the science of political economy. Thus any selection satisfying these interests must contain large excerpts from the *Grundrisse* and the *Theories of Surplus Value*, which, together with *Capital*, constitute the centrepiece of Marx's work.

To avoid overloading the text with footnotes I have added an annotated name index as well as a full subject index.

NOTE

Most of the extracts have been previously published elsewhere. In those which I translated myself, I have made a few minor alterations for the present collection.

I am grateful to my friends Nick Caswell and Gay Sharp for helping with the Index.

I. The Early Writings
1837–1844

1. Letter to his Father

On leaving school Marx spent a year at the provincial University of Bonn, where he thoroughly absorbed the prevalent Romantic mood before moving to the University of Berlin. Seventeen letters from his father to Marx have been preserved, but this is the only one from Marx in reply. It was written on 10 November 1837, when Marx was nineteen years old, after he had spent just over a year in the Law Faculty at Berlin. Usually Marx's letters to his father were short, so the length of this one indicates its importance. The letter recounts the evolution of his ideas during the previous year and criticizes them from his newly won Hegelian standpoint, the Hegelian philosophy being the one then dominant in Berlin. What attracted Marx to Hegel, after the romanticism of his year at Bonn and his brief enthusiasm for the idealism of Kant and Fichte, was the bridge he conceived Hegel to have built between what is and what ought to be. In this letter he touches on many of the themes that were to run right through his work: historical consciousness, for example, an attempt to situate himself within an evolving process, and the desire, following Hegel, to find the identity of the real and the rational.

Dear Father,

There are moments in one's life that represent the limit of a period and at the same time point clearly in a new direction.

In such a period of transition we feel ourselves compelled to consider the past and present with the eagle eye of thought in order to come to a realization of our actual position. Yes, History itself likes this sort of stock-taking and introspection which often make it look as though it were going backwards or standing still, whereas it is merely throwing itself into an armchair to understand itself and comprehend intellectually its own mental processes.

An individual, however, becomes lyrical at such moments, for every change is partly a swansong, partly an overture, to a new epic that is trying to find a form in brilliant colours that are not yet distinct. Yet we want to erect a memorial to our past experiences so that they may find again in our emotions the place that they have lost in our actions; and there could be no more sacred home for this memorial than the heart of parents, the mildest of judges, the most intimate sympathizers, a sun of love whose fire warms the inmost centre of our endeavours. How better could much that is disgraceful and blameworthy find forgiveness and excuse than when it appears as the result of an essentially necessary state of affairs, how else could the often untoward fall of chance and the mind's errors escape being thought the products of a deformed spirit?

So now that I am casting an eye back over the events of the year that I have lived here and thus answering, my dear father, your most precious letter from Ems, allow me to consider my situation (as I do life in general) as the result of an intellectual activity that finds expression on all sides—in science, art, and personal matters.

When I left you, a new world had just begun to exist for me, the world of love that was at first drunk with its own desire and hopeless. Even the journey to Berlin, which would otherwise have charmed me completely, excited me to an admiration of nature, and inflamed me with a zest for life, left me cold and even, surprisingly, depressed me; for the rocks that I saw were no rougher, no harsher than the emotions of my soul, the broad cities no more full of life than my blood, the tables of the inns no more overladen and indigestible than the stocks of fantasies that I carried with me, nor, finally, was any work of art as beautiful as Jenny.

When I arrived in Berlin I broke off all the connections that I had hitherto contracted, made rare and reluctant visits, and tried to steep myself in science and art.

Considering my state of mind then it was inevitable that lyric poetry should be my first project and certainly the pleasantest and readiest to hand. But my attitude and all my previous development made it purely idealistic. My heaven and art became a Beyond as distant as my love. Everything real began to dissolve and thus lose its finiteness, I attacked the present, feeling was expressed without moderation or form, nothing was natural, everything built of moonshine; I believed in a complete opposition between what is and what ought to be, and rhetorical reflections occupied the place of poetic thoughts, though there was perhaps also a certain warmth of emotion and desire for exuberance. These are the characteristics of all the poems of the first three volumes that Jenny received from me. The whole scope of a longing that sees no limits is expressed in many forms and broadens poetry out.

But poetry was to be, and had to be, only a sideline; I had to study jurisprudence and felt above all impelled to struggle with philosophy. Both were so interconnected that I examined Heineccius, Thibaut, and the sources completely uncritically like a schoolboy, and thus translated the first two books of Pandects into German and at the same time tried to elaborate a philosophy that would cover the whole field of law. As introduction I prefixed a few metaphysical propositions and continued this unhappy opus as far as public law, a work of almost three hundred pages.

Here the same opposition of 'is' and 'ought' which is the hallmark of idealism was the dominating and very destructive feature and engendered the following hopelessly mistaken division of the subject-matter: firstly came what I had so graciously christened the metaphysics of law, i.e. first principles, reflections, definitions distinct from all actual law and every actual form of law—just as you find in Fichte only here more modern and with less substance. This meant that from the outset the unscientific form of mathematical dogmatism where one circles round a subject, reasoning back and forth, without letting it unfold its own rich and living content, prevented any grasp of the truth. The mathematician constructs and proves the triangle, but it remains a pure abstraction in space and does not develop any further; you have to put it beside something else and then it takes up other positions and it is the juxtaposition of these different things that gives it different relationships and truths. Whereas in the practical expression of the living world of ideas in which law, the state, nature, and the whole of philosophy consist, the object itself must be studied in its own development, arbitrary divisions must not be introduced, and it is the ratio of the object itself which must develop out of its inner contradictions and find unity within itself.

The second part consisted of the philosophy of law, i.e. in accordance with the opinion I held at that time, a discussion of the development of ideas in positive Roman law, as though the development of the ideas of positive law (I don't mean in its purely finite terms) could ever be anything different from the formation of the concept of law which the first part should already have dealt with!

Moreover, I had further divided this section into formal and material legal doctrine, the first of which was to describe the pure form of the system in its consistent development, its divisions and range, while the second was to describe the self-incarnation of the form in its content. This was an error that I held in common with Herr v. Savigny, as I found out later in his learned work on property, only with the difference that he calls the formal definition of the idea 'the finding of the place that such and such a doctrine occupies in the (fictional) Roman system', and material 'the doctrine of the positive content that the Romans included in a concept thus defined', whereas I meant by form the necessary structure of the expressions of an idea and by matter the necessary quality of these expressions. The fault here was that I believed that the one could and must develop itself independently of the other and thus I did not obtain a true form but merely a desk into whose drawers I proceeded to pour sand. . . .

At the same time I translated Tacitus's *Germania* and Ovid's *Tristia* and began to learn English and Italian on my own, i.e. out of grammars, though I have not yet got anywhere with them. I also read Kelin's *Criminal Law* and his *Annals* and all the latest literature, though this latter only as a sideline. At the end of the term I again sought the dances of the Muses and the music of the Satyrs and in the last volume that I sent you the forced humour of *Scorpion and Felix* and the misconceived fantastic drama of *Oulanem* are shot through with idealism which finally changes completely, dissolving into purely formal art which has no objects to inspire it and no exciting progress of ideas.

And yet these last poems were the only ones in which suddenly, as though at the touch of a magic wand—oh! the touch was at first shattering—the kingdom of true poetry glittered opposite me like a distant fairy palace and all my creations dissolved into nothingness.

With these various occupations I had been forced during the first term to sit up through many nights, to fight through many a struggle and endure much excitement from within and without, and yet was not much richer at the end in spite of having deserted nature, art, and the world, and spurned friends. These thoughts were registered by my body and a doctor advised me to go to the country, and so for the first time I went through the whole of the long city and out of the gate to Stralow. I did not suspect that there my anaemic and languishing body would mature and acquire a robust strength.

A curtain had fallen, my holy of holies was rent asunder, and new gods had to be installed. I left behind the idealism which, by the way, I had nourished with that of Kant and Fichte, and came to seek the idea in the real itself. If the gods had before dwelt above the earth, they had now become its centre.

I had read fragments of Hegel's philosophy, but I did not care for its grotesque and rocky melody. Once again I wanted to dive off into the sea, but with the firm intention of finding the nature of the mind as necessary, concrete, and firmly established as that of physical nature, for I wanted to stop fencing and bring the pure pearls up to the sunlight.

I wrote a dialogue of about twenty-four pages entitled 'Cleanthes or the starting-point and necessary progress of philosophy'. Here art and science, which had become completely separate, regained to some extent their unity, and I vigorously set about the job itself, a philosophical and dialectical development of the divinity as it manifests itself as idea-in-itself, religion, nature, and history. My last sentence was the beginning of Hegel's system, and this work for whose sake I had made some acquaintance with natural science, Schelling, and history, which had caused me endless headaches and is written in so confused a manner (for it had actually to be a new logic) that I can now scarcely think myself back into it, this my dearest child, reared by moonlight, like a false siren delivers me into the arms of the enemy.

My vexation prevented me from thinking at all for several days and I ran like a madman around the garden beside the dirty waters of the Spree 'which washes souls and makes weak tea'. I even went on a hunting party with my landlord and rushed off to Berlin and wanted to embrace every old tramp I saw.

Soon afterwards I undertook only positive studies, the study of ownership by Savigny, Feuerbach and Grohlmann's *Criminal Law*, the *De Verhorum Significatione* of Cramer, Wenning-Ingenheim's *System of Pandects*, Mühlenbruch's *Doctrina Pandectarum*, which I am still working through, and finally a few titles from Lauterbach, *Civil Trials* and above all *Canon Law*, the first part of which—the *Concordia discordantium Canonum* of Gratian—I have read and excerpted almost entirely in the *corpus*, as also the supplement Lancelotti's *Institutiones*. Then I translated Aristotle's *Rhetoric* in part, read the *De Argumentis Scientiarum* of the famous Baco of Verulam, was very busy with Reimarus whose book *On the Instincts of Animals* I followed with delight, and also came across German law, though principally only in so far as I went through the capitularies of the Frankish kings and the letters that the Popes addressed to them. My vexation at Jenny's illness, my fruitless and failed intellectual endeavours, and my consuming anger at having to make my idol a view that I hated, made me ill, as I have already written to you, dear father. When I recovered I burnt all my poems and sketches for novels, etc., fancying that I could be completely free from them, which has at least not yet been disproved.

During my illness I had got to know Hegel from beginning to end, together with most of his disciples. Through several gatherings with friends in Stralow I obtained entrance into a graduate club among whose members were several university lecturers and the most intimate of my Berlin friends, Dr. Rutenberg. In the discussions here many contradictory views appeared and I attached myself ever more closely to the current philosophy that I had thought to escape. . . .

In the hope that by and by the clouds that surround our family will retreat and that I may be allowed to suffer and weep with you and perhaps give you tangible proofs of the deep and sincere sympathy and immeasurable love that I can often only express so badly; in the hope, too, that you, dearly beloved father, will take into consideration the often very disordered state of my mind, and forgive where my heart has seemed to err, overcome by my fighting spirit, and that you will soon be completely restored to health so that I may myself press you to my heart and tell you all.

Your ever loving son
Karl

Forgive, dear father, the illegible handwriting and bad style; it is almost four o'clock. The candle is burnt right down and my eyes are sore; a real anxiety has come over me and I will not be able to quieten the ghosts I have roused until I am near you again. Please give my love to dear, wonderful Jenny. I have already read her letter twelve times and I still find new delights. It is in every particular, including that of style, the most beautiful letter that I can imagine written by a woman.

BIBLIOGRAPHY

Original: *MEGA* I (i) 2, pp. 213–21.
Present translation: K. Marx, *The Early Texts*, ed. D. McLellan, Oxford, 1971, pp. 1–10.

Other translations: *Writings of the Young Marx on Philosophy and Society*, ed. L. Easton and
  K. Guddat, New York, 1967, pp. 40–50.
W. Glen-Doepel, in B. Delfgaauw, *The Young Marx*, London, 1967, pp. 135–47.
Commentaries: F. Mehring, *Karl Marx*, London, 1934, pp. 10–12.
W. Johnston, 'Marx's verses of 1836–7', *Journal of the History of Ideas*, April 1967.
D. McLellan, *Marx Before Marxism*, Penguin, 1972, pp. 66 ff.

2. Doctoral Thesis

In 1839, Marx, still in Berlin, began writing a doctoral thesis which would help him to get a job as a university lecturer. Pressure from his friend Bruno Bauer, quarrels with his family, his engagement to Jenny v. Westphalen and the consequent need to obtain a job quickly, made Marx present his thesis in a hurry and he obtained his degree *in absentia* from the University of Jena in April 1841. The title of the thesis was 'The Difference between Democritus' and Epicurus' Philosophy of Nature'. Many of Marx's Young Hegelian colleagues were interested in this post-Aristotelian Greek philosophy, both because it seemed to them to present the same problems as confronted post-Hegelian philosophy and also because it was the intellectual climate in which Christianity started—and religion was a constant subject of debate among the Young Hegelians.

The body of Marx's thesis is of little interest: it consists of a criticism of those who equated the natural philosophies of Democritus and Epicurus and a catalogue of the differences between these philosophies. Marx attacked Democritus' mechanistic determinism and praised Epicurus for introducing the idea of spontaneity into the movement of the atoms. The following extracts come from the far more interesting digressions interspersed between the notes and references and no doubt intended to be incorporated into a revised and enlarged version of the thesis for publication. Marx's starting-point in these digressions is Hegel's philosophy of history, which he intended to revise and push further. The extracts develop the problem raised by Marx's letter to his father: what is the philosopher's task after the seductive solution to the problem of the relationship of the real to the rational as offered by Hegel? Marx's answer, couched in rather obscure and abstract language, is that Hegel has to be radically rethought and put on a new basis, a basis that will involve the disappearance of philosophy at the same time as its realization. For the function of philosophy is to criticize existing reality and make the gap between the ideal and the real intolerable. Marx also introduces here the notion of *praxis*, though as yet in an idealistic form.

Preface

The form of this treatise would have been both more strictly scientific and also less pedantic in many of its developments had it not originally been intended to be a doctoral thesis. Extrinsic reasons have none the less persuaded me to have it printed in this form. Moreover, I think that I have here solved a problem in the history of Greek philosophy that has hitherto remained a mystery.

Specialists know that for the subject of this treatise there are no previous works that are at all useful. The blatherings of Cicero and Plutarch have been reproduced right up to the present time. Gassendi's exposition, which freed Epicurus from the interdict which the Church fathers and the Middle Ages, the period of unreason incarnate, had laid on him, is only an interesting stage. He tries to reconcile his Catholic conscience with his pagan science and Epicurus with the Church, which of course was a waste of effort. It is as though one wanted to put a Christian nun's habit on the serenely blooming body of a Greek Laïs. Gassendi has to learn too much philosophy from Epicurus for him to be in a position to teach us much about it.

This treatise should be considered as only the preliminary to a larger work in which I will describe in detail the cycle of Epicurean, Stoic, and Sceptic philosophies in their relationship to the whole of Greek speculation. The deficiencies in the form and so on of the present treatise will then disappear. . . .

Hegel has, it is true, by and large correctly described the general characteristics of these systems—but the admirably broad and bold plan of his history of philosophy, which really gave birth to the history of philosophy as a subject, made it impossible to enter into details; and also his conception of what he called 'speculative *par excellence*' prevented this giant of a thinker from recognizing the great importance that these systems have for the history of Greek philosophy and the Greek mind in general. These systems are the key to the true history of Greek philosophy. . . .

The reason for adding a critique of Plutarch's polemic against Epicurus' theology as an appendix was that this polemic does not stand in isolation, but represents a *genre* in that it strikingly conveys the attitude of the theological mind to philosophy.

My critique does not discuss, among other things, how completely false Plutarch's whole approach is when he calls philosophy before the bar of religion. On this subject, let a passage from David Hume suffice instead of any argument:

It is certainly a sort of insult for philosophy whose sovereign views should be recognized on all sides, when she is compelled on every occasion to defend herself because of her consequences and justify herself in the eyes of every art and science that is offended by her. One is put in mind of a king who is accused of high treason against his own subjects.

As long as a single drop of blood pulses in her world-conquering and totally free heart philosophy will continually shout at her opponents the cry of Epicurus: ʽἀσεβὴς δὲ οὐχ ὁ τοὺς τῶν πολλῶν θεοὺς ἀναιρῶν, ἀλλ ὁ τὰς τῶν πολλῶν δόξας θεοῖς προϑάπτων'. (The profane man is not the one who destroys the gods of the multitude but the one who foists the multitude's doctrines onto the gods.)

Philosophy makes no secret of it. The proclamation of Prometheus—ʽἁπλῷ λόγῳ τοὺς πάντας ἐχθαίρω θεούς' (in a word, I detest all the Gods)—is her own profession, her own slogan against all the gods of heaven and earth who do not recognize man's self-consciousness as the highest divinity. There shall be none other beside it.

But to the pitiful cowards who rejoice over the apparently worsening social position of philosophy she repeats what Prometheus said to the servant of the gods, Hermes: 'τῆς σῆς λατρείας τὴν ἐμὴν δυαπραξίαν, σαφῶς ἐπίστασ', οὐκ ἂν ἀλλάξαιμ' ἐγώ'. (Understand this well, I would not change my evil plight for your servility.)

Prometheus is the foremost saint and martyr in the philosopher's calendar.

From the Notes to the Dissertation

. . . As regards Hegel, too, it is pure ignorance on the part of his followers when they explain this or that aspect of his system as a compromise or something of the sort, i.e. when they pass a moral judgement on it. They forget that almost no time ago, as can plainly be shown them out of their own writings, they were fervent adherents to all aspects of his one-sidedness. . . .

It is conceivable that a philosopher should be guilty of this or that inconsistency because of this or that compromise; he may himself be conscious of it. But what he is not conscious of is that in the last analysis this apparent compromise is made possible by the deficiency of his principles or an inadequate grasp of them. So if a philosopher really has compromised it is the job of his followers to use the inner core of his thought to illuminate his own superficial expressions of it. In this way, what is a progress in conscience is also a progress in knowledge. This does not involve putting the conscience of a philosopher under suspicion, but rather construing the essential characteristics of his views, giving them a definite form and meaning, and thus at the same time going beyond them.

Moreover, I consider this unphilosophical evolution of a large part of the Hegelian school as a phenomenon that will always accompany a transition from discipline to freedom.

It is a psychological law that once the theoretical intellect has achieved freedom within itself it turns into practical energy and, emerging from the shadow kingdom of Amenthes as will, directs itself against the exterior reality of the world. (But it is important from the philosophical point of view to elaborate on these pages because from the precise manner of this transition we can draw conclusions as to the immanent characteristics and historical character of a philosophy. We see here its *curriculum vitae* reduced, so to speak, to its simplest expression, its quintessence.) But the *praxis* of philosophy is itself theoretical. It is the sort of critique that measures individual existing things by their essence and particular realities by the Idea. But this immediate realization of philosophy is fraught with contradictions in its innermost essence and it is its essence that appears in the phenomena and imprints its seal on them.

So long as philosophy as will goes forth against the world of appearances, the system is degraded to an abstract totality, i.e. it has become one side of the world with another side over against it. Its relation to the world is one of reflection. Being inspired with the desire to realize itself, there is a tension between it and other things. Its inner self-sufficiency and perfection are destroyed. What was an inner light becomes a consuming flame that turns outwards. As a consequence, the world's becoming philosophical coincides with philosophy's becoming worldly, the realization of philosophy coincides with its disappearance, the exterior battles of philosophy are against its own inner deficiencies; in the struggle it acquires precisely those defects against which it is fighting, and so only eliminates them by making them its own. Philosophy's opposite and enemy is always the same as philosophy itself, but with the factors reversed.

In the end this duality of the philosophical mind produces two schools completely opposed to one another, one of which, the liberal party as we may loosely call it, lays most emphasis on philosophy as a concept and principle, while the other holds fast to what are not concepts, to the real. This second school is positive philosophy. The activity of the first takes the form of a critique, i.e. philosophy turning itself against the exterior world; the activity of the latter is an attempt to philosophize, i.e. philosophical introspection. The second school sees the deficiency as immanent to philosophy, whereas the first sees it as a deficiency of the world that it is trying to make philosophical. Each of these parties does exactly what the other aims at and what it does not itself intend. But the first, in spite of its inner contradictions, is in general aware of its principle and aim. In the second, the inversion, the craziness, so to speak, is manifested in all its purity. As regards content, it is only the liberal party, because it is the party of the concept, which makes any real progress, whereas positive philosophy is only capable of requirements and tendencies whose form contradicts their meaning. . . .

. . . Just as in philosophy there are turning-points which in themselves develop to concreteness, gather its abstract principles into a totality, and so break off the unilinear process, so are there also moments in which philosophy turns its eyes towards the outside world, no longer merely comprehends but, as a practical being, spins intrigues with the world without intermediary. It comes forth from the shadowy kingdom of Amenthes and throws itself on the breast of the worldly sirens. It is the shrovetide of philosophy. Let her dress up like a dog as the Cynic does, put on the robes of a priest like the Alexandrine, or the fragrant spring garment of the Epicurean. It is essential for philosophy at this stage to put on actors' masks.

Just as in the story of the creation of mankind Deucalion threw stones behind him, so philosophy casts its eyes backwards (the bones of its mother are bright eyes), when its heart becomes strong enough to create a world; but like Prometheus, who stole fire from heaven and began to build houses and settle on the earth, so philosophy, which has evolved so as to impinge on the world, turns itself against the world that it finds. So now the Hegelian philosophy.

Philosophy has isolated itself so as to become a comolete and total world, and the nature of this totality is conditioned by philosophy's own development as is also the form that is taken by its transition to a practical relationship to reality. Thus there is a rift running through the totality of the world, and indeed this rift has been widened as far as possible in that intellectual existence has become free and attained to the richness of the universal. The heart-beat has become in a concrete manner in itself the distinction which the whole organism is. The rift in the world is not causal if its two sides are totalities. So the world that is opposed by a philosophy that is complete in itself is one that is rent asunder. Therefore, the activity of this philosophy appears too to be rent asunder and contradictory; its objective universality returns into the subjective forms of the individual minds in which it has its life. Normal harps will sound beneath any hand; those of Aeolus only when the storm strikes them. But we should not let ourselves be misled by the storm that follows a great, a world-philosophy.

Someone who does not appreciate this historical necessity must consequently deny that man could continue to live at all after a total philosophy, or else treat the dialectic of quantity as such as the highest category of conscious minds, and claim with some of our misguided Hegelians that mediocrity is the usual form in which absolute mind appears; but a mediocrity that gives itself out to be the normal appearance of the absolute has itself degenerated into boundlessness, namely a boundless pretension. Without this necessity it is impossible to understand how after Aristotle a Zeno, an Epicurus, even a Sextus Empiricus, how after Hegel the efforts of later philosophers, which are for the most part unmitigatedly deficient, could attain to the light of day.

In such times half-formed spirits have the opposite view to real commanders. They believe that they can make good their losses by reducing and dividing their forces and make a peace treaty with real needs, whereas Themistocles, when Athens was threatened with destruction, persuaded the Athenians to quit their city completely and found a new Athens on another element, the sea. Nor should we forget that the period that follows such catastrophes is an iron one, happy if it is marked by titanic struggles, lamentable if it is like the centuries that limp behind the great periods of art and busy themselves with imitating in wax, plaster, and copper what sprang from Carrara marble like Pallas Athene from the head of Zeus, father of the gods. But those periods are titanic that follow a total philosophy and its subjective forms of development, for the division that forms its unity is gigantic. Thus the Stoic, Epicurean, and Sceptic philosophies are followed by Rome. They are unhappy and iron for their gods are dead and the new goddess has as yet only the obscure form of fate, of pure light or of pure darkness. She still lacks the colours of the day. The root of the unhappiness, however, is that the soul of the period, the spiritual Monas, being sated with itself, shapes itself ideally on all sides in isolation and cannot recognize any reality that has come to fruition without it. Thus the happy aspect of this unhappy time lies in the subjective manner, the modality in which philosophy as subjective consciousness conceives its relation to reality.

Thus, for example, the Stoic, and Epicurean philosophies were the happiness of their time; thus the night-butterfly, when the universal sun has sunk, seeks the lamplight of a private person. . . .

BIBLIOGRAPHY

Original: MEGA I (i) 1, pp. 8 ff., 63 ff., 131 ff.
Present translation: K. Marx, *Early Texts*, pp. 12 ff.

Other translations: The body of the thesis is translated in: N. Livergood, *Marx's Philosophy of Action*, The Hague, 1967. There are extracts from the important digressions in: *Writings of the Young Marx*, ed. Easton and Guddat, pp. 51 ff.

Commentaries: C. Bailey, 'Karl Marx and Greek Atomism', *The Classical Quarterly*, Vol. 22, 1928.

N. Livergood, Introduction to work cited above.

D. McLellan, *Marx Before Marxism*, Penguin, 1972, pp. 74 ff.

H. Mins, 'Marx's Doctrinal Dissertation', *Science and Society*, 1948.

3. Articles for the *Rheinische Zeitung*

The seven extracts which follow were all written for the *Rheinische Zeitung*, a newspaper founded in January 1842 to defend the liberal interest of Rhineland industrialists against the Prussian Central government. Most of the Berlin Young Hegelians contributed articles during the first few months of the paper's existence, and it was natural that Marx, who had participated in the discussions preceding its foundation, should also contribute. His prospects of a university career evaporated when his friend Bruno Bauer was dismissed from his post for unorthodox teaching in March 1842, and Marx moved to the Rhineland to devote himself to full-time journalism. His contributions were so well received that in October 1842 he was offered the editorship. The paper was a great success, but its outspoken criticism of the government caused it to be suppressed in March 1843. During the year that he spent in journalism, Marx's views were in transition and reveal no systematic framework; it is of the essence of a polemicist to be eclectic, and Marx uses expressions and lines of argument drawn alike from Spinoza, Kant, and Hegel.

On the Freedom of the Press

The following extracts are taken from Marx's first contribution to the *Rheinische Zeitung*. Written in May 1842, it is a commentary on the debate in the Rhenish Parliament on the extent to which press publication of their proceedings should be allowed. The two extracts deal with religion as an ideological cloak and the role of law in society.

. . . Because the real situation of these gentlemen in the modern state bears no relation at all to the conception that they have of their situation, because they live in a world situated beyond the real world, and because in consequence their imagination holds the place of their head and their heart, they necessarily turn towards theory, being unsatisfied with practice, but it is towards the theory of the transcendent, i.e. religion. However, in their hands religion acquires a polemical bitterness impregnated with political tendencies and becomes, in a more or less conscious manner, simply a sacred cloak to hide desires that are both very secular and at the same time very imaginary.

Thus we shall find in our Speaker that he opposes a mystical/religious theory of his imagination to practical demands . . . and that to what is reasonable from the human point of view he opposes superhuman sacred entities and to the true sanctuary of ideas a vulgar point of view that is both arbitrary and unbelieving. . . .

Thus so far from a law on the press being a repressive measure directed against the repetition of a crime, the lack of a law dealing with the press should rather be seen as an exclusion of freedom of the press from the sphere of legal freedom, for legally recognized freedom exists in the state as law. Laws are as little repressive measures directed against freedom as the law of gravity is a repressive measure directed against movement, because although it keeps the heavenly bodies in perpetual motion yet it can also kill me if I wish to violate it and dance in the air. Laws are rather positive, bright and general norms in which freedom has attained to an existence that is impersonal, theoretical, and independent of the arbitrariness of individuals. A people's statute book is its Bible of freedom.

The law on the press is therefore the legal recognition of the freedom of the press. It is law because it is the positive existence of freedom. Thus it must always be present, even when it is never applied, as in North America, while censorship, like slavery, can never become legal, though it were a thousand times present as law.

There are no preventive laws at the present time. Law only prevents by forbidding. It becomes active law as soon as it is transgressed for it is only true law when in it the unconscious natural law of freedom becomes the conscious law of the state. Where law is true law, i.e. where it is the existence of freedom, it is the true existence of the freedom of man. Thus the laws cannot prevent man's actions, for they are the inner laws of life of his action itself, the conscious mirror-images of his life. Law thus retreats before man's life as a life of freedom, and only when his behaviour has actually shown that he has ceased to obey the natural law of freedom does the state law compel him to be free. Similarly physical law appears alien to me only when my life has ceased to be the life of these laws, when it is sick. Thus a preventive law is a meaningless contradiction. . . .

The Leading Article of the Kölnische Zeitung

This article was a reply to an attack on the 'new philosophical school' by Karl Hermes, editor of the *Kölnische Zeitung*, the big conservative rival of the *Rheinische Zeitung*. Hermes had accused the *Rheinische Zeitung* of intemperate discussion of religion. In the following extract Marx is concerned to show, in a rather Hegelian manner, that the state is not based on religion but on 'the rational character of freedom'.

. . . Either the Christian state corresponds to the concept of a state which is to be a realization of liberty according to reason, and in that case the only condition for a state's being Christian is that it should be rational and then it is sufficient to deduce the state from the rational character of human relationships, which is the job of philosophy. Or the state of rational freedom cannot be deduced from Christianity, and in that case you will yourselves admit that this deduction is not included in the attitude of Christianity, for it cannot wish for a bad state, and a state that is not a realization of rational freedom is a bad state.

However you answer this dilemma, you will have to agree that the construction of the state ought not to start from religion but from the rational character of freedom. Only the crassest ignorance would hold that this theory of the autonomous character that belongs to the concept of the state is the sudden fantasy of modern philosophers.

Philosophy has done with regard to politics what physics, mathematics, medicine, and each science have done in their respective sphere. Bacon of Verulam declared that theological physics was a virgin consecrated god and sterile: he emancipated physics from theology and it became fertile. You should no more ask a politician if he is a believer than you would put this question to a doctor. In the period that proceeds and immediately follows the great discovery by Copernicus of the true solar system, the law of gravity of the state was also discovered. Its centre of gravity was found to be in itself, and the different European governments tried to apply this discovery, with the superficiality of every first practical trial, in the system of the balance of powers. Similarly, first Machiavelli and Campenella, then later Hobbes, Spinoza, Hugo Grotius, through to Rousseau, Fichte, and Hegel began to consider the state through human eyes and deduced its natural laws from reason and experience and not from theology, just like Copernicus, who disregarded the fact that Joshua had ordered the sun to stop on Gabaon and the moon above the valley of Ajalon. Modern philosophy has only continued work that Heraclitus and Aristotle had already begun. Therefore your polemics are not against the reason of modern philosophy, they are against the ever new philosophy of reason. Of course, the ignorance which discovered for the first time yesterday or the day before in the *Rheinische Zeitung* or the *Kölnische Zeitung* the very old ideas on the state, this ignorance regards the ideas of history as sudden fantasies of isolated individuals because to it they are new and arrived overnight. This ignorance forgets that it assumes itself the old role of the doctor of the Sorbonne who thought it his duty publicly to accuse Montesquieu for having had the frivolity to declare that the supreme civic quality was political virtue and not religious virtue; it forgets that it assumes the role of Joachim Lange who denounced Wolff on the pretext that his doctrine of predestination would lead to the desertion of soldiers and thus the relaxation of a military discipline and in the end the dissolution of the state; finally it forgets that the Prussian civil code comes precisely from the philosophical school of that same 'Wolff' and the Code Napoléon not from the Old Testament but from the ideas of Voltaire, Rousseau, Condorcet, Mirabeau, Montesquieu, and the French Revolution. Ignorance is a demon, it is to be feared that it may yet play many a tragedy. The greatest Greek poets were right to represent it in terrible dramas of the royal families of Mycenae and Thebes in the form of a tragic destiny.

But if the previous professors of constitutional law have constructed the state from instincts either of ambition or sociability or even from reason, but from the individual's reason and not social reason, the profounder conception of modern philosophy deduces the state from the idea of the all. It considers the state as the great organism in which juridical, moral, and political liberties must be realized and in which each citizen, by obeying the laws of the state, only obeys the natural laws of his own reason, human reason. *Sapienti sat.* [A word to the wise]

Communism and the Augsburger Allgemeine Zeitung

This is Marx's first article as editor of the *Rheinische Zeitung*. The *Augsburger Allgemeine Zeitung* had accused the *Rheinische Zeitung* of communist sympathies. Moses Hess had been responsible for the paper's reprinting an article on workers' housing from a journal of Wilhelm Weitling and reporting the speeches of followers of Fourier delivered at a recent congress at Strasburg. The extract which follows is particularly interesting in that it reveals Marx's initially hostile reaction to French socialism.

. . . The *Rheinische Zeitung* does not even concede theoretical validity to communist ideas in their present form, let alone desire their practical realization, which it anyway finds impossible, and will subject these ideas to a fundamental criticism. If it had aims and capacities beyond well-polished phrases, the *Augsburger* would have perceived that books like those of Leroux and Considérant, and above all the acute work of Proudhon, cannot be criticized by superficial and transitory fancies but only after consistent and probing study. We have to take such theoretical works all the more seriously as we cannot agree with the *Augsburger* which finds the reality of communist ideas not in Plato but in its obscure acquaintance who, though being gifted in several lines of scientific research, sacrificed all the money he could lay his hands on and washed his comrades' plates and cleaned their boots according to the will of Father Enfantin. We are firmly convinced that the true danger does not lie in the practical attempt to carry out communist ideas but in their theoretical development; for practical attempts, even by the masses, can be answered with a cannon as soon as they become dangerous, but ideas that have overcome our intellect and conquered our conviction, ideas to which reason has riveted our conscience, are chains from which one cannot break loose without breaking one's heart; they are demons that one can overcome only by submitting to them. Yet the *Augsburger Zeitung* has never got to know the crisis of conscience caused by the rebellion of man's subjective desires against the objective insights of his own reason, for it has neither reason of its own, nor insights, nor even conscience. . . .

The Law on Thefts of Wood

This article also deals with a debate in the Rhenish parliament. A more stringent law on thefts of timber had been proposed. The gathering of dead wood had traditionally been unrestricted, but scarcities were caused by the agrarian crises of the 1820s and the growing needs of industry. The situation was getting out of hand: five-sixths of all prosecutions in Prussia dealt with wood, and the proportion was even higher in the Rhineland. So now it was being proposed that the keeper be the sole arbiter of an alleged offence and that he alone assess the damages. Marx's general view is that the state should defend customary law against the rapacity of the rich. It was the writing of this article that first directed Marx's attention to socioeconomic problems. He himself wrote in 1859 that 'the proceedings of the Rhenish parliament on thefts of wood . . . provided one of the first occasions for occupying myself with economic questions.'

. . . If every violation of property without differentiation or further definition is theft, would not private property be theft? Through my private property do I not exclude a third party from this property? And do I not thus violate his right to property? When you deny the distinction between essentially different types of the same crime, then you deny the crime as distinct from the law and you do away with the law itself, for every crime has a facet that is in common with the law. It is thus a fact as historical as it is rational that an undifferentiated harshness destroys all the effects of punishment for it has destroyed punishment as a consequence of the law. . . .

But we unpractical men lay claim, on behalf of the masses of the poor who have no political or social possessions, to what the learned and docile servants of the so-called historians found to be the true philosopher's stone which could form any impure pretension into the pure gold of right. We reclaim for poverty the right of custom, and moreover a right of custom which is not a local one but which is that of poverty in all lands. We go further and affirm that customary right by its nature can only be the right of the lowest and elementary mass of propertyless people.

Among the so-called customs of the privileged are understood customs against the law. The date of their birth falls in the period when the history of mankind formed a part of natural history and the Egyptian legend was proved true when all the gods concealed themselves in the form of animals. Mankind appears as disintegrated into particular animal races who are held together not by equality but by an inequality that regulates the laws. A universal lack of freedom requires laws that lack freedom, for whereas human law is the existence of freedom, animal law is the existence of a lack of freedom. . . .

The rights of aristocratic custom run counter by their content to the form of general law. They cannot be formed into laws because they are formulations of lawlessness. The fact that these customary rights are through their content in conflict with the form of law, i.e. its universality and necessity, proves that they are unjust customs and that, instead of being enforced in opposition to the law, they should be abrogated because of this opposition and even on occasion be punished. For no one stops behaving unjustly simply because this way of behaving is a custom, any more than the thieving son of a thief is excused by his family's idiosyncrasies. If a man behaves unjustly intentionally, then his intention should be punished, and if he behaves unjustly out of custom, then his custom should be punished as being a bad one. In an age of general law rational customary rights are nothing but the custom of legal rights, for rights do not cease to be customary once they have constituted themselves as law, but they do cease to be purely customary. For the law-abiding man law becomes his own custom whereas the man who does not abide by the law is constrained by it even though it is not his custom. Rights no longer depend on the chance of whether custom is rational for custom becomes rational, because rights are legal, because custom has become the custom of the state. . . .

It is unwillingly that we have followed this boring and stupid debate, but we thought it our duty to use an example to show what can be expected of an estates assembly motivated by particular interests, were it ever really called upon to legislate. . . .

We repeat once again that our estates have fulfilled their position as estates, but we are far from wishing to justify them thereby. The Rhinelander would have to triumph in them over the representative, and the man over the owner of the woods. Even in law it is not only the representation of particular interests but also the representation of the interest of the province that has been entrusted to them. However contradictory both these tasks may be, one should not hesitate in the case of a confrontation to sacrifice the representation of particular interests to that of the province. The feeling for right and law is the most important provincial characteristic of the Rhinelander; but it is self-evident that particular interests know no fatherland and no province either, no cosmopolitanism and no parochialism either.

Those imaginative writers who are pleased to find in romantic idealism a bottomless depth of character and a most fruitful source of peculiarly individual types of attitude in a representation of particular interests are quite wrong: such a representation destroys all natural and spiritual differences in that it enthrones in their place an immoral, foolish, and spiritless abstraction of limited content which is slavishly subordinate to a narrow consciousness. . . .

Letter to Arnold Ruge

This letter, written in November 1842, explains in detail the reasons for Marx's break with his Berlin Young Hegelian friends known as the *Freien*, or Free Men, whose contributions he eventually excluded from the newspaper. He had become increasingly estranged from his former colleagues whose extremism did not permit them, so he thought, to appreciate the difficulties involved in editing a Rhineland newspaper.

. . . A few days ago I received a letter from little Meyen, whose favourite category, and quite rightly, is 'ought', in which there was talk of my relationship: (1) to you and Herwegh; (2) to the Free Men; and (3) of the new principles of editing and the position *vis-à-vis* the government. I replied immediately and gave him my frank opinion of the deficiencies of their work, which finds freedom more in a licentious, sansculottish and thus convenient form than in a free, i.e. independent and profound content. I called for them to show less vague reasoning, fine-sounding phrases, conceited self-admiration, and more precision, more detail on concrete circumstances, and more knowledge of the subject. I explained that I held the smuggling into incidental theatre reviews etc. of communist and socialist dogmas, that is of a new world-view, to be unsuitable and indeed immoral, and that I desired quite a different and more profound discussion of communism if it were to be discussed at all. I then asked that religion should be criticized more within a critique of the political situation than the political situation within a critique of religion, because this approach fits better the nature of a newspaper and the education of the public; for religion has no content of its own and does not live from heaven but from earth and falls automatically with dissolution of the inverted reality whose theory it is. Finally I wished that, if philosophy were to be spoken of, there should be less trifling with the slogan 'atheism' (like children who assure anyone who will listen to them that they are not afraid of an ogre) and more presenting its content to the people. That's all. . . .

On the Estates Committees in Prussia

Marx here attacks an article in the *Augsburger Allgemeine Zeitung* that advocated the institution of the Estates Committees—national advisory committees whose membership was chosen on the basis of Estates. Frederick William IV had decided on their institution as an attempt to meet demands for popular representation. The extract below gives Marx's interpretation, again in a very Hegelian manner, of what 'representation' should mean.

. . . If this political self-reliance of particular interests were a necessity in the state, this would only be a symptom of its inner disease, just as the laws of nature say that an unhealthy body must break out in spots. One must opt for one of two points of view, either that the particular interests overstrain themselves, become alienated from the political spirit of the state and wish to limit the state, or the state concentrates itself in the government alone, and grants the limited spirit of the people as a recompense simply a sphere to ventilate its particular interests. Finally one could make a synthesis of both views. So if the desire for a representation of intelligence is to be meaningful, then we must interpret it as the desire for the conscious representation of the people's intelligence not in order to enforce individual needs against the state but to realize that its highest need is to make the state really its own creation, its own state. To be represented is in general something to be suffered; only the material, spiritless, dependent, insecure need representation; but no element in the state should be material, spiritless, dependent, insecure. Representation should not be conceived of as the representation of some stuff that is not the people itself, but only as its self-representation, as an action of state that only distinguishes itself by the universality of its content from the other manifestations of its political life. Representation must not be looked upon as a concession to defenceless weakness and powerlessness but as the self-conscious vitality of the strongest force. In a true state there is no landed property, no industry, no material stuff that can, as such elements, strike a bargain with the state; there are only spiritual powers and it is only in their resurrection in the state, in their political rebirth, that natural powers are capable of having a political voice. The state has spiritual nerves throughout the whole of nature, and it must appear at every point that not matter but form, not nature without the state but political nature, not the unfree object but the free man, dominates. . . .

Defence of the Moselle Correspondent

This article, too, engaged Marx's interest in economic questions. One of the correspondents of the *Rheinische Zeitung* had exposed the poverty of the wine-growers in the Moselle region. Challenged by the government, Marx defended his correspondent's conclusions. In his view, the problems were not so much the result of human intentions as of objective economic relationships; and a free press could help towards a resolution of such problems.

. . . When investigating political conditions, one is too easily tempted to neglect the objective character of the relationships and to explain everything from the wills of the persons acting. There are relationships, however, which determine both the actions of private persons and of individual authorities, and which are as independent of the will as breathing. If this objective standpoint is taken from the beginning, one will not presuppose an exclusively good or bad will on either side. Rather, one will observe relationships in which, at first, only persons appear to act; and as soon as it is proved that something was necessitated by circumstances, it will not be difficult to work out under which external conditions this thing actually had to come into being, and under which other conditions it could not have come about although a need for it was present. This can be determined with almost the same certainty as a chemist determines under which external conditions given substances will form a compound. . . .

BIBLIOGRAPHY

Original: MEW, Vol. 1: pp. 47 f. 58; pp. 103 f.; p. 108; pp. 113, 115, 116, 146 f.; Vol. 27, p. 412; *MEGA*, I, i (1), pp. 334 f.; *MEW*, Vol. 1, pp. 177.

Present translation: K. Marx, *Early Texts*, pp. 35 f.; pp. 41 f.; pp. 47 f.; pp. 49 f.; p. 53; pp. 56 f.; translated for present edition.

Other translations: Excerpts in *Writings of the Young Marx on Philosophy and Society*, ed. Easton and Guddat, pp. 96 ff.

Commentaries: D. Howard, *The Development of the Marxian Dialectic*, Carbondale, 1972, pp. 24 ff.

A. McGovern, 'Marx's First Political Writings: the *Rheinische Zeitung*—1842–43', in *Demythologising Marxism*, ed. F. Adelmann, The Hague, 1969.

D. McLellan, *Marx before Marxism*, Penguin, 1972, Ch. 4.

4. Critique of Hegel's 'Philosophy of Right'

On the suppression of the *Rheinische Zeitung* in March 1843, Marx moved to his future mother-in-law's house in Kreuznach where he stayed for the following six months, marrying Jenny in June. It was during this stay that he decided to get to grips with Hegel's political philosophy, a project he had had in mind for more than a year. What he produced was a 150-page manuscript in which he copied out Hegel's text paragraph by paragraph and interspersed it with lengthy critical comments. Marx intended to write up his manuscript for publication but left it unfinished.

Marx's general aim was to evaluate Hegel's political philosophy, which gave him scope to criticize existing political institutions and, more broadly, to discuss the question of the relationship of politics to economics. Of the four extracts which follow, the first is a general criticism of Hegel for starting with abstract ideas instead of with concrete reality; the second criticizes Hegel's defence of monarchy and proposes instead a democratic humanism strongly influenced by Feuerbach. The third extract gives a brilliant analysis of bureaucracy's tendency to form a state within a state; and the last proclaims universal suffrage as the solution to the split between civil society (i.e. society where men pursue their individual self-interest) and the political state (i.e. the sphere of man's collective, universal, or 'species' interests).

The manuscript was written when Marx's ideas were in a transient state: he had adopted the fundamental humanism of Feuerbach and, with it, Feuerbach's reversal of subject and predicate in the Hegelian dialectic. He considered it plain that the task ahead was the recovery by man of the social dimension of his nature that had been lost ever since the French Revolution levelled all citizens in the political state and thus accentuated the individualism of bourgeois society. It was clear to Marx that private property must cease to be the basis of social organization, but it is not obvious that he was arguing for its abolition, nor did he make clear the various roles of classes in the social evolution.

On Hegel's Dialectic

... 'The actual idea is Spirit which separates itself into the two ideal spheres of its concept, the family and civil society, which are its finite phase'—thus the separation of the state into family and civil society is ideal, i.e. necessary, is part of the essence of the state; the family and civil society are real parts of the state, real spiritual instances of will, they are the modes of being of the state; family and civil society make themselves into the state. They are the initiators. According to Hegel, they are, on the contrary, created by the actual idea; it is not their own life-process that unites them to the state, it is the life-process of the idea that has distinguished them from itself; indeed they are the finite phase of this idea; they owe their existence to another spirit than their own. They are definitions posited by a third party, not self-definitions. Therefore they too are defined as 'finiteness', as the 'actual idea's' own finitude. The aim of its being is not this being itself, for the idea separates off these presuppositions 'in order to leave its ideality and become explicit as finite actual spirit', that is, the political state cannot exist without the natural basis of the family and the artificial basis of civil society: they are its *conditio sine qua non* [essential condition]. However, the condition is put in the position of the conditioned, the determining of the determined, and the producer is in the position of the product of what it has itself produced. The actual idea only lowers itself to the 'finitude' of the family and civil society so as to transcend them and enjoy and produce its own infinity; thus (in order to attain its aim) it is to these spheres that spirit bestows the material of this its finite actuality (this? which? these spheres are its 'finite actuality', its 'material') i.e. human beings as a mass (the material of the state is here 'individuals, the mass', 'the state is composed by them', this composition is here expressed as an act of the idea, as an attribution that it performs with its own material; the fact is that the state originates in the mass as it exists as members of the family and civil society; speculation expresses this fact as an act of the idea, not as the idea of the mass, but as the act of a subjective idea, distinct from the fact itself), 'so that the function assigned to any given individual' (before only the assigning of individuals to the spheres of family and civil society was spoken of) 'is visibly mediated by circumstances, caprice etc.' Thus empirical reality is taken as it is; it is also declared to be rational, but it is rational not because of its own reason but because the empirical fact in its empirical existence has a meaning other than its own. The fact that served as a beginning is not conceived of as such but as a mystical result. The real becomes an appearance, but the idea has no other content than this appearance. Also the idea has no other aim than the logical one 'to become explicit as infinite actual spirit'. In this paragraph is set down the whole mystery of the philosophy of law and of Hegel's philosophy in general. . . .

On Democracy

. . . Democracy is the truth of monarchy, monarchy is not the truth of democracy. Monarchy is necessarily democracy as an inconsequence against itself, whereas the monarchical element in democracy is no inconsequence. Monarchy cannot, as democracy can, be understood in its own terms. None of the elements of democracy has a different meaning from its own meaning. Each is merely an element of the whole people. In monarchy one part determines the character of the whole. The whole constitution must modify itself in relation to the fixed point. Democracy is the constitution of the species. Monarchy is a variety and a bad one at that. Democracy is content and form. Monarchy ought only to be form, but it falsifies the content.

In monarchy the whole, the people, is subsumed under one of its modes of being, the political constitution; in democracy the constitution itself appears as only one determination, and the self-determination of the people at that. In monarchy we have the people of the constitution; in democracy we have the constitution of the people. Democracy is the solution to the riddle of all constitutions. Here the constitution is constantly, not only in itself and essentially but also in its existence and reality, brought back to its real basis, the real man, the real people, and set up as its own work. The constitution appears as what it is, the free product of man; one could say that this is valid in certain respects for constitutional monarchy also, but the specific difference of democracy is that in it the constitution is nothing more than one element in the being of the people, that the political constitution does not explicitly form the state.

Hegel starts from the state and makes man into the subjective aspect of the state; democracy starts from man and makes the state into objectified man. Just as religion does not make man, but man makes religion, so the constitution does not make the people, but the people makes the constitution. In a certain respect democracy has the same relation to all the other forms of state as Christianity has to all other forms of religion. Christianity is the religion *par excellence*, the essence of religion, deified man as a particular religion. Similarly democracy is the essence of all constitutions of the state, socialized man as a particular constitution of the state; it has the same relationship to other constitutions as the species has to its types, only that in this case, the species itself appears as a particular existence and thus over against existences that do not correspond to the essence, it appears as a particular type. Democracy is the Old Testament in relation to other political forms. Man is not there for the benefit of the law, but the law for the benefit of man; it is a human existence, whereas in other political forms man has only a legal existence. That is the fundamental character of democracy.

All other constructions of the state are a certain, definite, and particular form of the state. In democracy the formal principle is at the same time the material principle. Thus it is first the true unity of universal and particular. In a monarchy, for example, or in a republic, as merely a particular form of the state, political man has his particular existence beside the unpolitical, private man. Property, contract, marriage, civil society appear here (as Hegel develops it quite correctly for these abstract forms of the state, except that he thinks he is developing the idea of the state) as particular modes of being alongside the political state. They appear as the content, to which the political state has the relationship of organizing form, in fact merely a reason that is without content in itself, determining and limiting, now confirming, now denying. In democracy as the political state puts itself beside this content and distinguishes itself from it, it is itself only a particular content and a particular mode of existence of the people. In the monarchy, for

instance, this particular element, the political constitution, has the significance of a universal that dominates and determines all particulars. In democracy the state as a particular is only a particular and as a universal is a real
universal, i.e. is no particular characteristic distinguished from the rest of the
content. The manner in which the most recent French thinkers have conceived of this is that in a true democracy the political state disappears. This
is correct in so far as, *qua* political state and constitution, it is no longer valid
for the whole.

In all states that are not democracies, the state, the law, the constitution is
the dominant factor without really dominating, i.e. materially penetrating all
the other spheres that are not political. In a democracy the constitution, the
law, and the state itself are only a self-determination of the people and a
particular content of them in so far as it is a political constitution.

It is self-evident, moreover, that all forms of state have democracy as their
truth and therefore that they are untrue inasmuch as they are not democracies.

In the old states the political state formed the content of the state and
excluded the other spheres; the modern state is a compromise between the
political and the non-political state.

In democracy, the abstract state has ceased to be the dominant element.
The struggle between monarchy and republic is itself still only a struggle
inside the abstract form of the state. The *political* republic is democracy
inside the abstract form of the state. The abstract political form of democracy is therefore a republic; but here it ceases being only a political constitution.

Property and so on—in short the whole matter of law and the state—is
with little modification the same in North America as in Prussia. Thus there,
a republic is merely a form of the state as here the monarchy is. The content
of the state lies outside its constitution. So Hegel is right when he says: the
political state is the constitution, that is, the matter of the state is not
political. There is only an exterior identity here, a mutual determination. It
would be very difficult to construct the political state and the constitution
from the different elements of the people's life. It developed itself as universal reason over against the other spheres, as something beyond them. The
historical task consisted then in their re-vindication, but the particular
spheres did not realize here that their private essence coincides with the
other-worldly essence of the constitution or the political state, and that its
other-worldly being is nothing but the affirmation of their own alienation.
The political constitution was formerly the religious sphere, the religion of
the people's life, the heaven of its universality over against the earthly and
real existence. The political sphere was the only state sphere in the state, the
only sphere in which the content as well as the form was a content of the
species and the genuine universal; but at the same time this was in such a
manner that, because this sphere stood over against the others, its content

too became a formal and particular one. Political life in the modern state is the scholasticism of the people's life. Monarchy is the perfected expression of this alienation. Republicanism is its negative inside its own sphere. It is evident that the political constitution as such can only be elaborated where the private spheres have obtained an independent existence. Where trade and landed property are not free and have not yet been made independent, the political constitution also is unfree and still not independent. The Middle Ages were the democracy of unfreedom.

The abstraction of the state as such belongs only to the modern time, because the abstraction of private life also belongs only to modern times. The abstraction of the political state is a modern product.

In the Middle Ages there were serfs, feudal property, corporations of trade and of learned men, etc. This means that in the Middle Ages property, trade, society, and men were political; the material content of the state was delimited by its form; each private sphere had a political character or was a political sphere or politics formed the character of the private sphere. In the Middle Ages the political constitution was the constitution of private property, but only because the constitution of private property was the political constitution. In the Middle Ages the people's life and the state's life were identical. Man was the real principle of the state, but it was unfree man. So it is the democracy of unfreedom, perfected alienation. The abstract, reflected opposition only begins with the modern world. The Middle Ages embodied the real dualism and the modern time the abstract dualism. . . .

On Bureaucracy

. . . 'Bureaucracy' is the 'state formalism' of civil society. It is the 'state's consciousness', the 'state's will', the 'state's power' as a corporation and thus a particular, closed society within the state. (The 'general interest' can only maintain itself as a 'particular' *vis-à-vis* particulars, so long as the particular maintains itself as a 'general' *vis-à-vis* the general. Bureaucracy must thus safeguard the imaginary universality of the particular interest, the spirit of the corporation, in order to safeguard the imaginary particularity of the general interest, its own spirit. The state must be a corporation as long as the corporation wishes to be a state.) But bureaucracy wishes the corporation to be an imaginary power. Of course the individual corporation has this same desire for its particular interest against bureaucracy, but it desires bureaucracy against the other corporations, against other particular interest. Therefore bureaucracy, being the completion of the corporation, has the victory over the corporation which is the incomplete bureaucracy. It degrades this latter to an appearance or desires to degrade it to an appearance, but it desires that this appearance exist and believe in its own existence. The corporation is the attempt of civil society to become a state; but bureaucracy is the state that has really made itself into a civil society.

The 'state formalism' that bureaucracy is, is the 'state as formalism' and Hegel has described it as such a formalism. Because this 'state formalism' constitutes itself as a real power and comes to have a material content of its own, then it is self-evident that 'bureaucracy' is a web of 'practical illusions' or the 'illusion of the state'. The bureaucratic spirit is through and through a Jesuitical, theological spirit. The bureaucrats are the Jesuits and theologians of the state. Bureaucracy is the republic as priest.

Since it is of the essence of bureaucracy to be the 'state as formalism', so its aim implies this also. The real aim of the state thus appears to bureaucracy as an aim against the state. The spirit of bureaucracy is therefore the 'formal spirit of the state'. Thus it makes the 'formal spirit of the state' or the real lack of spirit by the state into a categorical imperative. Bureaucracy counts in its own eyes as the final aim of the state. Because 'it makes its 'formal' ends into its content, it enters into conflict everywhere with 'real' ends. It is therefore compelled to claim the formal for its content and its content as the formal. The aims of the state are transformed into the aims of the bureaux and the aims of the bureaux into the aims of the state. Bureaucracy is a circle from which no one can escape. Its hierarchy is a hierarchy of knowledge. The apex entrusts the lower circles with insight into the individual while the lower circles leave insight into the universal to the apex, so they deceive each other reciprocally.

Bureaucracy constitutes an imaginary state beside the real state and is the spiritualism of the state. Thus every object has a dual meaning, a real one and a bureaucratic one, just as knowledge is dual, a real and a bureaucratic (it is the same with the will). But the real thing is treated according to its bureaucratic essence, its other-worldly spiritual essence. Bureaucracy holds in its possession the essence of the state, the spiritual essence of society, it is its private property. The general spirit of bureaucracy is secret, mystery, safeguarded inside itself by hierarchy and outside by its nature as a closed corporation. Thus public political spirit and also political mentality appear to bureaucracy as a betrayal of its secret. The principle of its knowledge is therefore authority, and its mentality is the idolatry of authority. But within bureaucracy the spiritualism turns into a crass materialism, the materialism of passive obedience, faith in authority, the mechanism of fixed and formal behaviour, fixed principles, attitudes, traditions. As far as the individual bureaucrat is concerned, the aim of the state becomes his private aim, in the form of a race for higher posts, of careerism. Firstly he considers the real life as a material one, for the spirit of this life has its own separate existence in bureaucracy. Bureaucracy must therefore make it its job to render life as material as possible. Secondly, for himself life becomes material, i.e. in so far as it becomes an object of bureaucratic procedure, for his spirit is laid down for him, his aim lies outside himself, and his existence is the existence of the bureaucratic. The state only continues to exist as separate fixed spirits of bureaux whose connection is subordination and passive obedience. Real knowledge appears as devoid of content, as real life appears as dead, for this imaginary knowledge and this imaginary life count as essential. . . .

Independence and self-reliance in the political state . . . are achieved by private property whose apogee appears as inalienable landed property. Thus political dependence does not spring from the inner nature of the political state, it is no gift of the political state to its members, it is not the spirit that gives the state a soul. For the members of the political state receive their independence from a thing unconnected with the essence of the political state, a thing of abstract private law, from abstract private property. Political dependence is an accident of private property, not the substance of the political state. The political state and in it the legislative power, as we have seen, is the revealed mystery of the true worth and essence of the elements in the state. The significance that private property has in the political state is its essential, its true significance; the significance that difference in class has in the political state, is the essential significance of difference of class. Similarly the essence of princely power and of government appears in the 'legislative power'. It is here in the sphere of the political state, that the individual elements of the state relate to themselves as to the being of their species, their 'species-being'; for the political state is the sphere of their universal determination, their religious sphere. The political state is the mirror of truth for the different elements of the concrete state. . .

On Voting

. . . As we have seen, the state exists merely as political state. The totality of the political state is the legislature. To participate in the legislature is thus to participate in the political state and to prove and actualize one's existence as member of the political state, as member of the state. That all as individuals want to participate integrally in the legislature is nothing but the will of all to be actual (active) members of the state, or to give themselves a political existence, or to prove their existence as political and to effect it as such. We have further seen that the Estates are civil society as legislature, that they are its political existence. The fact, therefore, that civil society invades the sphere of legislative power *en masse*, and where possible totally, that actual civil society wishes to substitute itself for the fictional civil society of the legislature, is nothing but the drive of civil society to give itself political existence, or to make political existence its actual existence. The drive of civil society to transform itself into political society, or to make political society into the actual society, shows itself as the drive for the most fully possible universal participation in legislative power.

Here, quantity is not without importance. If the augmentation of the Estates is a physical and intellectual augmentation of one of the hostile forces—and we have seen that the various elements of the legislature oppose one another as hostile forces—then the question of whether all as individuals are members of the legislature or whether they should enter the legislature through deputies is the placing in question of the representative principle within the representative principle, i.e. within that fundamental conception of the political state which exists in constitutional monarchy. (1) The notion that the legislature is the totality of the political state is a notion of the abstraction of the political state. Because this one act is the sole political act of civil society, all should participate and want to participate in it at once. (2) All as individuals. In the Estates, legislative activity is not regarded as social, as a function of society, but rather as the act wherein the individuals first assume an actually and consciously social function, that is, a political function. Here the legislature is no derivative, no function of society, but simply its formation. This formation into a legislative power requires that all members of civil society regard themselves as individuals, that they actually face one another as individuals. The abstraction of 'being a member of the state' is their 'abstract definition', a definition that is not actualized in the actuality of their life.

There are two possibilities here: either the separation of the political state and civil society actually obtains, or civil society is actual political society. In the first case, it is impossible that all as individuals participate in the legislature, for the political state is an existent which is separated from civil society. On the one hand, civil society would abandon itself as such if all [its members] were legislators; on the other hand, the political state which stands over against it can tolerate it only if it has a form suitable to the standards of the state. In other words, the participation of civil society in the political state through deputies is precisely the expression of their separation and merely dualistic unity.

Given the second case, i.e. that civil society is actual political society, it is nonsense to make a claim which has resulted precisely from a notion of the political state as an existent separated from civil society, from the theological notion of the political state. In this situation, legislative power altogether loses the meaning of representative power. Here, the legislature is a representation in the same sense in which every function is representative. For example, the shoemaker is my representative in so far as he fulfills a social need, just as every definite social activity, because it is a species-activity, represents only the species; that is to say, it represents a determination of my own essence the way every man is the representative of the other. Here, he is representative not by virtue of something other than himself which he represents, but by virtue of what he is and does.

Legislative power is sought not for the sake of its content, but for the sake of its formal political significance. For example, executive power, in and for itself, has to be the object of popular desire much more than legislative power, which is the metaphysical political function. The legislative function is the will, not in its practical but in its theoretical energy. Here, the will should not preempt the law; rather, the actual law is to be discovered and formulated.

Out of this divided nature of the legislature—i.e. its nature as actual lawgiving function and at the same time representative, abstract-political function—stems a peculiarity which is especially prevalent in France, the land of political culture.

(We always find two things in the executive: the actual deed and the state's reason for this deed, as another actual consciousness, which in its total organization is the bureaucracy.)

The actual content of legislative power (so long as the prevailing special interests do not come into significant conflict with the *objectum quaestionis*) is treated very much *à part*, as a matter of secondary importance.

A question attracts particular attention only when it becomes political, that is to say, either when it can be tied to a ministerial question, and thus becomes a question of the power of the legislature over the executive, or when it is a matter of rights in general, which are connected with the political formalism. How has this phenomenon come about? Because the legislature is at the same time the representation of civil society's political existence; because in general the political nature of a question consists in its relationship to the various powers of the political state; and, finally, because the legislature represents political consciousness, which can manifest itself as political only in conflict with the executive. There is the essential demand that every social need, law, etc., be investigated and identified politically, that is to say, determined by the whole of the state in its social sense. But in the abstract political state this essential demand takes a new turn; specifically , it is given a formal change of expression in the direction of another power (content) besides its actual content. This is no abstraction of the French, but rather the inevitable consequence of the actual state's existing merely as the political state formalism examined above. The opposition within the representative power is the   ἐξοχὴν   (*par excellence*) political existence of the representative power. Within this representative constitution, however, the question under investigation takes a form other than that in which Hegel considered it. It is not a question of whether civil society should exercise legislative power through deputies or through all as individuals. Rather, it is a question of the extension and greatest possible universalization of voting, of active as well as passive suffrage. This is the real point of dispute in the matter of political reform, in France as well as in England.

Voting is not considered philosophically, that is, not in terms of its proper nature, if it is considered in relation to the crown or the executive. The vote is the actual relation of actual civil society to the civil society of the legis-lature, to the representative element. In other words, the vote is the immediate, the direct, the existing and not simply imagined relation of civil society to the political state. It therefore goes without saying that the vote is the chief political interest of actual civil society. In unrestricted suffrage, both active and passive, civil society has actually raised itself for the first time to an abstraction of itself, to political existence as its true universal and essential existence. But the full achievement of this abstraction is at once also the transcendence of the abstraction. In actually establishing its political existence as its true existence civil society has simultaneously established its civil existence, in distinction from its political existence, as inessential. And with the one separated, the other, its opposite, falls. Within the abstract political state the reform of voting advances the dissolution of this political state, but also the dissolution of civil society. . . .

BIBLIOGRAPHY

Original: MEGA, I (1) 1, pp. 401–553.

Present translation: First three extracts, Karl Marx, *Early Texts*, pp. 63–70; fourth extract Karl Marx, *Critique of Hegel's 'Philosophy of Right'*, ed. J. O'Malley, Cambridge, 1970, pp. 118 ff.

Other translations: Substantial excerpts in *Writings of the Young Marx*, ed. Easton and Guddat, pp. 151 ff. Complete translation: edition by J. O'Malley, cited above. K. Marx, *Early Writings*, ed. L. Colletti, New York and London, 1975.

Commentaries: S. Avineri, 'Marx's Critique of Hegel's "Philosophy of Right" in its Systematic Setting', *Cahiers de L'Isea*, August 1966.

S. Avineri, 'The Hegelian Origins of Marx's Political Thought', *The Review of Metaphysics*, 1967.

R. Berki, 'Perspectives in the Marxian Critique of Hegel's Political Philosophy', in *Hegel's Political Philosophy*, ed. Z. Pelczynski, Cambridge, 1971.

L. Dupré, *The Philosophical Foundations of Marxism*, New York, 1966, Ch. 4.

D. Howard, *The Development of the Marxian Dialectic*, Carbondale, 1972, pp. 48 ff.

J. Hyppolite, 'Marx's Critique of Hegel's Philosophy of the State', in J. Hyppolite, *Studies on Marx and Hegel*, London and New York, 1969.

D. McLellan, *Marx before Marxism*, Penguin, 1972, Ch. 5.

J. O'Malley, Introduction to edition cited above.

5. A Correspondence of 1843

With the demise of the *Rheinische Zeitung*, Marx began collaborating with Ruge on a Franco–German monthly, entitled *Deutsch–französische Jahrbücher*, to be published in Paris and to propagate a synthesis of German philosophy and French socialism. The *Jahrbücher* began with an exchange of letters between Marx, Ruge, Bakunin, and Feuerbach, which were designed to work out an ideological point of view for the journal. Marx wrote the two extracts below in May and September 1843 respectively. In the first he outlines his hopes for a union between intellectuals and 'suffering humanity'; in the second, drawing on Feuerbach's humanism and Hegel's views on reason in history, he criticizes contemporary forms of socialism and communism, and appeals for a reform of consciousness by exposing the roots of contemporary problems.

... The interior difficulties almost seem to be even greater than the exterior ones. For even though the 'whence' is not in doubt, yet all the more confusion reigns over the 'whither'. It is not only that a general anarchy has burst out among the reformers. Everyone will have to admit to himself that he has no exact view of what should happen. However, that is just the advantage of the new line that we do not anticipate the world dogmatically but wish to discover the new world by criticism of the old. For before, philosophers had the answer to all riddles lying in their desks and the stupid exterior world has only to open its mouth for the roasted pigeons of absolute knowledge to fly into it. Philosophy has become secularized and the most striking proof of this is that the philosophical mind itself is not merely in an exterior way drawn into the painful struggle but also in its inner nature. If our job is not building a future that will last for all ages, what we do have to accomplish now is all the more certain, I mean the reckless critique of all that exists, reckless in the sense that the critique is neither afraid of its own results nor of conflicting with the powers that be. ...

The existence of a suffering humanity which thinks and a thinking humanity which is oppressed must of necessity be disagreeable and unacceptable for the animal world of philistines who neither act nor think but merely enjoy.

On our side the old world must be brought right out into the light of day and the new one given a positive form. The longer that events allow thinking humanity time to recollect itself and suffering humanity time to assemble itself, the more perfect will be the birth of the product that the present carries in its womb. ...

And again the whole socialist principle is only one facet of the true reality of the human essence. We have just as much to take into account the other facet, the theoretical existence of man, and make religion, science, etc. the object of our critique. Moreover, we wish to have an effect on our contemporaries, and more particularly on our German contemporaries. The question is how to go about it. Two facts cannot be denied. Religion and politics are the twin subjects in which contemporary Germany is chiefly interested. We must start from these subjects in whatever state they are, and not oppose them with some ready-made system, like, for example, the *Voyage en Icarie* [a Utopian work by Etienne Cabet].

The reason has always existed, but not always in a rational form. Thus, the critic can start from any form that man's mind has taken, theoretical or practical, and develop out of the actual forms of existing reality the true reality as what it ought to be, that which is its aim. Now, as regards actual life, the political state, even where it is not yet consciously impregnated with socialist principles, contains in all its modern forms the demands of reason. Nor does it stop there. It presupposes everywhere the realization of reason. But in this way its ideals come into conflict everywhere with its real suppositions.

Thus, the social truth emerges everywhere out of this conflict of the political state with itself. As religion is the table of contents of the theoretical battles of mankind, so is the political state of its practical ones. So inside its republican form the political state expresses all social struggles, needs, and truths. We do not, therefore, sacrifice any of our principles when we make the exclusively political questions—for example, the difference between the estates and the representative system—the object of our critique. For this question really expresses in a political manner the difference between the lordship of man and the lordship of private property. Thus, the critic not only can but also must go into these political questions which, in the opinion of the crass socialists, are beneath all value. In that he develops the advantages of a representative system over one of the estates, he interests a large party in a practical manner. In that he raises the representative system from its political form to a universal one and thus gives force to its true and fundamental meaning, he compels this party to go beyond itself, for its victory implies its dissolution.

So there is nothing to stop us from making a critique of politics the starting-point of our critique, from taking part in party politics and so identifying ourselves with real battles. We do not then set ourselves opposite the world with a doctrinaire principle, saying: 'Here is the truth, kneel down here!' It is out of the world's own principles that we develop for it new principles. We do not say to her, 'Stop your battles, they are stupid stuff. We want to preach the true slogans of battle at you.' We merely show it what it is actually fighting about, and this realization is a thing that it must make its own even though it may not wish to.

The reform of consciousness consists solely in letting the world perceive its own consciousness by awaking it from dreaming about itself, in explaining to it its own actions. Our whole and only aim consists in putting religious and political questions in a self-conscious, human form, as is also the case in Feuerbach's critique of religion.

So our election cry must be: reform of consciousness not through dogmas, but through the analysis of mystical consciousness that is not clear to itself, whether it appears in a religious or political form. It will then be clear that the world has long possessed the dream of a thing of which it only needs to possess the consciousness in order really to possess it. It will be clear that the problem is not some great gap between the thoughts of the past and those of the future but the completion of thoughts of the past. Finally, it will be clear that humanity is not beginning a new work, but consciously bringing its old work to completion.

So we can summarize the tendency of our journal in one word: self-understanding (equals critical philosophy) by our age of its struggles and wishes. This is a task for the world and for us. It can only be the result of united forces. What is at stake is a confession, nothing more. To get its sins forgiven, humanity only needs to describe them as they are.

BIBLIOGRAPHY

Original: *MEGA*, I (1) 1, pp. 566, 572 ff.

Present translation: K. Marx, *Early Texts*, pp. 79 ff.

Other translations: *Writings of the Young Marx on Philosophy and Society*, ed. Easton and Guddat, pp. 210 ff. K. Marx, *Early Writings*, ed. L. Colletti, New York and London, 1975, pp. 199 ff.

Commentary: D. Howard, *The Development of the Marxian Dialectic*, Carbondale, 1972, pp. 84 ff.

D. McLellan, *Marx before Marxism*, Penguin, 1972, pp. 164 ff.

6. On the Jewish Question

Marx wrote this review article for the *Deutsch–französische Jahrbücher* in Kreuznach before he left for Paris in October 1843. Bruno Bauer had recently published two essays asserting that, in order to be able to live together, both Jews and Christians had to renounce what separated them—religion. Thus it was not only Jews but all men who needed emancipation. Civil rights were inconceivable in an absolute state with an established religion. Religious prejudice and religious separation would vanish when civil and religious castes and privileges were abolished and all men made 'equal' in the sense of the French Revolution or the American Constitution. Marx agrees with Bauer but complains that he has not gone far enough: Bauer subjects to criticism only the 'Christian state' and not the state as such, and thus fails to examine the relationship of political emancipation—that is, the granting of merely political rights—to human emancipation.

Marx goes on to point out that the mere disestablishment of religion does not abolish religious beliefs or the social ills that give rise to those beliefs, and cites the United States as an example. He then examines the relationship of the abstract political state to civil society, and demonstrates how this divides man into the 'citizen' (member of the universal state) and the 'bourgeois' (self-interested member of civil society). Turning from the rights of the citizen, Marx looks at the rights of man or natural rights, and criticizes their basic assumption that man is an essentially selfish creature. His solution is to abolish the gap between civil society and the state by making real in civil society the universal, communal, or 'species' essence of man inherent in the state.

I

On The Jewish Question

By Bruno Bauer

The German Jews seek emancipation. What sort of emancipation do they want? Civil, political emancipation. Bruno Bauer answers them: No one in Germany is politically emancipated. We ourselves are not free. How then could we liberate you? You Jews are egoists if you demand a special emancipation for yourselves as Jews. You ought to work as Germans for the political emancipation of Germany, and as men for the emancipation of mankind, and consider your particular sort of oppression and ignominy not as an exception to the rule but rather as a confirmation of it.

Or do the Jews want to be placed on an equal footing with Christian subjects? But in that case they recognize the Christian state as justified, and acquiesce in a regime of general enslavement. Why are they not pleased with their particular yoke when they are pleased with the general yoke? Why should the German interest himself in the emancipation of the Jews if the Jew does not interest himself in the liberation of the German?

The Christian state is only acquainted with privileges. In it the Jew possesses the privilege of being a Jew. As a Jew, he has rights that the Christian does not have. Why does he wish for rights that Christians enjoy and he does not have?

The wish of the Jew to be emancipated from the Christian state entails a demand that the Christian state should give up its religious prejudice. But does the Jew give up his own religious prejudice? Does he then have the right to demand of another that he forswear his religion? It is the very nature of the Christian state that prevents it from emancipating the Jew; but, adds Bauer, it is also the nature of the Jew that prevents his being emancipated. As long as the state is Christian and the Jew Jewish, the one is as incapable of bestowing emancipation as the other is incapable of receiving it.

The Christian state can only have its typical, i.e. privileged relationship to the Jew by permitting the separation of the Jew from the other subjects, but at the same time subjecting him to a pressure from the other separated spheres that is all the heavier since the Jew stands in religious opposition to the dominant religion. But likewise the Jew can only have a Jewish relationship to the state and treat it as alien to himself, for he opposes his own imaginary nationality to actual nationality, and his own imaginary law to actual law, fancies himself justified in separating himself from humanity, as a matter of principle takes no part in the movement of history, and waits on a destiny that has nothing in common with the destiny of mankind as a whole. He considers himself a member of the Jewish people and the Jewish people as the chosen people.

On what grounds then do you Jews seek emancipation? On account of your religion? But it is the mortal enemy of the state religion. As citizens? There are no citizens in Germany. As human beings? You are no more human beings than those to whom you appeal.

After a critical review of the way the question of Jewish emancipation was previously formulated and solved, Bauer frames the question in a new way. How, he asks, are they constituted, the Jew who is to be emancipated and the Christian state which is to do the emancipating? His answer consists in a critique of the Jewish religion; he analyses the religious opposition between Judaism and Christianity and explains the nature of the Christian state in a way that is bold, acute, witty, and thorough, and in a style as precise as it is pithy and energetic.

What, then, is Bauer's solution to the Jewish question and what is the result? To formulate a question is already to solve it. The critique of the Jewish question is the answer to it. Here is a resumé:

We must emancipate ourselves before we can emancipate others.

The most flexible form of the opposition between Christian and Jew is the religious opposition. How is an opposition to be done away with? By making it impossible. How does one make a religious opposition impossible? By abolishing religion. As soon as Jew and Christian recognize their opposed religions as merely different stages in the development of the human spirit, as different snake skins that history has cast off, and recognize man as the snake that used the skins for covering, then they will no longer be in religious opposition but only in a critical, scientific, human opposition. Science is thus their unity, and contradictions in science are solved by science itself.

The German Jew in particular suffers from the general lack of political emancipation and the pronounced Christianity of the state. In Bauer's opinion, however, the Jewish question has a general significance that is independent of specifically German circumstances. It is the question of the relationship of religion to the state, of the opposition between religious prejudice and political emancipation. Emancipation from religion is laid down as a precondition both for the Jew who desires to be politically emancipated and for the emancipating state which itself needs emancipation.

'Fine, people say (the Jew himself included), the Jew is not to be emancipated as a Jew, because he is a Jew, because he has universal human moral principles that are so outstanding; rather his Jewishness will take second place to his citizenship and he will be a citizen in spite of his being and remaining a Jew. In other words he is and remains a Jew in spite of his being a citizen and living in a condition similar to other men. For his narrow Jewish nature always in the end triumphs over his human and political obligations. The prejudice remains even though it is overcome by universal principles. But if it does remain then it would be more correct to say that it is the prejudice that overcomes everything else.

'The Jew would only be able to remain a Jew in the life of the state in a sophistical sense, that is, in appearance only; so if he wished to remain a Jew, the appearance would become what was essential and gain the upper hand. This means that his life in the state would become only an appearance or a momentary exception to the rule governing the real nature of things' ('The Capability of Present-day Jews and Christians for Liberation', *Twenty-One Sheets*, p. 57).

Let us listen, on the other hand, to how Bauer formulates the task of the state: 'France', it runs, 'has recently (Debate of the Chamber of Deputies for the 26th December 1840) given us apropos of the Jewish question a glimpse of a free life, as she does continually in all other political questions since the July Revolution. But she has revoked her freedom by law, thus declaring it to be a sham and on the other hand she has contradicted her free law by her actions' (*The Jewish Question*, p. 64).

'Universal freedom has not yet been established by law in France and the Jewish question still not solved because legal freedom, which consists in the equality of all citizens, is limited in practice since life is still dominated and divided by religious privileges, and this lack of freedom reacts on the law and forces it to agree to the division of citizens wo are in principle free, into oppressors and oppressed' (p. 65).

When, therefore, would the Jewish question in France be solved?

'The Jew, for example, would have had to cease being a Jew if he were to refuse to let his law stop him from fulfilling his duties to the state and his fellow citizens, for example, going to the Chamber of Deputies on the Sabbath and taking part in public debates. Any religious privilege at all, including, therefore, the monopoly of a privileged church, must be abolished and if some or many or even the overwhelming majority still believe themselves bound to fulfil their religious duties, then this must be allowed them as a purely private affair' (p. 65). 'Religion no longer exists when there is no longer a privileged religion. Take from religion its power of exclusion and it ceases to exist' (p. 66). 'Herr Martin du Nord was of the opinion that the proposal to omit the mention of Sunday in the law was equivalent to a motion declaring that Christianity had ceased to exist: a declaration that the abolition of the Sabbath law for the Jews would be equivalent to a proclamation of the dissolution of Judaism would be just as perfectly justified' (p. 71).

So Bauer requires on the one hand that the Jew give up Judaism and man in general give up religion in order to achieve civil emancipation. On the other hand it follows that for him the political abolition of religion is the equivalent of the abolition of all religion. The state that presupposes religion is not yet a true and real state. 'Of course religious ideas afford the state guarantees. But what state? What sort of state?' (p. 97).

It is here that Bauer's one-sided approach to the Jewish question appears.

It is in no way sufficient to inquire: Who should emancipate? Who should be emancipated? A proper critique would have a third question—*what sort of emancipation* is under discussion? What preconditions are essential for the required emancipation? It is only the critique of political emancipation itself that would be the final critique of the Jewish question itself that would be the final critique of the Jewish question and its true resolution into 'the general problems of the age'.

Bauer falls into contradictions because he does not formulate the question at this level. He poses conditions that are not grounded in the nature of political emancipation itself. He raises questions not contained within the problem and solves problems that leave his questions unanswered. Bauer says of the opponents of Jewish emancipation: 'Their one fault was that they presupposed the Christian state as the only true one and did not subject it to the same critique to which they subjected Judaism' (p. 3). Here Bauer's fault lies in the fact that he subjects only the Christian state to his critique, not 'the state as such'. That he does not investigate the relationship of political to human emancipation and thus poses conditions that are only explicable by supposing an uncritical confusion of political emancipation and universal human emancipation. Bauer asks the Jews: Does your standpoint give you the right to seek political emancipation? But we ask the reverse question: Has the standpoint of political emancipation the right to require from the Jews the abolition of Judaism and from all men the abolition of religion?

The Jewish question always presents itself differently according to the state in which the Jew lives. In Germany, where there is no political state, no state as such, the Jewish question is a purely theological one. The Jew finds himself in religious opposition to the state which recognizes Christianity as its foundation. This state is a professed theologian. Criticism is here criticism of theology, a two-sided criticism of Christian and of Jewish theology. But we are still always moving inside theology however critically we may be moving.

In France, which is a constitutional state, the Jewish question is a question of constitutionalism, a question of the incompleteness of political emancipation. Since here the appearance of a state religion is retained although in an empty and self-contradictory formula, namely that of the religion of the majority, the relationship of the Jew to the state contains the appearance of a religious or theological opposition.

It is in the North American states—or at least a part of them—that the Jewish question loses its theological importance for the first time and becomes a really *secular* question. It is only where the political state exists in its complete perfection that the relationship of the Jew and of the religious man in general to the political state, and thus the relationship of religion to the state, can stand out in all its peculiarities and purity. The criticism of this relationship ceases to be a theological criticism as soon as the state ceases to have a theological attitude to religion, as soon as it adopts the attitude of a state towards religion, i.e. a political attitude. Criticism then becomes a criticism of the political state. At this point, where criticism ceases to be theological, Bauer's criticism ceases to be critical. 'There is in America neither state religion nor a religion declared to be that of the majority, nor pre-eminence of any one way of worship over another. The state is stranger to all forms of worship' (G. de Beaumont, *Mary or Slavery in the U.S. . . .,* Paris, 1835, p. 214). There are even some North American states where 'the constitution does not impose religious belief and practice as a condition of political rights' (loc. cit., p. 225). And yet 'people in the U.S. do not believe that a man without religion can be an honest man' (loc. cit., p. 224). Yet North America is the land of religiosity *par excellence* as Beaumont, Tocqueville, and the Englishman Hamilton all aver with one voice. But the North American states are serving here only as an example. The question is: what is the relationship of complete political emancipation to religion? The fact that even in the land of completed political emancipation we find not only the existence of religion but a living existence full of freshness and strength furnishes us with the proof that the existence of religion does not contradict or impede the perfection of the state. But since the existence of religion is the existence of a defect, the source of this defect can only be sought in the nature of the state itself. Religion for us no longer has the force of a basis for secular deficiencies but only that of a phenomenon. Therefore we explain the religious prejudice of free citizens by their secular prejudice. We do not insist that they must abolish their religious limitation in order to abolish secular limitations. We insist that they abolish their religious limitations as soon as they abolish their secular limitations. We do not change secular questions into theological ones. We change theological questions into secular ones. History has for long enough been resolved into superstition: we now resolve superstition into history. The question of the relationship of political emancipation to religion becomes for us a question of the relationship of political emancipation to human emancipation. We criticize the religious weakness of the political state by criticizing the secular construction of the political state without regard to its religious weaknesses. We humanize the opposition of the state to a particular religion, Judaism for example, into the opposition of the state to particular secular elements, and the opposition of the state to religion in general into the opposition of the state to its own presuppositions in general.

The political emancipation of the Jew, the Christian, and religious man in general implies the emancipation of the state from Judaism, Christianity, and religion in general. The state as state emancipates itself from religion in the manner peculiar to its own nature by emancipating itself from the state religion, i.e. by not recognizing, as a state, any religion, by affirming itself simply as a state. Political emancipation is not the completed and consistent form of religious emancipation because political emancipation is not the completed and consistent form of human emancipation.

The limitations of political emancipation are immediately evident in the fact that a state can liberate itself from a limitation without man himself being truly free of it and the state can be a free state without man himself being a free man. Bauer himself tacitly admits this when he poses the following condition for political emancipation: 'Every single religious privilege, including the monopoly of a privileged church, must be abolished. If several or more or even the overwhelming majority of people still felt obliged to fulfil their religious duties, this practice should be left to them as a completely private matter.' Therefore the state can have emancipated itself from religion, even when the overwhelming majority of people is still religious. And the overwhelming majority does not cease to be religious simply because its religion is private.

But the attitude of the state, especially the free state, to religion is merely the attitude of the men who make up the state to religion. It follows from this that man liberates himself from an impediment through the medium of the state and politically by entering into opposition with himself and getting round this impediment in an abstract, limited, and partial manner. It follows also that when man liberates himself politically, he liberates himself by means of a detour, through the medium of something else, however neces-

sary that medium may be. It follows finally that man, even when he proclaims himself an atheist through the intermediary of the state, i.e. when he
proclaims the state to be atheist, still retains his religious prejudice, just
because he recognizes himself only by a detour and by the medium of
something else. Religion is precisely the recognition of man by detour
through an intermediary. The state is the intermediary between man and his
freedom. As Christ is the intermediary onto whom man unburdens all his
divinity, all his religious bonds, so the state is the mediator onto which he
transfers all his Godlessness and all his human liberty.

The political elevation of man above religion shares all the deficiencies
and all the advantages of political elevation in general. The state as state
annuls private property, for example, as soon as man declares in a political
manner that private property is abolished, as soon as he abolishes the requirement of a property qualification for active and passive participation at
elections, as has happened in many North American states. Hamilton inter-
prets this fact from the political standpoint quite correctly: 'the 'masses have
thus gained a victory over the property owners and monied classes'. Is
private property not abolished ideally speaking when the non-owner has
become the lawgiver for the owner? The census is the last political form of
recognizing private property.

And yet the political annulment of private property has not only not
abolished private property, it actually presupposes it. The state does away
with difference in birth, class, education, and profession in its own manner
when it declares birth, class, education, and profession to be unpolitical
differences, when it summons every member of the people to an equal
participation in popular sovereignty without taking the differences into consideration, when it treats all elements of the people's real life from the point
of view of the state. Nevertheless the state still allows private property,
education, and profession to have an effect in their own manner, that is as
private property, as education, as profession, and make their particular
natures felt. Far from abolishing these factual differences, its existence rests
on them as a presupposition, it only feels itself to be a political state and
asserts its universality by opposition to these elements. Therefore Hegel
defines the relationship of the political state to religion quite rightly when he
says: 'In order for the state to come into existence as the self-knowing ethical
actuality of spirit, it is essential that it should be distinct from the form of
authority and of faith. But this distinction emerges only in so far as divisions
occur within the ecclesiastical sphere itself. It is only in this way that the
state, above the particular churches, has attained to the universality of
thought—its formal principle—and is bringing this universality into existence.' [1942, p. 173] Of course! only thus does the state build its universality over and above its particular elements.

The perfected political state is by its nature the species-life of man in
opposition to his material life. All the presuppositions of this egoistic life

continue to exist in civil society outside the sphere of the state, but as proper to civil society. When the political state has achieved its true completion, man leads a double life, a heavenly one and an earthly one, not only in thought and consciousness but in reality, in life. He has a life both in the political community, where he is valued as a communal being, and in civil society, where he is active as a private individual, treats other men as means, degrades himself to a means, and becomes the plaything of alien powers. The political state has just as spiritual an attitude to civil society as heaven has to earth. It stands in the same opposition to civil society and overcomes it in the same manner as religion overcomes the limitations of the profane world, that is, it must likewise recognize it, reinstate it, and let itself once more be dominated by it. Man in the reality that is nearest to him, civil society, is a profane being. Here where he counts for himself and others as a real individual, he is an illusory phenomenon. In the state, on the other hand, where man counts as a species-being, he is an imaginary participant in an imaginary sovereignty, he is robbed of his real life and filled with an unreal universality.

The conflict with his citizenship and with other men as members of the community in which man as an adherent of a particular religion finds himself can be reduced to the secular division between political state and civil society. For man as a bourgeois 'life in the state is only an apparent and momentary exception to the essential rule'. [In this passage Marx uses 'bourgeois' to mean a member of civil society, and 'citizen' to mean an individual with political rights.] Of course the bourgeois, like the Jew, only remains in the life of the state sophistically speaking, just as the citizen only sophistically remains a Jew or bourgeois; but this sophism is not a personal matter. It is a sophism of the political state itself. The difference between the religious man and the citizen is the difference between the trader and the citizen, between the labourer and the citizen, between the property owner and the citizen, between the living individual and the citizen. The opposition to the political man in which the religious man finds himself is the same opposition in which the bourgeois finds himself to the citizen and the member of civil society to his political lion's skin.

This secular strife to which the Jewish question can in the last analysis be reduced—the relationship of the political state to its presuppositions, whether these be material elements like private property or intellectual like education, religion, the conflict between general and private interests, the rift between the political state and the civil society—these secular oppositions are left intact by Bauer while he polemicizes against their religious expressions. 'It is precisely the same need which is the basis of civil society, ensures its continued existence, and guarantees its necessity that also exposes its existence to perpetual dangers, sustains an unsure element within it, produces the continuing oscillating mixture of wealth and poverty, need and superfluity, and in general creates change' (p. 8).

Compare the whole section entitled 'Civil Society' (pp. 8–9), which is drafted from the main points of Hegel's 'Philosophy of Right'. Civil society in its opposition to the political state is recognized as necessary because the political state is recognized as necessary.

Political emancipation is of course a great progress. Although it is not the final form of human emancipation in general, it is nevertheless the final form of human emancipation inside the present world order. It is to be understood that I am speaking here of real, practical emancipation.

Man emancipates himself politically from religion by banishing it from the field of public law and making it a private right. Religion is no longer the spirit of the state where man behaves, as a species-being in community with other men albeit in a limited manner and in a particular form and a particular sphere: religion has become the spirit of civil society, the sphere of egoism, the *bellum omnium contra omnes* [war of all against all]. Its essence is no longer in community but in difference. It has become the expression of separation of man from his common essence, from himself and from other men, as it was originally. It is still only the abstract recognition of a particular perversion, private whim, and arbitrariness. For example, the infinite splintering of religion in North America already gives it the exterior form of a purely individual affair. It is shoved away into the crowd of private interests and exiled from the common essence as such. But we should not be deceived about the limitations of political emancipation. The separation of man into a public and a private man, the displacement of religion from the state to civil society is not a stage but the completion of political emancipation, which thus does not abolish or even try to abolish the actual religiosity of man.

The decomposition of man into Jew and citizen, protestant and citizen, religious man and citizen, this decomposition is no trick played upon political citizenship, no avoidance of political emancipation. It is political emancipation itself, the political manner of emancipating oneself from religion. Of course, in times when the political state is born violently as such out of civil society, when man's self-liberation tries to complete itself in the form of political self-liberation, the state must go as far as abolishing, destroying religion, but only in the same way as it goes as far as abolishing private property, at the most, by declaring a maximum, by confiscation or a progressive tax, or in the same way as it abolishes life, by the guillotine. In moments of particular self-consciousness political life tries to suppress its presuppositions, civil society and its elements, and to constitute itself as the real, harmonious life of man. However, this is only possible through violent opposition to its own conditions, by declaring the revolution to be permanent. The political drama therefore ends necessarily with the restoration of religion, private property, and all the elements of civil society, just as war ends with peace.

Indeed, it is not the so-called Christian state, that one that recognizes

Christianity as its basis, as the state religion, and thus adopts an exclusive attitude to other religions, that is the perfected Christian state, but rather the atheist state, the democratic state, the state that downgrades religion to the other elements of civil society. If the state is still a theologian, makes an official confession of the Christian faith, and does not yet dare to declare itself a state, then it has not yet succeeded in expressing its human basis, of which Christianity in the transcendental expression, in a secular, human form, in its reality as a state. The so-called Christian state is quite simply the non-state because it is only the human background of Christianity and not Christianity itself that can be translated into real human achievements.

The so-called Christian state is the Christian negation of the state, but in no way the state realization of Christianity. The state that still recognizes Christianity in the form of a religion, does not yet recognize it in a political form because it still has a religious attitude to religion, that is, it is not the real elaboration of the human basis of religion because it still accepts the unreal, the imaginary form of this human kernel. The so-called Christian state is the imperfect state and the Christian religion serves as a supplement and a sanctification of its imperfection. Religion therefore necessarily becomes a means for the state, and the state is one of hypocrisy. There is a great difference between the perfect state counting religion as one of its presuppositions because of the deficiencies in the general essence of the state, and the imperfect state declaring religion to be its foundation because the deficiencies in its particular existence make it a deficient state. In the latter case religion becomes imperfect politics. In the former the imperfection of even a perfect politics shows itself in religion. The so-called Christian state needs the Christian religion in order to complete itself as a state. The democratic state, the true state, does not need religion for its political completion. Rather it can abstract from religion, because it realizes the human foundations of religion in a secular manner. The so-called Christian state, on the other hand, has a political attitude towards religion and a religious attitude towards politics. When it degrades the forms of the state to an appearance, then it degrades religion just as surely to an appearance.

In order to explain this opposition, we shall examine Bauer's model of the Christian state, a model that derives from a study of the Christian Germanic state.

'In order to prove', says Bauer, 'the impossibility or non-existence of a Christian State, people have often recently pointed to the sayings in the Gospel which the present state does not only not follow, but cannot even begin to follow if it does not wish to bring about its complete dissolution as a state.' 'But the matter is not dealt with so easily. What do those sayings in the Gospel demand? Supernatural self-denial, subjection to the authority of revelation, disregard of the state, abolition of secular relationships. But the Christian state demands and performs all this. It has made the spirit of the Gospel its own, and if it does not repeat it in the same words that the Gospel uses, that is only because it expresses this spirit in political forms, that is, in forms that are certainly borrowed from the nature of the state and this world but which, in the religious rebirth that they must experience, are degraded to an appearance. Its disregard of the state is realized and completed through the political institutions' (p. 55).

Bauer now further develops the theme of how the people in a Christian state are merely non-people, have no more will of their own, and have their true existence in their leader to whom they are subject and who is nevertheless alien to them in origin and nature since he is God-given and arrived at without their own co-operation; Bauer also explains how the laws of this people are not their own work but direct revelations; how the supreme leader needs privileged intermediaries with his own people and the masses; how the masses themselves disintegrate into a number of particular groups formed and defined by chance which differentiate themselves through their interests, particular passions and prejudices, and obtain as a privilege the permission mutually to exclude each other, etc. (p. 56).

But Bauer himself says: 'Politics, if it is to be nothing but religion, cannot be politics; any more than dishwashing, if it has the force of a religious practice, should be treated as a household matter' (p. 108). In the Christian Germanic state, however, religion is a 'household matter' just as 'household matters' are religious. In the Christian Germanic state the dominance of religion is the religion of dominance.

The separation of the 'spirit of the Gospel' from the 'letter of the Gospel' is an irreligious act. The state which lets the Gospel speak political words, in words different from the Holy Spirit, commits sacrilege in its own religious eyes if not in the eyes of men. The state that recognizes Christianity as its highest norm and the Bible as its Magna Carta must be met with the words of the Holy Scripture, for every word of Scripture is holy. Both this state and the dregs of humanity on which it is based arrive at a painful contradiction that is insurmountable from the point of view of religious consciousness, if it has pointed out to it those sayings of the Gospel with which it 'does not conform and cannot conform unless it wishes to dissolve itself entirely'. And why does it not wish to dissolve itself entirely? It can give neither itself nor others an answer to this question. In its own consciousness the Christian state is an ideal whose realization is unattainable. It can only convince itself of its own existence by lies and so remains for ever an object of self-doubt, an insufficient, problematic object. Thus criticism is fully justified when it forces the state that appeals to the Bible into a crazed state of mind where it no longer knows whether it is an imagination or a reality, where the infamy of its worldly ends for which religion serves as a cloak arrives at an insoluble conflict with the honesty of its religious consciousness which views the final aim of the world as religion. This state can only pacify its inner uneasiness by becoming a myrmidon of the Catholic Church. In the face of the Catholic Church, which declares secular powers to be its bondsmen, the state is as powerless as is the secular power which affirms itself to be dominant over the religious spirit.

In the so-called Christian state it is alienation that is important, not man himself. The man who is important, the king, is a being specifically differentiated from other men (which is itself a religious conception), who is in direct contact with heaven and God. The relationships that hold sway here are ones of faith. The religious spirit is thus not yet really secularized.

But the religious spirit can never really be secularized. For what is it but the unsecular form of a stage in the development of the human spirit? The religious spirit can only be secularized in so far as the stage in the development of the human spirit whose religious expression it is emerges and constitutes itself in its secular form. This happens in the democratic state. The foundation of this state is not Christianity but the human foundation of Christianity. Religion remains as the ideal, unsecular consciousness of its members, because it is the ideal form of the stage of human development that is realized in this state.

What makes the members of the political state religious is the dualism between their individual life and their species-life, between life in civil society and political life, their belief that life in the state is the true life even though it leaves untouched their individuality. Religion is here the spirit of civil society, the expression of separation and distance of man from man. What makes a political democracy Christian is the fact that in it man, not only a single man but every man, counts as a sovereign being; but it is man as he appears uncultivated and unsocial, man in his accidental existence, man as he comes and goes, man as he is corrupted by the whole organization of our society, lost to himself, sold, given over to the domination of inhuman conditions and elements—in a word, man who is no longer a real species-being. The fantasy, dream, and postulate of Christianity, the sovereignty of man, but of man as an alien being separate from actual man, is present in democracy as a tangible reality and is its secular motto.

The religious and theological consciousness has all the more religious and theological force in the complete democracy as it is without political significance and earthly aims. It is the affair of minds that are shy of the world, the expression of a limited understanding, the product of arbitrariness and fantasy, a really other-worldly life. Christianity achieves here the practical expression of its significance of a universal religion in that it groups together the most different opinions in the form of Christianity and even more because it does not lay on others the requirements of Christianity, but only a religion in general, any religion (compare the above mentioned work of Beaumont). The religious consciousness revels in richness of religious opposition and religious diversity.

Thus we have shown that political emancipation from religion leaves religion intact even though it is no longer a privileged religion. The contradiction with his citizenship in which the adherent of a particular religion finds himself is only a part of the general secular contradiction between the political state and civil society. The perfect Christian state is the one that recognizes itself as a state and abstracts from the religion of its members. The emancipation of the state from religion is not the emancipation of actual man from religion.

So we do not say to the Jews, as Bauer does: you cannot be emancipated politically without emancipating yourselves radically from Judaism. Rather we say to them: because you can be politically emancipated without completely and consistently abandoning Judaism, this means that political emancipation itself is not human emancipation. If you Jews wish to achieve political emancipation without achieving human emancipation, then the incompleteness and contradiction does not only lie in you, it lies in the nature and category of political emancipation. If you are imprisoned within this category, then you are sharing in something common to everyone. Just as the state is evangelizing when it, although a state, has a Christian attitude to Jews, so the Jew is acting politically when he, although a Jew, requests civil rights.

But if a man, although a Jew, can be politically emancipated and acquire civil rights, can he claim and accept human rights? Bauer denies it.

The question is whether the Jew as such, i.e. the Jew who himself admits that his true nature compels him to live in eternal separation from others, is capable of accepting universal human rights and bestowing them on others.

The concept of human rights was first discovered by the Christian world in the previous century. It is not innate in man, but won in a struggle against the historical traditions in which man has hitherto been educated. Thus human rights are not a gift of nature, no dowry but the prize of the struggle against the accident of birth and against privileges that history transmitted from generation to generation up to the present time. They are the result of culture and only to be possessed by the man who has won and merited them.

Can the Jew really take possession of them? As long as he is a Jew the limited nature which makes him a Jew must gain the upper hand over the human nature that should bind him as a man to other men and must separate him off from non-Jews. He declares through this separation that the particular nature that makes him a Jew is his true and highest nature, before which his human nature must give way.

In the same way, the Christian as Christian cannot grant human rights [pp. 19, 20'.

. According to Bauer man must sacrifice the 'privilege of belief' in order to be able to receive general human rights. Let us discuss for a moment the so-called human rights, human rights in their authentic form, the form they have in the writings of their discoverers, the North Americans and French! These human rights are partly political rights that are only exercised in community with other men. Their content is formed by participation in the common essence, the political essence, the essence of the state. They fall under the category of political freedom, under the category of civil rights, which, as we have seen, in no way presuppose the consistent and positive abolition of religion, nor, consequently, of Judaism. It remains to discuss the other part of human rights, the rights of man, in so far as they differ from the rights of the citizen.

Among them are freedom of conscience, the right to exercise a chosen religion. The privilege of belief is expressly recognized either as a human right, or as a consequence of one of the human rights, freedom.

Declaration of the Rights of Man and of the Citizen, 1791, Article 10: 'No one should be molested because of his opinions, not even religious ones'. In the first section of the constitution of 1791 'the liberty of every man to practise the religion to which he adheres' is guaranteed as human right. *The Declaration of the Rights of Man . . . 1793* counts among human rights, in Article 7, 'the free exercise of religious practice'. Indeed, concerning the right to publish one's thoughts and opinions, to hold assemblies and practise one's religion, it goes as far as to say: 'the necessity of announcing these rights supposes either the present or the recent memory of despotism'. Compare the constitution of 1795, Section 14, Article 354.

Constitution of Pennsylvania, Article 9, Paragraph 3: 'All men have a natural and indefeasible right to worship Almighty God according to the dictates of their own consciences: no man can of right be compelled to attend, erect or support a place of worship, or to maintain any ministry, against his consent; no human authority can, in any case whatever, control or interfere with the rights of conscience.'

Constitution of New Hampshire, Articles 5 & 6: 'Among the natural rights, some are in their very nature unalienable. . . . Of this kind are rights of conscience' (Beaumont loc. cit., pp. 213, 214).

The incompatibility of religion with the rights of man is so far from being evident in the concept of the rights of man, that the right to be religious, to be religious in one's own chosen way, to practise one's chosen religion is expressly counted as one of the rights of man. The privilege of faith is a universal right of man.

The rights of man are as such differentiated from the right of the citizen. Who is the 'man' who is different from the 'citizen?' No one but the member of civil society. Why is the member of civil society called 'man', simply man, and why are his rights called the rights of man? How do we explain this fact? From the relationship of the political state to civil society, from the nature of political emancipation.

Above all we notice the fact that the so-called rights of man, the rights of man as different from the rights of the citizen are nothing but the rights of the member of civil society, i.e. egoistic man, man separated from other men and the community. The most radical constitution, the constitution of 1793, can say:

Declaration of the Rights of Man . . ., Article 2. These rights etc. (natural and imprescriptable rights) are: equality, liberty, security, property.

What does liberty consist of?

Article 6: 'Liberty is the power that belongs to man to do anything that does not infringe on the right of someone else' or according to the declaration of the rights of man of 1791 'liberty consists in the power of doing anything that does not harm others'.

Thus freedom is the right to do and perform what does not harm others. The limits within which each person can move without harming others is defined by the law, just as the boundary between two fields is defined by the fence. The freedom in question is that of a man treated as an isolated monad and withdrawn into himself. Why is the Jew, according to Bauer, incapable of receiving the rights of man? 'So long as he is a Jew the limited nature that makes him a Jew will get the upper hand over the human nature that should unite him as a man to other men and will separate him from the non-Jew.' But the right of man to freedom is not based on the union of man with man, but on the separation of man from man. It is the right to this separation, the rights of the limited individual who is limited to himself.

The practical application of the rights of man to freedom is the right of man to private property.

What does the right of man to property consist in?

Article 16 (Constitution of 1793): 'The right of property is the right which belongs to all citizens to enjoy and dispose at will of their goods and revenues, the fruit of their work and industry.'

Thus the right of man to property is the right to enjoy his possessions and dispose of the same arbitrarily, without regard for other men, independently from society, the right of selfishness. It is the former individual freedom together with its latter application that forms the basis of civil society. It leads man to see in other men not the realization but the limitation of his own freedom. Above all it proclaims the right of man 'to enjoy and dispose at will of his goods, his revenues and fruits of his work and industry'.

There still remain the other rights of man, equality and security.

Equality, here in its non-political sense, is simply the counterpart of the liberty described above, namely that each man shall without discrimination be treated as a self-sufficient monad. The constitution of 1795 defines the concept of this equality, in conformity with this meaning, thus:

Article 3 (Constitution of 1795): 'Equality consists of the fact that the law is the same for all, whether it protects or punishes.'

And security?

Article 8 (Constitution of 1793): 'Security consists in the protection afforded by society to each of its members for the conservation of his person, rights, and property.'

Security is the highest social concept of civil society, the concept of the police. The whole of society is merely there to guarantee to each of its members the preservation of his person, rights, and property. It is in this sense that Hegel calls civil society the 'state of need and of reason'.

The concept of security does not allow civil society to raise itself above its egoism. Security is more the assurance of egoism.

Thus none of the so-called rights of man goes beyond egoistic man, man as he is in civil society, namely an individual withdrawn behind his private interests and whims and separated from the community. Far from the rights of man conceiving of man as a species-being, species-life itself, society, appears as a framework exterior to individuals, a limitation of their original self-sufficiency. The only bond that holds them together is natural necessity, need and private interest, the conservation of their property and egoistic person.

It is already paradoxical that a people that is just beginning to free itself, to tear down all barriers between different sections of the people and form a political community, should solemnly proclaim (Declaration of 1791) the justification of egoistic man separated from his fellow men and the community. Indeed, this proclamation is repeated at a moment when only the most heroic devotion can save the nation, and is therefore peremptorily demanded, at a moment when the sacrifice of all the interests of civil society is raised to the order of the day and egoism must be punished as a crime (*Declaration of the Rights of Man . . . 1793*). This fact appears to be even more paradoxical when we see that citizenship, the political community, is degraded by the political emancipators to a mere means for the preservation of these so-called rights of man, that the citizen is declared to be the servant of egoistic man, the sphere in which man behaves as a communal being is degraded below the sphere in which man behaves as a partial being, finally that it is not man as a citizen but man as a bourgeois who is called the real and true man.

'The aim of every political association is the conversation of the natural and imprescriptible rights of man' (*Declaration of the Rights of Man . . . 1791*, Article 2). 'Government is instituted to guarantee man the enjoyment of his natural and imprescriptible rights' (*Declaration of the Rights of Man . . . 1791*, Article 1). So even in the moments of youthful freshness and enthusiasm raised to fever pitch by the pressure of circumstances, political life is declared to be a mere means whose end is the life of civil society. It is true that its revolutionary practice is in flagrant contradiction with its theory. While, for example, security is declared to be a right of man, the violation of the privacy of correspondence is publicly inserted in the order of the day. While the 'unlimited freedom of the press' (Constitution of 1793, Article 122) is guaranteed as a consequence of the right of man to individual freedom, the freedom of the press is completely destroyed, for 'the liberty of the press must not be permitted when it compromises public liberty' ('The Young Robespierre' in Buchez and Roux, *Parliamentary History of the French Revolution*, vol. 28, p. 159). This means then that the right of man to freedom ceases to be a right as soon as it enters into conflict with political life, whereas, according to the theory, political life is only the guarantee of the rights of man, the rights of individual man, and so must be given up as soon as it contradicts its end, these rights of man. But the practice is only the exception and the theory is the rule. Even though one were to treat the revolutionary practice as the correct version of the relationship, the riddle still remains to be solved of why, in the minds of the political emancipators, the relationship is turned upside-down and the end appears as the means and the means as the end. This optical illusion of their minds would always be the same riddle, although it would then be a psychological and theoretical riddle.

The riddle has a simple solution.

Political emancipation is at the same time the dissolution of the old society on which rests the sovereign power, the essence of the state alienated from the people. Political revolution is the revolution of civil society. What was the character of the old society? One word characterizes it. Feudalism. The old civil society had a directly political character. The elements of civil life, like, for example, property or the family or the type and manner of work, were, in the form of seigniorial right, estates, and corporations, raised to the level of elements of state life. They defined in this form the relationship of the single individual to the state as a whole, that is, his political relationship, the relationship of separation and exclusion from the other parts of society. For this sort of organization of the people's life did not turn property or work into social elements but completed their separation from the state as a whole, and made them into particular societies within society. But the vital functions and conditions of life in civil society was still political even though political in the feudal sense, that is, they excluded the individual from the states as a whole. They turned the particular relationship of the corporation to the totality of the state, into his own general relationship to the life of the people, as it turned his particular civil occupation into his general occupation and situation. As a consequence of this organization the unity of the state— the mind, will, and authority of this state unity, the power of the state in general—equally appears necessarily as the particular affair of a lord and servants who are cut off from the people.

The political revolution overthrew this feudal power and turned state affairs into affairs of the people; it turned the state into a matter of general concern, i.e. into a true state; it necessarily destroyed all estates, corporations, guilds, privileges which were so many expressions of the separation of the people from the community. The political revolution thus abolished the political character of civil society. It shattered civil society with its simple parts, on the one hand into individuals, on the other hand into the material and spiritual elements that make up the life experience and civil position of these individuals. It unfettered the political spirit that had, as it were, been split, cut up, and drained away into the various cul-de-sacs of feudal society. The political revolution collected this spirit together after its dispersion, freed it from its confusion with civil life, and set it up as the sphere that was common to all, the general affair of the people in ideal independence from the other particular elements of civil life. Particular professions and ranks sank to a merely individual importance. They were no longer the relationship of individuals to the state as a whole. Public affairs as such became the general affair of each individual and politics was a general occupation.

But the perfection of the idealism of the state was at the same time the perfection of the materialism of civil society. The shaking off of the political yoke entailed the shaking off of those bonds that had kept the egoistic spirit of civil society fettered. Political emancipation entailed the emancipation of civil society from politics, from even the appearance of a general content.

Feudal society was dissolved into its basis, into man. But into the man that was its true basis, egoistic man. This man, the member of civil society, is the basis, the presupposition of the political state. He is recognized by it as such in the rights of man.

But the freedom of egoistic man and the recognition of this freedom is the recognition of the unimpeded movement of the spiritual and material elements that go to make up its life.

Man was therefore not freed from religion; he received freedom of religion. He was not freed from property; he received freedom of property. He was not freed from the egoism of trade; he received freedom to trade.

The formation of the political state and the dissolution of civil society into independent individuals, who are related by law just as the estate and corporation men were related by privilege, is completed in one and the same act. Man as member of civil society, unpolitical man, appears necessarily as natural man. The rights of man appear as natural rights, because self-conscious activity is concentrated upon political action. Egoistic man is the passive, given results of the dissolved society, an object of immediate certainty and thus a natural object. Political revolution dissolves civil life into its component parts, without revolutionizing and submitting to criticism these parts themselves. Its attitude to civil society, to the world of need, to work, private interests, private law is that they are the foundation of its existence, its own presupposition that needs no further proof, and thus its natural basis. Finally, man as a member of civil society counts for true man, for man as distinct from the citizen, because he is man in his sensuous, individual, immediate existence, while political man is only the abstract fictional man, man as an allegorical or moral person. This man as he actually is, is only recognized in the form of the egoistic individual, and the true man only in the form of the abstract citizen.

The abstraction of the political man is thus correctly described by Rousseau: 'He who dares to undertake the making of a people's institutions ought to feel himself capable, so to speak, of changing human nature, of transforming each individual, who is by himself a complete and solitary whole, into part of a greater whole from which he in a manner receives his life and being; of altering man's constitution for the purpose of strengthening it; and of substituting a partial and moral existence of the physical and independent existence nature has conferred on us all. He must, in a word, take away from man his own resources and give him instead new ones alien to him, and incapable of being made use of without the help of other men.'

All emancipation is bringing back man's world and his relationships to man himself.

Political emancipation is the reduction of man, on the one hand to a member of civil society, an egoistic and independent individual, on the other hand to a citizen, a moral person.

The actual individual man must take the abstract citizen back into himself and, as an individual man in his empirical life, in his individual work and individual relationships become a species-being; man must recognize his own forces as social forces, organize them, and thus no longer separate social forces from himself in the form of political forces. Only when this has been achieved will human emancipation be completed.

II

The Capacity of Present-day Jews and Christians to Become Free

By BRUNO BAUER

This is the form that Bauer gives to the question of the relationship of the Jewish and Christian religions to each other and to criticism. Their relationship to criticism is their relationship to 'the capacity to become free'.

The conclusion is: 'the Christian has only one barrier to surmount, his religion, in order to give up religion altogether', and thus to become free; 'the Jew, on the other hand, has not only to break with his Jewish nature but also with the development and completion of his religion, a development that has remained alien to him' (p. 71). So Bauer here turns the question of Jewish emancipation into a purely religious question. The theological problem of who has the better prospect of getting to heaven, Jew or Christian, is repeated in the enlightened form: which of the two is more capable of emancipation? And the question is no longer: which gives freedom, Judaism or Christianity? It is rather the reverse: which gives more freedom: the negation of Judaism or the negation of Christianity.

If they wish to become free, then the Jew should not profess Christianity, but the dissolution of Christianity, the general dissolution of religion, i.e. the Enlightenment, criticism and its result, free humanity (p. 70).

It is still a profession that is in question for the Jews, but no longer the profession of Christianity but of the dissolution of Christianity.

Bauer demands of the Jews that they break with the essence of the Christian religion, a demand that, as he admits himself, does not proceed from the development of the Jewish essence.

It was to be predicted that when Bauer at the end of his *The Jewish Question* conceived of Judaism as merely the crude religious criticism of Christianity and thus only saw in it a religious significance, the emancipation of the Jews would turn into a philosophico-theological act.

Bauer understands the ideal, abstract essence of the Jew, his religion, to be his whole essence. He concludes therefore quite rightly: 'The Jew contributes nothing to humanity when he neglects his limited law', when he abolishes the whole of his Judaism (p. 65).

According to this, the relationship of Jews and Christians is as follows: the sole interest of Christians in Jewish emancipation is a general human and theoretical interest. Judaism is a fact that must offend the religious eye of the Christian. As soon as his eye ceases to be religious then this fact ceases to offend. The emancipation of the Jew is in itself no task for the Christian.

The Jew, on the other hand, has not only his own task in order to achieve his liberation but also the Christian's task, the 'Critique of the Synoptics' and 'The Life of Jesus', etc., to get through.

They must look to it themselves: they will create their own destiny; but history does not allow itself to be mocked (p. 71).

We attempt to break the theological conception of the problem. The question of the Jews' capacity for emancipation is changed for us into the question what particular social element needs to be overcome in order to abolish Judaism? For the fitness of the present-day Jew for emancipation is bound up with the relationship of Judaism to the emancipation of the contemporary world. And this relationship stems necessarily from the position of Judaism in the contemporary enslaved world.

Let us discuss the actual secular Jew, not the sabbath Jew as Bauer does, but the everyday Jew.

Let us look for the secret of the Jew not in his religion, but let us look for the secret of religion in the actual Jew.

What is the secular basis of Judaism? Practical need, selfishness.

What is the secular cult of the Jew? Haggling. What is his secular god? Money.

Well then, an emancipation from haggling and money, from practical, real Judaism would be the self-emancipation of our age.

An organization of society that abolished the presupposition of haggling, and thus its possibility, would have made the Jew impossible. His religious consciousness would dissolve like an insipid vapour into the real live air of society. On the other hand: if the Jew recognizes this practical essence of his as null and works for its abolition, he is working for human emancipation with his previous development as a basis and turning himself against the highest practical expression of human self-alienation.

Thus we recognize in Judaism a general contemporary anti-social element which has been brought to its present height by a historical development which the Jews zealously abetted in its harmful aspects and which now must necessarily disintegrate.

In the last analysis the emancipation of the Jews is the emancipation of humanity from Judaism.

The Jew has already emancipated himself in a Jewish manner.

The Jew who in Vienna, for example, is only tolerated, controls through the power of his money the fate of the whole empire. The Jew who may be without rights in the smallest of the German states decides the destiny of Europe. While the corporations and guilds turn a deaf ear to the Jew or do not yet favour him, their bold and selfish industry laughs at medieval institutions (Bruno Bauer, *The Jewish Question*, p. 114).

This is no isolated fact. The Jew has emancipated himself in a Jewish manner, not only annexing the power of money but also because through him and also apart from him money has become a world power and the practical spirit of the Jew has become the practical spirit of the Christian people. The Jews have emancipated themselves in so far as the Christians have become Jews.

'The pious and politically free inhabitant of New England', Captain Hamilton informs us, 'is a sort of Laocoön, who does not even make the least effort to free himself from the snakes that enlace him.'

Mammon is their idol, they adore him not with their lips alone but with all of the strength of their body and soul. In their eyes the world is nothing but a Stock Exchange and they are convinced that here on earth their only vocation is to become richer than other men. The market has conquered all their other thoughts, and their one relaxation consists in bartering objects. When they travel they carry, so to speak, their wares or their display counter about with them on their backs and talk of nothing but interest and profit. If they lost sight for a moment of their own business, this is merely so that they can pry into someone else's.

Indeed, the practical dominance of Judaism over the Christian world has reached its unambiguous, normal expression in North America. Here even the announcing of the gospel, the Christian pulpit, has become an article of trade, and the bankrupt gospel merchant becomes like the evangelist who has become rich in business.

A man such as you see at the head of a respectable congregation began by being a merchant—his trade fell off, so he became a minister; another began as a priest, but as soon as he had a certain sum of money at his disposal, he left the pulpit for business. In the eyes of many, the religious ministry is a real industrial career (Beaumont, loc. cit. pp. 185–6).

According to Bauer it is a hypocritical state of affairs when in theory political rights are denied the Jew, while in practice he possesses a monstrous power and exercises on a large scale a political influence that is limited on a small scale (*The Jewish Question*, p. 114).

The contradiction between the practical political power of the Jew and his political rights is the general contradiction between politics and the power of money. Whereas the first ideally is superior to the second, in fact it is its bondsman.

Judaism has maintained itself alongside Christianity not only as a religious critique of Christianity, not only as an incarnate doubt about the religious provenance of Christianity, but just as much because the practical Jewish spirit, Judaism or commerce, has maintained itself, and even reached its highest development, in Christian society. The Jew who is a particular member of civil society is only the particular appearance of the Judaism of civil society.

Judaism has maintained itself not in spite of, but because of, history.

From its own bowels civil society constantly begets Judaism.

What was the implicit and explicit basis of the Jewish religion? Practical need, egoism.

The monotheism of the Jew is therefore in reality the polytheism of many needs, a polytheism that makes even the lavatory an object of divine law. Practical need, or egoism, is the principle of civil society and appears as such in all its purity as soon as civil society has completely given birth to the political state. The god of practical need and selfishness is money.

Money is the jealous god of Israel before whom no other god may stand. Money debases all the gods of man and turns them into commodities. Money is the universal, self-constituted value of all things. It has therefore robbed the whole world, human as well as natural, of its own values. Money is the alienated essence of man's work and being, this alien essence dominates him and he adores it.

The god of the Jews has been secularized and has become the god of the world. Exchange is the actual god of the Jew. His god is only the illusion of exchange.

The view of nature that has obtained under the domination of private property and money is the actual despising and degrading of nature. It does really exist in the Jewish religion, but only in imagination.

In this sense Thomas Münzer declares it intolerable 'that all creation has been made into property: the fish in the water, the bird in the air, the off-spring of the earth—creation, too, must become free'.

What lies abstract in the Jewish religion, a contempt for theory, art, history, man as an end in himself, is the actual, conscious standpoint, the virtue of the money man. The species-relationship itself, the relationship of man to woman, etc., becomes an object of commerce! Woman is bartered.

The imaginary nationality of the Jew is the nationality of the merchant, of the money man in general.

The baseless and irrational law of the Jew is only the religious caricature of morality and law in general, the purely formal rights with which the world of selfishness surrounds itself.

Here, too, the highest relationship of man is the legal relationship, the relationship to laws that are not valid for him because they are the laws of his own will and essence, but because they are the masters and deviations from them are avenged.

Jewish Jesuitry, the same practical Jesuitry that Bauer points out in the Talmud, is the relationship of the world of selfishness to the dominant laws whose crafty circumvention forms the chief art of this world.

If the affairs of the world were to be conducted within the limits of its laws, this would entail the continual supersession of these laws.

Judaism could not develop itself any further theoretically as a religion because the attitude of practical need is narrow by nature and exhausted in a few traits.

The religion of practical need could, by its nature, find its completion not in theory but in practice, for this latter is its true form.

Judaism could not create a new world; it could only draw the new creations and relationships of the world into the sphere of its own industry, because practical need, whose spirit is selfishness, is passive and does not really extend itself, but finds itself extended by the progress of social circumstances.

Judaism reaches its apogee with the completion of civil society; but civil society first reaches its completion in the Christian world. Only under the domination of Christianity, which made all national, natural, moral, and theoretical relationships exterior to man, could civil society separate itself completely from the life of the state, tear asunder all the species-bonds of man, put egoism and selfish need in the place of these species-bonds and dissolve man into a world of atomistic individuals with hostile attitudes towards each other.

Christianity had its origin in Judaism. It has dissolved itself back into Judaism.

The Christian was from the beginning the theorizing Jew; the Jew is therefore the practical Christian, and the practical Christian has become the Jew again. Christianity has overcome real Judaism in appearance only. It was too gentlemanly, too spiritual, to remove the crudeness of practical need other than by raising it into the blue heavens.

Christianity is the sublime thought of Judaism; Judaism is the vulgar practical application of Christianity. But this practical application could only become universal after Christianity as the perfect religion had completed, in a theoretical manner, the self-alienation of man from himself and from nature.

Only then could Judaism attain general domination and make externalized man and externalized nature into alienable, saleable objects, a prey to the slavery of egoistic need and the market.

Selling is the practice of externalization. As long as man is imprisoned within religion, he only knows how to objectify his essence by making it into an alien, imaginary being. Similarly, under the domination of egoistic need he can only become practical, only create practical objects by putting his products and his activity under the domination of an alien entity and lending them the significance of an alien entity—money.

In its perfected practice the Christian egoism concerning the soul necessarily changes into the Jewish egoism concerning the body; heavenly need becomes earthly, and the subjectivism becomes selfishness. We explain the tenacity of the Jew not by his religion, but by the human basis of his religion, practical need, egoism.

Because the true essence of the Jew has been realized and secularized in civil society, it could not convince the Jew of the unreality of his religious essence which is merely the ideal perception of practical need. Thus it is not only in the Pentateuch or the Talmud that we find the essence of the contemporary Jew: we find it in contemporary society, not as an abstract but as a very empirical essence, not as the limitation of the Jew but as the Jewish limitations of society.

As soon as society manages to abolish the empirical essence of Judaism, the market and its presuppositions, the Jew becomes impossible, for his mind no longer has an object, because the subjective basis of Judaism, practical need, has become humanized, and because the conflict of man's individual, material existence with his species-existence has been superseded.

The social emancipation of the Jew implies the emancipation of society from Judaism.

BIBLIOGRAPHY

Original: *MEGA*, I (1) 1, pp. 576 ff.

Present translation: K. Marx, *Early Texts*, pp. 85 ff.

Other translations: K. Marx, *A World without Jews*, ed. D. Runes, New York, 1959.
K. Marx, *Early Writings*, ed. T. Bottomore, pp. 3 ff.; *Writings of the Young Marx on Philosophy and Society*, ed. Easton and Guddat, pp. 216 ff.
K. Marx, *Early Writings*, ed. L. Colletti, New York and London, 1975, pp. 211 ff.

Commentaries: S. Avineri, 'Marx and Jewish Emancipation', *Journal of the History of Ideas*, 1964.
E. Block, 'Man and Citizen according to Marx', in *Socialist Humanism*, ed. E. Fromm, London, 1967.
L. Dupré, *The Philosophical Foundations of Marxism*, New York, 1966, Ch. 5.
D. Howard, *The Development of the Marxian Dialectic*, Carbondale, 1972, pp. 94 ff.
J. Maguire, *Marx's Paris Writings: an Analysis*, Dublin, 1972, pp. 14 ff.
D. McLellan, *Marx before Marxism*, Penguin, 1972, Ch. 6.
D. Runes, Introduction to edition cited above.

7. Towards a Critique of Hegel's *Philosophy of Right:* Introduction

The following article was first published in the *Deutsch–französische Jahrbücher*. It was intended originally as an introduction to Marx's unfinished manuscript on Hegel's *Philosophy of Right*, and was written very early in 1844 under the impact of Marx's first few months in Paris, where he had moved in October 1843. The article begins with a series of brilliant epigrams on religion—a subject Marx considers adequately dealt with by his Young Hegelian colleagues. In the main body of the article Marx describes, in the first part, the current situation in Germany: intellectually very advanced owing to the ideas of Hegel and the Young Hegelians, but politically very backward. In the second part, he discusses the revolutionary possibilities latent in this contrast. From an analysis of the French Revolution Marx draws the optimistic conclusion that Germany is much more fitted than France for a radical revolution. His reasons are the absence of a strong middle class in Germany and the possibility of an explosive union between German philosophy and the 'class with radical chains'. This is the first occasion on which Marx proclaims his adherence to the cause of the proletariat: he had been in close contact with working-class revolutionaries since his arrival in Paris.

As far as Germany is concerned, the criticism of religion is essentially complete, and the criticism of religion is the presupposition of all criticism.

The profane existence of error is compromised as soon as its heavenly *oratio pro aris et focis* [prayer for hearth and home] is refuted. Man has found in the imaginary reality of heaven where he looked for a superman only the reflection of his own self. He will therefore no longer be inclined to find only the appearance of himself, the non-man, where he seeks and must seek his true reality.

The foundation of irreligious criticism is this: man makes religion, religion does not make man. Religion is indeed the self-consciousness and self-awareness of man who either has not yet attained to himself or has already lost himself again. But man is no abstract being squatting outside the world. Man is the world of man, the state, society. This state, this society, produces religion's inverted attitude to the world, because they are an inverted world themselves. Religion is the general theory of this world, its encyclopaedic compendium, its logic in popular form, its spiritual *point d'honneur*, its enthusiasm, its moral sanction, its solemn complement, its universal basis for consolation and justification. It is the imaginary realization of the human essence, because the human essence possesses no true reality. Thus, the struggle against religion is indirectly the struggle against the world whose spiritual aroma is religion.

Religious suffering is at the same time an expression of real suffering and a protest against real suffering. Religion is the sigh of the oppressed creature, the feeling of a heartless world, and the soul of soulless circumstances. It is the opium of the people.

The abolition of religion as the illusory happiness of the people is the demand for their real happiness. The demand to give up the illusions about their condition is a demand to give up a condition that requires illusion. The criticism of religion is therefore the germ of the criticism of the valley of tears whose halo is religion.

Criticism has plucked the imaginary flowers from the chains not so that man may bear chains without any imagination or comfort, but so that he may throw away the chains and pluck living flowers. The criticism of religion disillusions man so that he may think, act, and fashion his own reality as a disillusioned man come to his senses; so that he may revolve around himself as his real sun. Religion is only the illusory sun which revolves around man as long as he does not revolve around himself.

It is therefore the task of history, now the truth is no longer in the beyond, to establish the truth of the here and now. The first task of philosophy, which is in the service of history, once the holy form of human self-alienation has been discovered, is to discover self-alienation in its unholy forms. The criticism of heaven is thus transformed into the criticism of earth, the criticism of religion into the criticism of law, and the criticism of theology into the criticism of politics.

The following exposition—a contribution to this task—does not deal with the original but with its copy, the German philosophy of the state and law. The only reason is that it is dealing with Germany.

If we wanted to start with the German *status quo* itself, the result would still be an anachronism even if one did it in the only adequate way, i.e. negatively. Even the denial of our political present is already a dusty fact in the historical lumber-room of modern peoples. Even if I deny powdered wigs, I still have unpowdered wigs. If I deny the situation in the Germany of 1843, I am according to French reckoning, scarcely in the year 1789, still less at the focal point of the present.

Indeed, German history can congratulate itself on following a path that no people in the historial firmament have taken before it and none will take after. For we have shared the restorations of the modern peoples, without sharing their revolutions. We have had restorations, firstly because other peoples dared to make a revolution, and secondly because other peoples suffered a counter-revolution; once because our masters were afraid and once because they were not afraid. With our shepherds at our head, we continually found ourselves in the company of freedom only once—on the day of its burial.

There is a school that justifies the abjectness of today by the abjectness of yesterday, a school which declares every cry of the serf against the knout to be rebellious as long as the knout is an aged, historical knout with a pedigree, a school to which history, like the God of Isreael to his servant Moses, only shows its posterior. This school, the Historical School of Law, would have invented German history had it not been itself an invention of that history. It is a Shylock, but a servile Shylock that for every pound of flesh cut from the heart of the people swears upon its bond, its historical, Christian–Germanic bond.

Easy-going enthusiasts, on the other hand, Germanophiles by blood and liberal by reflection, look for the history of freedom beyond our history in the primeval Teutonic forests. But how is the history of our freedom different from the history of the wild boar's freedom if it is only to be found in the forests? Moreover, it is well known that the forest echoes back the same words that are shouted into it. So peace to the primeval German forests!

But war on the situation in Germany! Of course. It is below the level of history and below any criticism, but it remains an object of criticism just as the criminal, though below the level of humanity, yet remains an object of the executioner. In its struggle against this situation criticism is no passion of the head, it is the head of passion. It is no surgical knife, it is a weapon. Its object is its enemy that it does not aim to refute, but to annihilate. For the spirit of this situation is refuted already. In itself it is not an object worthy of thought but an existence as despicable as it is despised. Criticism does not itself need to arrive at an understanding with this object for it is already clear about it. It no longer pretends to be an end in itself but only a means. The essential feeling that animates it is indignation, its essential task is denunciation.

The point is to describe the counter-pressures of all social spheres, a general passive discontent, a narrowness that both recognizes and yet misconceives itself, all contained in the framework of a government that lives from the preservation of all mediocrities and is itself nothing but mediocrity in government.

What a charade! Society is infinitely divided into a multiplicity of races which stand opposed to each other with petty antipathies, bad consciences, and a brutal mediocrity. It is precisely the ambivalent and suspicious attitude to each other that leads their masters to treat them all without distinction, although with different formalities, as persons whose existence has been granted as a favour. And even the fact that they are dominated, ruled, and possessed they must recognize and profess as a favour of heaven! And on the other side are the masters themselves whose greatness is in inverse proportion to their number.

The criticism that tackles this state of affairs is engaged in a hand to hand battle, and in a hand to hand battle it does not matter whether the opponent is equally noble, well born, and interesting—the point is to hit him. The point is not to allow the Germans a moment of self-deceit or resignation. We must make the actual oppression even more oppressive by making them conscious of it, and the insult even more insulting by publicizing it. We must describe every sphere of German society as the disgrace of German society, we must force these petrified relationships to dance by playing their own tune to them! So as to give them courage, we must teach the people to be shocked by themselves. We are thus fulfilling an inevitable need of the German people and the needs that spring from the true character of a people are the final bases of its satisfaction.

Even for modern peoples this struggle against the narrow content of the German *status quo* is not without interest; for the German *status quo* is the unabashed consummation of the *ancien régime* and the *ancien régime* is the hidden deficiency of the modern state. The struggle against the German political present is the struggle against the past of modern peoples, and they are still burdened with reminiscences from this past. It is instructive for them to see the *ancien régime*, that played tragedy in their history, play comedy as a German ghost. Its history was tragic so long as it was the established power in the world and freedom was a personal fancy; in a word, so long as it believed and had to believe, in its own justification. So long as the *ancien régime*, as the existing world order, was struggling with a world that was just beginning, then there was on its part a universal historical error, but not a personal one. Its demise was therefore tragic.

The present German regime, on the other hand, an anachronism in flagrant contradiction to all generally recognized axioms, the nullity of the *ancien régime* exhibited for all the world to see, only imagines that it believes in itself, and requires this imagination from the rest of the world. If it believed in its own nature, would it try to hide it under the appearance of an alien nature and seek its salvation in hypocrisy and sophistry? The modern *ancien régime* is only the comedian of a world order whose real heroes are dead. History is thorough and passes through many stages when she carries a worn-out form to burial. The last stage of a world-historical form is its comedy. The gods of Greece who had already been mortally wounded in the *Prometheus Bound* tragedy of Aeschylus, had to die once more a comic death in the dialogues of Lucian. Why does history follow this course? So that mankind may take leave of its past joyfully. It is this joyful political function that we vindicate for the political powers of Germany.

But as soon as modern socio-political reality is submitted to criticism, as soon, that is, as criticism raises itself to the level of truly human problems, it finds itself outside the German *status quo*, or it would conceive of its object at a level below its object. An example: the relationship of industry and the world of wealth in general to the political world is one of the chief problems of modern times. In what form does this problem begin to preoccupy the Germans? Under the form of protectionism, a system of prohibitions and a national economy. German chauvinism has left men for matter, and so one fine morning our cotton knights and iron heroes found themselves changed into patriots. So people are beginning in Germany to recognize the interior sovereignty of monopoly by according it an exterior sovereignty. Thus we are now starting to begin in Germany when France and England are beginning to end. The old and rotten state of affairs against which these countries are in theoretical rebellion and which they only tolerate as one tolerates chains, is in Germany greeted like the rising dawn of a beautiful future that scarcely dares to pass from artful theory to implacable practice. While in France and England the problem is: political economy or domination of wealth by society, in Germany it is: national economy or domination of nationality by private property. Thus in France and England the problem is to abolish monopoly that has progressed to its final consequences; in Germany the problem is to progress as far as the final consequences of monopoly. There the problem is to find a solution, here it is to provoke a collision. This is a sufficient example of the German form of modern problems, an example of how our history, like a raw recruit, has so far only had a job of performing trivial historical drill after everyone else.

So if developments in Germany as a whole did not go beyond German political development, a German could no more take part in contemporary problems than can a Russian. But if the single individual is not bound by the limits of his nation, the whole nation is even less liberated by the liberation of an individual. Scythians made no progress at all towards a Greek culture because Greece counted a Scythian as one of her philosophers.

Happily we Germans are no Scythians.

As the ancient peoples have experienced their pre-history in imagination, in mythology, so we Germans have experienced our future history in thought, in philosophy. We are philosophical contemporaries without being historical ones. German philosophy is the ideal prolongation of German history. So if, instead of criticizing the incomplete works of our real history, we criticize the posthumous works of our ideal history, philosophy, then our criticism will be at the centre of the question of which the present age says: that is the question. What in developed peoples is the practical conflict with the modern state institutions, in Germany, where these institutions do not even exist, it is a critical conflict with the philosophical reflection of these institutions.

The German philosophy of law and of the state is the only theory in German history that stands *al pari* [on an equal footing] with the official modern present. The German people must therefore add this dream history to its existing circumstances and submit to criticism not only these existing circumstances but at the same time their abstract continuation. Its future can limit itself neither to the immediate negation of its real political and juridical circumstances nor to the immediate completion of its ideal political and juridical circumstances, for it has the immediate negation of its real situation in its ideal circumstances and has already almost left behind the immediate completion of its philosophy by looking at neighbouring peoples. Thus the practical political party in Germany is justified in demanding the negation of philosophy. Their error consists not in their demand, but in being content with the demand that they do not and cannot really meet. They believe that they can complete that negation by turning their back on philosophy and murmuring at her with averted head some vexatious and banal phrases. Their limited vision does not count philosophy as part of German reality or even fancies that it is beneath the level of German practice and the theories that serve it. You demand that we start from the real seeds of life, but forget that until now the real seed of the German people has only flourished inside its skull. In a word: you cannot transcend philosophy without realizing it.

The same error but with inverted factors is committed by the theoretical party that originates in philosophy.

It saw in the present struggle nothing but the critical struggle of philosophy with the German world and did not reflect that previous philosophy itself has belonged to this world and is its completion, albeit in ideas. It was critical with regard to its opposite but not to itself, for it started from the presuppositions of philosophy and remained content with the results thus obtained. Or else it presented demands and conclusions got from elsewhere as the demands and conclusions of philosophy, although these, supposing them to be well founded, can only be obtained by the negation of previous philosophy, of philosophy as philosophy. We reserve for later a more detailed description of this party. Its principal fault can be summed up thus: it thought it could realize philosophy without transcending it.

The criticism of the German philosophy of the state and of law which was given its most consistent, richest, and final version by Hegel, is both the critical analysis of the modern state and of the reality that depends upon it and also the decisive denial of the whole previous method of the German political and legal mind, whose principal and most general expression, raised to the level of a science, is precisely the speculative philosophy of law itself. Only Germany could give rise to a speculative philosophy of law, this abstract and exuberant thought of the modern state whose reality remains in the beyond, even though this is only beyond the Rhine. And inversely the German ideology of the modern state that abstracts from actual men was only possible because, and in so far as, the modern state itself abstracts from actual men or satisfies the whole man in a purely imaginary way. In politics the Germans have thought what other people have done. Germany was their theoretical conscience. The abstraction and conceit of its thought has always been in step with the reality of its narrow and trivial situation. So if the *status quo* of the German political system expresses the consummation of the *ancien régime*, the completion of the thorn in the flesh of the modern state, then the *status quo* of the German political consciousness expresses the incompletion of the modern state, the defectiveness of its very flesh.

Even were it only a decided opponent of the previous methods of the German political mind, the criticism of the speculative philosophy of law cannot end with itself, but in tasks for which there is only one solution: *praxis*.

This is the question: can Germany attain to a *praxis* that will be equal to her principles, i.e. can she attain to revolution that will not only raise her to the official level of modern peoples but to the human level that is the immediate future of these peoples?

The weapon of criticism cannot, of course, supplant the criticism of weapons; material force must be overthrown by material force. But theory, too, will become material force as soon as it seizes the masses. Theory is capable of seizing the masses as soon as its proofs are *ad hominem* and its proofs are *ad hominem* as soon as it is radical. To be radical is to grasp the matter by the root. But for man the root is man himself. The manifest proof of the radicalism of German theory and its practical energy is that it starts from the decisive and positive abolition of religion. The criticism of religion ends with the doctrine that man is the highest being for man, that is, with the categorical imperative to overthrow all circumstances in which man is humiliated, enslaved, abandoned, and despised, circumstances best described by the exclamation of a Frenchman on hearing of an intended tax on dogs: Poor dogs! They want to treat you like men!

Even historically speaking, theoretical emancipation has a specifically practical significance for Germany. For Germany's revolutionary past is theoretical, it is the Reformation. Once it was the monk's brain in which the revolution began, now it is in the philosopher's.

Certainly, Luther removed the servitude of devotion by replacing it by the servitude of conviction. He destroyed faith in authority by restoring the authority of faith. He turned priests into laymen by turning laymen into priests. He liberated man from exterior religiosity by making man's inner conscience religious. He emancipated the body from chains by enchaining the heart.

But even though Protestantism was not the true solution, it formulated the problem rightly. The question was now no longer the battle of the layman with the exterior priest, it was the battle with his own interior priest, his priestly nature. Protestantism by turning laymen into priests emancipated the lay popes, the princes, together with their clergy, the privileged and the philistines. Similarly philosophy, by turning the priestly Germans into men, will emancipate the people. But just as emancipation did not stop with the princes, so it will not stop with the secularization of goods involved in the spoliation of the church that was above all practised by hypocritical Prussia. The peasants' war, the most radical event in German history, failed then because of theology. Today, when theology itself has failed, the most unfree event in German history, our *status quo*, will be wrecked on philosophy. On the day before the Reformation Germany was the most unconditional servant of Rome; on the day before its revolution it is the unconditional servant of less than Rome, of Prussia and Austria, cabbage squires and philistines.

However, there appears to be a major obstacle to a radical German revolution. For revolutions need a passive element, a material basis. A theory will only be realized in a people in so far as it is the realization of what it needs. Will the enormous gulf between the demands of German thought and the replies of German actuality match the same gulf that exists between civil society and the state, and within civil society itself? Will theoretical needs immediately become practical ones? It is not enough that thought should tend towards reality, reality must also tend towards thought.

But Germany has not scaled the intermediary stages of political emancipation at the same time as modern peoples. Even the stages that she has passed beyond theoretically have not yet been reached in practice. How can she with one perilous leap not only go beyond her own barriers but also beyond the barriers of modern peoples, barriers which must in reality appear to her as a desirable liberation from her real barriers. A radical revolution can only be a revolution of radical needs, whose presuppositions and breeding-ground seem precisely to be lacking.

Germany, it is true, has only accompanied the development of modern peoples by the abstract activity of thought, without taking an active part in the real struggles of this development. But, on the other hand, it has shared the sufferings of this development without sharing its joys and its partial satisfactions. Abstract activity on the one hand is matched by abstract suffering on the other. Germany will therefore one fine morning find herself on a level with European decadence before it has ever stood on the level of European emancipation. She could be compared with a fetishist who suffers from the maladies of Christianity.

If we consider first the German governments, we find that the conditions of the age, the situation of Germany, the outlook of German culture, and finally their own happy instincts drive them to combine the civilized deficiencies of the modern political world, whose advantages we do not possess, with the barbaric deficiencies of the *ancien régime*, which we enjoy to the full. Thus Germany must participate more and more in the unreason, if not in the reason, even of forms of state that go beyond her present *status quo*. Is there, for example, a country in the world that shares so naïvely as so-called constitutional Germany all the illusions of the constitutional state without sharing its realities? Or did it not have to be the brain wave of a German regime to link the terrors of censorship with those of the French September laws which presuppose freedom of the press? In the Roman Pantheon the gods of all nations were to be found, and in the Holy Roman German Empire are to be found the sins of all forms of state. This eclecticism will reach a height as yet unsuspected: this is guaranteed by the politico-aesthetic gourmandizing of a German king who thinks to play all the roles of monarchy, the feudal as well as the bureaucratic, the absolute as well as the constitutional, the autocratic as well as the democratic, in his own person if not through the person of the people, and for himself if not for the people. Germany is the political deficiencies of the present constituted into a world of their own and as such will not be able to overthrow specifically German barriers without overthrowing the general barriers of the political present.

It is not the radical revolution that is a Utopian dream for Germany, not universal human emancipation; it is the partial, purely political revolution, the revolution which leaves the pillars of the house still standing. What is the basis of a partial, purely political revolution? It is that a part of civil society emancipates itself and attains to universal domination, that a particular class undertakes the general emancipation of society from its particular situation. This class frees the whole of society, but only under the presupposition that the whole of society is in the same situation as this class, that it possesses, or can easily acquire, for example, money and education.

No class in civil society can play this role without arousing a moment of enthusiasm in itself and among the masses. It is a moment when the class fraternizes with society in general and dissolves itself into society; it is identified with society and is felt and recognized as society's general representative. Its claims and rights are truly the claims and rights of society itself of which it is the real social head and heart. A particular class can only vindicate for itself general supremacy in the name of the general rights of society. Revolutionary energy and intellectual selfconfidence alone are not enough to gain this position of emancipator and thus to exploit politically all spheres of society in the interest of one's own sphere. So that the revolution of a people and the emancipation of a particular class of civil society may coincide, so that one class can stand for the whole of society, the deficiency of all society must inversely be concentrated in another class; a particular class must be a class that rouses universal scandal and incorporates all limitations; a particular social sphere must be regarded as the notorious crime of the whole society, so that the liberation of this sphere appears as universal self-liberation. So that one class *par excellence* may appear as the class of liberation, another class must inversely be the manifest class of oppression. The universally negative significance of the French nobility and clergy determined the universally positive significance of the class nearest to them and opposed to them: the bourgeoisie.

But not only is every particular class in Germany lacking in the consistency, insight, courage, and boldness that could mark it as the negative representative of society; they are also lacking in what breadth of mind that can identify, even if only for a moment, with the mind of the people, that genius that can infuse material force with political power, that revolutionary zeal that can throw at its adversary the defiant words: I am nothing and I should be all. The principal element in the honest morality of not only individual Germans but also of classes is that modest egoism that parades its narrowness and lets it be used against itself. Thus the relationship of the different spheres of German society to each other is not dramatic but epic. For each begins to be conscious of itself and to take up a position near the others with its particular claims, not as soon as it is oppressed but as soon as the conditions of the time without any co-operation create a lower social stratum which they in their turn can oppress. Even the moral self-awareness of the German middle class rests simply on the consciousness of being the representative of philistine mediocrity of all other classes. It is thus not only the German kings that ascend the throne *mal à propos* [inopportunely]; it is also every sphere of civil society that is defeated before it has celebrated its victory, that has developed its own limitations before it has overcome the limitations that confront it, that shows its narrow-mindedness before it can show its generosity. The result is that even the opportunity for an important role is past before it was to hand and that, as soon as that class begins to struggle with the class above it, it is engaged in a struggle with the class below. Thus the princes are fighting against the king, the bureaucracy against the nobility, the bourgeoisie against all of them, while the proletariat is already beginning its fight against the bourgeoisie. The middle class scarcely dares to conceive of emancipation from its own point of view, and already the development of social circumstances and the progress of political theory declare this point of view itself to be antiquated or at least problematical.

In France it is enough that one should be something in order to wish to be all. In Germany one must be nothing, if one is to avoid giving up everything. In France partial emancipation is the basis of universal emancipation; in Germany universal emancipation is a *conditio sine qua non* of every partial emancipation. In France it is the reality, in Germany the impossibility, of a gradual liberation that must give birth to total freedom. In France every class of the people is politically idealistic and is not primarily conscious of itself as a particular class but as a representative of general social needs. The role of emancipator thus passes in a dramatic movement to different classes of the French people until it comes to the class which no longer realizes social freedom by presupposing certain conditions that lie outside mankind and are yet created by human society, but which organizes the conditions of human existence by presupposing social freedom. In Germany, on the contrary, where practical life is as unintellecutal as intellectual life is unpractical, no class of civil society has the need for, or capability of achieving, universal emancipation until it is compelled to by its immediate situation, by material necessity and its own chains.

So where is the real possibility of a German emancipation?

We answer: in the formation of a class with radical chains, a class in civil society that is not a class of civil society, of a social group that is the dissolution of all social groups, of a sphere that has a universal character because of its universal sufferings and lays claim to no particular right, because it is the object of no particular injustice but of injustice in general.

This class can no longer lay claim to a historical status, but only to a human one. It is not in a one-sided opposition to the consequences of the German political regime, it is in total opposition to its presuppositions. It is, finally, a sphere that cannot emancipate itself without emancipating itself from all other spheres of society and thereby emancipating these other spheres themselves. In a word, it is the complete loss of humanity and thus can only recover itself by a complete redemption of humanity. This dissolution of society, as a particular class, is the proletariat.

The proletariat is only beginning to exist in Germany through the invasion of the industrial movement. For it is not formed by the poverty produced by natural laws but by artificially induced poverty. It is not made up of the human masses mechanically oppressed by the weight of society, but of those who have their origin in society's brutal dissolution and principally the dissolution of the middle class, although, quite naturally, its ranks are gradually swelled by natural poverty and Germano-Christian serfdom.

When the proletariat proclaims the dissolution of the hitherto existing world order, it merely declares the secret of its own existence, since it is in fact the dissolution of this order. When it demands the negation of private property it is only laying down as a principle for society what society has laid down as a principle for the proletariat, what has already been incorporated in itself without its consent as the negative result of society. The proletarian thus finds that he has in relation to the world of the future the same right as the German king in relation to the world of the past, when he calls the people *his* people as he might call a horse *his* horse. When the king declares the people to be his private property he is only confirming that the private property owner is king.

As philosophy finds in the proletariat its material weapons, so the proletariat finds in philosophy its intellectual weapons, and as soon as the lightning of thought has struck deep into the virgin soil of the people, the emancipation of the Germans into men will be completed.

Let us summarize our results:

The only liberation of Germany that is practically possible is the liberation from the theoretical standpoint that declares man to be the highest being for man. In Germany emancipation from the Middle Ages is only possible as an emancipation from the partial overcoming of the Middle Ages. In Germany no form of slavery can be broken without every form of slavery being broken. Germany is thorough and cannot make a revolution without its being a thorough one. The emancipation of Germany is the emancipation of man. The head of this emancipation is philosophy, its heart is the proletariat. Philosophy cannot realize itself without transcending the proletariat, the proletariat cannot transcend itself without realizing philosophy.

When all interior conditions are fulfilled, the day of German resurrection will be heralded by the crowing of the Gallic cock.

BIBLIOGRAPHY

Original: *MEGA*, I (1) 1, pp. 607 ff.

Present translation: K. Marx, *Early Texts*, pp. 115 ff.

Other translations: K. Marx and F. Engels, *On Religion*, Moscow, 1957, pp. 41 ff.
K. Marx, *Early Writings*, ed. T. Bottomore, London, 1963, pp. 43 ff.; *Writings of the Young Marx on Philosophy and Society*, ed. Easton and Guddat, pp. 249 ff.
K. Marx, *Critique of Hegel's 'Philosophy of Right'*, ed. J. O'Malley, Cambridge, 1970, pp. 131 ff.
K. Marx, *Early Writings*, ed. L. Colletti, New York and London, 1975, pp. 243 ff.

Commentaries: L. Dupré, *The Philosophical Foundations of Marxism*, New York, 1966, Ch. 5.
D. Howard, *The Development of the Marxian Dialectic*, Carbondale, 1972, pp. 111 ff.
J. Maguire, *Marx's Paris Writings: an Analysis*, Dublin, 1972, pp. 31 ff.
D. McLellan, *Marx before Marxism*, Penguin, 1962, Ch. 6.
E. Olssen, 'Marx and the Resurrection', *Journal of the History of Ideas*, 1968.
R. Tucker, *Philosophy and Myth in Karl Marx*, Cambridge, 1961, pp. 106 ff.

8. Economic and Philosophical Manuscripts

These manuscripts, the most important of Marx's early writings, were written in the summer of 1844. They represent Marx's first draft of his 'Economics'—the project to which he was to devote the rest of his life. The manuscripts fall into four main groups: firstly, there is a passage on alienated labour—the most finished and readily comprehensible of the manuscripts, in which Marx details the ways in which the worker's relationship to his product result in his alienation. Secondly, in the manuscript headed 'Private Property and Communism', Marx outlines his view of communist man and society. In the third section he discusses the relationship of capitalism to human needs; and in the final section he gives what is probably his fullest account of his view of Hegel's dialectic, praising him for having discovered man's world-creating capacities, but criticizing his abstract, philosophical portrayal.

Marx intended to write up this work for publication, but other problems distracted him. When they were first published in 1932, they were thought by many to portray a humanist and even an existentialist Marx—very different from the Marx of the later writings—and this discrepancy gave rise to a protracted debate on the continuity or discontinuity of Marx's thought. The 1844 manuscripts certainly show him under the influence of Feuerbach's humanism (though Marx's interest in politics, economics, and even history was foreign to Feuerbach), and he was soon to distance himself considerably from Feuerbach's ideas. Nevertheless, many of the positions taken up by Marx in 1844 were still present in the *Grundrisse* and even in *Capital*.

Preface

I have announced in the *Deutsch–französische Jahrbücher* a forthcoming critique of legal and political science in the form of a critique of Hegel's philosophy of right. While I was working on the manuscript for publication it became clear that it was quite inopportune to mix criticism directed purely against speculation with that of other and different matters, and that this mixture was an obstacle to the development of my line of thought and to its intelligibility. Moreover, the condensation of such rich and varied subjects into a single work would have permitted only a very aphoristic treatment, and, on the other hand, such an aphoristic presentation would have created the appearance of an arbitrary systematization. I will therefore present one after another a critique of law, morality, politics, etc. in different independent brochures and then finally in a separate work try to show the connection of the whole and the relationship of the parts to each other and end with a criticism of the elaboration of the material by speculative philosophy.

Therefore in the present work the connection of political economy with the state, law, morality, civil life etc. is only dealt with in so far as political economy itself professes to deal with these subjects.

I do not need to reassure the reader who is familiar with political economy that my results have been obtained through a completely empirical analysis founded on a conscientious and critical study of political economy.

It is self-evident that apart from the French and English socialists I have also used the works of German socialists. However, the substantial and original German works in this field can be reduced—apart from Weitling's work—to the articles published by Hess in the *Twenty-One Sheets* and to Engels's *Sketch of a Critique of Political Economy* in the *Deutsch-französische Jahrbücher*, where I also outlined the first elements of the present work in a completely general way.

Apart from these writers who have treated political economy in a critical manner, positive criticism in general, including therefore the positive German criticism of political economy, owes its true foundation to the discoveries of Feuerbach. The petty jealousy of some and the real anger of others seems to have instigated a veritable conspiracy of silence against his *Philosophy of the Future* and *Theses for the Reform of Philosophy* in *Anekdota*, although they are used tacitly.

The first positive humanist and naturalist criticism dates from Feuerbach. The less bombastic they are, the more sure, deep, comprehensive, and lasting is the effect of Feuerbach's works, the only ones since Hegel's *Phenomenology* and *Logic* to contain a real theoretical revolution.

I considered the final chapter of the present work, 'The Critical Analysis of the Hegelian Dialectic and Philosophy in General', to be absolutely necessary. This is in contradistinction to the critical theologians of our time who have not completed any such task. This deficiency of theirs is inevitable, for even the critical theologian remains a theologian, and thus must either begin with definite presuppositions of philosophy regarded as an authority or, if the process of criticism and the discoveries of someone else have made him doubt his philosophical presuppositions, he abandons them in a cowardly and unjustified manner, abstracts from them, and only proclaims his slavery to them and vexation at this slavery in a negative, unconscious, and sophistical way.

The reason for his purely negative and unconscious expression is partly that he constantly repeats an assurance of the purity of his own criticism and partly that he wishes to avert the eye of the observer and his own eye from the fact that criticism must necessarily come to terms with its birth-place, the Hegelian dialectic and German philosophy in general.

However much theological criticism was at the beginning of the movement the really progressive stage, on close examination it is in the last analysis nothing but the apogee and the result old transcendent philosophy, particularly the Hegelian, pushed to theological caricature. The justice meted out by history is interesting in that theology, which was always the fly in the philosophical ointment, is now called to represent in itself the negative dissolution of philosophy, its process of decomposition. I shall demonstrate this historical nemesis in detail on another occasion.

On the other hand, the extent to which Feuerbach's discoveries about the essence of philosophy still necessitate a critical treatment of the philosophical dialectic, at least to serve as proof, will become apparent from my development of the subject.

Alienated Labour

We started from the presuppositions of political economy. We accepted its vocabulary and its laws. We presupposed private property, the separation of labour, capital, and land, and likewise of wages, profit, and ground rent; also division of labour; competition; the concept of exchange value, etc. Using the very words of political economy we have demonstrated that the worker is degraded to the most miserable sort of commodity; that the misery of the worker is in inverse proportion to the power and size of his production; that the necessary result of competition is the accumulation of capital in a few hands, and thus a more terrible restoration of monopoly; and that finally the distinction between capitalist and landlord, and that between peasant and industrial worker disappears and the whole of society must fall apart into the two classes of the property owners and the propertyless workers.

Political economy starts with the fact of private property, it does not explain it to us. It conceives of the material process that private property goes through in reality in general abstract formulas which then have for it a value of laws. It does not understand these laws, i.e. it does not demonstrate how they arise from the nature of private property. Political economy does not afford us any explanation of the reason for the separation of labour and capital, of capital and land. When, for example, political economy defines the relationship of wages to profit from capital, the interest of the capitalist is the ultimate court of appeal, that is, it presupposes what should be its result. In the same way competition enters the argument everywhere. It is explained by exterior circumstances. But political economy tells us nothing about how far these exterior, apparently fortuitous circumstances are merely the expression of a necessary development. We have seen how it regards exchange itself as something fortuitous. The only wheels that political economy sets in motion are greed and war among the greedy, competition.

It is just because political economy has not grasped the connections in the movement that new contradictions have arisen in its doctrines, for example, between that of monopoly and that of competition, freedom of craft and corporations, division of landed property and large estates. For competition, free trade, and the division of landed property were only seen as fortuitous circumstances created by will and force, not developed and comprehended as necessary, inevitable, and natural results of monopoly, corporations, and feudal property.

So what we have to understand now is the essential connection of private property, selfishness, the separation of labour, capital, and landed property, of exchange and competition, of the value and degradation of man, of monopoly and competition, etc.—the connection of all this alienation with the money system.

Let us not be like the political economist who, when he wishes to explain something, puts himself in an imaginary original state of affairs. Such an original stage of affairs explains nothing. He simply pushes the question back into a grey and nebulous distance. He presupposes as a fact and an event what he ought to be deducing, namely the necessary connection between the two things, for example, between the division of labour and exchange. Similarly, the theologian explains the origin of evil through the fall, i.e. he presupposes as an historical fact what he should be explaining.

We start with a contemporary fact of political economy:

The worker becomes poorer the richer is his production, the more it increases in power and scope. The worker becomes a commodity that is all the cheaper the more commodities he creates. The depreciation of the human world progresses in direct proportion to the increase in value of the world of things. Labour does not only produce commodities; it produces itself and the labourer as a commodity and that to the extent to which it produces commodities in general.

What this fact expresses is merely this: the object that labour produces, its product, confronts it as an alien being, as a power independent of the producer. The product of labour is labour that has solidified itself into an object, made itself into a thing, the objectification of labour. The realization of labour is its objectification. In political economy this realization of labour appears as a loss of reality for the worker, objectification as a loss of the object or slavery to it, and appropriation as alienation, as externalization.

The realization of labour appears as a loss of reality to an extent that the worker loses his reality by dying of starvation. Objectification appears as a loss of the object to such an extent that the worker is robbed not only of the objects necessary for his life but also of the objects of his work. Indeed, labour itself becomes an object he can only have in his power with the greatest of efforts and at irregular intervals. The appropriation of the object appears as alienation to such an extent that the more objects the worker produces, the less he can possess and the more he falls under the domination of his product, capital.

All these consequences follow from the fact that the worker relates to the product of his labour as to an alien object. For it is evident from this presupposition that the more the worker externalizes himself in his work, the more powerful becomes the alien, objective world that he creates opposite himself, the poorer he becomes himself in his inner life and the less he can call his own. It is just the same in religion. The more man puts into God, the less he retains in himself. The worker puts his life into the object and this means that it no longer belongs to him but to the object. So the greater this activity, the more the worker is without an object. What the product of his labour is, that he is not. So the greater this product the less he is himself. The externalization of the worker in his product implies not only that his labour becomes an object, an exterior existence but also that it exists outside him, independent and alien, and becomes a self-sufficient power opposite him, that the life that he has lent to the object affronts him, hostile and alien.

Let us now deal in more detail with objectification, the production of the worker, and the alienation, the loss of the object, his product, which is involved in it.

The worker can create nothing without nature, the sensuous exterior world. It is the matter in which his labour realizes itself, in which it is active, out of which and through which it produces.

But as nature affords the means of life for labour in the sense that labour cannot live without objects on which it exercises itself, so it affords a means of life in the narrower sense, namely the means for the physical subsistence of the worker himself.

Thus the more the worker appropriates the exterior world of sensuous nature by his labour, the more he doubly deprives himself of the means of subsistence, firstly since the exterior sensuous world increasingly ceases to be an object belonging to his work, a means of subsistence for his labour; secondly, since it increasingly ceases to be a means of subsistence in the direct sense, a means for the physical subsistence of the worker.

Thus in these two ways the worker becomes a slave to his object: firstly he receives an object of labour, that is he receives labour, and secondly, he receives the means of subsistence. Thus it is his object that permits him to exist first as a worker and secondly as a physical subject. The climax of this slavery is that only as a worker can he maintain himself as a physical subject and it is only as a physical subject that he is a worker.

(According to the laws of political economy the alienation of the worker in his object is expressed as follows: the more the worker produces the less he has to consume, the more values he creates the more valueless and worthless he becomes, the more formed the product the more deformed the worker, the more civilized the product, the more barbaric the worker, the more powerful the work the more powerless becomes the worker, the more cultured the work the more philistine the worker becomes and more of a slave to nature.)

Political economy hides the alienation in the essence of labour by not considering the immediate relationship between the worker (labour) and production. Labour produces works of wonder for the rich, but nakedness for the worker. It produces palaces, but only hovels for the worker; it produces beauty, but cripples the worker; it replaces labour by machines but throws a part of the workers back to a barbaric labour and turns the other part into machines. It produces culture, but also imbecility and cretinism for the worker.

The immediate relationship of labour to its products is the relationship of the worker to the objects of his production. The relationship of the man of means to the objects of production and to production itself is only a consequence of this first relationship. And it confirms it. We shall examine this other aspect later.

So when we ask the question: what relationship is essential to labour, we are asking about the relationship of the worker to production.

Up to now we have considered only one aspect of the alienation or externalization of the worker, his relationship to the products of his labour. But alienation shows itself not only in the result, but also in the act of production, inside productive activity itself. How would the worker be able to affront the product of his work as an alien being if he did not alienate himself in the act of production itself? For the product is merely the summary of the activity of production. So if the product of labour is externalization, production itself must be active externalization, the externalization of activity, the activity of externalization. The alienation of the object of labour is only the résumé of the alienation, the externalization in the activity of labour itself.

What does the externalization of labour consist of then?

Firstly, that labour is exterior to the worker, that is, it does not belong to his essence. Therefore he does not confirm himself in his work, he denies himself, feels miserable instead of happy, deploys no free physical and intellectual energy, but mortifies his body and ruins his mind. Thus the worker only feels a stranger. He is at home when he is not working and when he works he is not at home. His labour is therefore not voluntary but compulsory, forced labour. It is therefore not the satisfaction of a need but only a means to satisfy needs outside itself. How alien it really is is very evident from the fact that when there is no physical or other compulsion, labour is avoided like the plague. External labour, labour in which man externalizes himself, is a labour of self-sacrifice and mortification. Finally, the external character of labour for the worker shows itself in the fact that it is not his own but someone else's, that it does not belong to him, that he does not belong to himself in his labour but to someone else. As in religion the human imagination's own activity, the activity of man's head and his heart, reacts independently on the individual as an alien activity of gods or devils, so the activity of the worker is not his own spontaneous activity. It belongs to another and is the loss of himself.

The result we arrive at then is that man (the worker) only feels himself freely active in his animal functions of eating, drinking, and procreating, at most also in his dwelling and dress, and feels himself an animal in his human functions.

Eating, drinking, procreating, etc. are indeed truly human functions. But in the abstraction that separates them from the other round of human activity and makes them into final and exclusive ends they become animal.

We have treated the act of alienation of practical human activity, labour, from two aspects. (1) The relationship of the worker to the product of his labour as an alien object that has power over him. This relationship is at the same time the relationship to the sensuous exterior world and to natural objects as to an alien and hostile world opposed to him. (2) The relationship of labour to the act of production inside labour. This relationship is the relationship of the worker to his own activity as something that is alien and does not belong to him; it is activity that is passivity, power that is weakness, procreation that is castration, the worker's own physical and intellectual energy, his personal life (for what is life except activity?) as an activity directed against himself, independent of him and not belonging to him. It is self-alienation, as above it was the alienation of the object.

We now have to draw a third characteristic of alienated labour from the two previous ones.

Man is a species-being not only in that practically and theoretically he makes both his own and other species into his objects, but also, and this is only another way of putting the same thing, he relates to himself as to the present, living species, in that he relates to himself as to a universal and therefore free being.

Both with man and with animals the species-life consists physically in the fact that man (like animals) lives from inorganic nature, and the more universal man is than animals the more more universal is the area of inorganic nature from which he lives. From the theoretical point of view, plants, animals, stones, air, light, etc. form part of human consciousness, partly as objects of natural science, partly as objects of art; they are his intellectual inorganic nature, his intellectual means of subsistence, which he must first prepare before he can enjoy and assimilate them. From the practical point of view, too, they form a part of human life and activity. Physically man lives solely from these products of nature, whether they appear as food, heating, clothing, habitation, etc. The universality of man appears in practice precisely in the universality that makes the whole of nature into his inorganic body in that it is both (i) his immediate means of subsistence and also (ii) the material object and tool of his vital activity. Nature is the inorganic body of a man, that is, in so far as it is not itself a human body. That man lives from nature means that nature is his body with which he must maintain a constant interchange so as not to die. That man's physical and intellectual life depends on nature merely means that nature depends on itself, for man is a part of nature.

While alienated labour alienates (1) nature from man, and (2) man from himself, his own active function, his vital activity, it also alienates the species from man; it turns his species-life into a means towards his individual life. Firstly it alienates species-life and individual life, and secondly in its abstraction it makes the latter into the aim of the former which is also conceived of in its abstract and alien form. For firstly, work, vital activity, and productive life itself appear to man only as a means to the satisfaction of a need, the need to preserve his physical existence. But productive life is species-life. It is life producing life. The whole character of a species, its generic character, is contained in its manner of vital activity, and free conscious activity is the species-characteristic of man. Life itself appears merely as a means to life.

The animal is immediately one with its vital activity. It is not distinct from it. They are identical. Man makes his vital activity itself into an object of his will and consciousness. He has a conscious vital activity. He is not immediately identical to any of his characterizations. Conscious vital activity differentiates man immediately from animal vital activity. It is this and this alone that makes man a species-being. He is only a conscious being, that is, his own life is an object to him, precisely because he is a species-being. This is the only reason for his activity being free activity. Alienated labour reverses the relationship so that, just because he is a conscious being, man makes his vital activity and essence a mere means to his existence.

The practical creation of an objective world, the working-over of inorganic nature, is the confirmation of man as a conscious species-being, that is, as a being that relates to the species as to himself and to himself as to the species. It is true that the animal, too, produces. It builds itself a nest, a dwelling, like the bee, the beaver, the ant, etc. But it only produces what it needs immediately for itself or its offspring; it produces one-sidedly whereas man produces universally; it produces only under the pressure of immediate physical need, whereas man produces freely from physical need and only truly produces when he is thus free; it produces only itself whereas man reproduces the whole of nature. Its product belongs immediately to its physical body whereas man can freely separate himself from his product. The animal only fashions things according to the standards and needs of the species it belongs to, whereas man knows how to produce according to the measure of every species and knows everywhere how to apply its inherent standard to the object; thus man also fashions things according to the laws of beauty.

Thus it is in the working over of the objective world that man first really affirms himself as a species-being. This production is his active species-life. Through it nature appears as his work and his reality. The object of work is therefore the objectification of the species-life of man; for he duplicates himself not only intellectually, in his mind, but also actively in reality and thus can look at his image in a world he has created. Therefore when alienated labour tears from man the object of his production, it also tears from him his species-life, the real objectivity of his species and turns the advantage he has over animals into a disadvantage in that his inorganic body, nature, is torn from him.

Similarly, in that alienated labour degrades man's own free activity to a means, it turns the species-life of man into a means for his physical existence.

Thus consciousness, which man derives from his species, changes itself through alienation so that species-life becomes a means for him.

Therefore alienated labour:

(3) makes the species-being of man, both nature and the intellectual faculties of his species, into a being that is alien to him, into a means for his individual existence. It alienates from man his own body, nature exterior to him, and his intellectual being, his human essence.

(4) An immediate consequence of man's alienation from the product of his work, his vital activity and his species-being, is the alienation of man from man. When man is opposed to himself, it is another man that is opposed to him. What is valid for the relationship of a man to his work, of the product of his work and himself, is also valid for the relationship of man to other men and of their labour and the objects of their labour.

In general, the statement that man is alienated from his species-being, means that one man is alienated from another as each of them is alienated from the human essence.

The alienation of man and in general of every relationship in which man stands to himself is first realized and expressed in the relationship with which man stands to other men.

Thus in the situation of alienated labour each man measures his relationship to other men by the relationship in which he finds himself placed as a worker.

We began with a fact of political economy, the alienation of the worker and his production. We have expressed this fact in conceptual terms: alienated, externalized labour. We have analysed this concept and thus analysed a purely economic fact.

Let us now see further how the concept of alienated, externalized labour must express and represent itself in reality.

If the product of work is alien to me, opposes me as an alien power, whom does it belong to then?

If my own activity does not belong to me and is an alien, forced activity to whom does it belong then?

To another being than myself.

Who is this being?

The gods? Of course in the beginning of history the chief production, as for example, the building of temples etc. in Egypt, India, and Mexico was both in the service of the gods and also belonged to them. But the gods alone were never the masters of the work. And nature just as little. And what a paradox it would be if, the more man mastered nature through his work and the more the miracles of the gods were rendered superfluous by the miracles of industry, the more man had to give up his pleasure in producing and the enjoyment in his product for the sake of these powers.

The alien being to whom the labour and the product of the labour belongs, whom the labour serves and who enjoys its product, can only be man himself. If the product of labour does not belong to the worker but stands over against him as an alien power, this is only possible in that it belongs to another man apart from the worker.

If his activity torments him it must be a joy and a pleasure to someone else. This alien power above man can be neither the gods nor nature, only man himself.

Consider further the above sentence that the relationship of man to himself first becomes objective and real to him through his relationship to other men. So if he relates to the product of his labour, his objectified labour, as to an object that is alien, hostile, powerful, and independent of him, this relationship implies that another man is the alien, hostile, powerful, and independent master of this object. If he relates to his own activity as to something unfree, it is a relationship to an activity that is under the domination, oppression, and yoke of another man.

Every self-alienation of man from himself and nature appears in the relationship in which he places himself and nature to other men distinct from himself. Therefore religious self-alienation necessarily appears in the relationship of layman to priest, or, because here we are dealing with a spiritual world, to a mediator, etc. In the practical, real world, the self-alienation can only appear through the practical, real relationship to other men. The means through which alienation makes progress are themselves practical. Through alienated labour, then, man creates not only his relationship to the object and act of production as to alien and hostile men; he creates too the relationship in which other men stand to his production and his product and the relationship in which he stands to these other men. Just as he turns his production into his own loss of reality and punishment and his own product into a loss, a product that does not belong to him, so he creates the domination of the man who does not produce over the production and the product. As he alienates his activity from himself, so he hands over to an alien person an activity that does not belong to him.

Up till now we have considered the relationship only from the side of the worker and we will later consider it from the side of the non-worker.

Thus through alienated, externalized labour the worker creates the relationship to this labour of a man who is alien to it and remains exterior to it. The relationship of the worker to his labour creates the relationship to it of the capitalist, or whatever else one wishes to call the master of the labour. Private property is thus the product, result, and necessary consequence of externalized labour, of the exterior relationship of the worker to nature and to himself.

Thus private property is the result of the analysis of the concept of externalized labour, i.e. externalized man, alienated work, alienated life, alienated man.

We have, of course, obtained the concept of externalized labour (externalized life) from political economy as the result of the movement of private property. But it is evident from the analysis of this concept that, although private property appears to be the ground and reason for externalized labour, it is rather a consequence of it, just as the gods are originally not the cause but the effect of the aberration of the human mind, although later this relationship reverses itself.

It is only in the final culmination of the development of private property that these hidden characteristics come once more to the fore, in that firstly it is the product of externalized labour and secondly it is the means through which labour externalizes itself, the realization of this externalization.

This development sheds light at the same time on several' previously unresolved contradictions.

1. Political economy starts from labour as the veritable soul of production, and yet it attributes nothing to labour and everything to private property. Proudhon has drawn a conclusion from this contradiction that is favourable to labour and against private property. But we can see that this apparent contradiction is the contradiction of alienated labour with itself and that political economy has only expressed the laws of alienated labour.

We can therefore also see that wages and private property are identical: for wages, in which the product, the object of the labour, remunerates the labour itself, are just a necessary consequence of the alienation of labour. In the wage system the labour does not appear as the final aim but only as the servant of the wages. We will develop this later and for the moment only draw a few consequences.

An enforced raising of wages (quite apart from other difficulties, apart from the fact that, being an anomaly, it could only be maintained by force) would only mean a better payment of slaves and would not give this human meaning and worth either to the worker or to his labour.

Indeed, even the equality of wages that Proudhon demands only changes the relationship of the contemporary worker to his labour into that of all men to labour. Society is then conceived of as an abstract capitalist.

Wages are an immediate consequence of alienated labour and alienated labour is the immediate cause of private property. Thus the disappearance of one entails also the disappearance of the other.

2. It is a further consequence of the relationship of alienated labour to private property that the emancipation of society from private property, etc., from slavery, is expressed in its political form by the emancipation of the workers. This is not because only their emancipation is at stake but because general human emancipation is contained in their emancipation. It is contained within it because the whole of human slavery is involved in the

relationship of the worker to his product and all slave relationships are only modifications and consequences of this relationship.

Just as we have discovered the concept of private property through an analysis of the concept of alienated, externalized labour, so all categories of political economy can be deduced with the help of these two factors. We shall recognize in each category of market, competition, capital, money, only a particular and developed expression of these first two fundamental elements.

However, before we consider this structure let us try to solve two problems:

1. To determine the general essence of private property as it appears as a result of alienated labour in its relationship to truly human and social property.

2. We have taken the alienation and externalization of labour as a fact and analysed this fact. We now ask, how does man come to externalize, to alienate his labour? How is this alienation grounded in human development? We have already obtained much material for the solution of this problem, in that we have turned the question of the origin of private property into the question of the relationship of externalized labour to the development of human history. For when we speak of private property we think we are dealing with something that is exterior to man. When we speak of labour, then we are dealing directly with man. This new formulation of the problem already implies its solution.

To take point 1, the general nature of private property and its relationship to truly human property.

Externalized labour has been broken down into two component parts that determine each other or are only different expressions of one and the same relationship. Appropriation appears as alienation, as externalization, and externalization as appropriation, and alienation as true enfranchisement. We have dealt with one aspect, alienated labour as regards the worker himself, that is, the relationship of externalized labour to itself. As a product and necessary result of this relationship we have discovered the property relationship of the non-worker to the worker and his labour.

As the material and summary expression of alienated labour, private property embraces both relationships, both that of the worker to his labour, the product of his labour and the non-worker, and that of the non-worker to the worker and the product of his labour.

We have already seen that for the worker who appropriates nature through his work, this appropriation appears as alienation, his own activity as activity for and of someone else, his vitality as sacrifice of his life, production of objects as their loss to an alien power, an alien man: let us now consider the relationship that this man, who is alien to labour and the worker, has to the worker, to labour and its object.

The first remark to make is that everything that appears in the case of the worker to be an activity of externalization, of alienation, appears in the case of the non-worker to be a state of externalization, of alienation.

Secondly, the real, practical behaviour of the worker in production and towards his product (as a state of mind) appears in the case of the non-worker opposed to him as theoretical behaviour. Thirdly, the non-worker does everything against the worker that the worker does against himself but he does not do against himself what he does against the worker.

Let us consider these three relationships in more detail. . . . [The manuscript breaks off unfinished here.]

Private Property and Communism

The overcoming of self-alienation follows the same course as self-alienation itself. At first, private property is considered only from its objective aspect, but still with labour as its essence. The form of its existence is therefore capital, that is to be abolished 'as such' (Proudhon). Or else the source of the harmfulness of private property, its alienation from human existence, is thought of as consisting in the particular type of labour, labour which is levelled down, fragmented, and therefore unfree. This is the view of Fourier, who like the physiocrats also considered agriculture as labour *par excellence*. Saint-Simon on the other hand declares industrial labour to be the essential type, and demands as well exclusive rule by industrialists and the improvement of the condition of the workers. Finally, communism is the positive expression of the overcoming of private property, appearing first of all as generalized private property. In making this relationship universal communism is:

1. In its original form only a generalization and completion of private property. As such it appears in a dual form: firstly, it is faced with such a great domination of material property that it wishes to destroy everything that cannot be possessed by everybody as private property; it wishes to abstract forcibly from talent, etc. It considers immediate physical ownership as the sole aim of life and being. The category of worker is not abolished but extended to all men. The relationship of the community to the world of things remains that of private property. Finally, this process of opposing general private property to private property is expressed in the animal form of opposing to marriage (which is of course a form of exclusive private property) the community of women where the woman becomes the common property of the community. One might say that the idea of the community of women reveals the open secret of this completely crude and unthinking type of communism. Just as women pass from marriage to universal prostitution, so the whole world of wealth, that is the objective essence of man, passes from the relationship of exclusive marriage to the private property owner to the relationship of universal prostitution with the community. By

systematically denying the personality of man this communism is merely the consistent expression of private property which is just this negation. Universal envy setting itself up as a power is the concealed form of greed which merely asserts itself and satisfies itself in another way. The thoughts of every private property owner as such are at least turned against those richer than they as an envious desire to level down. This envious desire is precisely the essence of competition. Crude communism is only the completion of this envy and levelling down to a preconceived minimum. It has a particular and limited standard. How little this abolition of private property constitutes a real appropriation is proved by the abstract negation of the whole world of culture and civilization, a regression to the unnatural simplicity of the poor man without any needs who has not even arrived at the stage of private property, let alone got beyond it.

In this theory the community merely means a community of work and equality of the wages that the communal capital, the community as general capitalist, pays out. Both sides of the relationship are raised to a sham universality, labour being the defining characteristic applied to each man, while capital is the universality and power of society.

The infinite degradation in which man exists for himself is expressed in his relationship to woman as prey and servant of communal lust; for the secret of this relationship finds an unambiguous, decisive, open, and unveiled expression in the relationship of man to woman and the conception of the immediate and natural relationship of the sexes. The immediate, natural, and necessary relationship of human being to human being is the relationship of man to woman. In this natural relationship of the sexes man's relationship to nature is immediately his relationship to man, and his relationship to man is immediately his relationship to nature, his own natural function. Thus, in this relationship is sensuously revealed and reduced to an observable fact how far for man his essence has become nature or nature has become man's human essence. Thus, from this relationship the whole cultural level of man can be judged. From the character of this relationship we can conclude how far man has become a species-being, a human being, and conceives of himself as such; the relationship of man to woman is the most natural relationship of human being to human being. Thus it shows how far the natural behaviour of man has become human or how far the human essence has become his natural essence, how far his human nature has become nature for him. This relationship also shows how far the need of man has become a human need, how far his fellow men as men have become a need, how far in his most individual existence he is at the same time a communal being.

The first positive abolition of private property, crude communism, is thus only the form in which appears the ignominy of private property that wishes to establish itself as the positive essence of the community.

2. The second form of communism is: (*a*) still political in nature, whether

democratic or despotic; (*b*) with the abolition of the state, but still incomplete and still under the influence of private property, i.e. the alienation of man. In both forms communism knows itself already to be the reintegration or return of man into himself, the abolition of man's self-alienation. But since it has not yet grasped the positive essence of private property or the human nature of needs, it is still imprisoned and contaminated by private property. It has understood its concept, but not yet its essence.

3. Thirdly, there is communism as the positive abolition of private property and thus of human self-alienation and therefore the real reappropriation of the human essence by and for man. This is communism as the complete and conscious return of man conserving all the riches of previous development for man himself as a social, i.e. human being. Communism as completed naturalism is humanism and as completed humanism is naturalism. It is the genuine solution of the antagonism between man and nature and between man and man. It is the true solution of the struggle between existence and essence, between objectification and self-affirmation, between freedom and necessity, between individual and species. It is the solution to the riddle of history and knows itself to be this solution.

The whole movement of history, therefore, both as regards the real engendering of this communism, the birth of its empirical existence, and also as regards its consciousness and thought, is the consciously comprehended process of its becoming. On the other hand, the communism that is still incomplete seeks an historical proof for itself in what already exists by selecting isolated historical formations opposed to private property. It tears isolated phases out of the movement (Cabet, Villegardelle, etc. in particular ride this hobby horse) and sserts them as proofs of its historical pedigree. But all it succeeds in showing is that the disproportionately larger part of this movement contradicts its assertions and that if it has ever existed, it is precisely its past being that refutes its pretension to essential being.

We can easily see how necessary it is that the whole revolutionary movement should find not so much its empirical as its theoretical basis in the development of private property and particularly the economic system.

This material, immediately sensuous, private property, is the material, sensuous expression of man's alienated life. Its movement of production and consumption is the sensuous revelation of the movement of all previous production, i.e. the realization or reality of man. Religion, family, state, law, morality, science, and art are only particular forms of production and fall under its general law. The positive abolition of private property and the appropriation of human life is therefore the positive abolition of all alienation, thus the return of man out of religion, family, state, etc. into his human, i.e. social being. Religious alienation as such occurs only in man's interior consciousness, but economic alienation is that of real life, and its abolition therefore covers both aspects. It is obvious that the movement begins differently with different peoples according to whether the actual conscious life of the people is lived in their minds or in the outer world, is an ideal or a real life. Communism begins immediately with atheism (Owen) and atheism is at first still very far from being communism, for this atheism is still rather an abstraction.[1] The philanthropy of atheism is therefore at first only an abstract philosophical philanthropy whereas that of communism is immediately real and directly orientated towards action.

We have seen how, presupposing the positive supercession of private property, man produces man, himself and other men; how also the object, which is the direct result of his personal activity, is at the same time his own existence for other men and their existence for him. In the same way, both the materials of his labour and man its author are the result and at the same time the origin of the movement. (And private property is historically necessary precisely because there must be this origin.) So the general character of the whole movement is a social one; as society produces man as man, so it is produced by man. Activity and enjoyment are social both in their content and in their mode of existence; they are social activity and social enjoyment. The human significance of nature is only available to social man; for only to social man is nature available as a bond with other men, as the basis of his own existence for others and theirs for him, and as the vital element in human reality; only to social man is nature the foundation of his own human existence. Only as such has his natural existence become a human existence and nature itself become human. Thus society completes the essential unity of man and nature, it is the genuine resurrection of nature, the accomplished naturalism of man and the accomplished humanism of nature.

Social activity and social enjoyment by no means exist only in the form of a directly communal activity and directly communal enjoyment. But communal activity and enjoyment, i.e. activity and enjoyment that is expressed and confirmed in the real society of other men, will occur everywhere where this direct expression of sociability arises from the content of the activity or enjoyment and corresponds to its nature.

But even if my activity is a scientific one, etc., an activity that I can seldom perform directly in company with other men, I am still acting socially since I am acting as a man. Not only the material of my activity—like language itself for the thinker—is given to me as a social product, my own existence is social activity; therefore what I individually produce, I produce individually for society, conscious of myself as a social being.

My universal consciousness is only the theoretical form whose living form is the real community, society, whereas at the present time universal consciousness is an abstraction from real life and as such turns into its enemy. Thus the activity of my universal consciousness is as such my theoretical existence as a social being.

[1] Prostitution is only a particular expression of the general prostitution of the worker, and because prostitution is a relationship which includes both the person prostituted and the person prostituting—whose baseness is even greater—thus the capitalist, too, etc. is included within this category. (Footnote by Marx.)

It is above all necessary to avoid restoring society as a fixed abstraction opposed to the individual. The individual is the social being. Therefore, even when the manifestation of his life does not take the form of a communal manifestation performed in the company of other men, it is still a manifestation and confirmation of social life. The individual and the species-life of man are not different, although, necessarily, the mode of existence of individual life is a more particular or a more general mode of species-life or the species-life is a more particular or more general individual life.

Man confirms his real social life in his species-consciousness and in his thought he merely repeats his real existence just as conversely his species-being is confirmed in his species-consciousness and exists for itself in its universality as a thinking being.

However much he is a particular individual (and it is precisely his particularity that makes him an individual and a truly individual communal being) man is just as much the totality, the ideal totality, the subjective existence of society as something thought and felt. Man exists also in reality both as the contemplation and true enjoyment of social existence and as the totality of human manifestations of life.

Thus thought and being are indeed distinct, but at the same time they together form a unity.

Death appears as the harsh victory of the species over the particular individual and seems to contradict their unity; but the particular individual is only a determinate species-being and thus mortal.

4. Private property is only the sensuous expression of the fact that man is both objective to himself and, even more, becomes a hostile and inhuman object to himself, that the expression of his life entails its externalization, its realization becomes the loss of its reality, an alien reality. Similarly the positive supersession of private property, that is, the sensuous appropriation by and for man of human essence and human life, of objective man and his works, should not be conceived of only as direct and exclusive enjoyment, as possession and having. Man appropriates his universal being in a universal manner, as a whole man. Each of his human relationships to the world— seeing, hearing, smell, tasting, feeling, thinking, contemplating, feeling, willing, acting, loving—in short all the organs of his individuality, just as the organs whose form is a directly communal one, are in their objective action, or their relation to the object, the appropriation of this object. The appropriation of human reality, their relationship to the object, is the confirmation of human reality. It is therefore as manifold as the determinations and activities of human nature. It is human effectiveness and suffering, for suffering, understood in the human sense, is an enjoyment of the self for man.

Private property has made us so stupid and narrow-minded that an object is only ours when we have it, when it exists as capital for us or when we directly possess, eat, drink, wear, inhabit it, etc. in short, when we use it. Yet private property itself in its turn conceives of all these direct realizations of property merely as means of life, and the life which they serve is that of private property, labour, and capitalization.

Thus all physical and intellectual senses have been replaced by the simple alienation of all these senses, the sense of having. Man's essence had to be reduced to this absolute poverty, so it might bring forth out of itself its own inner riches. (On the category of having see Hess in *Twenty-One Sheets.*)

The supersession of private property is therefore the complete emancipation of all human senses and qualities, but it is this emancipation precisely in that these senses and qualities have become human, both subjectively and objectively. The eye has become a human eye when its object has become a social, human object produced by man and destined for him. Thus in practice the senses have become direct theoreticians. They relate to the thing for its own sake but the thing itself is an objective human relationship to itself and to man and vice versa. (I can in practice only relate myself humanly to an object if the object relates itself humanly to man.) Need and enjoyment have thus lost their egoistic nature and nature has lost its mere utility in that its utility has become human utility.

In the same way I can appropriate the senses and enjoyment of other men. Apart, then, from these immediate organs, social organs are constituted in the form of society: thus, for example, direct social activity with others is an organ of the manifestation of life and a manner in which to appropriate human life.

It is evident that the human eye enjoys things differently from the crude, inhuman eye, the human ear differently from the crude ear, etc.

We have seen that man does not lose himself in his object provided that it is a human object or objective humanity. This is only possible if it becomes a social object for him and he himself becomes a social being, while society becomes a being for him in this object.

Therefore in so far as generally in society reality becomes the reality of man's faculties, human reality, and thus the reality of his own faculties, all objects become for him the objectification of himself. They are objects that confirm and realize his individuality, his own objects, i.e. he becomes an object himself. How they become his own depends on the nature of the object and the faculty that corresponds to it. For it is just the distinctness of this relationship that constitutes the specific real mode of affirmation. The eye perceives an object differently from the ear and the object of the eye is different from that of the ear. What makes each faculty distinct is just its particular essence and thus also the particular mode of its objectification, of its objectively real, living being. Thus man is affirmed in the objective world not only in thought but through all his senses.

Just as society that is being born finds all of the material for its cultural formation through the development of private property with its material and intellectual wealth and poverty, so society when formed produces man in the whole wealth of its being, man rich in profound and manifold sensitivity as its constant reality.

It can be seen how subjectivism and objectivism, spiritualism and materialism, activity and passivity lose their opposition and thus their existence as opposites only in a social situation; it can be seen how the solution of theoretical opposition is only possible in a practical way, only through the practical energy of man, and their solution is thus by no means an exercise in epistemology but a real problem of life that philosophy could not solve just because it conceived of it as a purely theoretical task.

It can be seen how the history of industry and its previous objective existence is an open book of man's faculties and his psychology available to view. It was previously not conceived of in its connection with man's essence but only under the exterior aspect of utility, because man, moving inside the sphere of alienation, could only apprehend religion as the generalized existence of man, or history in its abstract and universal form of politics, art, literature, etc., as the reality of human faculties and the human species-act. In everyday material industry (which one can just as well consider as a part of that general development, or the general development can be considered as a particular part of industry, because all human activity has hitherto been labour, i.e. industry, self-alienated activity) we have the objectified faculties of man before us in the form of sensuous, alien, utilitarian objects, in the form of alienation. A psychology for which this book, and therewith the most tangible and accessible part of history, remains closed cannot become a genuine science with a real content. What should one think of a science whose preconceptions disregarded this large field of man's labour and which is not conscious of its incompleteness even though so broad a wealth of man's labour means nothing to it apart, perhaps, from what can be expressed in a single word 'need', and 'common need'?

The natural sciences have developed an enormous activity and appropriated an ever-increasing amount of material. However, philosophy has remained as alien to them as they to philosophy. The momentary union was only an imaginary illusion. The wish was there, but the ability lacking. Historians themselves only afford natural science a passing glance, as making for enlightenment, utility, and isolated great discoveries. But natural science by means of industry has penetrated human life all the more effectively, changed its form and prepared for human emancipation, even though in the first place it lead to complete dehumanization. Industry is the real historical relationship of nature, and therefore of natural science, to man. If then it is conceived of as the open revelation of human faculties, then the human essence of nature or the natural essence of man will also be understood. Natural science will then lose its one-sidedly materialist, or rather idealistic, orientation and become the basis of human science as it has already, though in an alienated form, become the basis of actual human life. And to have one basis for life and another for science would be in itself a falsehood. Nature as it is formed in human history—the birth process of human society—is the real nature of man and thus nature as fashioned by industry is true anthropological nature, though in an alienated form.

Sense-experience (see Feuerbach) must be the basis of all science. Science is only real science when it starts from sense-experience in the dual form of sense perception and sensuous need, in other words when it starts from nature. The whole of history is a preparation for 'man' to become the object of sense perception and for needs to be the needs of 'man as man'. History itself is the real part of natural history, of nature's becoming man. Natural science will later comprise the science of man just as much as the science of man will embrace natural science; they will be one single science.

Man is the direct object of natural science; for directly sensuous nature for man is man's sense-experience (the expressions are identical) in the shape of other men presented to him in a sensuous way. For it is only through his fellow man that his sense-experience becomes human for him. But nature is the direct object of the science of man. Man's first object—man himself—is nature, sense-experience; and particular human sensuous faculties are only objectively realized in natural objects and can only attain to self-knowledge in the science of nature in general. The elements of thought itself, the element of the vital manifestation of thought, language, is sensuous in character. The social reality of nature and human natural science or the natural science of man are identical expressions.

It can be seen how the wealthy man and the plenitude of human need take the place of economic wealth and poverty. The wealthy man is the man who needs a complete manifestation of human life and a man in whom his own realization exists as an inner necessity, as a need. Not only the wealth of man but also his poverty contain equally, under socialism, a human and therefore social meaning. Poverty is the passive bond that lets man feel his greatest wealth, his fellow man, as a need. The domination of the objective essence within me, the sensuous eruption of my essential activity is the passion that here becomes the activity of my essence.

5. A being only counts itself as independent when it stands on its own feet and it stands on its own feet as long as it owes its existence to itself. A man who lives by grace of another considers himself a dependent being. But I live completely by the grace of another when I owe him not only the maintenance of my life but when he has also created my life, when he is the source of my life. And my life necessarily has such a ground outside itself if it is not my own creation. The idea of creation is thus one that it is very difficult to drive out of the minds of people. They find it impossible to conceive of nature and man existing through themselves since it contradicts all the evidences of practical life.

The idea of the creation of the world received a severe blow from the science of geogeny, the science which describes the formation and coming

into being of the earth as a process of self-generation. Spontaneous generation is the only practical refutation of the theory of creation.

Now it is easy to say to the single individual what Aristotle already said:
you are engendered by your father and your mother and so in your case it is
the mating of two human beings, a human species-act, that has produced the
human being. You see, too, that physically also man owes his existence to
man. So you must not only bear in mind the aspect of the infinite regression
and ask further: who engendered my father and his grandfather, etc., you
must also grasp the circular movement observable in that progression whereby man renews himself by procreation and thus always remains the subject.
But you will answer: I grant you this circular movement but then grant me
the progression that pushes me ever further backwards until I ask, who
created the first man and the world as a whole? I can only answer you: your
question itself is a product of abstraction. Ask yourself how you come to ask
such a question; ask yourself whether your question is not put from a
standpoint that I cannot accept because it is an inverted one. Ask yourself
whether that progress exists as such for rational thought. When you inquire
about the creation of the world and man, then you abstract from man and
the world. You suppose them non-existent and yet require me to prove to
you that they exist. I say to you: give up your abstraction and you will give
up your question, or if you wish to stick to your abstraction then be consistent, and if you think of man and the world as non-existent then think of
yourself as non-existent, also, for you too are a part of the world and man.
Do not think, do not ask me questions, for immediately you think and ask,
your abstraction from the being of nature and man has no meaning. Or are
you such an egoist that you suppose everything to be nothing and yet wish to
exist yourself?

You can reply to me: I do not wish to suppose the nothingness of the
world and so on; I am only asking you about their origins, as I ask an
anatomist about the formation of bones, etc.

But since for socialist man what is called world history is nothing but the
creation of man by human labour and the development of nature for man, he
has the observable and irrefutable proof of his self-creation and the process
of his origin. Once the essential reality of man in nature, man as the existence of nature for man, and nature for man as the existence of man, has
become evident in practical life and sense experience, then the question of an
alien being, of a being above nature and man—a question that implies an
admission of the unreality of nature and man—has become impossible in
practice. Atheism, as a denial of this unreality, has no longer any meaning,
for atheism is a denial of God and tries to assert through this negation the
existence of man; but socialism as such no longer needs this mediation; it
starts from the theoretical and practical sense-perception of man and nature
as the true reality. It is the positive self-consciousness of man no longer
mediated through the negation of religion, just as real life is the positive

reality of man no longer mediated through communism as the negation of private property. Communism represents the positive in the form of the negation of the negation and thus a phase in human emancipation and rehabilitation, both real and necessary at this juncture of human development. Communism is the necessary form and dynamic principle of the immediate future, but communism is not as such the goal of human development, the form of human society.

Critique of Hegel's Dialectic and General Philosophy

6. Perhaps this is the place to make some remarks towards an understanding and justification of my attitude to Hegel's dialectic in general and in particular its elaboration in the *Phenomenology* and *Logic* and finally about its relationship to the modern critical movement.

Modern German criticism was so busy with the content of what it had inherited, and its progress, though imprisoned within its material, was so forceful that there developed a completely uncritical attitude to the method of criticism and a total unawareness of the apparently merely formal but in fact essential question: where do we stand now concerning Hegel's dialectic? The unawareness of the relationship of modern criticism to Hegel's philosophy in general and his dialectic in particular was so great that critics like Strauss [and Bruno Bauer] are completely imprisoned within Hegel's logic, the former completely and the latter at least implicitly in his *Synoptics* (where, in opposition to Strauss, he replaces the substance of 'abstract nature' with the 'self-consciousness' of abstract man) and even in his *Christianity Revealed*. [Bruno Bauer, *Das entdeckte Christentum*, Zurich and Winterthur, 1843.] Thus we read for example in *Christianity Revealed*: 'as though self-consciousness, in positing the world, that which is different, and in producing itself in what it produces, since it then suppresses the difference between its product and itself and is only itself in the productive movement, did not have its purpose in this movement' [Bruno Bauer, op. cit., p. 113] etc. Or: 'they (the French materialists) have not yet appreciated that the movement of the universe only becomes really explicit and achieves unity with itself in the movement of self consciousness'. [Bruno Bauer, op. cit., pp. 114 f.] Not only do these expressions not differ in their vocabulary from the Hegelian conception, they even repeat it literally.

How little during the process of criticism (Bauer in his *Synoptics*) an awareness was shown of its relationship to Hegel's dialectic and how little this awareness grew even after the process of material criticism is shown by Bauer when in his *Good Cause of Freedom* [Bruno Bauer, *Die gute Sache der Freiheit und meine eigene Angelegenheit*, Zurich and Winterthur, 1842, pp. 193 ff.] he brushes aside the indiscreet question of Herr Gruppe 'what will now happen to logic?' by referring him to future critics.

But now Feuerbach, both in his *Theses* in the *Anekdota* and in more detail in his *Philosophy of the Future*, has radically reversed the old dialectic and philosophy and the criticism that was unable to accomplish this itself, has seen it done by someone else, and proclaimed itself pure, decisive, and absolute criticism that has a clear vision of itself. [Marx refers to Bruno Bauer's *Allgemeine Literatur Zeitung*, Charlottenburg, 1844.] This criticism has now in its spiritual pride reduced the whole movement of history to the relationship between itself and the rest of the world that by contrast falls into the category of 'the mass', and dissolved all dogmatic oppositions into the single dogmatic opposition between its own cleverness and the stupidity of the world, between the critical Christ and the 'rabble' of humanity. It has daily and hourly demonstrated its own excellence in comparison with the stupidity of the mass and has finally announced the critical last judgement in that the day is approaching when the whole of fallen humanity will be assembled before it, be divided into groups, and each particular rabble receive its certificate of poverty. The critical school has published its superiority to human sentiments and to the world over which, enthroned in superior solitude, it lets echo from time to time from its sarcastic lips the laughter of the Olympian gods. After all these entertaining antics of idealism (Young Hegelianism), whose death agony takes the form of criticism, it has not once breathed a word of the necessity of a critical debate with its own source, Hegel's dialectic; indeed it has not even been capable of giving any criticism of Feuerbach's dialectic. It is thoroughly devoid of self-criticism.

Feuerbach is the only person to have a serious and critical relationship to the Hegelian dialectic and to have made real discoveries in this field; in short, he has overcome the old philosophy. The greatness of his achievement and the unpretentious simplicity with which Feuerbach presents it to the world are in a strikingly opposite inverse ratio.

Feuerbach's great achievement is:

1. To have proved that philosophy is nothing but religion conceptualized and rationally developed; and thus that it is equally to be condemned as another form and mode of existence of human alienation.

2. To have founded true materialism and real science by making the social relationship of 'man to man' the basic principle of his theory.

3. To have opposed to the negation of the negation that claims to be the absolute positive, the positive that has its own self for foundation and basis.

This is how Feuerbach explains Hegel's dialectic and thus justifies his taking the positive knowledge afforded by the senses as his starting-point:

Hegel starts from the alienation of substance (in logical terms: infinity, abstract universality), from the absolute and unmoved abstraction, i.e. in popular language, he starts from religion and theology.

Secondly, he supersedes the infinite and posits the actual, the perceptible, the real, the finite, and the particular. (Philosophy as supersession of religion and theology.)

Thirdly, he supersedes the positive in its turn and reinstates the abstraction, the infinite. Reinstatement of religion and theology.

Thus Feuerbach conceives of the negation of the negation only as an internal contradiction of philosophy, as philosophy that affirms theology (transcendence, etc.) after it has just denied it and thus affirms it in opposition to itself.

The positing of self-affirmation and self-confirmation present in the negation of the negation is not considered to be an independent affirmation since it is not yet sure of itself, still burdened with its opposite, doubtful of itself and thus needing proof, and not demonstrated by the fact of its own existence. It is therefore contrasted directly with affirmation that is verified by the senses and based on itself.

But Hegel has nevertheless discovered an expression of the historical movement that is merely abstract, logical, and speculative in that he conceived of the positive aspect of the negation of the negation as the sole, unique positive and the negative aspect in it as the only true self-affirming act of all being. This history is not yet the real history of man as a presupposed subject but only the history of the act of creation and the origin of man. We shall explain both the abstract form of this movement in Hegel and also what differentiates it from modern criticism and the same process in Feuerbachs' *Essence of Christianity*; or rather, we explain the critical form of the process that is still uncritical in Hegel.

Let us look at Hegel's system. We must begin with Hegel's *Phenomenology*, the true birthplace and secret of Hegel's philosophy.

PHENOMENOLOGY

(A) SELF-CONSCIOUSNESS

I. *Consciousness*

    (*a*) Sense certainty or the 'this' and meaning.

    (*b*) Perception or the thing with its properties and illusion.

    (*c*) Power and understanding, phenomena and the super-sensible world.

II. *Self-consciousness*. The truth of the certainty of oneself.

    (*a*) Dependence and independence of self-consciousness. Mastery and servitude.

    (*b*) Freedom of self-consciousness. Stoicism and scepticism. The unhappy consciousness.

III. *Reason*. Certainty and truth of reason.

    (*a*) Observational reason: observation of nature and self-consciousness.

    (*b*) Realization of rational self-consciousness through itself. Pleasure and necessity. The law of the heart and the madness of self-conceit. Virtue and the way of the world.

(*c*) Individuality which is real in and for itself. Legislative reason. Reason as testing laws.

(B) SPIRIT

  I. True spirit: customary morality.

 II. Self-alienated spirit, culture.

III. Spirit sure of itself, morality.

(C) RELIGION

  Natural religion, the religion of art, revealed religion.

(D) ABSOLUTE KNOWLEDGE

Hegel's *Encyclopaedia* begins with logic, with pure speculative thought, and ends with absolute knowledge, with the philosophical or absolute, i.e. super-human, abstract mind that is self-conscious and self-conceiving. Similarly the whole of the *Encyclopaedia* is nothing but the extended being of philosophical mind, its self-objectification; and philosophical mind is nothing but the alienated mind of the world conceiving of itself and thinking inside its self-alienation, i.e. abstractly. Logic is the money of the mind, the speculative thought-value of man and nature, their essence which has become indifferent to any real determination and thus unreal. It is externalized thought that abstracts from nature and real man, abstract thought. The exterior character of this abstract thought . . . nature as it exists for this abstract thought. Nature is exterior to it and represents its loss of itself. And mind understands nature as exterior, as an abstract thought, but as alien abstract thought. Finally spirit, which is thought returning to its birthplace and which, as anthropological, phenomenological, psychological, moral, and artistic–religious spirit, only considers itself valid when it finally discovers and affirms itself as absolute knowledge in absolute, i.e. abstract, spirit, receives its conscious and appropriate existence. For its real mode of existence is abstraction.

Hegel has committed a double error.

The first is most evident in the *Phenomenology*, the birthplace of the Hegelian philosophy. When he considers, for example, wealth and the power of the state as beings alienated from man's being, this happens only in their conceptual form. . . . They are conceptual beings and thus simply an alienation of pure, i.e. abstract, philosophical thought. The whole process therefore ends with absolute knowledge. What these objects are alienated from and what they affront with their pretention to reality, is just abstract thought. The philosopher, who is himself an abstract form of alienated man, sets himself up as the measure of the alienated world. The whole history of externalization and the whole recovery of this externalization is therefore nothing but the history of the production of abstract, i.e. absolute, thought—logical, speculative thought. Alienation, which thus forms the real interest of this externalization and its supersession is the opposition inside thought itself of the implicit and the explicit, of consciousness and self-consciousness, of object and subject, that is, it is the opposition inside thought itself of abstract thought and sensuous reality or real sensuous experience.

All other oppositions and their movements are only the appearance, the cloak, the exoteric form of these two opposites that alone are interesting and which give meaning to other, profane contradictions. What is supposed to be the essence of alienation that needs to be transcended is not that man's being objectifies itself in an inhuman manner in opposition to itself, but that it objectifies itself in distinction from, and in opposition to, abstract thought.

The appropriation of man's objectified and alienated faculties is thus firstly only an appropriation that occurs in the mind, in pure thought, i.e. in abstraction. It is the appropriation of these objects as thoughts and thought processes. Therefore in the *Phenomenology* in spite of its thoroughly negative and critical appearance and in spite of the genuine criticism, often well in advance of later developments, that is contained within it, one can already see concealed as a germ, as a secret potentiality, the uncritical positivism and equally uncritical idealism of Hegel's later works, this philosophical dissolution and restoration of existing empirical reality. Secondly, the vindication of the objective world for man (for example, the knowledge that sense perception is not abstract sense perception but human sense perception; that religion, wealth, etc., are only the alienated reality of human objectification, of human faculties put out to work, and therefore only the way to true human reality), this appropriation or the insight into this process, appears in Hegel in such a way that sense perception, religion, state power, etc., are spiritual beings; for spirit alone is the true essence of man and the true form of spirit is thinking spirit, logical, speculative spirit. The human character of nature and of historically produced nature, the product of man, appears as such in that they are products of abstract mind, and thus phases of mind, conceptual beings. The *Phenomenology* is thus concealed criticism that is still obscure to itself and mystifying; but in so far as it grasps the alienation of man, even though man appears only in the form of mind, it contains all the elements of criticism concealed, often already prepared and elaborated in a way that far surpasses Hegel's own point of view. The 'unhappy consciousness', the 'honest consciousness', the struggle of the 'noble and base consciousness' etc. etc., these single sections contain the elements, though still in an alienated form, of a criticism of whole spheres like religion, the state, civil life, etc. Just as the essence, the object, appears as a conceptual being, so the subject is always consciousness or self-consciousness, or rather the object only appears as abstract consciousness, man only as self-consciousness. Thus the different forms of alienation that occur are only different forms of consciousness and self-consciousness. Since the abstract consciousness that the object is regarded as being, is only in itself a phase in the differentiation of self-consciousness, the result of the process is the identity of consciousness and self-consciousness, absolute knowledge, the process of abstract thought that is no longer outward looking but only takes place inside itself. In other words, the result is the dialectic of pure thought.

Therefore the greatness of Hegel's *Phenomenology* and its final product, the dialectic of negativity as the moving and creating principle, is on the one hand that Hegel conceives of the self-creation of man as a process, objectification as loss of the object, as externalization and the transcendence of this externalization. This means, therefore, that he grasps the nature of labour and understands objective man, true, because real, man as the result of his own labour. The real, active relationship of man to himself as a species-being or the manifestation of himself as a real species-being, i.e. as a human being, is only possible if he uses all his species powers to create (which is again only possible through the co-operation of man and as a result of history), if he relates himself to them as objects, which can only be done at first in the form of alienation.

We shall now describe in detail the one-sidedness and limitations of Hegel using as a text the final chapter of the *Phenomenology* on absolute knowledge, the chapter which contains both the quintessence of the *Phenomenology*, its relationship to speculative dialectic, and also Hegel's attitude to both and to their interrelations.

For the moment we will only say this in anticipation: Hegel adopts the point of view of modern economics. He conceives of labour as the self-confirming essence of man. He sees only the positive side of labour, not its negative side. Labour is the means by which man becomes himself inside externalization or as externalized man. The only labour that Hegel knows and recognizes is abstract, mental labour. Thus Hegel conceives of what forms the general essence of philosophy, the externalization of man who knows himself or externalized science that thinks itself, as the essence of labour and can therefore, in contrast to previous philosophy, synthesize its individual phases and present his philosophy as the philosophy. What other philosophers have done—to conceive of single phases of nature and man's life as phases of self-consciousness, indeed of abstract self-consciousness—this Hegel knows by doing philosophy. Therefore his science is absolute.

Let us now proceed to our subject.

Absolute knowledge. Last chapter of the *Phenomenology*.

The main point is that the object of consciousness is nothing but self-consciousness or that the object is only objectified self-consciousness, self-consciousness as object. (Positing that man = consciousness.)

It is necessary therefore to overcome the objects of consciousness. Objectivity as such is considered to be an alien condition not fitting man's nature and self-consciousness. Thus the reappropriation of the objective essence of man, which was produced as something alien and determined by alienation, not only implies the transcendence of alienation, but also of objectivity. This means that man is regarded as a non-objective, spiritual being.

Hegel describes the process of the overcoming of the object of consciousness as follows:

The object does not only show itself as returning into the Self: that is according to Hegel the one-sided conception of this process. Man is equated with Self. But the Self is only man abstractly conceived and produced by abstraction. It is the Self that constitutes man. His eye, his ear, etc., take their nature from his Self; each of his faculties belongs to his Self. But in that case it is quite false to say: self-consciousness has eyes, ears, and faculties. Self-consciousness is rather a quality of human nature, of the human eye, etc., human nature is not a quality of self-consciousness.

The Self, abstracted and fixed for itself, is man as abstract egoist, egoism raised to its pure abstraction in thought (we will return to this point later).

For Hegel, the human essence, man, is the same as self-consciousness. All alienation of man's essence is therefore nothing but the alienation of self-consciousness. The alienation of self-consciousness is not regarded as the expression of the real alienation of man's essence reflected in knowledge and thought. The real alienation (or the one that appears to be real) in its inner concealed essence that has first been brought to the light by philosophy, is nothing but the appearance of the alienation of the real human essence, self-consciousness. The science that comprehends this is therefore called 'phenomenology'. Thus all reappropriation of the alienated objective essence appears as an incorporation into self-consciousness. Man making himself master of his own essence is only self-consciousness making itself master of objective essence. The return of the object into the Self is therefore the reappropriation of the object.

Universally expressed, the overcoming of the object of consciousness implies:

1. That the object presents itself to consciousness as about to disappear.

2. That it is the externalization of self-consciousness that creates 'thingness'.

3. That this externalization has not only a negative but also a positive significance.

4. That this significance is not only implicit and for us but also for self-consciousness itself.

5. For self-consciousness, the negative aspect of the object or its self-supersession has a positive significance, or, in other words, it knows the nullity of the object because it externalizes itself, for in this externalization it posits itself as object or establishes the object as itself, in virtue of the indivisible unity of being for itself.

6. At the same time, this other phase is also present that self-

consciousness has just as much superseded and re-absorbed this alienation and objectivity and thus is at home in its other being as such.

7. This is the movement of consciousness and consciousness is therefore the totality of its phases.

8. Similarly, consciousness must have related itself to the object in all its determinations, and have conceived it in terms of each of these determinations. This totality of determinations makes the object intrinsically a spiritual being, and it becomes truly so for consciousness by the perception of every one of these determinations as the Self, or by what was earlier called the spiritual attitude towards them.

Concerning 1. That the object as such presents itself to the consciousness as about to disappear is the above-mentioned return of the object into the Self.

Concerning 2. The externalization of self-consciousness posits 'thingness'. Because man is equated with self-consciousness, his externalized objective essence or 'thingness' is equated with externalized self-consciousness and 'thingness' is posited through this externalization. ('Thingness' is what is an object for man, and the only true object for him is the object of his essence or his objectified essence. Now since it is not real man as such—and therefore not nature, for man is only human nature—that is made the subject, but only self-consciousness, the abstraction of man, 'thingness' can only be externalized self-consciousness.) It is quite understandable that a natural, living being equipped and provided with objective, i.e. material faculties should have real, natural objects for the object of its essence and that its self-alienation should consist in the positing of the real, objective world, but as something exterior to it, not belonging to its essence and overpowering it. There is nothing incomprehensible and paradoxical in that. Rather the opposite would be paradoxical. It is equally clear that a self-consciousness, i.e. its externalization can only posit 'thingness', i.e. only an abstract thing, a thing of abstraction and no real thing. It is further clear that 'thingness' is not something self-sufficient and essential in contrast to self-consciousness, but a mere creation established by it. And what is established is not self-confirming, but only confirms the act of establishment which has for a moment, but only a moment, crystallized its energy into a product and in appearance given it the role of an independent and real being.

When real man of flesh and blood, standing on the solid, round earth and breathing in and out all the powers of nature, posits his real objective faculties, as a result of his externalization, as alien objects, it is not the positing that is the subject; it is the subjectivity of objective faculties whose action must therefore be an objective one. An objective being has an objective effect and it would not have an objective effect if its being did not include an objective element. It only creates and posits objects because it is posited by objects, because it is by origin natural. Thus in the act of positing it does not degenerate from its 'pure activity' into creating an object; its objective product only confirms its objective activity, its activity as an activity of an objective, natural being.

We see here how consistent naturalism or humanism is distinguished from both idealism and materialism and constitutes at the same time their unifying truth. We see also how only naturalism is capable of understanding the process of world history.

Man is a directly natural being. As a living natural being he is on the one hand equipped with natural vital powers and is an active, natural being. These powers of his are dispositions, capacities, instincts. On the other hand, man as a natural, corporeal, sensuous, objective being is a passive, dependent, and limited being, like animals and plants, that is, the objects of his instincts are exterior to him and independent of him and yet they are objects of his need, essential objects that are indispensable for the exercise and confirmation of his faculties. The fact that man is an embodied, living, real, sentient objective being means that he has real, sensuous objects as the objects of his life-expression. In other words, he can only express his being in real, sensuous objects. To be objective, natural and sentient and to have one's object, nature and sense outside oneself or oneself to be object, nature and sense for a third person are identical. Hunger is a natural need; so it needs a natural object outside itself to satisfy and appease it. Hunger is the objective need of a body for an exterior object in order to be complete and express its being. The sun is the object of the plant, an indispensable object that confirms its life, just as the plant is the object of the sun in that it is the expression of the sun's life-giving power and objective faculties.

A being that does not have its nature outside itself is not a natural being and has no part in the natural world. A being that has no object outside itself is not an objective being. A being that is not itself an object for a third being has no being for its object, i.e. has no objective relationships and no objective existence.

A non-objective being is a non-being.

Imagine a being which is neither itself an object nor has an object. Firstly, such a being would be the only being, there would be no being outside it, it would exist solitary and alone. For as soon as there are objects outside myself, as soon as I am not alone, I am something distinct, a different reality from the object outside me. Thus for this third object, I am a reality different from it, i.e. its object. Thus an object that is not the object of another being supposes that no objective being exists. As soon as I have an object, this object then has me as an object. But a non-objective being is an unreal, non-sensuous being that is only thought of, i.e. an imaginary being, a being of abstraction. To be sensuous, i.e. to be real, is to be an object of sense, a sensuous object, thus to have sensuous objects outside oneself, to have objects of sense perception. To be sentient is to suffer.

Man as an objective, sentient being is therefore a suffering being, and, since he is a being who feels his sufferings, a passionate being. Passion is man's faculties energetically striving after their object.

But man is not only a natural being, he is a human natural being. This means that he is a being that exists for himself, thus a species-being that must confirm and exercise himself as such in his being and knowledge. Thus human objects are not natural objects as they immediately present themselves nor is human sense, in its purely objective existence, human sensitivity and human objectivity. Neither nature in its objective aspect nor in its subjective aspect is immediately adequate to the human being. And as everything natural must have an origin, so man too has his process of origin, history, which can, however, be known by him and thus is a conscious process of origin that transcends itself. History is the true natural history of man. (We shall return to this point later.)

Thirdly, since the positing of 'thingness' is itself only an appearance, an act that contradicts the essence of pure activity, it must again be transcended and 'thingness' be denied.

Concerning 3, 4, 5, 6. (3). This externalization of consciousness has not only negative but also positive significance and (4) this significance is not only implicit and for us, but also for self-consciousness itself. (5) For self-consciousness the negative aspect of the object or its self-transcendence has a positive significance or in other words it knows the nullity of the object because it externalizes itself, for in this externalization it knows itself as object, or in virtue of the indivisible unity of being for itself, establishes the object for itself. (6) At the same time, this other phase is also present in the process, namely that self-consciousness has just as much superseded and reabsorbed this alienation and objectivity and thus is at home in its other being as such.

We have already seen that the appropriation of the alienated objective essence or the supersession of objectivity regarded as alienation, which must progress from indifferent strangeness to a really inimical alienation, means for Hegel at the same time, or even principally, the supersession of objectivity, since what offends self-consciousness in alienation is not the determinate character of the object but its objective character. The object is thus a negative, self-annulling being, a nullity. This nullity has for consciousness not only a negative but also a positive meaning, for this nullity of the object is precisely the self-confirmation of its non-objectivity and abstraction. For consciousness itself the nullity of the object has a positive significance because it knows this nullity, objective being as its own self-externalization; because it knows that this nullity only exists through its self-externalization

. . .

The way that consciousness is and that something is for it, is knowledge. Knowledge is its only act. Thus something exists for it in so far as it knows this something. Knowing is its only objective relationship. It knows the nullity of the object, i.e. that the object is not distinct from itself, the non-being of the object for itself, because it recognizes the object as its own self-externalization. In other words, it knows itself, knows knowledge as object, because the object is only the appearance of an object, a mirage, that essentially is nothing but knowledge itself that opposes itself to itself and is thus faced with a nullity, something that has no objectivity outside knowledge. Knowing knows that in so far as it relates itself to an object it is only exterior to itself, alienates itself. It knows that it only appears to itself as an object or that what appears to it as an object is only itself.

On the other hand, says Hegel, there is implied this other aspect: that consciousness has equally superseded this externalization and objectivity and taken it back into itself and thus is at home in its other being as such.

In this discussion we have assembled all the illusions of speculation.

Firstly, self-consciousness at home in its other being as such. It is, therefore, if we here abstract from the Hegelian abstraction and substitute man's self-consciousness for self-consciousness, at home in its other being as such. This implies, for one thing, that consciousness, knowing as knowing, thinking as thinking, pretends to be directly the opposite of itself, sensuous reality, life; it is thought overreaching itself in thought (Feuerbach). This aspect is entailed in so far as consciousness as mere consciousness is not offended by alienated objectivity but by objectivity as such.

The second implication is that in so far as self-conscious man has recognized the spiritual world (or the general spiritual mode of existence of his world) as self-externalization and superseded it, he nevertheless confirms it again in this externalized form and declares it to be his true being, restores it, pretends to be at home in his other being as such. Thus, for example, after the supersession of religion and the recognition of it as the product of self-alienation, man nevertheless finds himself confirmed in religion as such. Here is the root of Hegel's false positivism or his merely apparent criticism. This is what Feuerbach has characterized as the positing, negation, and restoration of religion or theology, although it should be understood to have a wider application. Thus reason finds itself at home in unreason as such. Man who has recognized that he has been leading an externalized life in law, politics, etc. leads his true human life in this externalized life as such. Thus the true knowledge and the true life is the self-affirmation and self-contradiction in contradiction with itself and with the knowledge and the nature of the object.

So there can be no more question of a compromise on Hegel's part with religion, the state, etc., for this falsehood is the falsehood of his very principle.

If I know religion as externalized human self-consciousness, then what I know in it as religion is not my self-consciousness, but the confirmation in it of my externalized self-consciousness. Thus I know that the self-consciousness that is part of my own self is not confirmed in religion, but rather in the abolition and supersession of religion.

Therefore, in Hegel the negation of the negation is not the confirmation of true being through the negation of apparent being. It is the confirmation of apparent being or self-alienated being in its denial or the denial of this apparent being as a being dwelling outside man and independent of him, its transformation into a subject.

Therefore supersession plays a very particular role in which negation and conservation are united.

Thus for example, in Hegel's *Philosophy of Right*, private right superseded equals morality, morality superseded equals the family, the family superseded equals civil society, civil society superseded equals the state, and state superseded equals world history. In reality private right, morality, family, civil society, state, etc. remain, only they become 'phases', modes of men's existence, which have no validity in isolation but which dissolve and create themselves. They are mere phases in the process. In their real existence, this moving being of theirs is concealed. It only comes to the fore and is revealed in thought, in philosophy, and therefore my true religious existence is my existence in the philosophy of religion, my true political existence is my existence in the philosophy of law, my true natural existence is in the philosophy of nature, my true artistic existence is in the philosophy of art, my true human existence is my existence in philosophy. Similarly, the true existence of religon, the state, nature, and art is the philosophy of religion, the state, nature, and art. But if the philosophy of religion, etc. is the only true existence of religion, then it is only as a philosopher of religion that I am really religious and so I deny real religiousness and really religious men. But at the same time I confirm them, too, partly inside my own existence or inside the alienated existence that I oppose to them (for this is merely their philosophical expression), partly in their peculiar and original form, for I count them as only apparently other being, as allegories, forms of their own true existence (i.e my philosophical existence), concealed by sensuous veils.

Similarly, quality superseded equals quantity, quantity superseded equals measure, measure superseded equals essence, essence superseded equals appearance, appearance superseded equals reality, reality superseded equals concept, concept superseded equals objectivity, objectivity superseded equals the absolute idea, absolute idea superseded equals nature, nature superseded equals subjective spirit, subjective spirit superseded equals ethical objective spirit, ethical objective spirit superseded equals art, art superseded equals religion, religion superseded equals absolute knowledge.

On the one hand, this supersession is a supersession of something thought, and thus private property as thought is superseded in the thought of morality. And because this thought imagines that it is directly the opposite of itself, sensuous reality, and thus also that its action is sensuous, real action, this supersession in thought that lets its object remain in reality believes it has really overcome it. On the other hand, since the object has now become for it a phase in its thought process, it is therefore regarded in its real existence as being a self-confirmation of thought, of self-consciousness and abstraction.

From one angle, therefore, the being that Hegel transcends in philosophy is not actual religion, state, nature, but religion as itself already an object of knowledge, dogmatics; and similarly with jurisprudence, political science, natural science. From this angle, therefore, he stands in opposition both to actual being and to direct, non-philosophical science or to the non-philosophical conception of this being. Thus he contradicts current conceptions.

From another angle the man who is religious, etc. can find his final confirmation in Hegel.

Let us consider, within the framework of alienation, positive phases of the Hegelian dialectic.

(a) Supersession as an objective movement absorbing externalization. This is the insight expressed within alienation of the reappropriation of objective being through the supersession of its alienation. It is the alienated insight into the real objectification of man, into the real appropriation of his objective essence through the destruction of the alienated character of the objective world, through its supersession in its alienated character of the objective world, through its supersession in its alienated existence. In the same way, atheism as the supersession of God is the emergence of theoretical humanism, and communism as the supersession of private property is the indication of real human life as man's property, which is also the emergence of practical humanism. In other words, atheism is humanism mediated with itself through the supersession of religion, and communism is humanism mediated with itself through the supersession of private property. Only through the supersession of this mediation, which is, however, a necessary pre-condition, does positive humanism that begins with itself come into being.

But atheism and communism are no flight, no abstraction, no loss of the objective world engendered by man or of his faculties that have created his objectivity, no poverty-stricken regression to unnatural and underdeveloped simplicity. They are rather the first real emergence and genuine realization of man's essence as something actual.

Thus in considering the positive side of self-referring negation (although still in an alienated form) Hegel conceives of the alienation of man's self and his being, the loss of his object and his reality, as self-discovery, manifestation of being, objectification, realization. In short, Hegel conceives, inside his abstraction, labour to be the self-engendering act of man, his relation to himself as an alien being and the manifestation of his own being as something alien, as the emergence of the species-consciousness and species-life.

In Hegel, apart from, or rather as a consequence of, the inversion we have already described, this act appears as merely formal because it is abstract, and the human essence itself is only regarded as an abstract, thinking being, as self-consciousness.

Secondly, because the conception is formal and abstract, the supersession of externalization becomes a confirmation of externalization. In other words, for Hegel the process of self-creation and self-objectification as self-externalization and self-alienation is the absolute and therefore final manifestation of human life which has itself for aim, is at peace with itself and has attained its true nature.

Therefore, this movement in its abstract form as dialectic is regarded as true human life, and because it is still an abstraction, an alienation of human life, it is viewed as a divine process, but the divine process of man, a process gone through by his absolute, pure, abstract being separated from himself.

Thirdly, this process must have an agent, a subject; but the subject only comes into being as the result; this result, the subject knowing itself as absolute self-consciousness, is therefore God, absolute spirit, the idea that knows and manifests itself. Real man and real nature become mere predicates or symbols of this hidden, unreal man and unreal nature. The relationship of subject and predicate to each other is thus completely inverted: a mystical subject–object or subjectivity reaching beyond the object, absolute subject as a process (it externalizes itself, returns to itself from its externalization and at the same time re-absorbs its externalization); a pure and unceasing circular movement within itself. . . .

On Money

What I have thanks to money, what I pay for, i.e. what money can buy, that is what I, the possessor of the money, am myself. My power is as great as the power of money. The properties of money are my—(its owner's)—properties and faculties. Thus what I am and what I am capable of is by no means determined by my individuality. I am ugly, but I can buy myself the most beautiful women. Consequently I am not ugly, for the effect of ugliness, its power of repulsion, is annulled by money. As an individual, I am lame, but money can create twenty-four feet for me; so I am not lame; I am a wicked, dishonest man without conscience or intellect, but money is honoured and so also is its possessor. Money is the highest good and so its possessor is good. Money relieves me of the trouble of being dishonest; so I am presumed to be honest. I may have no intellect, but money is the true mind of all things and so how should its possessor have no intellect? Moreover he can buy himself intellectuals and is not the man who has power over intellectuals not more intellectual than they? I who can get with money everything that the human heart longs for, do I not possess all human capacities? Does not my money thus change all my incapacities into their opposite?

If money is the bond that binds me to human life, that binds society to me and me to nature and men, is not money the bond of all bonds? Can it not tie and untie all bonds? Is it not, therefore, also the universal means of separation? It is the true agent both of separation and of union, the galvano-chemical power of society.

Shakespeare brings out two particular properties of money:

1. It is the visible god-head, the transformation of all human and natural qualities into their opposites, the general confusion and inversion of things; it makes impossibilities fraternize.

2. It is the universal whore, the universal pander between men and peoples.

The inversion and confusion of all human and natural qualities, the fraternization of impossibilities, this divine power of money lies in its being the externalized and self-externalizing species-being of man. It is the externalized capacities of humanity.

What I cannot do as a man, thus what my individual faculties cannot do, this I can do through money. Thus money turns each of these faculties into something that it is not, i.e. into its opposite.

If I long for a meal, or wish to take the mail coach because I am not strong enough to make the journey on foot, then money procures the meal and the mail coach for me. This means that it changes my wishes from being imaginary, and translates them from their being in thought, imagination, and will into a sensuous, real being, from imagination to life, from imaginary being to real being. The truly creative force in this mediation is money.

Demand also exists for the man who has no money but his demand is simply an imaginary entity that has no effective existence for me, for a third party or for other men and thus remains unreal and without an object. The difference between a demand that is based on money and effective and one that is based on my needs, passions, wishes, etc. and is ineffective is the difference between being and thought, between a representation that merely exists within me and one that is outside me as a real object.

If I have no money for travelling, then I have no need, no real and self-realizing need, to travel. If I have a vocation to study, but have no money for it, then I have no vocation to study, no effective, genuine vocation. If, on the contrary, I do not really have a vocation to study, but have the will and the money, then I have an effective vocation thereto. Money is the universal means and power, exterior to man, not issuing from man as man or from human society as society, to turn imagination into reality and reality into mere imagination. Similarly it turns real human and natural faculties into mere abstract representations and thus imperfections and painful imaginings, while on the other hand it turns the real imperfections and imaginings, the really powerless faculties that exist only in the imagination of the individual, into real faculties and powers. This description alone suffices to make money the universal inversion of individualities that turns them into their opposites and gives them qualities at variance with their own. As this perverting power, money then appears as the enemy of man and social bonds that pretend to self-subsistence. It changes fidelity into infidelity, love into hate, hate into love, virtue into vice, vice into virtue, slave into master, master into slave, stupidity into intelligence and intelligence into stupidity.

Since money is the existing and self-affirming concept of value and confounds and exchanges all things, it is the universal confusion and exchange of all things, the inverted world, the confusion and exchange of all natural and human qualities.

He who can buy courage is courageous though he be a coward. Because money can be exchanged not for a particular quality, for a particular thing or human faculty, but for the whole human and actual objective world, from the point of view of its possessor it can exchange any quality for any other, even contradictory qualities and objects; it is the fraternization of incompatibles and forces contraries to embrace.

If you suppose man to be man and his relationship to the world to be a human one, then you can only exchange love for love, trust for trust, etc. If you wish to appreciate art, then you must be a man with some artistic education; if you wish to exercise an influence on other men, you must be a man who has a really stimulating and encouraging effect on other men. Each of your relationships to man—and to nature—must be a definite expression of your real individual life that corresponds to the object of your will. If you love without arousing a reciprocal love, that is, if your love does not as such produce love in return, if through the manifestation of yourself as a loving person you do not succeed in making yourself a beloved person, then your love is impotent and a misfortune. . . .

BIBLIOGRAPHY

Original: *MEGA*, I (3), pp. 29 ff.

Present translation: K. Marx, *Early Texts*, pp. 131 ff.

Other translations: K. Marx, *Economic and Philosophical Manuscripts of 1844*, trans. M. Milligan, Moscow, 1959.

K. Marx, *Early Writings*, ed. T. Bottomore, London, 1963, pp. 66 ff.; *Writings of the Young Marx on Philosophy and Society*, ed. L. Easton and K. Guddatt, New York, 1967, pp. 284 ff.

K. Marx, *Early writings*, ed. L. Colletti, New York and London, 1975, pp. 279 ff.

Commentaries: H. Adams, *Karl Marx in his Earlier Writings*, 2nd edn., London, 1965, Ch. 8.

L. Althusser, *For Marx*, London, 1970.

D. Braybrooke, 'Diagnosis and Remedy in Marx's Doctrine of Alienation', *Social Research*, autumn 1968.

L. Colletti, Introduction to edition cited above.

L. Dupré, *The Philosophical Foundations of Marxism*, New York, 1966, pp. 120 ff.

L. Easton, 'Alienation and History in the Early Marx', *Philosophy and Phenomenological Research*, Dec. 1961.

E. Fromm, *Marx's Concept of Man*, New York, 1961.

D. Howard, *The Development of the Marxian Dialectic*, Carbondale, 1972, pp. 141 ff.

H. Koren, *Marx and the Authentic Man*, Duquesne, 1967.

K. Löwith, 'Self-alienation in the Early Writings of Marx', *Social Research*, 1954.

J. Maguire, *Marx's Paris Writings*, Dublin, 1972, pp. 41 ff.

D. McLellan, *Marx before Marxism*, Penguin, 1972, Ch. 7.

J. O'Neill, 'The Concept of Estrangement in the Early and Later Writings of Karl Marx', *Philosophy and Phenomenological Research*, Sept. 1964, reprinted in: *Sociology as a Skin Trade*, London, 1972.

F. Pappenheim, *The Alienation of Modern Man*, New York, 1959.

R. Schacht, *Alienation*, New York, 1970, Ch. 3.

R. Tucker, *Philosophy and Myth in Karl Marx*, Cambridge, 1961, Part 3.

9. Letter to Ludwig Feuerbach

This letter, published only recently, shows how enthusiastic Marx was about Feuerbach's ideas during the period when he was composing the *Economic and Philosophical Manuscripts*.

Dear Sir,

I take the opportunity presented to me of sending you an article of mine, in which are sketched some elements of my critical philosophy of law which I had already finished once but then subjected to a new rewriting so as to be generally intelligible. I lay no particular value on this article, but I am glad to find an opportunity of being able to assure you of the exceptional respect and—allow me the word—love that I have for you. Your *Philosophy of the Future* and *Essence of Faith* are, in spite of their limited scope, of more weight than the whole of contemporary German literature put together.

In these writings you have—whether intentionally I do not know—given a philosophical basis to socialism, and the communists, too, have similarly understood these works in that sense. The unity of man with man based on the real differences between men, the concept of human species transferred from the heaven of abstraction to the real earth, what is this other than the concept of society! . . .

BIBLIOGRAPHY

Original: MEW, Vol. 27, p. 425.
Present translation: K. Marx, *Early Texts*, p. 184.
Other translations: None.
Commentaries: on Feuerbach, see further:
  E. Kamenka, *The Philosophy of Ludwig Feuerbach*, London, 1970.

10. On James Mill

The following extract comes from the notebooks which Marx kept when he was writing the *Economic and Philosophical Manuscripts*. He copied into them extracts from the classical economists with critical comments. By far the most interesting is the following commentary on James Mill, in which Marx describes the dehumanizing effects of the credit system and exchange in capitalist society; in opposition to this Marx outlines the relationships of production that would exist in communist society. The extract below forms, in some ways, a positive counterpart to the negative account of alienation in the *Economic and Philosophical Manuscripts*.

... Credit is the economic judgement on the morality of a man. In credit, instead of metal or paper, man himself has become the mediator of exchange, only not as man, but as the existence of capital and interest. Thus the medium of exchange has certainly returned and been transferred to man, but only because man has been transferred outside him and himself taken on a material form. Money is not transcended in man inside the credit relationship but man himself has been changed into money or money become incarnate in him. Human individuality, human morality has itself become both an article of commerce and the material in which money exists. Instead of money, paper is my own personal being, my flesh and blood, my social value and status, the material body of the spirit of money. Credit no longer analyses money value into money but into human flesh and the human heart. This is because all progress and inconsequence inside a false system produces the worst regression and the worst and basest consequences. Inside the system of credit man's nature alienated from itself confirms itself, under the appearance of an extreme economic recognition of man, in a double way: (1) the antithesis between capitalist and worker and between large and small capitalists becomes even greater in that credit is given only to him who already has and is a new opportunity for the rich man to accumulate; or in that the poor man either confirms or denies his whole existence according to the arbitrary will and judgement that the rich man passes on him, and sees his whole existence depend upon this arbitrariness. (2) The reciprocal dissimulation, hypocrisy, and pretended sanctity is forced to a culmination so that the man who has no credit not only has the simple judgement passed on him that he is poor, but also the moral judgement that he possesses no trust or recognition, thus that he is a social pariah, a bad man. Also the poor man undergoes this humiliation in addition to his privation and has to make a humiliating request for credit to the rich man. (3) Because of this completely ideal existence of money, man cannot detect what is counterfeit in any material other than his own person and must himself become counterfeit, obtain credit by stealth and lies, and this credit relationship both on the side of the man who trusts and of the man who needs trust, becomes an object of commerce, an object of mutual deception and misuse. Here there is still plain in all its clarity the mistrust that is the basis of economic trust, the mistrustful consideration of whether to give credit or not, the spying into the secrets of the private life, etc. of the person seeking credit; the betrayal of temporary difficulties in order to ruin a rival through the sudden shaking of his credit. The whole system of bankruptcy, ghost companies, etc. . . . . in state credit, the state occupies exactly the same place as man does above . . . his play with papers of state shows how he has become a plaything of businessmen. (4) The credit system has its final completion in the banking system. The creation of bankers, the state domination of the banks, the concentration of capital in these hands, this economic Areopagus of the nation, is the worthy completion of the world of money. In that the moral recognition of a man, trust in the state, etc. in the credit system take the form of credit, the secret that is contained in the lie of moral recognition, the immoral baseness of this morality, and the hypocrisy and egoism in that trust of the state come to the fore and show themselves for what they really are.

Exchange, both of human activity within production itself and also of human products with each other, is equivalent to species-activity and species-enjoyment whose real, conscious, and true being is social activity and social enjoyment. Since human nature is the true communal nature of man, men create and produce their communal nature by their natural action; they produce their social being which is no abstract, universal power over against single individuals, but the nature of each individual, his own activity, his own life, his own enjoyment, his own wealth. Therefore this true communal nature does not originate in reflection, it takes shape through the need and egoism of individuals, i.e. it is produced directly by the effect of their being. It is not dependent on man whether this communal being exists or not; but so long as man has not recognized himself as man and has not organized the world in a human way, this communal nature appears in the form of alienation, because its subject, man, is a self-alienated being. Men, not in the abstract, but as real, living, particular individuals, are this nature. It is, therefore, as they are. Therefore, to say that man alienates himself is the same as to say that the society of this alienated man is a caricature of his real communal nature, his true species-life, that therefore his activity appears to him as a suffering, his own creation appears as an alien power, his wealth as poverty, the natural tie that binds him to other men appears as an unnatural tie and the separation from other men as his true being; his life appears as sacrifice of life, the realization of his essence as a loss of the reality of his life, his production as a production of his own nothingness, his power over the object as the power of the object over him, and he himself, the master of his creation, appears as its slave.

Economics conceives of the communal nature of man, or his self-affirming human nature, the mutual completion that leads to the species-life, to the truly human life, under the form of exchange and commerce. Society, says Destutt de Tracy, is a series of mutual exchanges. It is precisely this movement of mutual integration. Society, says Adam Smith, is a commercial society and each of its members is a tradesman.

We can see how economics rigidifies the alienated form of social intercourse, as the essential, original form that corresponds to man's nature.

Economics, as does the actual process, starts from the relationship of man to man, as that of private property owner to private property owner. If man as private property owner is presupposed, i.e man as an exclusive owner who keeps his personality and distinguishes himself from other men by means of this exclusive property (private property is his personal, peculiar, and thus essential being), then the loss or surrender of private property is an externalization of man and of private property itself. Here we shall only take up the last point. If I hand over my private property to another, then it ceases to be mine; it becomes a thing that is independent of me lying outside my control, exterior to me. Thus I externalize my private property. Thus in relation to myself I turn it into externalized private property. But I only turn it into an externalized thing, I only abolish my personal relationship to it, I give it back to the elementary powers of nature when I only externalize it with reference to myself. It only becomes externalized private property when it ceases to be my private property without thereby ceasing altogether to be private property, i.e. when it enters into the same relationship with another man apart from me that it had to me, in a word, when it becomes the private property of another man. Violence excepted, how would I come to externalize my private property in favour of another man? Economics correctly answers; out of necessity, out of need. The other man is also a private property owner, but he owns another thing that I want and cannot and will not do without, that appears to me as necessary to the completion of my being and the realization of my essence.

The tie that binds the two private property owners to each other is the specific nature of the object that is the stuff of their private property. The desire and need for both these objects shows each of the private property owners and makes them realize that he has another essential relationship to objects apart from that of private property, that he is not the particular being that he imagined but a total being whose needs in relation also to the products of another's work are an inner property. For the need of a thing is the most evident and irrefutable proof that the thing belongs to my essence, that its being is for me, its property is the property and peculiarity of my essence. Thus, owners of private property are driven to give up their private property but to do so in such a way that it confirms private property at the same time, or to give up the private property inside the relationship of private property. Thus each externalizes a part of his private property to the other. The social connection or relationship between the two owners of private property is therefore a reciprocal externalization, the relationship of externalization supposed on both sides or externalization as the relationship between both owners, whereas in private property by itself the externalization is only in relation to oneself, i.e. one-sided.

Thus exchange or trade is the social species-act, the communal nature, the social commerce and integration of man inside private property and thus the exterior, externalized species-act. This is the reason that it appears as trade. This is also the reason that it is the opposite of the social relationship.

Through this reciprocal externalization or alienation of private property, private property itself gets into the position of externalized private property. For firstly it has ceased to be the product of labour, the exclusive, peculiar personality of its owner. This latter has externalized it, it has left the owner whose product it was and has acquired a personal meaning for the man whose product it is not. It has lost its personal meaning for its owner. Secondly, it has been related to another piece of private property and made equivalent to it. It has been replaced by the private property of a different nature, as it itself replaces private property of a different nature. Thus, on both sides private property appears as representing private property of a different nature. It appears as the equivalent of another natural product, and both sides are interrelated in such a way that each represents the being of the other and both relate to each other as substitutes for themselves and the other. The being of private property has therefore as such become a substitute, and an equivalence. Instead of possessing a direct self-identity it is only a relation with something else. As an equivalence, its being is no longer its own. It has therefore become a value and most directly an exchange value. Its existence as value is different from its immediate existence; it is exterior to its specific being, an externalized aspect of itself; it is only a relative existence of the same.

We must keep for another time a more precise definition of the nature of this value and also of the process by which it turns into a price.

If the relationship of exchange is presupposed, labour immediately becomes labour of wages. The condition of alienated labour reaches its culmination in that: (1) on one side wage labour and the product of the worker stands in no direct relationship to his need or the nature of his work, but is determined on both sides by social combinations hostile to the worker; and (2) the person who buys the product is not a producer himself but exchanges for it what another has produced. In the crude form of externalized private property, barter, each of the private property owners has produced under the impulse of direct need, of his situation, and of the material available to him. Therefore, each exchanges with the other only the superfluity of his own production. Labour was, of course, the direct source of his existence, but at the same time it was also the affirmation of his individual existence. Through exchange his labour has become partly a source of gain. Its aim and its nature have become different. The product is produced for value, exchange-value, equivalency, and no longer because of its direct, personal connection with the producer. The more varied the production becomes, so the producer's needs are more varied while his activity becomes more one-sided and his labour can more and more be characterized as wage-labour, until finally it is purely this and it becomes quite accidental and inessential whether the producer has the immediate enjoyment of a product that he personally needs and also whether the very activity of his labour enables him to enjoy his personality, realize his natural capacities and spiritual aims.

In wage-labour is contained: (1) the alienation and disconnection between labour and the man who labours; (2) the alienation and disconnection between labour and its object; (3) that the worker is governed by social needs that are alien and do violence to him: he subjects himself to them out of egoistic need and necessity and they only have significance for him as a means of satisfying his want just as to them he appears as a slave of his needs; (4) that to the worker the purpose of his activity seems to be the maintenance of his individual life and what he actually does is regarded as a means; his life's activity is in order to gain the means to live.

Thus, the greater and more elaborate appears the power of society inside the private property relationship, the more egoistic, antisocial, and alienated from his own essence becomes man.

Thus, the mutual exchange of the products of human activity appears as commerce and barter: similarly mutual completion and exchange of activity itself appears as division of labour which makes of man an extremely abstract being, a machine etc., and leads to an abortion of his intellectual and physical faculties.

It is precisely the unity of human labour that is viewed only as its division because man's social being only comes into existence as its opposite and in its alienated form. Division of labour grows with civilization.

Within the presuppositions of the division of labour the product, the material of private property, is considered by the individual more and more as an equivalent. Moreover, just as he no longer exchanges what is superfluous and the object of his production can simply be immaterial to him, so he does not exchange his product for what is immediately necessary to him. The equivalent exists as a money equivalent which is the immediate result of wage labour and the medium of exchange (see above).

The complete domination of the alienated thing over man is fully manifested in money, the complete indifference both with regard to the nature of the material and the specific nature of the private property, and to the personality of the private property owner.

What was domination of person over person is now the general domination of the thing over the person, of the product over the producer. Just as the characteristic externalization of private property lay in the equivalent and value, so money is the existence of this externalization that is sensuous and objective to itself.

It is self-evident that political economy can only grasp this whole development as a fact, a result of the fortuitous force of circumstances.

The separation of work from itself = the separation of the worker from the capitalist = the separation of work from capital whose original form separates into landed property and moveable property . . . the original characteristic of private property is monopoly; and so as soon as it provides itself with a political constitution, it is that of monopoly. The completion of monopoly is competition. For the political economist production, consumption, and their mediator, exchange or distribution, are separate. The separation of production and consumption of activity and enjoyment in different individuals and in the same individual is the separation of work from its object and from itself as enjoyment. Distribution is the self-confirming power of private property. The separation of work, capital, and private property from each other and similarly the separation of work from work, of capital from capital and landed property from landed property, and finally the separation of labour from wages, of capital from profit and profit from rent, and lastly of landed property from ground rent permits self-alienation to appear both in its own form and in that of mutual alienation.

Man—and this is the basic presupposition of private property—only produces in order to have. The aim of production is possession. Not only does production have this utilitarian aim; it also has a selfish aim; man produces only his own exclusive possession. The object of his production is the objectification of his immediate, selfish need. Thus, in this savage and barbaric condition man's production is measured, is limited by the extent of his immediate need whose immediate content is the object produced.

Thus, in these circumstances man no longer produces according to his immediate needs. His need is limited by his production. Demand and supply are thus exactly coterminous. His production is measured by his need. In this case there is no exchange or it is reduced to the exchange of his labour against the product of his labour and this exchange is the hidden form or kernel of real exchange.

As soon as exchange exists, production goes beyond the immediate limits of possession. But over-production does not leave selfish need behind. It is rather an indirect way of satisfying a need that can only be objectified in the production of another and not in this production. Production has become the source of wages, wage-labour. Thus, while in the first situation need is the measure of production, in the second situation production or rather the ownership of the product is the measure of how far needs can be satisfied.

I have produced for myself and not for you, as you have produced for yourself and not for me. You are as little concerned by the result of my production in itself as I am directly concerned by the result of your production. That is, our production is not a production of men for men as such, that is, social production. Thus, as a man none of us is in a position to be able to enjoy the product of another. We are not present to our mutual products as men. Thus, neither can our exchange be the mediating movement which confirms that my product is for you, because it is an objectification of your own essence, your need. For what links our productions together is not the human essence. Exchange can only set in motion and activate the attitude that each of us has to his own product and thus to the product of another. Each of us sees in his own product only his own selfish needs objectified, and thus in the product of another he only sees the objectification of another selfish need independent and alien to him.

Of course as man you have a human relationship to my product; you have a need for my product. Therefore, it is present to you as an object of your desires and will. But your need, your desires, your will are powerless with regard to my product. This means, therefore, that your human essence, which as such necessarily has an intrinsic relationship to my production, does not acquire power and property over my production, for the peculiarity and power of the human essence is not recognized in my production. They are more a fetter that makes you depend on me because they manoeuvre you into a position of dependence on my product. Far from being the means of affording you power over my production, they are rather the means of giving me power over you.

When I produce more of an object than I myself directly require, my over-production is calculated and refined according to your need. It is only in appearance that I produce more of this object. In reality I produce another object, the object of your production that I count on exchanging for my surplus, an exchange that I have already completed in my thought. The social relationship in which I stand to you, my work for your need, is also a mere appearance and similarly our mutual completion is a mere appearance for which mutual plundering serves as a basis. An intention to plunder and deceive is necessarily in the background, for since our exchange is a selfish one both on your side and on mine, and since each selfishness tries to overcome the other person's, of necessity we try to deceive each other. Of course, the measure of the power that I gain for my object over yours needs your recognition in order to become a real power. But our mutual recognition of the mutual power of our objects is a battle in which he conquers who has more energy, strength, insight, and dexterity. If I have enough physical strength, I plunder you directly. If the kingdom of physical strength no longer holds sway, then we seek to deceive each other and the more dextrous beats the less. Who defeats whom is an accident as far as the totality of the relationship is concerned. The ideal intended victory is with both sides, i.e. each has, in his own judgement, defeated the other.

Thus, exchange is brought about necessarily on both sides by the object of each man's production and possession. The ideal relationship to the mutual objects of our reproduction is of course our mutual need. But the actual and true relationship, the one that has a real effect, is simply the mutual ex-clusive possession of mutual production. What gives your need of my things a value, a worth, and an effect for me is only your object, the equivalent of my object. Thus, our mutual product is therefore the means, the mediation, the instrument, the recognized power of our mutual need of each other. Your demand, therefore, and the equivalent of your possession are expres-sions that have the same meaning and value for me, and your demand is only effective and therefore becomes meaningful when it has effect and meaning in relation to me. Simply as a man without this instrument your demand is for you an unsatisfied desire, and for me a non-existent imagining. Thus, as a man you stand in no relationship to my object, because I myself have no human relationship to it. But the true power over an object is the means and thus we mutually regard our products as the power of each over another and over himself. This means that our own product has reared up against us. It seemed to be our property, but in reality we are its property. We ourselves are excluded from true property because our property excludes other men.

The only intelligible language that we speak to one another consists in our objects in their relationships to one another. We would not understand a human speech and it would remain ineffective; on the one hand it would be seen and felt as an entreaty or a prayer and thus as a humiliation and therefore used with shame and a feeling of abasement, while on the other side it would be judged brazen and insane and as such rejected. Our mutual alienation from the human essence is so great that the direct language of this essence seems to us to be an affront to human dignity, and in contrast the alienated language of the values of things seems to be the language that justifies a self-reliant and self-conscious human dignity.

Of course in your eyes your product is an instrument, a means to be able to control my product and thus to satisfy your needs. But in my eyes it is the aim of our exchange. For me you are only an instrumental means for the production of this object, that is an end for me while you yourself conversely have the same relationship to my object. But: (1) each of us really acts as the other sees him. You have really made yourself into the means, the instrument, the producer of your own object in order to gain power over mine; (2) your object is to you only the perceivable cloak, the hidden form of my object; for what its production means and expresses is: power to pur-chase my object. So actually you have for yourself become a means and instrument of your object of which your desire is a slave, and you have performed the service of a slave so that the object of your desires shall no more afford you its charity. If this mutual enslavement to an object at the beginning of the process appears now as in a relationship of lordship and slavery, that is only the crude and open expression of our true relationship.

Our mutual value is for us the value of our mutual products. Thus, man himself is for us mutually worthless.

Supposing that we had produced in a human manner; each of us would in his production have doubly affirmed himself and his fellow men. I would have: (1) objectified in my production my individuality and its peculiarity and thus both in my activity enjoyed an individual expression of my life and also in looking at the object have had the individual pleasure of realizing that my personality was objective, visible to the senses and thus a power raised beyond all doubt. (2) In your enjoyment or use of my product I would have had the direct enjoyment of realizing that I had both satisfied a human need by my work and also objectified the human essence and therefore fashioned for another human being the object that met his need. (3) I would have been for you the mediator between you and the species and thus been acknowledged and felt by you as a completion of your own essence and a necessary part of yourself and have thus realized that I am confirmed both in your thought and in your love. (4) In my expression of my life I would have fashioned your expression of your life, and thus in my own activity have realized my own essence, my human, my communal essence.

In that case our products would be like so many mirrors, out of which our essence shone.

Thus, in this relationship what occurred on my side would also occur on yours.

If we consider the different stages as they occur in our supposition:

My work would be a free expression of my life, and therefore a free enjoyment of my life. Presupposing private property, my work is an alienation of my life, because I work in order to live, to furnish myself with the means of living. My work is not my life.

Secondly: In work the peculiarity of my individuality would have been affirmed since it is my individual life. Work would thus be genuine, active property. Presupposing private property, my individuality is so far externalized that I hate my activity: it is a torment to me and only the appearance of an activity and thus also merely a forced activity that is laid upon me through an exterior, arbitrary need, not an inner and necessary one.

My labour can only appear in my object as what it is. It cannot appear as what it essentially is not. Therefore, it appears still as merely the expression of my loss of self and my powerlessness that is objective, observable, visible, and therefore beyond all doubt. . . .

BIBLIOGRAPHY

Original: *MEGA*, I (3), pp. 425 ff.

Present translation: K. Marx, *Early Texts*, pp. 192 ff.

Other translations: *Writings of the Young Marx on Philosophy and Society*, ed. L. Easton and K. Guddatt, New York, 1967, pp. 270 ff.

K. Marx, *Early Writings*, ed. L. Colletti, New York and London, 1975, pp. 259 ff.

Commentaries: D. McLellan, 'Marx's Concept of the Unalienated Society', *Review of Politics*, 1969.

11. Critical Remarks on the Article: 'The King of Prussia and Social Reform'

The *Deutsch–französische Jahrbücher* stopped publication after its first number, partly owing to financial difficulties, but also owing to differences of opinion between its two co-editors, Marx and Ruge. While Marx moved to communism, Ruge remained a liberal democrat. In the following article Marx made public his break with Ruge. In the summer of 1844 several thousand weavers in Silesia had smashed the machinery that threatened their livelihood and been suppressed with great brutality. In his article criticizing Frederick William IV's paternalist attitude, Ruge asserted that no social revolt could succeed in Germany since political consciousness was extremely underdeveloped there and social revolutions sprang from political revolutions. Continuing themes from the *Jewish Question*, Marx claimed that political consciousness was not sufficient to produce a revolution and that, on the contrary, any genuine revolution would be a social one in which the state would be abolished.

. . . From the political point of view the state and any organization of society are not two distinct things. The state is the organization of society. In so far as the state admits the existence of social abuses, it seeks their origin either in natural laws that no human power can control or in the private sector which is independent of it or in the inadequacy of the administration that depends on the state. Thus, England sees misery as founded on the natural law according to which population must always outstrip the means of subsistence. On the other hand, it explains pauperism by the cussedness of the poor, as the King of Prussia explains it by the unchristian spirit of the rich and the Convention by the counter-revolutionary and suspicious attitude of the property owners. Therefore, England punishes the poor, the King of Prussia exhorts the rich, and the Convention beheads the property owners.

In short, all states look for the causes in accidental or intended faults of administration, and therefore seek the remedy for its evils in administrative measures. Why? Simply because the administration is the organizing activity of the state.

The state cannot abolish the contradiction which exists between the role and good intentions of the administration on the one hand and the means at its disposal on the other without abolishing itself, for it rests on this contradiction. It rests on the contrast between public and private life, on the contrast between general and particular interests. The administration must therefore limit itself to a formal and negative activity, for its power ceases just where civil life and work begin. Indeed, in the face of the consequences that spring from the unsocial nature of this civil life, this private property, this commerce, this industry, this reciprocal plundering of different civil groups, in face of these consequences, impotence is the natural law of the administration. For this tearing apart, this baseness, this slavery of civil society is the natural basis on which the modern state rests, as the civil society of slavery was the natural basis on which the classical state rested. The existence of the state and the existence of slavery are inseparable. The classical state and classical slavery—frank and open class oppositions—were not more closely forged together than the modern state and modern world of haggling, hypocritical, Christian oppositions. If the modern state wished to do away with the impotence of its administration, it would have to do away with the contemporary private sphere for it only exists in contrast to the private sphere. But no living being believes that the defects of its specific existence are grounded in what is essential to its own life, but in circumstances exterior to its life. Suicide is unnatural. So the state cannot believe in the intrinsic impotence of its administration, i.e. of itself. It can appreciate only its formal, accidental defects, and try to remedy them. And if these modifications are fruitless then it thinks that social evils are a natural imperfection independent of man, a law of God, or else that the will of individuals is too perverted to be able to match the good intentions of the administration. And what perverse individuals! They complain about the government whenever it limits freedom and yet require the government to prevent the necessary consequences of this freedom!

The more powerful the state, and thus the more political a country is, the less is it inclined to look in the state itself, that is in the present organization of society whose active, self-conscious, and official expression is the state, for the cause of social evils, and thus understand their general nature. Political intelligence is political just because it thinks inside the limits of politics. The sharper and livelier it is the less capable it is of comprehending social evils. The classical period of political intelligence is the French Revolution. Far from seeing the source of social defects in the state, the heroes of the French Revolution see in social defects the source of political misfortunes. Thus Robespierre sees in extremes of poverty and riches only an impediment to pure democracy. So he wishes to establish a general Spartan frugality. The principle of politics is the will. The more one-sided, and thus the more perfect political intelligence is, the more it believes in the omnipotence of the will, the blinder it is to the natural and intellectual limits of the will, and thus the more incapable it is of discovering the sources of social evils. No further explanation is necessary to refute the stupid hope of the Prussian that it is the vocation of 'the political intelligence to discover the root of social misery in Germany'. . . .

But do not all revolts without exception break out in the wretched isolation of man from his common essence? Is not isolation a necessary presupposition of any revolt? Would the revolution of 1789 have taken place without the wretched isolation of the French bourgeois from their common essence? Their very aim was to do away with their isolation.

But the common essence from which the worker is isolated is a common essence of quite a different reality and compass from the political collectivity. This collectivity from which his own work separates him is life itself, physical and intellectual life, human morality, human activity, human enjoyment, human essence. The human essence is the true collectivity of man. And so isolation from this essence is out of all proportion more universal, insupportable, terrifying, and full of contradictions than isolation from the political collectivity, the abolition of this isolation or even a partial reaction and revolt against it is all the more immeasurable as man is more immeasurable than the citizen and human life than political life. An industrial revolt can therefore be as partial as it likes; it contains within it a universal soul: a political revolt can be as universal as it likes, even under the most colossal form it conceals a narrow spirit.

The Prussian ends his essay with the following fitting sentence:

A social revolution without a political soul (i.e. with an organizing intelligence operating from the standpoint of the whole) is impossible.

We have seen. A social revolution, even though it be limited to a single industrial district, involves from the standpoint of the whole, because it is a human protest against a dehumanized life, because it starts from the standpoint of the single, real individual, because the collectivity against whose separation from himself the individual reacts is the true collectivity of man, the human essence. The political soul of revolution consists on the contrary in a tendency of the classes without political influence to end their isolation from the top positions in the state. Their standpoint is that of the state, an abstract whole, that only exists through a separation from real life, that is inconceivable without the organized opposition, the general concept of humanity and its individual existence. Thus a revolution with a political soul also organizes, in conformity with its limited and double nature, a ruling group in society to society's detriment.

We wish to confide in the Prussian what a 'social revolution with a political soul' is; we entrust him at the same time with the secret that he is never able even in his stylish phrases to raise himself above the narrow political standpoint.

A 'social revolution with a political soul' is either a contradiction in terms, if the Prussian understands by social revolution a social revolution in opposition to a political one and nevertheless gives the social revolution a political soul instead of a social one. Or a 'social revolution with a political soul' is just a paraphrase for what used to be called a political revolution, is simply a revolution. Every revolution is social in as far as it destroys the old society. Every revolution is political in so far as it destroys the old power.

Let the Prussian choose between paraphrase and nonsense! A political revolution with a social soul is as rational as a social revolution with a political soul is paraphrastic or nonsensical. Revolution in general—the overthrow of the existing power and dissolution of previous relationships—is a political act. Socialism cannot be realized without a revolution. But when its organizing activity begins, when its peculiar aims, its soul, come forward, then socialism casts aside its political cloak.

This long development was necessary to tear apart the web of errors that lie hidden in a single newspaper column. Not all readers have the education and time necessary to give an account of such literary charlatanism. Thus, does not the anonymous Prussian have a duty *vis-à-vis* the reading public to forgo for the time being all writing on political and social matters and declamations on the situation in Germany, and rather begin with a conscientious attempt to clarify his own situation?

BIBLIOGRAPHY

Original: *MEGA*, I (3), pp. 5 ff.

Present translation: K. Marx, *Early Texts*, pp. 213 ff.

Other translations: *Writings of the Young Marx on Philosophy and Society*, ed. L. Easton and K. Guddatt, New York, 1967, pp. 338 ff.

Commentaries: K. Marx, *Early Writings*, ed. L. Colletti, New York and London, 1975, pp. 81 ff.
D. Howard, *The Development of the Marxian Dialectic*, Carbondale, 1972, pp. 135 ff.
D. McLellan, *Marx before Marxism*, Penguin, 1972, pp. 204 ff.
F. Mehring, *Karl Marx*, London, 1936, pp. 78 ff.

II. The Materialist Conception of History
1844–1847

12. *The Holy Family*

Shortly after finishing the *Economic and Philosophical Manuscripts*, Marx met Engels, who was passing through Paris on his way home from Manchester. The two became firm friends and established, as Engels put it later, their 'complete agreement in all theoretical fields'. On the basis of this agreement, they decided to publish a pamphlet that would finally discredit the ideas of the Young Hegelians and principally of Bruno Bauer. Engels wrote fifteen pages or so and then left Paris, while Marx typically expanded the 'pamphlet' into a fair-sized book which was published in February 1845 with the ironic title (referring to Bruno Bauer and his brothers) *The Holy Family*.

The book is extremely discursive, being a criticism of random articles in the *Allgemeine Literatur-Zeitung* edited by Bruno and Edgar Bauer. Much of it lacks permanent interest as it consists of hair-splitting arguments on unimportant topics. This is particularly so of the section dealing with Bauer's comments on Eugène Sue's enormous Gothic novel *The Mysteries of Paris*. In the more important parts of the book, extracted below, Marx defends Proudhon against Bauer's attacks. Marx praises Proudhon as the first thinker to have questioned the existence of private property and demonstrated its inhuman effects on society; Marx extrapolates from this to describe the self-alienation of man produced by the dialectical antagonism between wealth and the proletariat. After criticizing Hegel's speculative idealism, Marx attacks Bauer's view that 'the Spirit' is the progressive force in history rather than the movement of the masses—a view that had led him to an extremely conservative political position. A corollary of Bauer's general views was the assertion that the French Revolution had gone wrong as soon as the masses had become enthusiastic about its principles. In the final extract, Marx traces the intellectual pedigree of socialism back to the French materialist philosophers of the eighteenth century.

The Holy Family is a polemical rather than a constructive work and so contains no systematic presentation of Marx's ideas. Marx termed his standpoint 'real humanism' and was obviously still under the influence of Feuerbach. At the same time the book is the first of Marx's writings to show clearly the imprint of the materialist conception of history that was to be presented so strikingly in *The German Ideology*.

On Proudhon

As the first criticism of any science necessarily finds itself under the influence of the premisses of the science it is fighting against, so Proudhon's treatise *Qu'est-ce que la propriété?* is the criticism of political economy from the standpoint of political economy. We need go no deeper into the juridical part of the book, which criticizes law from the standpoint of law, for our main interest is the criticism of political economy. Proudhon's treatise will therefore be outstripped by a criticism of political economy, including Proudhon's conception of political economy. This work became possible only after Proudhon's own work, just as Proudhon's criticism supposed the physiocrats' criticism of the mercantile system, Adam Smith's criticism of the physiocrats, Ricardo's criticism of Adam Smith, and the works of Fourier and Saint-Simon.

All treatises on political economy take private property for granted. This basic premiss is for them an incontestable fact admitting of no further investigation, nay more, a fact which is spoken about only 'accidentally', as Say naïvely admits. But Proudhon makes a critical investigation—the first resolute, pitiless, and at the same time scientific investigation—of the foundation of political economy, private property. This is the great scientific progress he made, a progress which revolutionizes political economy and first makes a real science of political economy possible. Proudhon's treatise *Qu'est-ce que la propriété?* is as important for modern political economy as Sieyes's work *Qu'est-ce que le tiers état?* for modern politics.

Proudhon does not consider the further forms of private property, e.g. wages, trade, value, price, money, etc., as forms of private property in themselves, as they are considered, for example, in *Deutsch–französische Jahrbücher* (see *Notes for a Critique of Political Economy*, by F. Engels), but uses these economic premisses as an argument against economists; this is fully in keeping with his historically justified standpoint to which we referred above.

Accepting the relations of private property as human and reasonable, political economy moves in permanent contradiction to its basic premiss, private property, a contradiction analogous to that of theology, which, continually giving a human interpretation to religious conceptions, is by this very fact in constant conflict with its basic premiss, the superhuman character of religion. Thus, in political economy wages appear at the beginning as the proportional share of the product due to labour. Wages and profit on capital stand in a friendly, mutually favourable, apparently most human relationship to each other. Afterwards it turns out that they stand in the most hostile relationship, in inverse proportion to each other. Value is determined at the beginning in an apparently reasonable way by the cost of production of an object and its social usefulness. Later it turns out that value is determined quite fortuitously and that it does not need to bear any relation to cost of production or social usefulness. The magnitude of wages is determined at the beginning by free agreement between the free worker and the free capitalist. Later it turns out that the worker is compelled to allow the capitalist to determine it, just as the capitalist is compelled to fix it as low as possible. Freedom of the contracting parties has been supplanted by compulsion. The position is the same in trade and all other political–economic relations. The economists themselves occasionally feel these contradictions, the discussions of which is the main content of the struggle between them. When, however, the economists become conscious of these contradictions, they themselves attack private property in one of its particular forms as the falsifier of what are in themselves (i.e. in their imagination) reasonable wages, in itself reasonable value, in itself reasonable trade. Adam Smith, for instance, occasionally polemizes against the capitalists, Destutt de Tracy against the bankers, Simonde de Sismondi against the factory system, Ricardo against landed property, and nearly all modern economists against the non-industrial capitalists, in whom property appears as a mere consumer.

Thus, as an exception—when they attack some special abuse—the economists occasionally stress the semblance of humanity in economic relations, while sometimes, and as often as not, they take these relations precisely in their marked difference from the human, in their strictly economic sense. They stagger about within that contradiction completely unaware of it.

Proudhon puts an end to this unconsciousness once for all. He takes the human semblance of the economic relations seriously and sharply opposes it to their inhuman reality. He forces them to be in reality what they imagine themselves to be, or, to be more exact, to give up their own idea of themselves and confess their real inhumanity. He is therefore consistent when he represents as the falsifier of economic relations not this or that particular kind of private property as other economists do, but private property taken in its entirety. He does all that a criticism of political economy from the standpoint of political economy can do.

Herr Edgar, who wishes to characterize the standpoint of the treatise *Qu'est-ce que la propriété?*, naturally does not say a word of political economy or of the distinctive character of that treatise, which is precisely that it has made the essence of private property the vital question of political economy and jurisprudence. This is all self-evident for Critical Criticism. Proudhon, it says, has done nothing new by his negation of private property. He has only divulged one of Critical Criticism's close secrets.

'Proudhon', Herr Edgar continues immediately after his characterizing translation, 'therefore finds something Absolute, an eternal foundation in history, a god that guides mankind—justice.'

Proudhon's treatise, written in French in 1840, does not adopt the standpoint of German development in 1844. It is Proudhon's standpoint, a standpoint which is shared by countless diametrically opposed French writers and therefore gives Critical Criticism the advantage of having characterized the most contradictory standpoints with a single stroke of the pen. Incidentally, to settle with this Absolute in history one has only to apply logically the law formulated by Proudhon himself, that of the implementation of justice by its negation. If Proudhon does not carry logic that far, it is only because he had the misfortune of being born a Frenchman, not a German.

Alienation and the Proletariat

... By investigating 'the whole as such' to find the conditions for its existence, Critical Criticism is searching in the genuine theological manner, outside the whole, for the conditions for its existence. Critical speculation moves outside the object which it pretends to deal with. The whole contradiction is nothing but the movement of both its sides, and the condition for the existence of the whole lies in the very nature of the two sides. Critical Criticism dispenses with the study of this real movement which forms the whole in order to be able to declare that it, Critical Criticism as the calm of knowledge, is above both extremes of the contradiction, and that its activity, which has made the 'whole as such', is now alone in a position to abolish the abstraction of which it is the maker.

Proletariat and wealth are opposites; as such they form a single whole. They are both forms of the world of private property. The question is what place each occupies in the antithesis. It is not sufficient to declare them two sides of a single whole.

Private property as private property, as wealth, is compelled to maintain itself, and thereby its opposite, the proletariat, in existence. That is the positive side of the contradiction, self-satisfied private property.

The proletariat, on the other hand, is compelled as proletariat to abolish itself and thereby its opposite, the condition for its existence, what makes it the proletariat, i.e. private property. That is the negative side of the contradiction, its restlessness within its very self, dissolved and self-dissolving private property.

The propertied class and the class of the proletariat present the same human self-alienation. But the former class finds in this self-alienation its confirmation and its good, its own power: it has in it a semblance of human existence. The class of the proletariat feels annihilated in its self-alienation; it sees in it its own powerlessness and the reality of an inhuman existence. In the words of Hegel, the class of the proletariat is in abasement indignation at that abasement, an indignation to which it is necessarily driven by the contradiction between its human nature and its condition of life, which is the outright, decisive, and comprehensive negation of that nature.

Within this antithesis the private owner is therefore the conservative side, the proletarian, the destructive side. From the former arises the action of preserving the antithesis, from the latter, that of annihilating it.

Indeed private property, too, drives itself in its economic movement towards its own dissolution, only, however, through a development which does not depend on it, of which it is unconscious and which takes place against its will, through the very nature of things; only inasmuch as it produces the proletariat as proletariat, that misery conscious of its spiritual and physical misery, that dehumanization conscious of its dehumanization and therefore self-abolishing. The proletariat executes the sentence that private property pronounced on itself by begetting the proletariat, just as it carries out the sentence that wage-labour pronounced on itself by bringing forth wealth for others and misery for itself. When the proletariat is victorious, it by no means becomes the absolute side of society, for it is victorious only by abolishing itself and its opposite. Then the proletariat disappears as well as the opposite which determines it, private property.

When socialist writers ascribe this historic role to the proletariat, it is not, as Critical Criticism pretends to think, because they consider the proletarians as gods. Rather the contrary. Since the abstraction of all humanity, even of the semblance of humanity, is practically complete in the full-grown proletariat; since the conditions of life of the proletariat sum up all the conditions of life of society today in all their inhuman acuity; since man has lost himself in the proletariat, yet at the same time has not only gained theoretical consciousness of that loss, but through urgent, no longer disguisable, absolutely imperative need—that practical expression of necessity— is driven directly to revolt against that inhumanity; it follows that the proletariat can and must free itself. But it cannot free itself without abolishing the conditions of its own life. It cannot abolish the conditions of its own life without abolishing all the inhuman conditions of life of society today which are summed up in its own situation. Not in vain does it go through the stern but steeling school of labour. The question is not what this or that proletarian, or even the whole of the proletariat at the moment considers as its aim. The question is what the proletariat is, and what, consequent on that being, it will be compelled to do. Its aim and historical action is irrevocably and obviously demonstrated in its own life situation as well as in the whole organization of bourgeois society today. There is no need to dwell here upon the fact that a large part of the English and French proletariat is already conscious of its historic task and is constantly working to develop that consciousness into complete clarity.

On Idealist Philosophy

The mystery of the Critical presentation of the *Mystères de Paris* is the mystery of speculative Hegelian construction. Once Herr Szeliga has proclaimed 'degeneracy within civilization' and rightlessness in the state 'Mysteries', i.e. has dissolved them in the category 'Mystery', he lets 'Mystery' begin its speculative career. A few words will suffice to characterize speculative construction in general; Herr Szeliga's treatment of the *Mystères de Paris* will give the application in detail.

If from real apples, pears, strawberries, and almonds I form the general idea 'Fruit', if I go further and imagine that my abstract idea 'Fruit', derived from real fruit, is an entity existing outside me, is indeed the true essence of the pear, the apple, etc.; then, in the language of speculative philosophy I am declaring that 'Fruit' is the substance of the pear, the apple, the almond, etc. I am saying, therefore, that to be a pear is not essential to the pear, that to be an apple is not essential to the apple; that what is essential to these things is not their real being, perceptible to the senses, but the essence that I have extracted from them and then foisted on them, the essence of my idea— 'Fruit'. I therefore declare apples, pears, almonds, etc. to be mere forms of existence, *modi*, of 'Fruit'. My finite understanding supported by my senses does, of course, distinguish an apple from a pear and a pear from an almond; but my speculative reason declares these sensuous differences unessential, indifferent. It sees in the apple the same as in the pear, and in the pear the same as in the almond, namely 'Fruit'. Particular real fruits are no more than semblances whose true essence is 'the Substance'—'Fruit'.

By this method one attains no particular wealth of definition. The miner-alogist whose whole science consisted in the statement that all minerals are really 'Mineral' would be a mineralogist only in his imagination. For every mineral the speculative mineralogist says 'Mineral' and his science is reduced to repeating that word as many times as there are real minerals.

Having reduced the different real fruits to the one fruit of abstraction— 'Fruit', speculation must, in order to attain some appearance of real content, try somehow to find its way back from 'Fruit', from Substance to the different profane real fruits, the pear, the apple, the almond, etc. It is as hard to produce real fruits from the abstract idea 'Fruit' as it is easy to produce this abstract idea from real fruits. Indeed it is impossible to arrive at the opposite of an abstraction without relinquishing the abstraction.

The speculative philosopher therefore relinquishes the abstraction 'Fruit', but in a speculative, mystical fashion—with the appearance of not relin-quishing it. Thus he rises above his abstraction only in appearance. He argues like this:

If apples, pears, almonds, and strawberries are really nothing but 'Substance', 'Fruit', the question arises: Why does 'Fruit' manifest itself to me sometimes as an apple, sometimes as a pear, sometimes as an almond? Why this appearance of diversity which so strikingly contradicts my speculative conception of 'Unity'; 'Substance'; 'Fruit'?

This, answers the speculative philosopher, is because fruit is not dead, undifferentiated, motionless, but living, self-differentiating, moving. The diversity of profane fruits is significant not only to my sensuous understand-ing, but also to 'Fruit' itself and to speculative reasoning. The different profane fruits are different manifestations of the life of the one 'Fruit'; they are crystallizations of 'Fruit' itself. In the apple 'Fruit' gives itself an apple-like existence, in the pear a pear-like existence. We must therefore no longer say as from the standpoint of Substance: a pear is 'Fruit', an apple is 'Fruit', an almond is 'Fruit', but 'Fruit' presents itself as a pear, 'Fruit' presents itself as an apple, 'Fruit' presents itself as an almond; and the differences which distinguish apples, pears, and almonds from one another are the self-differentiations of 'Fruit' making the particular fruits subordinate members of the life-process of 'Fruit'. Thus 'Fruit' is no longer a contentless, undifferentiated unity; it is oneness as allness, as 'totalness' of fruits, which constitute an 'organic ramified series'. In every member of that series 'Fruit' gives itself a more developed, more explicit existence, until it is finally the 'summary' of all fruits and at the same time living unity which contains all those fruits dissolved in itself just as much as it produces them from within itself, as, for instance, all the limbs of the body are constantly dissolved in blood and constantly produced out of the blood.

We see that if the Christian religion knows only one Incarnation of God, speculative philosophy has as many incarnations as there are things, just as it has here in every fruit an incarnation of the 'Substance', of the Absolute 'Fruit'. The main interest for the speculative philosopher is therefore to produce the existence of the real profane fruits and to say in some mysterious way that there are apples, pears, almonds, and raisins. But the apples, pears, almonds, and raisins that we get in the speculative world are nothing but semblances of apples, semblances of pears, semblances of almonds, and semblances of raisins; they are moments in the life of 'Fruit', that abstract being of reason, and therefore themselves abstract beings of reason. Hence what you enjoy in speculation is to find all the real fruits there, but as fruits which have a higher mystic significance, which are grown out of the ether of your brain and not out of the material earth, which are incarnations of 'Fruit', the Absolute Subject. When you return from the abstraction, the preternatural being of reason 'Fruit', to real natural fruits, you give, contrariwise, the natural fruits a preternatural significance and transform them into so many abstractions. Your main interest is then to point out the unity of 'Fruit' in all the manifestions of its life—the apple, the pear, the almond—that is, the mystical interconnection between these fruits, how in each one of them 'Fruit' develops by degrees and necessarily progresses, for instance, from its existence as a raisin to its existence as an almond. The value of profane fruits no longer consists in their natural qualities but in their speculative quality which gives each of them a definite place in the life-process of 'Absolute Fruit'.

The ordinary man does not think he is saying anything extraordinary when he states that there are apples and pears. But if the philosopher expresses those existences in the speculative way he says something extraordinary. He works a wonder by producing the real natural being, the apple, the pear, etc., out of the unreal being of reason 'Fruit', i.e. by creating those fruits out of his own abstract reason, which he considers as an Absolute Subject outside himself, represented here as 'Fruit'. And in every existence which he expresses he accomplishes an act of creation.

It goes without saying that the speculative philosopher accomplishes this constant creation only by representing universally known qualities of the apple, the pear, etc., which exist in reality, as definitions discovered by him; by giving the names of the real things to what abstract reason alone can create, to abstract formulas of reason, finally, by declaring his own activity, by which he passes from the idea of an apple to the idea of a pear, to be the self-activity of the Absolute Subject, 'Fruit'.

In the speculative way of speaking, this operation is called comprehending the substance as the subject, as an inner process, as an Absolute Person and that comprehension constitutes the essential character of Hegel's method.

These preliminary remarks were necessary to make Herr Szeliga intelligible. After thus far dissolving real relations, e.g. right and civilization, in the category of mysteries and thereby making 'Mystery' a substance, he now rises to the real speculative Hegelian height and transforms 'Mystery' into self-existing subject incarnating itself in real situations and persons so that the manifestations of its life are countesses, marquises, *grisettes* [street girls], porters, notaries, and charlatans, love intrigues, balls, wooden doors, etc. Having produced the category 'Mystery' out of the real world, he produces the real world out of that category.

The mysteries of speculative construction in Herr Szeliga's presentation will be all the more visibly disclosed as he has an indisputable double advantage over Hegel. First, Hegel has the sophistic mastery of being able to present as a process of the imagined being of reason itself, of the Absolute Subject, the process by which the philosopher goes by sensory perception and imagination from one object to another. Besides, Hegel very often gives a real presentation, embracing the thing itself, within the speculative presentation. This real reasoning within the speculative reasoning misleads the reader into considering the speculative reasoning as real and the real as speculative.

With Herr Szeliga both these difficulties vanish. His dialectics have no hypocrisy or pretence. He performs his tricks with the most laudable honesty and the most sincere straightforwardness. But then he nowhere develops any real content, so that his speculative construction is free from all disturbing complications, from all ambiguous disguises, and appeals to the eye in its naked beauty. In Herr Szeliga we also see a brilliant illustration of how speculation on the one hand apparently freely creates its object *a priori* out of itself, and on the other hand, for the very reason that it wishes to get rid by sophistry of its reasonable and natural dependence on the object, falls into the most unreasonable and unnatural bondage to the object whose most accidental and individual attributes it is obliged to construe as absolutely necessary and general.

The Idealist View of History

. . . Just as according to old teleologists plants exist to be eaten by animals and animals by men, history exists in order to serve as the act of consumption of theoretical eating—proving. Man exists so that history may exist and history exists so that the proof of truths may exist. In that Critically trivialized form we have the repetition of the speculative wisdom that man exists and that history exists so that truth may be brought to self-consciousness.

That is why history, like truth, becomes a person apart, a metaphysical subject of which real human individuals are but the bearers. That is why Absolute Criticism uses expressions like these:

'History will not be joked at . . . history has exerted its greatest efforts to . . . history has been engaged . . . what would be the purpose of history? . . . history provides the explicit proof; history proposes truths,' etc.

If, as Absolute Criticism affirms, history has so far been occupied with only a few such truths—the simplest of all—which in the end are self-evident, 'this indigence to which previous human experiences were reduced proves first of all only Absolute Criticism's own indigence. From the unCritical standpoint the result of history is, on the contrary, that the most complicated truth, the quintessence of all truth, man, understands himself in the end by himself.'

'But truths', Absolute Criticism continues to argue, 'which seem to the mass to be so crystal clear that they are understood of themselves from the start . . . and that the mass deems proof superfluous, are not worth history supplying explicit proof of them; they constitute no part whatever of the problem which history is engaged in solving.'

In its holy zeal against the mass Absolute Criticism flatters it in the most refined way. If a truth is crystal clear because it seems crystal clear to the mass; if history's attitude to truths depends on the opinion of the mass, the opinion of the mass is absolute, infallible, it is law for history, and history proves only what the mass does not consider as crystal clear, what therefore needs proof. It is the mass, therefore, that prescribes history's 'task' and 'occupation'.

Absolute Criticism speaks of 'truths which are understood of themselves from the start'. In its Critical naïveness it invents an absolute 'from the start' and an abstract, immutable 'Mass'. There is just as little difference, in the eyes of Criticism, between the 'from the start' of the sixteenth-century mass and the 'from the start' of the nineteenth-century mass as between those masses themselves. It is precisely a feature of a truth which has become true and obvious and is understood of itself that it 'is understood of itself from the start'. Absolute Criticism's polemic against truths which are understood of themselves from the start is a polemic against truths which, in general, 'are understood of themselves'.

A truth which is understood of itself has lost its salt, its meaning, its value for Absolute Criticism as for divine dialectics. It has become flat, like stale water. On the one hand, therefore, Absolute Criticism proves everything which is understood of itself and, besides, many things which have the luck

of being incomprehensible and will therefore never be understood of themselves. On the other hand it considers as understood of itself everything
which needs some proof. Why? Because it is understood of itself that real
problems are not understood of themselves.

As 'Truth', like history, is an ethereal subject separate from the material
mass, it addresses itself not to the empirical man but to the 'innermost of the
soul'; in order to be 'truly apprehended' it does not act on his vulgar body,
which may live in the bowels of an English basement or at the top of a
French block of poky flats; it 'drags' on and on 'through' his idealistic
intestines. Absolute Criticism does certify that 'the mass' has so far in its
own way, i.e. superficially, been touched by the truths that history has been
so gracious as to 'propose'; 'but at the same time it prophesies that the
attitude of the mass to historical progress will completely change.'

It will not be long before the mysterious meaning of this Critical prophecy
is 'crystal clear' to us.

'All great actions of previous history', we are told, 'were failures from the
start and had no effective success because the mass became interested in and
enthusiastic over them; in other words they were bound to come to a pitiful
end because the idea in them was such that it had to be satisfied with a
superficial comprehension and therefore to rely on the approbation of the
mass.'

It seems that comprehension ceases to be superficial when it suffices for,
corresponds to, an idea. It is only for appearance's sake that Herr Bruno
brings out a relation between an idea and its comprehension, as it is also only
for appearance's sake that he brings out a relation between unsuccessful
historical action and the mass. If, therefore, Absolute Criticism condemns
something as being 'superficial', it is simply previous history whose actions
and ideas were those of the 'masses'. It rejects massy history to replace it by
Critical history (see Herr Jules Faucher on English problems of the day).
According to previous unCritical history, i.e. history not conceived in the
sense of Absolute Criticism, it must further be precisely distinguished to
what extent the mass was 'interested' in aims and to what extent it was
'enthusiastic' over them. The 'idea' always disgraced itself in so far as it
differed from the 'interest'. On the other hand it is easy to understand that
every massy 'interest' asserting itself historically goes far beyond its real
limits in the 'idea' or 'imagination' when it first came on the scene and is
confused with human interest in general. This illusion constitutes what
Fourier calls the tone of each historical epoch. The interest of the bourgeoisie in the 1789 Revolution, far from having been a 'failure', 'won' everything and had 'effective success', however much the 'pathos' of it evaporated
and the 'enthusiastic' flowers with which that interest adorned its cradle
faded. That interest was so powerful that it vanquished the pen of Marat,
the guillotine of the Terror, and the sword of Napoleon as well as the
crucifix and the blue blood of the Bourbons. The Revolution was a 'failure'

only for the mass which did not find in the political 'idea' the idea of its real 'interest', whose real life-principle did not therefore coincide with the life-principle of the Revolution; the mass whose real conditions for emancipation were substantially different from the conditions within which the bourgeoisie could emancipate itself and society. If the revolution, which can exemplify all great historical 'actions', was a failure, it was so because the mass whose living conditions it did not substantially go beyond was an exclusive, limited mass, not an all-embracing one. If it was a failure it was not because it aroused the 'enthusiasm' and 'interest' of the mass, but because the most numerous part of the mass, the part most greatly differing from the bourgeoisie, did not find its real interest in the principle of the revolution, had no revolutionary principle of its own, but only an 'idea', and hence only an object of momentary enthusiasm and only apparent exaltation.

With the thoroughness of the historical action the size of the mass whose action it is will therefore increase. In Critical history, according to which in historical actions it is not a matter of the active mass, of empirical action, or of the empiric interest of that action but rather only of 'an idea' 'in them', affairs must naturally take a different course.

'In the mass,' Criticism teaches us, 'not somewhere else, as its former liberal spokesmen believed, is the true enemy of the spirit to be found.'

The enemies of progress outside the mass are precisely those products of self-debasement, self-rejection, and self-estrangement of the mass which have been endowed with independent being and a life of their own. The mass therefore rises against its own deficiency when it rises against the independently existing products of its self-debasement just as man, turning against the existence of God, turns against his own religiosity. But as those practical self-estrangements of the mass exist in the real world in an outward way, the mass must fight them in an outward way. It must by no means consider these products of its self-estrangement as mere ideal fancies, mere estrangements of self-consciousness, and must not wish to abolish material estrangement by a purely inward spiritual action. As early as 1789 Loustalot's journal gave the motto:

> The great appear great in our eyes
> Only because we kneel.
> Let us rise!

But to rise it is not enough to do so in thought and to leave hanging over our real sensual head the real palpable yoke that cannot be subtilized away with ideas. Yet Absolute Criticism has learnt from Hegel's *Phenomenology* at least the art of changing real objective chains that exist outside me into mere ideal, mere subjective chains existing in me, and thus to change all exterior palpable struggles into pure struggles of thought.

It is on this Critical transformation that the pre-established harmony between Critical Criticism and the censorship is based. From the Critical point of view the writer's fight against the censor is not a fight of 'man against man'. The censor is nothing but my own tact personified for me by the solicitous police, my own tact struggling against my tactlessness and un-Criticalness. The struggle of the writer with the censor is only apparently, only in the eyes of wicked sensuality, anything else than the interior struggle of the writer with himself. In so far as the censor is a real individual different from myself, a police official who mishandles the product of my mind by applying an external standard which has nothing to do with the matter in question; he is but a massy imagination, an un-Critical figment of the brain. When Feuerbach's *Theses on the Reform of Philosophy* were prohibited by the censor, it was not the official barbarity of the censor that was to blame but the lack of refinement of Feuerbach's *Theses*. 'Pure' Criticism, unsullied by mass or matter, also has in the censor a purely 'ethereal' form, free from any massy reality.

Absolute Criticism has declared the 'mass' to be the true enemy of the spirit. This it develops as follows:

'The spirit now knows where to look for its only adversary—in the self-deception and the pithlessness of the mass.'

Absolute Criticism proceeds from the dogma of the absolute competency of the 'spirit'. Furthermore, it proceeds from the dogma of the extramundane existence of the spirit, i.e. of its existence outside the mass of humanity. Finally it transforms 'the spirit', 'progress', on the one hand, and the 'mass', on the other, into fixed beings, into concepts, and relates them one to the other in that form as given invariable extremes. It does not occur to Absolute Criticism to investigate the 'spirit' itself, to find out whether it is not its own spiritualistic nature, its airy pretensions that justify 'the phrase', 'self-deception' and 'pithlessness'. The spirit, on the contrary, is absolute, but unfortunately at the same time it continually falls into spiritlessness: it continually calculates without the master, hence it must necessarily have an adversary that intrigues against it. That adversary is the mass.

The position is the same with 'progress'. In spite of 'progress's' pretensions, continual retrogressions and circular movements are to be observed. Not suspecting that the category 'Progress' is completely empty and abstract, Absolute Criticism is so profound as to recognize 'progress' as being absolute and to explain retrogression by supposing a 'personal adversary' of progress, the mass. As 'the mass' is nothing but the 'opposite of the spirit', of progress, of 'Criticism', it can also be defined only by that imaginary opposition; outside that opposition all that Criticism can say about the meaning and the existence of the mass is the senseless, because completely undefined:

'The mass, in the sense in which the "word" also embraces the so-called educated world.'

'Also' and 'so-called' are enough for its Critical definition. The 'Mass' is therefore distinct from the real masses and exists as the 'Mass' only for 'Criticism'.

All communist and socialist writers proceeded from the observation that, on the one hand, even the most favourable brilliant deeds seemed to remain without brilliant results, to end in trivialities, and, on the other, all progress of the spirit had so far been progress against the mass of mankind, driving it to an ever more dehumanized predicament. They therefore declare 'progress' (see Fourier) to be an inadequate abstract phrase; they assumed (see Owen among others) a fundamental flaw in the civilized world; that is why they submitted the real bases of contemporary society to incisive criticism. To this communist criticism corresponded immediately in practice the movement of the great mass against which history had so far developed. One must be acquainted with the studiousness, the craving for knowledge, the moral energy, and the unceasing urge for development of the French and English workers to be able to form an idea of the human nobleness of that movement.

How infinitely profound 'Absolute Criticism' must be to have in face of these intellectual and practical facts, but a one-sided conception of only one aspect of the relationship—the continual foundering of the spirit—and, vexed at this, to seek besides an adversary of the 'Spirit' and find it in the 'Mass'. In the end all this great Critical discovery comes to tautology. According to Criticism, the spirit has so far had a limit, an obstacle, in other words, an adversary, because it has had an adversary. Who, then, is the adversary of the Spirit? Spiritlessness. For the mass is defined only as the 'opposite' of the spirit, as spiritlessness or to take more precise definitions of spiritlessness, 'indolence', 'superficiality', 'self-complacency'. What a fundamental advantage over the communist writers it is not to have traced spiritlessness, indolence, superficiality, and self-complacency to their origin but to have branded them morally and exposed them as the opposite of the spirit, of progress! If these qualities are proclaimed qualities of the Mass, as of a subject still distinct from them, that distinction is nothing but a Critical semblance of distinction. Only in appearance has Absolute Criticism a definite concrete subject besides abstract qualities of spiritlessness, indolence, etc., for the 'Mass' in the Critical conception is nothing but those abstract qualities, another word for them, a fantastic personification of them.

Meanwhile, the relation between 'spirit and mass' has still a hidden sense which will be completely revealed in the course of the reasoning. We only indicate it here. That relation discovered by Herr Bruno is, in fact, nothing but a Critically caricatural realization of Hegel's conception of history; this, in turn, is nothing but the speculative expression of the Christian Germanic dogma of the opposition between spirit and matter, between God and the world. This opposition is expressed in history, in the very world of man, in only a few chosen individuals opposed as the active spirit to the rest of mankind, as the spiritless mass, as matter.

Hegel's conception of history assumes an Abstract or Absolute Spirit which develops in such a way that mankind is a mere mass bearing it with a varying degree of consciousness or unconsciousness. Within empiric, exoteric history he therefore has a speculative, esoteric history develop. The history of mankind becomes the history of the abstract spirit of mankind, a spirit beyond all man!

Parallel with this doctrine of Hegel's there developed in France that of the Doctrinairians, proclaiming the sovereignty of reason in opposition to the sovereignty of the people in order to exclude the masses and rule alone. This was quite consistent. If the activity of real mankind is nothing but the activity of a mass of human individuals then abstract generality, Reason, the Spirit must contrariwise have an abstract expression restricted to a few individuals. It then depends on the situation and imaginative power of each individual whether he will pass for a representative of that 'spirit'.

In Hegel the Absolute Spirit of history already treats the mass as material and finds its true expression only in philosophy. But with Hegel, the philosopher is only the organ through which the creator of history, the Absolute Spirit, arrives at self-consciousness by retrospection after the movement has ended. The participation of the philosopher in history is reduced to this retrospective consciousness, for real movement is accomplished by the Absolute Spirit unconsciously, so that the philosopher appears *post festum* [after the event].

Hegel is doubly inconsistent: first because, while declaring that philosophy constitutes the Absolute Spirit's existence he refuses to recognize the real philosophical individual as the Absolute Spirit; secondly, because according to him the Absolute Spirit makes history only in appearance. . . .

The Jewish Question Revisited

. . . As industrial activity is not abolished by the abolition of the privileges of the trades, guilds, and corporations, but, on the contrary, real industry begins only after the abolition of these privileges; as ownership of the land is not abolished when privileges of land ownership are abolished, but, on the contrary, begins its universal movement with the abolition of privileges and the free division and free alienation of land; as trade is not abolished by the abolition of trade privileges but finds its true materialization in free trade; so religion develops in its practical universality only where there is no privileged religion (cf. the North American States).

The modern 'public system', the developed modern state, is not based, as Criticism thinks, on a society of privileges, but on a society in which privileges are abolished and dissolved; on developed civil society based on the vital elements which were still politically fettered in the privilege system and have been set free. Here 'no privileged exclusivity' stands opposed either to any other exclusivity or to the public system. Free industry and free trade abolish privileged exclusivity and thereby the struggle between the privileged exclusivities. In its place they set man free from privilege—which isolates from the social whole but at the same time joins in a narrower exclusivity—man, no longer bound to other men even by the semblance of common ties. Thus they produce the universal struggle of man against man, individual against individual. In the same way civil society as a whole is this war among themselves of all those individuals no longer isolated from the others by anything else but their individuality, and the universal uncurbed movement of the elementary forces of life freed from the letters of privilege. The contradiction between the democratic representative state and civil society is the perfection of the classic contradiction between public common-wealth and slavedom. In the modern world each one is at the same time a member of slavedom and of the public commonwealth. Precisely the slavery of civil society is in appearance the greatest freedom because it is in appear-ance the perfect independence of the individual. Indeed, the individual considers as his own freedom the movement, no longer curbed or fettered by a common tie or by man, the movement of his alienated life elements, like property, industry, religion, etc.; in reality, this is the perfection of his slavery and his inhumanity. Right has here taken the place of privilege.

It is therefore only here, where we find no contradiction between free theory and the practical import of privilege, but, on the contrary, the prac-tical abolition of privilege, free industry, free trade, etc., conforming to 'free theory', where the public system is not faced with any privileged exclusivity, where the contradiction expounded by Criticism is abolished; here only do we find the accomplished modern state.

Here reigns the reverse of the law which Herr Bauer, in connection with the debates in the French Chamber, formulated in perfect agreement with Monsieur Martin (*du Nord*):

'As Monsieur Martin (*du Nord*) saw in the motion not to mention Sunday in the law a motion declaring that Christianity had ceased to exist, with the same right, and a completely warranted right, the declaration that the law of the Sabbath is no longer binding on the Jews would be the declaration of the dissolution of Judaism.'

It is just the opposite in the developed modern state. The state declares that religion, like the other elements of civil life, only begins to exist in its full scope when the state declares it to be non-political and thus leaves it to itself. To the dissolution of the political existence of these elements, for example, the dissolution of property by the abolition of the property qualifi-cation for electors, the dissolution of religion by the abolition of the state church, to this very proclamation of their civil death corresponds their most vigorous life, which henceforth obeys its own laws undisturbed and develops to its full scope.

Anarchy is the law of civil society emancipated from disjointing privileges, and the anarchy of civil society is the basis of the modern public system, just as the public system is in turn the guarantee of that anarchy. To the same extent as the two are opposed to each other they also determine each other.

It is clear how capable Criticism is of assimilating the 'new'. But if we remain within the bounds of 'pure Criticism' the question arises: Why did Criticism not conceive as a universal contradiction the contradiction that it disclosed in connection with the debates in the French Chamber, although in its own opinion that is what 'should have been' done?

'That step was, however, then impossible—not only because . . . not only because . . . but also because without that last remnant of interior involvement with its opposite criticism was impossible and could not have come to the point from which it had only one step to make.'

It was impossible . . . because . . . it was impossible! Criticism affirms moreover, that the fateful 'one step' necessary to 'come to the point from which it had only one step to make' was impossible. Who will dispute that? In order to come to a point from which there is only 'one step' to make, it is absolutely impossible to make still that 'one step' that leads beyond the point beyond which there is still 'one step'.

All's well that ends well! At the end of the encounter with the Mass, who is hostile to Criticism's *Die Judenfrage*, 'Criticism' admits that its conception of 'the rights of man', its 'appraisal of religion in the French Revolution', the 'free political essence it pointed to occasionally in concluding its considerations', in a word, that the 'whole time of the French Revolution was no more nor no less for Criticism than a symbol—that is to say, not the time of the revolutionary actions of the French in the exact and prosaic sense, but a symbol, only a fantastic expression of the figures which it saw at the end.' We shall not deprive Criticism of the consolation that when it erred politically it did so only at the 'conclusion' and at the 'end' of its work. A well-known drunkard used to console himself with the thought that he was never drunk before midnight.

On the Jewish question Criticism has indisputably continually won ground from the enemy. In No. 1 of *Die Judenfrage* the treatise of 'Criticism' defended by Herr Bauer was still absolute and revealed the 'true' and 'general' significance of the Jewish question. In No. 2 Criticism had neither the 'will' nor the 'right' to go beyond Criticism. In No. 3 it had still to make 'one step' but that step was 'impossible'—because it was 'impossible'. It was not its 'will or right' but its involvement in its 'opposite' that prevented it from making that 'one step'. It would have liked to clear the last obstacle, but unfortunately there was a last remnant of Mass on its Critical seven-league boots.

The French Revolution

... The limitedness of the Mass forced 'the Spirit', 'Criticism', Herr Bauer, to consider the French Revolution not as the time of the revolutionary endeavours of the French in the 'prosaic sense' but 'only' as the 'symbol and fantastic expression' of the Critical figments of his own brain. Criticism does penance for its 'oversight' by submitting the Revolution to a further examination. At the same time it punishes the seducer of its innocence—'the Mass'—by communicating to it the results of that 'further examination'.

'The French Revolution was an experiment which still belonged entirely to the eighteenth century.'

The chronological truth that an experiment of the eighteenth century like the French Revolution is still entirely an experiment of the eighteenth century and not, for example, an experiment of the nineteenth seems 'still entirely' to be one of those truths 'which are understood of themselves from the start'. But in the terminology of Criticism, which is very prejudiced against 'crystal-clear' truths, a truth like that is called an 'examination' and therefore naturally has its place in a 'further examination of the revolution'.

'The ideas which the French Revolution gave rise to did not, however, lead beyond the system that it wanted to abolish by force.'

Ideas can never lead beyond an old-world system but only beyond the ideas of the old-world system. Ideas cannot carry anything out at all. In order to carry out ideas men are needed who dispose of a certain practical force. In its literal sense the Critical sentence is therefore another example of a truth that is understood of itself, that is, another 'examination'.

Undeterred by this examination, the French Revolution brought forth ideas which led beyond the ideas of the entire old-world system. The revolutionary movement which began in 1789 in the *Cercle social*, which in the middle of its course had as its chief representatives Leclerc and Roux and which finally was temporarily defeated with Babœuf's conspiracy, brought forth the communist idea which Babœuf's friend Buonarroti reintroduced into France after the Revolution of 1830. This idea, consistently developed, is the idea of the new world system.

'After the Revolution had therefore' (!) 'abolished feudal barriers in the life of the people, it was compelled to satisfy the pure egotism of the nation and to fan it itself, and, on the other hand, to curb it by its necessary complement, the recognition of a supreme being, that higher confirmation of the general state system, the functions of which is to hold together the individual self-seeking atoms.'

The egotism of the nation is the natural egotism of the general state system, as opposed to the egotism of the feudal estates. The supreme being is the higher confirmation of the general state system, that is, again the nation. Nevertheless, the supreme being is supposed to curb the egotism of the nation, that is, of the general state system! A really Critical task, to curb egotism by means of its confirmation and even of its religious confirmation, i.e. by recognizing that it is superhuman and therefore cannot be curbed by man! The creators of the supreme being were not aware of this, their Critical intention.

Monsieur Buchez, who supports national fanaticism with religious fanaticism, understands his hero Robespierre better.

Rome and Greece were ruined by nationalism. Criticism therefore says nothing specific about the French Revolution when it says that nationalism was its downfall, just as it says nothing about the nation when it defines its egotism as pure. This pure egotism appears rather to be a very dark one, natural and adulterated with flesh and blood when compared, for example, with Fichte's 'ego'. But if, in contrast to the egotism of the feudal estates its purity is only relative, no 'further examination of the revolution' was needed to see that the egotism which has a nation as its content is more general or purer than that which has as its content a particular estate or a particular corporation.

Criticism's explanations on the general state system are no less instructive. They are confined to saying that the general system must hold together the separate self-seeking atoms.

Speaking exactly and in the prosaic sense, the members of civil society are not atoms. The specific property of the atom is that it has no properties and is therefore not connected with beings outside it by any relations determined by its own natural necessity. The atom has no needs, it is self-sufficient; the world outside it is absolute vacuum, i.e. it is contentless, senseless, meaningless, just becuas the atom has all its fullness in itself. The egotistic individual in civil society may in his non-sensuous imagination and lifeless abstraction inflate himself to the size of an atom, i.e. to an unrelated, self-sufficient, wantless, absolutely full, blessed being. Unblessed sensuous reality does not bother about his imagination; each of his senses compels him to believe in the existence of the world and the individuals outside him and even his profane stomach reminds him every day that the world outside him is not empty, but is what really fills. Every activity and property of his being, every one of his vital urges becomes a need, a necessity, which his self-seeking transforms into seeking for other things and human beings outside him. But as the need of one individual has no self-understood sense for the other egotistic individual capable of satisfying that need, and therefore no direct connection with its satisfaction, each individual has to create that connection; it thus becomes the intermediary between the need of another and the object of that need. Therefore it is natural necessity, essential human properties, however alienated thay may seem to be, and interest that hold the members of civil society together: civil, not political life is their real tie. It is therefore not the state that holds the atoms of civil society together, but the fact that they are atoms only in imagination, in the heaven of their fancy, but in reality beings tremendously different from atoms, in other words, not divine egoists, but egotistic human beings. Only political superstition today imagines that social life must be held together by the state whereas in reality the state is held together by civil life.

French Materialism and the Origins of Socialism

... 'Spinozism dominated the eighteenth century in its later French variety, which made matter into substance, as well as in deism, which conferred on matter a more spiritual name. ... Spinoza's French school and the supporters of deism were but two sects disputing over the true meaning of his system. ... The simple fate of this Enlightenment was its sinking into romanticism after being obliged to surrender to the reaction which began after the French movement.'

That is what Criticism says.

To the Critical history of French materialism we shall oppose a brief outline of its profane, massy history. We shall admit with due respect the abyss between history as it really happened and history as it happened according to the decree of 'Absolute Criticism', the creator equally of the old and of the new. And finally, obeying the prescriptions of Criticism, we shall make the 'Why?', 'Whence?', and 'Whither?' of Critical history the 'objects of a persevering study'.

'Speaking exactly and in the prosaic sense', the French Enlightenment of the eighteenth century, in particular French materialism, was not only a struggle against the existing political institutions and the existing religion and theology; it was just as much an open struggle against metaphysics of the seventeenth century, and against all metaphysics, in particular that of Descartes, Malebranche, Spinoza, and Leibnitz. Philosophy was opposed to metaphysics as Feuerbach, in his first decisive attack on Hegel, opposed sober philosophy to drunken speculation. Seventeenth-century metaphysics, beaten off the field by the French Enlightenment, to be precise, by French materialism of the eighteenth century, was given a victorious and solid restoration in German philosophy, particularly in speculative German philosophy of the nineteenth century. After Hegel linked it in so masterly a fashion with all subsequent metaphysics and with German idealism and founded a metaphysical universal kingdom, the attack on speculative metaphysics and metaphysics in general again corresponded, as in the eighteenth century, to the attack on theology. It will be defeated for ever by materialism which has now been perfected by the work of speculation itself and coincides with humanism. As Feuerbach represented materialism in the theoretical domain, French and English socialism and communism in the practical field represent materialism which now coincides with humanism.

'Speaking exactly and in the prosaic sense', there are two trends in French materialism; one traces its origin to Descartes, the other to Locke. The latter is mainly a French development and leads direct to socialism. The former, mechanical materialism, merges with what is properly French natural science. The two trends cross in the course of development. We have no need here to go deep into French materialism, which comes direct from Descartes, any more than into the French Newton school or the development of French natural science in general.

We shall therefore just note the following:

Descartes in his physics endowed matter with self-creative power and conceived mechanical motion as the act of its life. He completely separated his physics from his metaphysics. Within his physics matter is the only substance, the only basis of being and of knowledge.

Mechanical French materialism followed Descartes's physics in opposition to his metaphysics. His followers were by profession anti-metaphysicists, i.e. physicists.

The school begins with the physician Leroy, reaches its zenith with the physician Cabanis, and the physician Lamettrie is its centre. Descartes was still living when Leroy, like Lamettrie in the eighteenth century, transposed the Cartesian structure of animals to the human soul and affirmed that the soul is a modus of the body and ideas are mechanical motions. Leroy even thought Descartes had kept his real opinion secret. Descartes protested. At the end of the eighteenth century Cabanis perfected Cartesian materialism in his treatise: *Rapport du Physique et du Moral de l'homme*.

Cartesian materialism still exists today in France. It had great success in mechanical natural science which, 'speaking exactly and in the prosaic sense', will be least of all reproached with romanticism.

Metaphysics of the seventeenth century, represented in France by Descartes, had materialism as its antagonist from its very birth. It personally opposed Descartes in Gassendi, the restorer of epicurean materialism. French and English materialism was always closely related to Democritus and Epicurus. Cartesian metaphysics had another opponent in the English materialist Hobbes. Gassendi and Hobbes were victorious over their opponent long after their death when metaphysics was already officially dominant in all French schools.

Voltaire observed that the indifference of Frenchmen to the disputes between Jesuits and Jansenists in the eighteenth century was due less to philosophy than to Law's financial speculation. And, in fact, the downfall of seventeenth-century metaphysics can be explained by the materialistic theory of the eighteenth century only as far as that theoretical movement itself is explained by the practical nature of French life at the time. That life was turned to the immediate present, worldly enjoyment and worldly interests, the earthly world. Its anti-theological, anti-metaphysical, and materialistic practice demanded corresponding anti-theological, anti-metaphysical, and materialistic theories. Metaphysics had in practice lost all credit. Here we have only to indicate briefly the theoretical process.

In the seventeenth century metaphysics (cf. Descartes, Leibnitz, and others) still had an element of positive, profane content. It made discoveries in mathematics, physics, and other exact sciences which seemed to come within its pale. This appearance was done away with as early as the beginning of the eighteenth century. The positive sciences broke off from it and determined their own separate fields. The whole wealth of metaphysics was reduced to beings of thought and heavenly things, althojgh this was the very time when real beings and carthly things began to be the centre of all interest. Metaphysics had gone stale. In the very year in which Malebranche and Arnauld, the last great French metaphysicians of the seventeenth century, died, Helvetius and Condillac were born.

The man who deprived seventeenth-century metaphysics of all credit in the domain of theory was Pierre Bayle. His weapon was scepticism which he forged out of metaphysics' own magic formulas. He at first proceeded from Cartesian metaphysics. As Feuerbach was driven by the fight against speculative theology to the fight against speculative philosophy precisely because he recognized in speculation the last prop of theology, because he had to force theology to turn back from pretended science to coarse, repulsive faith, so Bayle too was driven by religious doubt to doubt about metaphysics which was the support of that faith. He therefore critically investigated metaphysics from its very origin. He became its historian in order to write the history of its death. He mainly refuted Spinoza and Leibnitz.

Pierre Bayle did not only prepare the reception of materialism and the philosophy of common sense in France by shattering metaphysics with his scepticism. He heralded atheistic society, which was soon to come to existence, by proving that a society consisting only of atheists is possible, that an atheist can be a respectable man and that it is not by atheism but by superstition and idolatry that man debases himself.

To quote the expression of a French writer, Pierre Bayle was 'the last metaphysician in the seventeenth-century sense of the word and the first philosopher in the sense of the eighteenth century'.

Besides the negative refutation of seventeenth-century theology and metaphysics, a positive, anti-metaphysical system was required. A book was needed which would systematize and theoretically justify the practice of life of the time. Locke's treatise on the origin of human reason came from across the Channel as if in answer to a call. It was welcomed enthusiastically like a long-awaited guest.

To the question: Was Locke perchance a follower of Spinoza? 'Profane' history may answer:

Materialism is the son of Great Britain by birth. Even Britain's scholastic Duns Scotus wondered: 'Can matter think?'

In order to bring about that miracle he had recourse to God's omnipotence, i.e. he forced theology itself to preach materialism. In addition he was a nominalist. Nominalism is a main component of English materialism and is in general the first expression of materialism.

The real founder of English materialism and all modern experimental science was Bacon. For him natural science was true science and physics based on perception was the most excellent part of natural science. Anaxagoras with his *homoeomeria* [doctrine of simple substances] and Democritus with his atoms are often the authorities he refers to. According to his teaching the senses are infallible and are the source of all knowledge. Science is experimental and consists in applying a rational method to the data provided by the senses. Induction, analysis, comparison, observation, and experiment are the principal requisites of rational method. The first and most important of the inherent qualities of matter is motion, not only mechanical and mathematical movement, but still more impulse, vital life-spirit, tension, or, to use Jacob Böhme's expression, the throes of matter. The primary forms of matter are the living, individualizing forces of being inherent in it and producing the distinctions between the species.

In Bacon, its first creator, materialism contained latent and still in a naïve way the germs of all-round development. Matter smiled at man with poetical sensuous brightness. The aphoristic doctrine itself, on the other hand, was full of the inconsistencies of theology.

In its further development materialism became one-sided. Hobbes was the one who systematized Bacon's materialism. Sensuousness lost its bloom and became the abstract sensuousness of the geometrician. Physical motion was sacrificed to the mechanical or mathematical, geometry was proclaimed the principal science. Materialism became hostile to humanity. In order to overcome the anti-human incorporeal spirit in its own field, materialism itself was obliged to mortify its flesh and become an ascetic. It appeared as a being of reason, but it also developed the implacable logic of reason.

If man's senses are the source of all his knowledge, Hobbes argues, proceeding from Bacon, then conception, thought, imagination, etc., are nothing but phantoms of the material world more or less divested of its sensuous form. Science can only give a name to these phantoms. One name can be applied to several phantoms. There can even be names of names. But it would be a contradiction to say, on the one hand, that all ideas have their origin in the world of the senses and to maintain, on the other hand, that a word is more than a word, that besides the beings represented, which are always individual, there exist also general beings. An incorporeal substance is just as much a nonsense as an incorporeal body. Body, being, substance, are one and the same real idea. One cannot separate the thought from matter which thinks. Matter is the subject of all changes. The word infinite is meaningless unless it means the capacity of our mind to go on adding without end. Since only what is material is perceptible, knowable, nothing is known of the existence of God. I am sure only of my own existence. Every human passion is a mechanical motion ending or beginning. The objects of impulses are what is called good. Man is subject to the same laws as nature; might and freedom are identical.

Hobbes systematized Bacon, but did not give a more precise proof of his basic principle that our knowledge and our ideas have their source in the world of the senses.

Locke proved the principle of Bacon and Hobbes in his essay on the origin of human reason.

Just as Hobbes did away with the theistic prejudices in Bacon's material-ism, so Collins, Dodwell, Coward, Hartley, Priestley, and others broke down the last bounds of Locke's sensualism. For materialists, at least, deism is no more than a convenient and easy way of getting rid of religion.

We have already mentioned how opportune Locke's work was for the French. Locke founded the philosophy of bon sens, of common sense, i.e. he said indirectly that no philosopher can be at variance with the healthy human senses and reason based on them.

Locke's immediate follower, Condillac, who also translated him into French, at once opposed Locke's sensualism to seventeenth-century metaphysics. He proved that the French had quite rightly rejected metaphy-sics as the mere bungling of fancy and theological prejudice. He published a refutation of the systems of Descartes, Spinoza, Leibnitz, and Malebranche.

In his *Essai sur l'origine des connaissances humaines* he expounded Locke's ideas and proved that not only the soul, but the senses too, not only the art of creating ideas, but also the art of sensuous perception are matters of experience and habit. The whole development of man therefore depends on education and environment. It was only by eclectic philosophy that Condillac was ousted from the French schools.

The difference between French and English materialism follows from the difference between the two nations. The French imparted to English mater-ialism wit, flesh and blood, and eloquence. They gave it the temperament and grace that it lacked. They civilized it.

In Helvetius, who also based himself on Locke, materialism became really French. Helvetius conceived it immediately in its application to social life (Helvetius, *De l'homme, de ses facultés intellectuelles et de son éducation*). Sensuous qualities and self-love, enjoyment and correctly understood per-sonal interests are the bases of morality. The natural equality of human intelligence, the unity of progress of reason and progress of industry, the natural goodness of man, and the omnipotence of education are the main points in his system.

In Lamettrie's works we find a combination of Descartes's system and English materialism. He makes use of Descartes's physics in detail. His *Man Machine* is a treatise after the model of Descartes's beast-machine. The physical part of Holbach's *Système de la nature, ou des lois du monde physique et du monde moral* is also a result of the combination of French and English materialism, while the moral part is based substantially on the moral of

Helvetius. Robinet (*De la Nature*), the French materialist who had the most connection with metaphysics and was therefore praised by Hegel, refers explicitly to Leibnitz.

We need not dwell on Volney, Dupuis, Diderot, and others any more than on the physiocrats, having already proved the dual origin of French materialism from Descartes's physics and English materialism, and the opposition of French materialism to seventeenth-century metaphysics and to the metaphysics of Descartes, Spinoza, Malebranche, and Leibnitz. The Germans could not see this opposition before they came into the same opposition with speculative metaphysics.

As Cartesian materialism merges into natural science proper, the other branch of French materialism leads direct to socialism and communism.

There is no need of any great penetration to see from the teaching of materialism on the original goodness and equal intellectual endowment of men, the omnipotence of experience, habit, and education, and the influence of environment on man, the great significance of industry, the justification of enjoyment, etc., how necessarily materialism is connected with communism and socialism. If man draws all his knowledge, sensation, etc., from the world of the senses and the experience gained in it, the empirical world must be arranged so that in it man experiences and gets used to what is really human and that he becomes aware of himself as man. If correctly understood interest is the principle of all moral, man's private interest must be made to coincide with the interest of humanity. If man is unfree in the materialist sense, i.e. is free not through the negative power to avoid this or that, but through the positive power to assert his true individuality, crime must not be punished in the individual, but the anti-social source of crime must be destroyed, and each man must be given social scope for the vital manifestation of his being. If man is shaped by his surroundings, his surroundings must be made human. If man is social by nature, he will develop his true nature only in society, and the power of his nature must be measured not by the power of separate individuals but by the power of society.

This and similar propositions are to be found almost literally even in the oldest French materialists. This is not the place to assess them. *Fable of the Bees, or Private Vices Made Public Benefits*, by Mandeville, one of the early English followers of Locke, is typical of the social tendencies of materialism. He proves that in modern society vice is indispensable and useful. This was by no means an apology of modern society.

Fourier proceeds immediately from the teaching of the French materialists. The Babouvists were coarse, uncivilized materialists, but mature communism too comes directly from French materialism. The latter returned to its mother country, England, in the form Helvetius gave it. Bentham based his system of correctly understood interest on Helvetius's moral, and Owen proceeded from Bentham's system to found English communism. Exiled to England, the Frenchman Cabet came under the influence of communist ideas there and on his return to France became the most popular, although the most superficial, representative of communism. Like Owen, the more scientific French communists, Dezamy, Gay, and others, developed the teaching of materialism as the teaching of real humanism and the logical basis of communism. . . .

BIBLIOGRAPHY

Original: MEW, Vol. 2, pp. 1 ff.

This translation: The Holy Family, trans. R. Dixon, Moscow, 1956, pp. 45 ff., 78 ff., 105 ff., 156 ff., 167 ff. Reproduced by kind permission of Lawrence and Wishart Ltd.

Other translations: Extracts in Writings of the Young Marx on Philosophy and Society, ed. L. Easton and K. Guddatt, pp. 361 ff.

Commentaries: H. Adams, Karl Marx in his Earlier Writings, 2nd edn., London, 1965, Ch. 9.

G. Cohen, 'Bourgeois and Proletarians', Journal of the History of Ideas, Apr. 1968.

L. Dupré, The Philosophical Foundations of Marxism, New York, 1966, Ch. 5.

13. Theses on Feuerbach

Shortly after finishing *The Holy Family*, Marx was compelled to leave Paris. He settled in Brussels for the next three years and continued his reading of economics. Although Marx had always been critical of Feuerbach to some extent, he now felt too closely identified with him and jotted down in his notebooks, probably in April 1845, the following eleven points in which he summarized his disagreements with Feuerbach. They show clearly how Marx's materialism is differentiated from all forms of static or mechanical materialism, and thus throw light on the meaning of terms such as 'objectivity' or 'science' in connection with Marx.

I

The chief defect of all hitherto existing materialism (that of Feuerbach included) is that the thing, reality, sensuousness, is conceived only in the form of the object or of contemplation, but not as sensuous human activity, practice, not subjectively. Hence, in contradistinction to materialism, the active side was developed abstractly by idealism—which, of course, does not know real, sensuous activity as such. Feuerbach wants sensuous objects, really distinct from the thought objects, but he does not conceive human activity itself as objective activity. Hence, in *Das Wesen des Christentums*, he regards the theoretical attitude as the only genuinely human attitude, while practice is conceived and fixed only in its dirty-judaical manifestation. Hence he does not grasp the significance of 'revolutionary', of 'practical-critical', activity.

II

The question whether objective truth can be attributed to human thinking is not a question of theory but is a practical question. Man must prove the truth, i.e. the reality and power, the this-sidedness of his thinking in practice. The dispute over the reality or non-reality of thinking that is isolated from practice is a purely scholastic question.

III

The materialist doctrine concerning the changing of circumstances and up-bringing forgets that circumstances are changed by men and that it is essential to educate the educator himself. This doctrine must, therefore, divide society into two parts, one of which is superior to society.

The coincidence of the changing of circumstances and of human activity or self-changing can be conceived and rationally understood only as revolutionary practice.

IV

Feuerbach starts out from the fact of religious self-alienation, of the duplication of the world into a religious world and a secular one. His work consists in resolving the religious world into its secular basis. But that the secular basis detaches itself from itself and establishes itself as an independent realm in the clouds can only be explained by the cleavages and self-contradictions within this secular basis. The latter must, therefore, in itself be both understood in its contradiction and revolutionized in practice. Thus, for instance, after the earthly family is discovered to be the secret of the holy family, the former must then itself be destroyed in theory and in practice.

V

Feuerbach, not satisfied with abstract thinking, wants contemplation; but he does not conceive sensuousness as practical, human-sensuous activity.

VI

Feuerbach resolves the religious essence into the human essence. But the human essence is no abstraction inherent in each single individual. In its reality it is the ensemble of the social relations.

Feuerbach, who does not enter upon a criticism of this real essence, is consequently compelled:

1. To abstract from the historical process and to fix the religious sentiment as something by itself and to presuppose an abstract—isolated—human individual.

2. Essence, therefore, can be comprehended only as 'genus', as an internal, dumb generality which naturally unites the many individuals.

VII

Feuerbach, consequently, does not see that the 'religious sentiment' is itself a social product, and that the abstract individual whom he analyses belongs to a particular form of society.

VIII

All social life is essentially practical. All mysteries which lead theory to mysticism find their rational solution in human practice and in the comprehension of this practice.

IX

The highest point reached by contemplative materialism, that is, materialism which does not comprehend sensuousness as practical activity, is the contemplation of single individuals and of civil society.

X

The standpoint of the old materialism is civil society; the standpoint of the new is human society, or social humanity.

XI

The philosophers have only interpreted the world, in various ways; the point is to change it.

BIBLIOGRAPHY

Original: *MEW*, Vol. 3, pp. 533 ff.

Present translation: *The German Ideology*, trans. S. Ryazanskaya, Moscow, 1964, pp. 659 ff. Reproduced by kind permission of Lawrence and Wishart Ltd.

Other translations: *Writings of the Young Marx on Philosophy and Society*, ed. L. Easton and K. Guddatt, pp. 400 ff.

S. Hook, *From Hegel to Marx*, Ann Arbor, 1962, pp. 273 ff.

K. Marx, *Selected Writings in Sociology and Social Philosophy*, ed. T. Bottomore and M. Rubel, London, 1956, pp. 67 ff.

Commentaries: E. Bloch, *On Karl Marx*, New York, 1971, Ch. 4.

S. Hook, edition cited above, pp. 272 ff.

N. McInnes, *The Western Marxists*, London, 1973, Ch. 1.

N. Rotenstreich, *Basic Problems of Marx's Philosophy*, New York, 1965, pp. 27 ff.

14. *The German Ideology*

Marx and Engels declared *The German Ideology* to have been written 'to settle accounts with our former philosophical views'. It is no coincidence that the largest sections are devoted to Feuerbach and Stirner: *The Holy Family* was to have been their last publication on the subject of Young Hegelianism, but Stirner had published, in November 1844, *The Ego and its Own*, an anarcho-existentialist statement that branded Marx and Engels as disciples of Feuerbach and attracted a lot of attention in Germany. Marx therefore felt obliged to deal with Feuerbach and Stirner as a preliminary to his economic work. There was also a section on the 'true socialist' followers of Feuerbach who wished to base socialism on an ethical ideal. The book also had the practical political aim of clarifying socialist principles for the net of Communist Correspondence Committees that Marx and Engels had founded, and which were to become one of the ingredients of the Communist League.

By far the most important part of the book is the first section. This was nominally concerned with Feuerbach but in fact is an extensive description and definition of the newly worked-out materialist conception of history. Marx and Engels begin by making fun of the philosophical pretensions of the Young Hegelians; the main body of this section is then divided into three parts: a general statement of the historical and materialist approach in contrast to that of the Young Hegelians, a historical analysis employing this method, and an account of its immediate future—a communist revolution. The section on Stirner, on the other hand, takes up more than two-thirds of the book and is extremely tedious, its acres of diatribe being only rarely relieved by the few perspicacious comments extracted below.

From any standpoint on Marx's works, *The German Ideology* is one of his major achievements. Cutting through the cloudy metaphysics of so much Young Hegelian and even 'true socialist' writing, it sets out the materialist conception of history with a force and in a detail that Marx never afterwards surpassed. In spite of strenuous efforts, Marx and Engels did not succeed in finding a publisher for their manuscript and left it 'to the gnawing of the mice'. It was first published in 1932.

Preface

Hitherto men have constantly made up for themselves false conceptions about themselves, about what they are and what they ought to be. They have arranged their relationships according to their ideas of God, of normal man, etc. The phantoms of their brains have got out of their hands. They, the creators, have bowed down before their creations. Let us liberate them from the chimeras, the ideas, dogmas, imaginary beings under the yoke of which they are pining away. Let us revolt against the rule of thoughts. Let us teach men, says one, to exchange these imaginations for thoughts which correspond to the essence of man; says the second, to take up a critical attitude to them; says the the third, to knock them out of their heads; and—existing reality will collapse.

These innocent and childlike fancies are the kernel of the modern Young Hegelian philosophy, which not only is received by the German public with horror and awe, but is announced by our philosophic heroes with the solemn consciousness of its cataclysmic dangerousness and criminal ruthlessness. The first volume of the present publication has the aim of uncloaking these sheep, who take themselves and are taken for wolves; of showing how their bleating merely imitates in a philosophic form the conceptions of the German middle class; how the boasting of these philosophic commentators only mirrors the wretchedness of the real conditions in Germany. It is its aim to debunk and discredit the philosophic struggle with the shadows of reality, which appeals to the dreamy and muddled German nation.

Once upon a time a valiant fellow had the idea that men were drowned in water only because they were possessed with the idea of gravity. If they were to knock this notion out of their heads, say by stating it to be a superstition, a religious concept, they would be sublimely proof against any danger from water. His whole life long he fought against the illusion of gravity, of whose harmful results all statistics brought him new and manifold evidence. This honest fellow was the type of the new revolutionary philosophers in Germany. . . .

The Premises of the Materialist Method

The premises from which we begin are not arbitrary ones, not dogmas, but real premises from which abstraction can only be made in the imagination. They are the real individuals, their activity and the material conditions under which they live, both those which they find already existing and those produced by their activity. These premises can thus be verified in a purely empirical way.

The first premise of all human history is, of course, the existence of living human individuals. Thus the first fact to be established is the physical organization of these individuals and their consequent relation to the rest of nature. Of course, we cannot here go either into the actual physical nature of man, or into the natural conditions in which man finds himself—geological, oro-hydrographical, climatic, and so on. The writing of history must always set out from these natural bases and their modification in the course of history through the action of men.

Men can be distinguished from animals by consciousness, by religion, or anything else you like. They themselves begin to distinguish themselves from animals as soon as they begin to produce their means of subsistence, a step which is conditioned by their physical organization. By producing their means of subsistence men are indirectly producing their actual material life.

The way in which men produce their means of subsistence depends first of all on the nature of the actual means of subsistence they find in existence and have to reproduce. This mode of production must not be considered simply as being the production of the physical existence of the individuals. Rather it is a definite form of activity of these individuals, a definite form of expressing their life, a definite mode of life on their part. As individuals express their life, so they are. What they are, therefore, coincides with their production, both with *what* they produce and with *how* they produce. The nature of individuals thus depends on the material conditions determining their production.

This production only makes its appearance with the increase of population. In its turn this presupposes the intercourse of individuals with one another. The form of this intercourse is again determined by production.

The relations of different nations among themselves depend upon the extent to which each has developed its productive forces, the division of labour, and internal intercourse. This statement is generally recognized. But not only the relation of one nation to others, but also the whole internal structure of the nation itself depends on the stage of development reached by its production and its internal and external intercourse. How far the productive forces of a nation are developed is shown most manifestly by the degree to which the division of labour has been carried. Each new productive force, in so far as it is not merely a quantitative extension of productive forces already known (for instance the bringing into cultivation of fresh land), causes a further development of the division of labour.

The division of labour inside a nation leads at first to the separation of industrial and commercial from agricultural labour, and hence to the separation of town and country and to the conflict of their interests. Its further development leads to the separation of commercial from industrial labour. At the same time, through the division of labour inside these various branches there develop various divisions among the individuals co-operating in definite kinds of labour. The relative position of these individual groups is determined by the methods employed in agriculture, industry, and commerce (patriarchalism, slavery, estates, classes). These same conditions are to be seen (given a more developed intercourse) in the relations of different nations to one another.

The various stages of development in the division of labour are just so many different forms of ownership, i.e. the existing stage in the division of labour determines also the relations of individuals to one another with reference to the material, instrument, and product of labour.

The first form of ownership is tribal ownership. It corresponds to the undeveloped stage of production, at which a people lives by hunting and fishing, by the rearing of beasts, or, in the highest stage, agriculture. In the latter case it presupposes a great mass of uncultivated stretches of land. The division of labour is at this stage still very elementary and is confined to a further extension of the natural division of labour existing in the family. The social structure is, therefore, limited to an extension of the family; patriarchal family chieftains, below them the members of the tribe, finally slaves. The slavery latent in the family only develops gradually with the increase of population, the growth of wants, and with the extension of external relations, both of war and of barter.

The second form is the ancient communal and State ownership which proceeds especially from the union of several tribes into a city by agreement or by conquest, and which is still accompanied by slavery. Beside communal ownership we already find movable, and later also immovable, private property developing, but as an abnormal form subordinate to communal ownership. The citizens hold power over their labouring slaves only in their community, and on this account alone, therefore, they are bound to the form of communal ownership. It is the communal private property which compels the active citizens to remain in this spontaneously derived form of association over against their slaves. For this reason the whole structure of society based on this communal ownership, and with it the power of the people, decays in the same measure as, in particular, immovable private property evolves. The division of labour is already more developed. We already find the antagonism of town and country; later the antagonism between those states which represent town interests and those which represent country interests, and inside the towns themselves the antagonism between industry and maritime commerce. The class relation between citizens and slaves is now completely developed.

With the development of private property, we find here for the first time the same conditions which we shall find again, only on a more extensive scale, with modern private property. On the one hand, the concentration of private property, which began very early in Rome (as the Licinian agrarian law proves) and proceeded very rapidly from the time of the civil wars and especially under the Emperors; on the other hand, coupled with this, the transformation of the plebeian small peasantry into a proletariat, which, however, owing to its intermediate position between propertied citizens and slaves, never achieved an independent development.

The third form of ownership is feudal or estate property. If antiquity started out from the town and its little territory, the Middle Ages started out from the country. This differing starting-point was determined by the sparseness of the population at that time, which was scattered over a large area and which received no large increase from the conquerors. In contrast to Greece and Rome, feudal development at the outset, therefore, extends over a much wider territory, prepared by the Roman conquests and the spread of agriculture at first associated with it. The last centuries of the declining Roman Empire and its conquest by the barbarians destroyed a number of productive forces; agriculture had declined, industry had decayed for want of a market, trade had died out or been violently suspended, the rural and urban population had decreased. From these conditions and the mode of organization of the conquest determined by them, feudal property developed under the influence of the Germanic military constitution. Like tribal and communal ownership, it is based again on a community; but the directly producing class standing over against it is not, as in the case of the ancient community, the slaves, but the enserfed small peasantry. As soon as feudalism is fully developed, there also arises antagonism towards the towns. The hierarchical structure of landownership, and the armed bodies of retainers associated with it, gave the nobility power over the serfs. This feudal organization was, just as much as the ancient communal ownership, an association against a subjected producing class; but the form of association and the relation to the direct producers were different because of the different conditions of production.

This feudal system of landownership had its counterpart in the towns in the shape of corporative property, the feudal organization of trades. Here property consisted chiefly in the labour of each individual person. The necessity for association against the organized robber barons, the need for communal covered markets in an age when the industrialist was at the same time a merchant, the growing competition of the escaped serfs swarming into the rising towns, the feudal structure of the whole country: these combined to bring about the guilds. The gradually accumulated small capital of individual craftsmen and their stable numbers, as against the growing population, evolved the relation of journeyman and apprentice, which brought into being in the towns a hierarchy similar to that in the country.

Thus the chief form of property during the feudal epoch consisted on the one hand of landed property with serf labour chained to it, and on the other of the labour of the individual with small capital commanding the labour of journeymen. The organization of both was determined by the restricted conditions of production—the small-scale and primitive cultivation of the land and the craft type of industry. There was little division of labour in the heyday of feudalism. Each country bore in itself the antithesis of town and country; the division into estates was certainly strongly marked; but apart from the differentiation of princes, nobility, clergy, and peasants in the country, and masters, journeymen, apprentices, and soon also the rabble of casual labourers in the towns, no division of importance took place. In agriculture it was rendered difficult by the strip-system, beside which the cottage industry of the peasants themselves emerged. In industry there was no division of labour at all in the individual trades themselves, and very little between them. The separation of industry and commerce was found already in existence in older towns; in the newer it only developed later, when the towns entered into mutual relations.

The grouping of larger territories into feudal kingdoms was a necessity for the landed nobility as for the towns. The organization of the ruling class, the nobility, had, therefore, everywhere a monarch at its head.

The fact is, therefore, that definite individuals who are productively active in a definite way enter into these definite social and political relations. Empirical observation must in each separate instance bring out empirically, and without any mystification and speculation, the connection of the social and political structure with production. The social structure and the State are continually evolving out of the life-process of definite individuals, but of individuals, not as they may appear in their own or other people's imagination, but as they really are, i.e. as they operate, produce materially, and hence as they work under definite material limits, presuppositions, and conditions independent of their will.

The production of ideas, of conceptions, of consciousness, is at first directly interwoven with the material activity and the material intercourse of men, the language of real life. Conceiving, thinking, the mental intercourse of men, appear at this stage as the direct efflux of their material behaviour. The same applies to mental production as expressed in the language of politics, laws, morality, religion, metaphysics, etc. of a people. Men are the producers of their conceptions, ideas, etc.—real, active men, as they are conditioned by a definite development of their productive forces and of the intercourse corresponding to these, up to its furthest forms. Consciousness can never be anything else than conscious existence, and the existence of men is their actual life-process. If in all ideology men and their circumstances appear upside-down as in a *camera obscura*, this phenomenon arises just as much from their historical life-process as the inversion of objects on the retina does from their physical life-process.

In direct contrast to German philosophy which descends from heaven to earth, here we ascend from earth to heaven. That is to say, we do not set out from what men say, imagine, conceive, nor from men as narrated, thought of, imagined, conceived, in order to arrive at men in the flesh. We set out from real, active men, and on the basis of their real life-process we demonstrate the development of the ideological reflexes and echoes of this life-process. The phantoms formed in the human brain are also, necessarily, sublimates of their material life-process, which is empirically verifiable and bound to material premises. Morality, religion, metaphysics, all the rest of ideology and their corresponding forms of consciousness, thus no longer retain the semblance of independence. They have no history, no development; but men, developing their material production and their material intercourse, alter, along with this their real existence, their thinking and the products of their thinking. Life is not determined by consciousness, but consciousness by life. In the first method of approach the starting-point is consciousness taken as the living individual; in the second method, which conforms to real life, it is the real living individuals themselves, and consciousness is considered solely as their consciousness.

This method of approach is not devoid of premises. It starts out from the real premises and does not abandon them for a moment. Its premises are men, not in any fantastic isolation and rigidity, but in their actual, empirically perceptible process of development under definite conditions. As soon as this active life-process is described, history ceases to be a collection of dead facts as it is with the empiricists (themselves still abstract), or an imagined activity of imagined subjects, as with the idealists.

Where speculation ends—in real life—there real, positive science begins: the representation of the practical activity, of the practical process of development of men. Empty talk about consciousness ceases, and real knowledge has to take its place. When reality is depicted, philosophy as an independent branch of knowledge loses its medium of existence. At the best its place can only be taken by a summing-up of the most general results, abstractions which arise from the observation of the historical development of men. Viewed apart from real history, these abstractions have in themselves no value whatsoever. They can only serve to facilitate the arrangement of historical material, to indicate the sequence of its separate strata. But they by no means afford a recipe or schema, as does philosophy, for neatly trimming the epochs of history. On the contrary, our difficulties begin only when we set about the observation and the arrangement—the real depiction —of our historical material, whether of a past epoch or of the present. The removal of these difficulties is governed by premisses which it is quite impossible to state here, but which only the study of the actual life-process and the activity of the individuals of each epoch will make evident. We shall select here some of these abstractions, which we use in contradistinction to the ideologists, and shall illustrate them by historical examples.

Since we are dealing with the Germans, who are devoid of premisses, we must begin by stating the first premiss of all human existence and, therefore, of all history, the premiss, namely, that men must be in a position to live in order to be able to 'make history'. But life involves before everything else eating and drinking, a habitation, clothing, and many other things. The first historical act is thus the production of the means to satisfy these needs, the production of material life itself. And indeed this is an historical act, a fundamental condition of all history, which today, as thousands of years ago, must daily and hourly be fulfilled merely in order to sustain human life. Even when the sensuous world is reduced to a minimum, to a stick as with Saint Bruno, it presupposes the action of producing the stick. Therefore in any interpretation of history one has first of all to observe this fundamental fact in all its significance and all its implications and to accord it its due importance. It is well known that the Germans have never done this, and they have never, therefore, had an earthly basis for history and consequently never an historian. The French and the English, even if they have conceived the relation of this fact with so-called history only in an extremely one-sided fashion, particularly as long as they remained in the toils of political ideology, have nevertheless made the first attempts to give the writing of history a materialistic basis by being the first to write histories of civil society, of commerce and industry.

The second point is that the satisfaction of the first need (the action of satisfying, and the instrument of satisfaction which has been acquired) leads to new needs; and this production of new needs is the first historical act. Here we recognize immediately the spiritual ancestry of the great historical wisdom of the Germans who, when they run out of positive material and when they can serve up neither theological nor political nor literary rubbish, assert that this is not history at all, but the 'prehistoric era'. They do not, however, enlighten us as to how we proceed from this nonsensical 'prehistory' to history proper; although, on the other hand, in their historical speculation they seize upon this 'prehistory' with especial eagerness because they imagine themselves safe there from interference on the part of 'crude facts', and, at the same time, because there they can give full rein to their speculative impulse and set up and knock down hypotheses by the thousand.

The third circumstance which, from the very outset, enters into historical development, is that men, who daily remake their own life, begin to make other men, to propagate their kind: the relation between man and woman, parents and children, the family. The family, which to begin with is the only social relationship, becomes later, when increased needs create new social relations and the increased population new needs, a subordinate one (except in Germany), and must then be treated and analysed according to the existing empirical data, not according to 'the concept of the family', as is the custom in Germany. These three aspects of social activity are not of course to be taken as three different stages, but just as three aspects or, to make it clear to the Germans, three 'moments', which have existed simultaneously since the dawn of history and the first men, and which still assert themselves in history today.

The production of life, both of one's own in labour and of fresh life in procreation, now appears as a double relationship: on the one hand as a natural, on the other as a social, relationship. By social we understand the co-operation of several individuals, no matter under what conditions, in what manner, and to what end. It follows from this that a certain mode of production, or industrial stage, is always combined with a certain mode of co-operation, or social stage, and this mode of co-operation is itself a 'productive force'. Further, that the multitude of productive forces accessible to men determines the nature of society, hence, that the 'history of humanity' must always be studied and treated in relation to the history of industry and exchange. But it is also clear how in Germany it is impossible to write this sort of history, because the Germans lack not only the necessary power of comprehension and the material but also the 'evidence of their senses', for across the Rhine you cannot have any experience of these things since history has stopped happening. Thus it is quite obvious from the start that there exists a materialistic connection of men with one another, which is determined by their needs and their mode of production, and which is as old as men themselves. This connection is ever taking on new forms, and thus presents a 'history' independently of the existence of any political or religious nonsense which in addition may hold men together.

Only now, after having considered four moments, four aspects of the primary historical relationships, do we find that man also possesses 'consciousness', but, even so, not inherent, not 'pure' consciousness. From the start the 'spirit' is afflicted with the curse of being 'burdened' with matter, which here makes its appearance in the form of agitated layers of air, sounds, in short, of language. Language is as old as consciousness, language is practical consciousness that exists also for other men, and for that reason alone it really exists for me personally as well; language, like consciousness, only arises from the need, the necessity, of intercourse with other men. Where there exists a relationship, it exists for me: the animal does not enter into 'relations' with anything, it does not enter into any relation at all. For the animal, its relation to others does not exist as a relation. Consciousness is, therefore, from the very beginning a social product, and remains so as long as men exist at all. Consciousness is at first, of course, merely consciousness concerning the immediate sensuous environment and consciousness of the limited connection with other persons and things outside the individual who is growing self-conscious. At the same time it is consciousness of nature, which first appears to men as a completely alien, all-powerful, and unassailable force, with which men's relations are purely animal and by which they are overawed like beasts; it is thus a purely animal consciousness of nature (natural religion) just because nature is as yet hardly modified historically. (We see here immediately that this natural religion or this particular relation of men to nature is determined by the form of society and vice versa. Here, as everywhere, the identity of nature and man appears in such a way that the restricted relation of men to nature determines their restricted relation to one another, and their restricted relation to one another determines men's restricted relation to nature.) On the other hand, man's consciousness of the necessity of associating with the individuals around him is the beginning of the consciousness that he is living in society at all. This beginning is as animal as social life itself at this stage. It is mere herd-consciousness, and at this point man is only distinguished from sheep by the fact that with him consciousness takes the place of instinct or that his instinct is a conscious one. This sheep-like or tribal consciousness receives its further development and extension through increased productivity, the increase of needs, and, what is fundamental to both of these, the increase of population. With these there develops the division of labour, which was originally nothing but the division of labour in the sexual act, then that division of labour which develops spontaneously or 'naturally' by virtue of natural predisposition (e.g. physical strength), needs, accidents, etc. etc. Division of labour only becomes truly such from the moment when a division of material and mental labour appears. (The first form of ideologists, priests, is concurrent.) From this moment onwards consciousness can really flatter itself that it is something other than consciousness of existing practice, that it really represents something without representing something real; from now on consciousness is in a position to emancipate itself from the world and to proceed to the formation of 'pure' theory, theology, philosophy, ethics, etc. But even if this theory, theology, philosophy, ethics, etc. comes into contradiction with the existing relations, this can only occur because existing social relations have come into contradiction with existing forces of production; this, moreover, can also occur in a particular national sphere of relations through the appearance of the contradiction, not within the national orbit, but between this national consciousness and the practice of other nations, i.e. between the national and the general consciousness of a nation (as we see it now in Germany).

Moreover, it is quite immaterial what consciousness starts to do on its own: out of all such muck we get only the one inference that these three moments, the forces of production, the state of society, and consciousness, can and must come into contradiction with one another, because the division of labour implies the possibility, nay the fact, that intellectual and material activity—enjoyment and labour, production and consumption—devolve on different individuals, and that the only possibility of their not coming into contradiction lies in the negation in its turn of the division of labour. It is self-evident, moreover, that 'spectres', 'bonds', 'the higher being', 'concept', 'scruple', are merely the idealistic, spiritual expression, the conception apparently of the isolated individual, the image of very empirical fetters and limitations, within which the mode of production of life and the form of intercourse coupled with it move.

Private Property and Communism

With the division of labour, in which all these contradictions are implicit, and which in its turn is based on the natural division of labour in the family and the separation of society into individual families opposed to one another, is given simultaneously the distribution, and indeed the unequal distribution, both quantitative and qualitative, of labour and its products, hence property: the nucleus, the first form of which lies in the family, where wife and children are the slaves of the husband. This latent slavery in the family, though still very crude, is the first property, but even at this early stage it corresponds perfectly to the definition of modern economists who call it the power of disposing of the labour-power of others. Division of labour and private property are, moreover, identical expressions: in the one the same thing is affirmed with reference to activity as is affirmed in the other with reference to the product of the activity.

Further, the division of labour implies the contradiction between the interest of the separate individual or the individual family and the communal interest of all individuals who have intercourse with one another. And indeed, this communal interest does not exist merely in the imagination, as the 'general interest', but first of all in reality, as the mutual interdependence of the individuals among whom the labour is divided. And finally, the division of labour offers us the first example of how, as long as man remains in natural society, that is, as long as a cleavage exists between the particular and the common interest, as long, therefore, as activity is not voluntarily, but naturally, divided, man's own deed becomes an alien power opposed to him, which enslaves him instead of being controlled by him. For as soon as the distribution of labour comes into being, each man has a particular, exclusive sphere of activity, which is forced upon him and from which he cannot escape. He is a hunter, a fisherman, a shepherd, or a critical critic, and must remain so if he does not want to lose his means of livelihood; while in communist society, where nobody was one exclusive sphere of activity but each can become accomplished in any branch he wishes, society regulates the general production and thus makes it possible for me to do one thing today and another tomorrow, to hunt in the morning, fish in the afternoon, rear cattle in the evening, criticize after dinner, just as I have a mind, without ever becoming hunter, fisherman, cowherd, or critic. This fixation of social activity, this consolidation of what we ourselves produce into an objective power above us, growing out of our control, thwarting our expectations, bringing to naught our calculations, is one of the chief factors in historical development up till now.

And out of this very contradiction between the interest of the individual and that of the community the latter takes an independent form as the State, divorced from the real interests of individual and community, and at the same time as an illusory communal life, always based, however, on the real ties existing in every family and tribal conglomeration—such as flesh and blood, language, division of labour on a larger scale, and other interests— and especially, as we shall enlarge upon later, on the classes, already determined by the division of labour, which in every such mass of men separate out, and of which one dominates all the others. It follows from this that all struggles within the State, the struggle between democracy, aristocracy, and monarchy, the struggle for the franchise, etc. etc. are merely the illusory forms in which the real struggles of the different classes are fought out among one another. Of this the German theoreticians have not the faintest inkling, although they have received a sufficient introduction to the subject in the *Deutsch–französische Jahrbücher* and *Die heilige Familie*. Further, it follows that every class which is struggling for mastery, even when its domination, as is the case with the proletariat, postulates the abolition of the old form of society in its entirety and of domination itself, must first conquer for itself political power in order to represent its interest in turn as the general interest, which immediately it is forced to do. Just because individuals seek only their particular interest, which for them does not coincide with their communal interest, the latter will be imposed on them as an interest 'alien' to them, and 'independent' of them, as in its turn a particular, peculiar 'general' interest; or they themselves must remain within this discord, as in democracy. On the other hand, too, the practical struggle of these particular interests, which constantly really run counter to the communal and illusory communal interests, makes practical intervention and control necessary through the illusory 'general' interest in the form of the State.

The social power, i.e. the multiplied productive force, which arises through the co-operation of different individuals as it is determined by the division of labour, appears to these individuals, since their co-operation is not voluntary but has come about naturally, not as their own united power, but as an alien force existing outside them, of the origin and goal of which they are ignorant, which they thus cannot control, which on the contrary passes through a peculiar series of phases and stages independent of the will and the action of man, nay even being the prime governor of these.

How otherwise could, for instance, property have had a history at all, have taken on different forms, and landed property, for example, according to the different premises given, have proceeded in France from parcellation to centralization in the hands of a few, in England from centralization in the hands of a few to parcellation, as is actually the case today? Or how does it happen that trade, which after all is nothing more than the exchange of products of various individuals and countries, rules the whole world through the relation of supply and demand—a relation which, as an English economist says, hovers over the earth like the Fates of the ancients, and with invisible hand allots fortune and misfortune to men, sets up empires and overthrows empires, causes nations to rise and to disappear—while with the abolition of the basis of private property, with the communistic regulation of production (and, implicit in this, the destruction of the alien relation between men and what they themselves produce), the power of the relation of supply and demand is dissolved into nothing, and men get exchange, production, the mode of their mutual relation, under their own control again?

This 'alienation' (to use a term which will be comprehensible to the philosophers) can, of course, only be abolished given two practical premises. For it to become an 'intolerable' power, i.e. a power against which men make a revolution, it must necessarily have rendered the great mass of humanity 'propertyless', and produced, at the same time, the contradiction of an existing world of wealth and culture, both of which conditions presuppose a great increase in productive power, a high degree of its development. And, on the other hand, this development of productive forces (which itself implies the actual empirical existence of men in their world-historical, instead of local, being) is an absolutely necessary practical premiss because without it want is merely made general, and with destitution the struggle for necessities and all the old filthy business would necessarily be reproduced; and furthermore, because only with this universal development of productive forces is a universal intercourse between men established, which produces in all nations simultaneously the phenomenon of the 'propertyless' mass (universal competition), makes each nation dependent on the revolutions of the others, and finally has put world-historical, empirically universal individuals in place of local ones. Without this, (1) communism could only exist as a local event; (2) the forces of intercourse themselves could not have developed as universal, hence intolerable powers: they would have remained home-bred conditions surrounded by superstition; and (3) each extension of intercourse would abolish local communism. Empirically, communism is only possible as the act of the dominant peoples 'all at once' and simultaneously, which presupposes the universal development of productive forces and the world intercourse bound up with communism. Moreover, the mass of propertyless workers—the utterly precarious position of labour-power on a mass scale cut off from capital or from even a limited satisfaction and, therefore, no longer merely temporarily deprived of work itself as a secure source of life—presupposes the world market through competition. The proletariat can thus only exist world-historically, just as communism, its activity, can only have a 'world-historical' existence. World-historical existence of individuals means existence of individuals which is directly linked up with world history.

Communism is for us not a state of affairs which is to be established, an ideal to which reality will have to adjust itself. We call communism the real movement which abolishes the present state of things. The conditions of this movement result from the premisses now in existence.

Communism and History

. . . In history up to the present it is certainly an empirical fact that separate individuals have, with the broadening of their activity into world-historical activity, become more and more enslaved under a power alien to them (a pressure which they have conceived of as a dirty trick on the part of the so-called universal spirit, etc.), a power which has become more and more enormous and, in the last instance, turns out to be the world market. But it is just as empirically established that, by the overthrow of the existing state of society by the communist revolution (of which more below) and the abolition of private property which is identical with it, this power, which so baffles the German theoreticians, will be dissolved; and that then the liberation of each single individual will be accomplished in the measure in which history becomes transformed into world history. From the above it is clear that the real intellectual wealth of the individual depends entirely on the wealth of his real connections. Only then will the separate individuals be liberated from the various national and local barriers, be brought into practical connection with the material and intellectual production of the whole world and be put in a position to acquire the capacity to enjoy this all-sided production of the whole earth (the creations of man). All-round dependence, this natural form of the world-historical co-operation of individuals, will be transformed by this communist revolution into the control and conscious mastery of these powers, which, born of the action of men on one another, have till now overawed and governed men as powers completely alien to them. Now this view can be expressed again in speculative-idealistic, i.e. fantastic, terms as 'self-generation of the species' ('society as the subject'), and thereby the consecutive series of interrelated individuals connected with each other can be conceived as a single individual, which accomplishes the mystery of generating itself. It is clear here that individuals certainly make one another, physically and mentally, but do not make themselves either in the nonsense of Saint Bruno, or in the sense of the 'Unique', of the 'made' man.

This conception of history depends on our ability to expound the real process of production, starting out from the material production of life itself, and to comprehend the form of intercourse connected with this and created by this mode of production (i.e. civil society in its various stages) as the basis of all history; and to show it in its action as State, to explain all the different theoretical products and forms of consciousness, religion, philosophy, ethics, etc., etc., and trace their origins and growth from that basis; by which means, of course, the whole thing can be depicted in its totality (and therefore, too, the reciprocal action of these various sides on one another). It has not, like the idealistic view of history, in every period to look for a category, but remains constantly on the real ground of history; it does not explain practice from the idea but explains the formation of ideas from material practice; and accordingly it comes to the conclusion that all forms and products of consciousness cannot be dissolved by mental criticism, by resolution into 'self-consciousness' or transformation into 'apparitions', 'spectres', 'fancies', etc., but only by the practical overthrow of the actual social relations which gave rise to this idealistic humbug; that not criticism but revolution is the driving force of history, also of religion, of philosophy and all other types of theory. It shows that history does not end by being resolved into 'self-consciousness' as 'spirit of the spirit', but that in it at each stage there is found a material result: a sum of productive forces, an historically created relation of individuals to nature and to one another, which is handed down to each generation from its predecessor; a mass of productive forces, capital funds and conditions, which, on the one hand, is indeed modified by the new generation, but also, on the other, prescribes for it its conditions of life and gives it a definite development, a special character. It shows that circumstances make men just as much as men make circumstances.

This sum of productive forces, capital funds, and social forms of intercourse, which every individual and generation finds in existence as something given, is the real basis of what the philosophers have conceived as 'substance' and 'essence of man', and what they have deified and attacked; a real basis which is not in the least disturbed, in its effect and influence on the development of men, by the fact that these philosophers revolt against it as 'self-consciousness' and the 'Unique'. These conditions of life, which different generations find in existence, decide also whether or not the periodically recurring revolutionary convulsion will be strong enough to overthrow the basis of the entire existing system. And if these material elements of a complete revolution are not present (namely, on the one hand the existing productive forces, on the other the formation of a revolutionary mass, which revolts not only against separate conditions of society up till then, but against the very 'production of life' till then, the 'total activity' on which it was based), then, as far as practical development is concerned, it is absolutely immaterial whether the idea of this revolution has been expressed a hundred times already, as the history of communism proves.

In the whole conception of history up to the present this real basis of history has either been totally neglected or else considered as a minor matter quite irrelevant to the course of history. History must, therefore, always be written according to an extraneous standard; the real production of life seems to be primeval history, while the truly historical appears to be separated from ordinary life, something superterrestrial. With this the relation of man to nature is excluded from history and hence the antithesis of nature and history is created. The exponents of this conception of history have consequently only been able to see in history the political actions of princes and States, religious and all sorts of theoretical struggles, and in particular in each historical epoch have had to share the illusion of that epoch. For instance, if an epoch imagines itself to be actuated by purely 'political' or 'religious' motives, although 'religion' and 'politics' are only forms of its true motives, the historian accepts this opinion. The 'idea', the 'conception' of the people in question about their real practice, is transformed into the sole determining, active force, which controls and determines their practice. When the crude form in which the division of labour appears with the Indians and Egyptians calls forth the caste system in their State and religion, the historian believes that the caste system is the power which has produced this crude social form. While the French and the English at least hold by the political illusion, which is moderately close to reality, the Germans move in the realm of the 'pure spirit', and make religious illusion the driving force of history. The Hegelian philosophy of history is the last consequence, reduced to its 'finest expression', of all this German historiography, for which it is not a question of real, nor even of political, interests, but of pure thoughts, which consequently must appear to Saint Bruno as a series of 'thoughts' that devour one another and are finally swallowed up in 'self-consciousness'. . . .

In reality and for the practical materialist, i.e. the communist, it is a question of revolutionizing the existing world, of practically attacking and changing existing things. When occasionally we find such views with Feuerbach, they are never more than isolated surmises and have much too little influence on his general outlook to be considered here as anything else than embryos capable of development. Feuerbach's 'conception' of the sensuous world is confined on the one hand to mere contemplation of it, and on the other to mere feeling; he says 'Man' instead of 'real historical man'. 'Man' is really 'the German'. In the first case, the contemplation of the sensuous world, he necessarily lights on things which contradict his consciousness and feeling, which disturb the harmony he presupposes, the harmony of all parts of the sensuous world and especially of man and nature. To remove this disturbance, he must take refuge in a double perception, a profane one which only perceives the 'flatly obvious' and a higher, philosophical, one which perceives the 'true essence' of things. He does not see how the sensuous world around him is not a thing given direct from all eternity, remaining ever the same, but the product of industry and of the state of society; and, indeed, in the sense that it is an historical product, the result of the activity of a whole succession of generations, each standing on the shoulders of the preceding one, developing its industry and its intercourse, modifying its social system according to the changed needs. Even the objects of the simplest 'sensuous certainty' are only given him through social development, industry, and commercial intercourse. The cherry-tree, like almost all fruit-trees, was, as is well known, only a few centuries ago transplanted by commerce into our zone, and therefore only by this action of a definite society in a definite age it has become 'sensuous certainty' for Feuerbach.

Incidentally, when we conceive things thus, as they really are and happened, every profound philosophical problem is resolved, as will be seen even more clearly later, quite simply into an empirical fact. For instance, the important question of the relation of man to nature (Bruno [Bauer] goes so far as to speak of 'the antitheses in nature and history' (p. 110), as though these were two separate 'things' and man did not always have before him an historical nature and a natural history) out of which all the 'unfathomably lofty works' on 'substance' and 'self-consciousness' were born, crumbles of itself when we understand that the celebrated 'unity of man with nature' has always existed in industry and has existed in varying forms in every epoch according to the lesser or greater development of industry, just like the 'struggle' of man with nature, right up to the development of his productive powers on a corresponding basis. Industry and commerce, production and the exchange of the necessities of life, themselves determine distribution, the structure of the different social classes, and are, in turn, determined by it as to the mode in which they are carried on; and so it happens that in Manchester, for instance, Feuerbach sees only factories and machines, where a hundred years ago only spinning-wheels and weaving-looms were to be seen, or in the Campagna of Rome he finds only pasture lands and swamps, where in the time of Augustus he would have found nothing but the vine-yards and villas of Roman capitalists. Feuerbach speaks in particular of the perception of natural science; he mentions secrets which are disclosed only to the eye of the physicist and chemist; but where would natural science be without industry and commerce? Even this 'pure' natural science is provided with an aim, as with its material, only through trade and industry, through the sensuous activity of men. So much is this activity, this unceasing sensuous labour and creation, this production, the basis of the whole sensuous world as it now exists, that, were it interrupted only for a year, Feuerbach would not only find an enormous change in the natural world, but would very soon find that the whole world of men and his own perceptive faculty, nay his own existence, were missing. Of course, in all this the priority of external nature remains unassailed, and all this has no application to the original men produced by *generatio aequivoca* [spontaneous generation]; but this differentiation has meaning only in so far as man is considered to be distinct from nature. For that matter, nature, the nature that preceded human history, is not by any means the nature in which Feuerbach lives, it is nature which today no longer exists anywhere (except perhaps on a few Australian coral islands of recent origin) and which, therefore, does not exist for Feuerbach.

Certainly Feuerbach has a great advantage over the 'pure' materialists in that he realizes how man too is an 'object of the senses'. But apart from the fact that he only conceives him as an 'object of the senses', not as 'sensuous activity', because he still remains in the realm of theory and conceives of men not in their given social connection, not under their existing conditions of life, which have made them what they are, he never arrives at the really existing active men, but stops at the abstraction 'man', and gets no further than recognizing 'the true, individual, corporeal man' emotionally, i.e. he knows no other 'human relationships' 'of man to man' than love and friend-ship, and even then idealized. He gives no criticism of the present conditions of life. Thus he never manages to conceive the sensuous world as the total living sensuous activity of the individuals composing it; and therefore when, for example, he sees instead of healthy men a crowd of scrofulous, over-worked, and consumptive starvelings, he is compelled to take refuge in the 'higher perception' and in the ideal 'compensation in the species', and thus to relapse into idealism at the very point where the communist materialist sees the necessity, and at the same time the condition, of a transformation both of industry and of the social structure.

As far as Feuerbach is a materialist he does not deal with history, and as far as he considers history he is not a materialist. With him materialism and history diverge completely, a fact which incidentally is already obvious from what has been said. . . .

The ideas of the ruling class are in every epoch the ruling ideas, i.e. the class which is the ruling material force of society is at the same time its ruling intellectual force. The class which has the means of material production at its disposal, has control at the same time over the means of mental production, so that thereby, generally speaking, the ideas of those who lack the means of mental production are subject to it. The ruling ideas are nothing more than the ideal expression of the dominant material relationships, the dominant material relationships grasped as ideas; hence of the relationships which make the one class the ruling one, therefore, the ideas of its dominance. The individuals composing the ruling class possess among other things consciousness, and therefore think. In so far, therefore, as they rule as a class and determine the extent and compass of an epoch, it is self-evident that they do this in its whole range, hence among other things rule also as thinkers, as producers of ideas, and regulate the production and distribution of the ideas of their age: thus their ideas are the ruling ideas of the epoch. For instance, in an age and in a country where royal power, aristocracy, and bourgeoisie are contending for mastery and where, therefore, mastery is shared, the doctrine of the separation of powers proves to be the dominant idea and is expressed as an 'eternal law' . . .

Our investigation hitherto started from the instruments of production, and it has already shown that private property was a necessity for certain industrial stages. In *industrie extractive* [raw materials industry] private property still coincides with labour; in small industry and all agriculture up till now property is the necessary consequence of the existing instruments of production; in big industry the contradiction between the instrument of production and private property appears for the first time and is the product of big industry; moreover, big industry must be highly developed to produce this contradiction. And thus only with big industry does the abolition of private property become possible.

In big industry and competition the whole mass of conditions of existence, limitations, biases of individuals, are fused together into the two simplest forms: private property and labour. With money every form of intercourse, and intercourse itself, is considered fortuitous for the individuals. Thus money implies that all previous intercourse was only intercourse of individuals under particular conditions, not of individuals as individuals. These conditions are reduced to two: accumulated labour or private property, and actual labour. If both or one of these ceases, then intercourse comes to a standstill. The modern economists themselves, e.g. Sismondi, Cherbuliez, etc., oppose 'association of individuals' to 'association of capital'.

On the other hand, the individuals themselves are entirely subordinated to the division of labour and hence are brought into the most complete dependence on one another. Private property, in so far as within labour itself it is opposed to labour, evolves out of the necessity of accumulation, and has still, to begin with, rather the form of the communality; but in its further development it approaches more and more the modern form of private property. The division of labour implies from the outset the division of the conditions of labour, of tools and materials, and thus the splitting-up of accumulated capital among different owners, and thus, also, the division between capital and labour, and the different forms of property itself. The more the division of labour develops and accumulation grows, the sharper are the forms that this process of differentiation assumes. Labour itself can only exist on the premiss of this fragmentation.

Thus two facts are here revealed. First the productive forces appear as a world for themselves, quite independent of and divorced from the individuals, alongside the individuals: the reason for this is that the individuals, whose forces they are, exist split up and in opposition to one another, while, on the other hand, these forces are only real forces in the intercourse and association of these individuals. Thus, on the one hand, we have a totality of productive forces, which have, as it were, taken on a material form and are for the individuals no longer the forces of the individuals but of private property, and hence of the individuals only in so far as they are owners of private property themselves. Never, in any earlier period, have the productive forces taken on a form so indifferent to the intercourse of individuals as individuals, because their intercourse itself was formerly a restricted one. On the other hand, standing over against these productive forces, we have the majority of the individuals from whom these forces have been wrested away, and who, robbed thus of all real life-content, have become abstract individuals, but who are, however, only by this fact put into a position to enter into relation with one another as individuals.

The only connection which still links them with the productive forces and with their own existence—labour—has lost all semblance of self-activity and only sustains their life by stunting it. While in the earlier periods self-activity and the production of material life were separated, in that they devolved on different persons, and while, on account of the narrowness of the individuals themselves, the production of material life was considered as a subordinate mode of self-activity, they now diverge to such an extent that altogether material life appears as the end, and what produces this material life, labour (which is now the only possible but, as we see, negative form of self-activity), as the means.

Thus things have now come to such a pass that the individuals must appropriate the existing totality of productive forces, not only to achieve self-activity, but also merely to safeguard their very existence. This appropriation is first determined by the object to be appropriated, the productive forces, which have been developed to a totality and which only exist within a universal intercourse. From this aspect alone, therefore, this appropriation must have a universal character corresponding to the productive forces and the intercourse.

The appropriation of these forces is itself nothing more than the development of the individual capacities corresponding to the material instruments of production. The appropriation of a totality of instruments of production is, for this very reason, the development of a totality of capacities in the individuals themselves.

This appropriation is further determined by the persons appropriating. Only the proletarians of the present day, who are completely shut off from all self-activity, are in a position to achieve a complete and no longer restricted self-activity, which consists in the appropriation of a totality of productive forces and in the thus postulated development of a totality of capacities. All earlier revolutionary appropriations were restricted; individuals, whose self-activity was restricted by a crude instrument of production and a limited intercourse, appropriated this crude instrument of production, and hence merely achieved a new state of limitation. Their instrument of production became their property, but they themselves remained subordinate to the division of labour and their own instrument of production. In all expropriations up to now, a mass of individuals remained subservient to a single instrument of production; in the appropriation by the proletarians, a mass of instruments of production must be made subject to each individual, and property to all. Modern universal intercourse can be controlled by individuals, therefore, only when controlled by all.

This appropriation is further determined by the manner in which it must be effected. It can only be effected through a union, which by the character of the proletariat itself can again only be a universal one, and through a revolution, in which, on the one hand, the power of the earlier mode of production and intercourse and social organization is overthrown, and, on the other hand, there develops the universal character and the energy of the proletariat, without which the revolution cannot be accomplished; and in which, further, the proletariat rids itself of everything that still clings to it from its previous position in society.

Only at this stage does self-activity coincide with material life, which corresponds to the development of individuals into complete individuals and the casting-off of all natural limitations. The transformation of labour into self-activity corresponds to the transformation of the earlier limited intercourse into the intercourse of individuals as such. With the appropriation of the total productive forces through united individuals, private property comes to an end. While previously in history a particular condition always appeared as accidental, now the isolation of individuals and the particular private gain of each man have themselves become accidental. . . .

Finally, from the conception of history we have sketched we obtain these further conclusions: (1) In the development of productive forces there comes a stage when productive forces and means of intercourse are brought into being, which, under the existing relationships, only cause mischief, and are no longer productive but destructive forces (machinery and money); and connected with this a class is called forth, which has to bear all the burdens of society without enjoying its advantages, which, ousted from society, is forced into the most decided antagonism to all other classes; a class which forms the majority of all members of society, and from which emanates the consciousness of the necessity of a fundamental revolution, the communist consciousness, which may, of course, arise among the other classes too through the contemplation of the situation of this class. (2) The conditions under which definite productive forces can be applied are the conditions of the rule of a definite class of society, whose social power, deriving from its property, has its practical-idealistic expression in each case in the form of the State; and, therefore, every revolutionary struggle is directed against a class, which till then has been in power. (3) In all revolutions up till now the mode of activity always remained unscathed and it was only a question of a different distribution of this activity, a new distribution of labour to other persons, while the communist revolution is directed against the preceding mode of activity, does away with labour, and abolishes the rule of all classes with the classes themselves, because it is carried through by the class which no longer counts as a class in society, is not recognized as a class, and is in itself the expression of the dissolution of all classes, nationalities, etc. within present society; and (4) Both for the production on a mass scale of this communist consciousness, and for the success of the cause itself, the alternation of men on a mass scale is necessary, an alteration which can only take place in a practical movement, a revolution; this revolution is necessary, therefore, not only because the ruling class cannot be overthrown in any other way, but also because the class overthrowing it can only in a revolution succeed in ridding itself of all the muck of ages and become fitted to found society anew. . . .

Communist Revolution

. . . Communism differs from all previous movements in that it overturns the basis of all earlier relations of production and intercourse, and for the first time consciously treats all natural premises as the creatures of hitherto existing men, strips them of their natural character and subjugates them to the power of the united individuals. Its organization is, therefore, essentially economic, the material production of the conditions of this unity; it turns existing conditions into conditions of unity. The reality, which communism is creating, is precisely the true basis for rendering it impossible that anything should exist independently of individuals, in so far as reality is only a product of the preceding intercourse of individuals themselves. Thus the communists in practice treat the conditions created up to now by production and intercourse as inorganic conditions, without, however, imagining that it was the plan or the destiny of previous generations to give them material, and without believing that these conditions were inorganic for the individuals creating them. The difference between the individual as a person and what is accidental to him is not a conceptual difference but a historical fact. This distinction has a different significance at different times—e.g. the estate as something accidental to the individual in the eighteenth century, the family more or less too. It is not a distinction that we have to make for each age, but one which each age makes itself from among the different elements which it finds in existence, and indeed not according to any theory, but compelled by material collisions in life. What appears accidental to the later age as opposed to the earlier—and this applies also to the elements handed down by an earlier age—is a form of intercourse which corresponded to a definite stage of development of the productive forces. The relation of the productive forces to the form of intercourse is the relation of the form of intercourse to the occupation or activity of the individuals. (The fundamental form of this activity is, of course, material, on which depend all other forms—mental, political, religious, etc. The various shaping of material life is, of course, in every case dependent on the needs which are already developed, and the production, as well as the satisfaction, of these needs is a historical process, which is not found in the case of a sheep or a dog, although sheep and dogs in their present form certainly, but *malgré eux* [in spite of themselves], are products of a historical process.) The conditions under which individuals have intercourse with each other, so long as the above-mentioned contradiction is absent, are conditions appertaining to their individuality, in no way external to them; conditions under which these definite individuals, living under definite relationships, can alone produce their material life and what is connected with it, are thus the conditions of their self-activity and are produced by this self-activity. The definite condition under which they produce thus corresponds, as long as the contradiction has not yet appeared, to the reality of their conditioned nature, their one-sided existence, the one-sidedness of which only becomes evident when the contradiction enters on the scene and thus exists for the later individuals. Then this condition appears as an accidental fetter, and the consciousness that it is a fetter is imputed to the earlier age as well.

These various conditions, which appear first as conditions of self-activity, later as fetters upon it, form in the whole evolution of history a coherent series of forms of intercourse, the coherence of which consists in this: in the place of an earlier form of intercourse, which has become a fetter, a new one is put, corresponding to the more developed productive forces and, hence, to the advanced mode of the self-activity of individuals—a form which in its turn becomes a fetter and is then replaced by another. Since these conditions correspond at every stage to the simultaneous development of the productive forces, their history is at the same time the history of the evolving productive forces taken over by each new generation, and is, therefore, the history of the development of the forces of the individuals themselves.

Since this evolution takes place naturally, i.e. is not subordinated to a general plan of freely combined individuals, it proceeds from various localities, tribes, nations, branches of labour, etc. each of which to start with develops independently of the others and only gradually enters into relation with the others. Furthermore, it takes place only very slowly; the various stages and interests are never completely overcome, but only subordinated to the prevailing interest and trail along beside the latter for centuries afterwards. It follows from this that within a nation itself the individuals, even apart from their pecuniary circumstances, have quite different developments, and that an earlier interest, the peculiar form of intercourse of which has already been ousted by that belonging to a later interest, remains for a long time afterwards in possession of a traditional power in the illusory community (State, law), which has won an existence independent of the individuals; a power which in the last resort can only be broken by a revolution. This explains why, with reference to individual points which allow of a more general summing-up, consciousness can sometimes appear further advanced than the contemporary empirical relationships, so that in the struggles of a later epoch one can refer to earlier theoreticians as authorities. . . .

. . . It follows from all we have been saying up till now that the communal relationship into which the individuals of a class entered, and which was determined by their common interests over against a third party, was always a community to which these individuals belonged only as average individuals, only in so far as they lived within the conditions of existence of their class—a relationship in which they participated not as individuals but as members of a class. With the community of revolutionary proletarians, on the other hand, who take their conditions of existence and those of all members of society under their control, it is just the reverse; it is as individuals that the individuals participate in it. It is just this combination of individuals (assuming the advanced stage of modern productive forces, of course) which puts the conditions of the free development and movement of individuals under their control—conditions which were previously abandoned to chance and had won an independent existence over against the separate individuals just because of their separation as individuals, and because of the necessity of their combination which had been determined by the division of labour, and through their separation had become a bond alien to them. Combination up till now (by no means an arbitrary one, such as is expounded for example in the *Contrat social*, but a necessary one) was an agreement upon these conditions, within which the individuals were free to enjoy the freaks of fortune (compare, e.g., the formation of the North American State and the South American republics). This right to the undisturbed enjoyment, within certain conditions, of fortuity and chance has up till now been called personal freedom. These conditions of existence are, of course, only the productive forces and forms of intercourse at any particular time. . . .

For the proletarians, on the other hand, the condition of their existence, labour, and with it all the conditions of existence governing modern society, have become something accidental, something over which they, as separate individuals, have no control, and over which no social organization can give them control. The contradiction between the individuality of each separate proletarian and labour, the condition of life forced upon him, becomes evident to him himself, for he is sacrificed from youth upwards and, within his own class, has no chance of arriving at the conditions which would place him in the other class.

Thus, while the refugee serfs only wished to be free to develop and assert those conditions of existence which were already there, and hence, in the end, only arrived at free labour, the proletarians, if they are to assert themselves as individuals, will have to abolish the very condition of their existence hitherto (which has, moreover, been that of all society up to the present), namely, labour. Thus they find themselves directly opposed to the form in which, hitherto, the individuals, of which society consists, have given themselves collective expression, that is, the State. In order, therefore, to assert themselves as individuals, they must overthrow the State. . . .

Egoism and Communism

How is that personal interests always develop, against the will of individuals, into class interests, into common interests which acquire independent existence in relation to the individual persons, and in their independence assume the form of general interests? How is it that as such they come into contradiction with actual individuals and in this contradiction, by which they are defined as general interests, they can be conceived by consciousness as ideal and even as religious, holy interests? How is it that in this process of private interests acquiring independent existence as class interests the personal behaviour of the individual is bound to undergo substantiation, alienation, and at the same time exists as a power independent of him and without him, created by intercourse, and becomes transformed into social relations, into a series of powers which determine and subordinate the individual, and which, therefore, appear in the imagination as 'holy' powers? If Sancho [Stirner] had only understood the fact that within the frameworks of definite modes of production, which, of course, are not dependent on the will, alien practical forces, which are independent not only of isolated individuals but even of all of them together, always come to stand above people—then he could be fairly indifferent as to whether this fact is presented in a religious form or distorted in the imagination of the egoist, for whom everything occurs in the imagination, in such a way that he puts nothing above himself. Sancho would then have descended from the realm of speculation into the realm of reality, from what people imagine they are to what they actually are, from what they imagine about themselves to how they act and are bound to act in definite circumstances. What seems to him a product of thought, he would have understood to be a product of life. . . .

Incidentally, even in the banal, petty-bourgeois German form in which Sancho perceives the contradiction of personal and general interests, he should have realized that individuals have always started out from themselves, and could not do otherwise, and that therefore both the aspects he noted are aspects of the personal development of individuals; both are equally engendered by the empirical conditions of life, both are only expressions of one and the same personal development of people and are therefore only in seeming contradiction to each other. . . .

Communism is simply incomprehensible to our saint because the communists do not put egoism against self-sacrifice or self-sacrifice against egoism, nor do they express this contradiction theoretically either in its sentimental or in its highflown ideological form; on the contrary, they demonstrate the material basis engendering it, with which it disappears of itself. The communists do not preach morality at all, such as Stirner preaches so extensively. They do not put to people the moral demand: love one another, do not be egoists, etc.; on the contrary, they are very well aware that egoism, just as much as self-sacrifice, is in definite circumstances a necessary form of the self-assertion of individuals. Hence, the communists by no means want . . . to do away with the 'private individual' for the sake of the 'general', self-sacrificing man. . . .

Theoretical communists, the only ones who have time to devote to the study of history, are distinguished precisely because they alone have discovered that throughout history the 'general interest' is created by individuals who are defined as 'private persons'. They know that this contradiction is only a seeming one because one side of it, the so-called 'general', is constantly being produced by the other side, private interest, and by no means opposes the latter as an independent force with an independent history—so that this contradiction is in practice always being destroyed and reproduced. Hence it is not a question of the Hegelian 'negative unity' of two sides of a contradiction, but of the materially determined destruction of the preceding materially determined mode of life of individuals, with the disappearance of which this contradiction together with its unity also disappears.

Power as the Basis of Right

. . . In actual history, those theoreticians who regarded power as the basis of right, were in direct contradiction to those who looked on will as the basis of right. . . . If power is taken as the basis of right, as Hobbes, etc. do, then right, law, etc. are merely the symptom, the expression of other relations upon which State power rests. The material life of individuals, which by no means depends merely on their 'will', their mode of production and form of intercourse, which mutually determine each other—this is the real basis of the State and remains so at all the stages at which division of labour and private property are still necessary, quite independently of the will of individuals. These actual relations are in no way created by the State power; on the contrary they are the power creating it. The individuals who rule in these conditions, besides having to constitute their power in the form of the State, have to give their will, which is determined by these definite conditions, a universal expression as the will of the State, as law—an expression whose content is always determined by the relations of this class, as the civil and criminal law demonstrates in the clearest possible way. . . . Their personal power is based on conditions of life which as they develop are common to many individuals, and the continuance of which they, as ruling individuals, have to maintain against others and, at the same time, maintain that they hold good for all. The expression of this will, which is determined by their common interests, is law. It is precisely because individuals who are independent of one another assert themselves and their own will, which on this basis is inevitably egoistical in their mutual relations, that self-denial is made necessary in law and right, self-denial in the exceptional case, and self-assertion of their interests in the average case (which, therefore, not they, but only the 'egoist in agreement with himself' regards as self-denial). The same applies to the classes which are ruled, whose will plays just as small a part in determining the existence of law and the State. For example, so long as the productive forces are still insufficiently developed to make competition superfluous, and therefore would give rise to competition over and over again, for so long the classes which are ruled would be wanting the impossible if they had the 'will' to abolish competition and with it the State and the law. Incidentally, too, it is only in the imagination of the ideologist that this 'will' arises before conditions have developed far enough to make its production possible. After conditions have developed sufficiently to produce it, the ideologist is able to imagine this will as being purely arbitrary and therefore as conceivable at all times and under all circumstances.

Like right, so crime, i.e. the struggle of the isolated individual against the prevailing conditions, is not the result of pure arbitrariness. On the contrary, it depends on the same conditions as that rule. The same visionaries who see in right and law the domination of some independently existing, general will can see in crime the mere violation of right and law. Hence the State does not exist owing to the ruling will, but the State which arises from the material mode of life of individuals has also the form of a ruling will. If the latter loses its domination, it means that not only has the will changed but also the material existence and life of the individuals, and only for that reason has their will changed. It is possible for rights and laws to be 'inherited', but in that case they are no longer ruling, but nominal, of which striking examples are furnished by the history of ancient Roman law and English law. We saw earlier how a theory and history of pure thought could arise among philosophers owing to the divorce between ideas and the individuals and their empirical relations which serve as the basis of these ideas. In the same way, here too one can divorce right from its real basis, whereby one obtains a 'ruling will' which in different epochs becomes modified in various ways and has its own, independent history in its creations, the laws. On this account, political and civil history becomes ideologically merged in a history of the rule of successive laws. This is the specific illusion of lawyers and politicians. . . .

Utilitarianism

. . . Hegel has already proved in his *Phänomenologie* how this theory of mutual exploitation, which Bentham expounded *ad nauseam*, could already at the beginning of the present century have been considered a phase of the previous one. Look at his chapter on 'The Struggle of Enlightenment with Superstition', where the theory of usefulness is depicted as the final result of enlightenment. The apparent stupidity of merging all the manifold relationships of people in the one relation of usefulness, this apparently metaphysical abstraction arises from the fact that, in modern bourgeois society, all relations are subordinated in practice to the one abstract monetary–commercial relation. This theory came to the fore with Hobbes and Locke at the same time as the first and second English revolutions, those first battles by which the bourgeoisie won political power. It is to be found even earlier, of course, among writers on political economy, as a tacit premiss. Political economy is the real science of this theory of utility; it acquires its true content among the physiocrats, since they were the first to treat political economy systematically. In Helvétius and Holbach one can already find an idealization of this doctrine, which fully corresponds to the attitude of opposition adopted by the French bourgeoisie before the revolution. In Holbach, all the activity of individuals in their mutual intercourse, e.g. speech, love, etc., is depicted as a relation of utility and utilization. Hence the actual relations that are presupposed here are speech, love, the definite manifestations of definite qualities of individuals. Now these relations are supposed not to have the meaning peculiar to them but to be the expression and manifestation of some third relation introduced in their place, the relation of utility or utilization. This paraphrasing ceases to be meaningless and arbitrary only when these relations have validity for the individual not on their own account, not as self-activity, but rather as disguises, though by no means disguises of the category of utilization, but of an actual third aim and relation which is called the relation of utility.

The verbal masquerade only has meaning when it is the unconscious or deliberate expression of an actual masquerade. In this case, the utility relation has a quite definite meaning, namely, that I derive benefit for myself by doing harm to someone else; further, in this case the use that I derive from some relation is in general alien to this relation, just as we saw above in connection with ability that from each ability a product alien to it was demanded, a relation determined by social relations—and this is precisely the relation of utility.

All this is actually the case with the bourgeois. For him only one relation is valid on its own account—the relation of exploitation; all other relations have validity for him only in so far as he can include them under this one relation, and even where he encounters relations which cannot be directly subordinated to the relation of exploitation, he does at least subordinate them to it in his imagination. The material expression of this use is money, the representative of the value of all things, people, and social relations. Incidentally, one sees at a glance that the category of 'utilization' is first of all abstracted from the actual relations of intercourse which I have with other people (but by no means from reflection and mere will) and then these relations are made out to be the reality of the category that has been abstracted from them themselves, a wholly metaphysical method of procedure. In exactly the same way and with the same justification, Hegel depicted all relations as relations of the objective spirit. Hence Holbach's theory is the historically justified philosophical illusion about the bourgeoisie just then developing in France, whose thirst for exploitation could still be described as a thirst for the full development of individuals in conditions of intercourse freed from the old feudal fetters. Liberation from the standpoint of the bourgeoisie, i.e. competition, was, of course, for the eighteenth century the only possible way of offering the individuals a new career for freer development. The theoretical proclamation of the consciousness corresponding to this bourgeois practice, the consciousness of mutual exploitation as the universal mutual relation of all individuals, was also a bold and open step forward, a mundane enlightenment as to the meaning of the political, patriarchal, religious, and sentimental embroidery of exploitation under feudalism, an embroidery which corresponded to the form of exploitation at that time and which was made into a system especially by the theoretical writers of the absolute monarchy. . . .

The advances made by the theory of utility and exploitation, its various phases, are closely connected with the various periods of development of the bourgeoisie. In the case of Helvétius and Holbach, the actual content of the theory never went much beyond paraphrasing the mode of expression of the writers at the time of the absolute monarchy. With them it was a different method of expression; it reflected not so much the actual fact but rather the desire to reduce all relations to the relation of exploitation, and to explain the intercourse of people from material needs and the ways of satisfying them. The problem was set. Hobbes and Locke had before their eyes both the earlier development of the Dutch bourgeoisie (both of them had lived for some time in Holland) and the first political actions by which the English bourgeoisie emerged from local and provincial limitations, as well as a comparatively highly developed stage of manufacture, overseas trade, and colonization. This particularly applies to Locke, who wrote during the first period of English economy, at the time of the rise of joint-stock companies, the Bank of England, and England's mastery of the seas. In their case, and particularly in that of Locke, the theory of exploitation was still directly connected with the economic content.

Helvétius and Holbach were confronted not only by English theory and the preceding development of the Dutch and English bourgeoisie, but also by the French bourgeoisie which was still struggling for its free development. The commercial spirit, universal in the eighteenth century, had especially in France taken possession of all classes in the form of speculation. The financial difficulties of the government and the resulting disputes over taxation occupied the attention of all France even at that time. In addition, Paris in the eighteenth century was the only world city, the only city where there was personal intercourse among individuals of all nations. These premisses, combined with the more universal character typical of Frenchmen in general, gave the theory of Helvétius and Holbach its peculiar universal colouring, but at the same time deprived it of the positive economic content that was still to be found among the English. The theory which for the English still was simply the registration of a fact becomes for the French a philosophical system. This generality devoid of positive content, such as we find it in Helvétius and Holbach, is essentially different from the substantial comprehensive view which is first found in Bentham and Mill. The former corresponds to the struggling, still undeveloped bourgeoisie, the latter to the ruling, developed bourgeoisie.

The content of the theory of exploitation that was neglected by Helvétius and Holbach was developed and systematized by the physiocrats—who worked at the same time as Holbach; but as they took as their basis the undeveloped economic relations of France where feudalism, under which landownership plays the chief role, was still not broken, they remained in thrall to the feudal outlook in so far as they declared landownership and land cultivation to be that productive force which determines the whole structure of society.

The theory of exploitation owes its further development in England to Godwin, and especially to Bentham, who gradually re-incorporated the economic content which the French had neglected, in proportion as the bourgeoisie succeeded in asserting itself both in England and in France. Godwin's *Political Justice* was written during the terror, and Bentham's chief works during and after the French Revolution and the development of large-scale industry in England. The complete union of the theory of utility with political economy is to be found, finally, in Mill.

At an earlier period political economy had been the subject of inquiry either by financiers, bankers, and merchants, i.e. in general by persons directly concerned with economic relations, or by persons with an all-round education like Hobbes, Locke, and Hume, for whom it was of importance as a branch of encyclopaedic knowledge. Thanks to the Physiocrats, political economy for the first time was raised to the rank of a special science and has been treated as such ever since. As a special branch of science it absorbed the other relations—political, juridical, etc.—to such an extent that it reduced them to economic relations. But it considered this subordination of all relations to itself only one aspect of these relations, and thereby allowed them for the rest an independent significance also outside political economy. The complete subordination of all existing relations to the relation of utility, and its unconditional elevation to be the sole content of all other relations, we find for the first time in Bentham, where, after the French Revolution and the development of large-scale industry, the bourgeoisie no longer appears as a special class, but as the class whose conditions of existence are those of the whole society.

When the sentimental and moral paraphrases, which for the French were the entire content of the utility theory, had been exhausted, all that remained for its further development was the question how individuals and relations were to be used, to be exploited. Meanwhile the reply to this question had already been given in political economy; the only possible step forward was by inclusion of the economic content. Bentham achieved this advance. But the idea had already been stated in political economy that the chief relations of exploitation are determined by production by and large, independently of the will of individuals who find them already in existence. Hence, no other field of speculative thought remained for the utility theory than the attitude of individuals to these important relations, the private exploitation of an already existing world by individuals. On this subject Bentham and his school indulged in lengthy moral reflections. Thereby the whole criticism of the existing world provided by the utility theory also moved within a narrow compass. Prejudiced in favour of the conditions of the bourgeoisie, it could criticize only those relations which had been handed down from a past epoch and were an obstacle to the development of the bourgeoisie. Hence, although the utility theory does expound the connection of all existing relations with economic relations it does so only in a restricted way.

From the outset the utility theory had the aspect of a theory of general utility, yet this aspect only became fraught with meaning when economic relations, especially division of labour and exchange, were included. With division of labour, the private activity of the individual becomes generally useful; Bentham's general utility becomes reduced to the same general utility that is operative in competition. By taking into account the economic relations of rent, profit, and wages, the definite exploitation relations of separate classes were introduced, since the manner of exploitation depends on the position in life of the exploiter. Up to this point the theory of utility was able to base itself on definite social facts; its further account of the manner of exploitation amounts to a mere recital of catechism phrases.

The economic content gradually turned the utility theory into a mere apologia for the existing state of affairs, an attempt to prove that under existing conditions the mutual relations of people today are the most advantageous and generally useful. It has this character among all modern economists. . . .

Artistic Talent under Communism

. . . He [Stirner] imagines that the so-called organizers of labour wanted to organize the entire activity of each individual, and yet it is precisely among them that a difference is drawn between directly productive labour, which has to be organized, and labour which is not directly productive. In regard to the latter, however, it was not their view, as Sancho imagines, that each should do the work of Raphael, but that anyone in whom there is a potential Raphael should be able to develop without hindrance. Sancho imagines that Raphael produced his pictures independently of the division of labour that existed in Rome at the time. If he were to compare Raphael with Leonardo da Vinci and Titian, he would know how greatly Raphael's works of art depended on the flourishing of Rome at that time, which occurred under Florentine influence, while the works of Leonardo depended on the state of things in Florence, and the works of Titian, at a later period, depended on the totally different development of Venice. Raphael as much as any other artist was determined by the technical advances in art made before him, by the organization of society and the division of labour in his locality, and, finally, by the division of labour in all the countries with which his locality had intercourse. Whether an individual like Raphael succeeds in developing his talent depends wholly on demand, which in turn depends on the division of labour and the conditions of human culture resulting from it.

In proclaiming the uniqueness of work in science and art, Stirner adopts a position far inferior to that of the bourgeoisie. At the present time it has already been found necessary to organize this 'unique' activity. Horace Vernet would not have had time to paint even a tenth of his pictures if he regarded them as works which 'only this Unique person is capable of producing'. In Paris, the great demand for vaudevilles and novels brought about the organization of work for their production, organization which at any rate yields something better than its 'unique' competitors in Germany. In astronomy, people like Arago, Herschel, Encke, and Bessel considered it necessary to organize joint observations and only after that obtained some fruitful results. In historical science, it is absolutely impossible for the 'Unique' to achieve anything at all, and in this field, too, the French long ago surpassed all other nations thanks to organization of labour. Incidentally, it is self-evident that all these organizations based on modern division of labour still lead only to extremely limited results, representing a step forward only compared with the previous narrow isolation. . . .

The exclusive concentration of artistic talent in particular individuals, and its suppression in the broad mass which is bound up with this, is a consequence of division of labour. If, even in certain social conditions, everyone were an excellent painter, that would not at all exclude the possibility of each of them being also an original painter, so that here too the difference between 'human' and 'unique' labour amounts to sheer nonsense. In any case, with a communist organization of society, there disappears the subordination of the artist to local and national narrowness, which arises entirely from division of labour, and also the subordination of the artist to some definite art, thanks to which he is exclusively a painter, sculptor, etc., the very name of his activity adequately expressing the narrowness of his professional development and his dependence on division of labour. In a communist society there are no painters but at most people who engage in painting among other activities. . . .

The Free Development of Individuals
in Communist Society

. . . The transformation of the individual relationship into its opposite, a merely material relationship, the distinction of individuality and chance by the individuals themselves, as we have already shown, is an historical proces., and at different stages of development assumes different, ever sharper, and more universal forms. In the present epoch, the domination of material conditions over individuals, and the suppression of individuality by chance, has assumed its sharpest and most universal form, thereby setting existing individuals a very definite task. It has set them the task of replacing the domination of circumstances and of chance over individuals by the domination of individuals over chance and circumstances. It has not, as Sancho imagines, put forward the demand that 'I should develop myself', which up to now every individual has done without Sancho's good advice; it has instead called for liberation from one quite definite mode of development. This task, dictated by present-day conditions, coincides with the task of the communist organization of society.

We have already shown above that the abolition of a state of things in which relationships become independent of individuals, in which individuality is subservient to chance and the personal relationships of individuals are subordinated to general class relationships, etc.—the abolition of this state of things is determined in the final analysis by the abolition of division of labour. We have also shown that the abolition of division of labour is determined by the development of intercourse and productive forces to such a degree of universality that private property and division of labour become fetters on them. We have further shown that private property can be abolished only on condition of an all-round development of individuals, because the existing character of intercourse and productive forces is an all-round one, and only individuals that are developing in an all-round fashion can appropriate them, i.e. can turn them into free manifestations of their lives. We have shown that at the present time individuals must abolish private property, because the productive forces and forms of intercourse have developed so far that, under the domination of private property, they have become destructive forces, and because the contradiction between the classes has reached its extreme limit. Finally, we have shown that the abolition of private property and of the division of labour is itself the union of individuals on the basis created by modern productive forces and world intercourse.

Within communist society, the only society in which the original and free development of individuals ceases to be a mere phrase, this development is determined precisely by the connection of individuals, a connection which consists partly in the economic prerequisites and partly in the necessary solidarity of the free development of all, and, finally, in the universal character of the activity of individuals on the basis of the existing productive forces. Here, therefore, the matter concerns individuals at a definite historical stage of development and by no means merely individuals chosen at random, even disregarding the indispensable communist revolution which itself is a general condition of their free development. The individuals' consciousness of their mutual relations will, of course, likewise become something quite different, and, therefore, will no more be the 'principle of love' or *dévouement*, than it will be egoism. . . .

BIBLIOGRAPHY

Original: *MEW*, Vol. 3, pp. 13 ff.

Present translation: K. Marx and F. Engels, *The German Ideology*, Moscow, 1964, pp. 23 ff., pp. 270 ff., pp. 357 ff., pp. 365 f., p. 441, p. 443. Reproduced by kind permission of Lawrence and Wishart Ltd.

Other translations: None.

Commentaries: H. Adams, *Karl Marx in his Earlier Writings*, 2nd edn., London, 1965, Ch. 10.

C. Arthur, Introduction to *The German Ideology*, London, 1971.

L. Dupré, *The Philosophical Foundations of Marxism*, New York, 1966, Ch. 6.

D. McLellan, *Karl Marx: His Life and Thought*, London and New York, 1973, pp. 142 ff.

Also the books in the General Bibliography by Acton, Bober, Evans, Jordan, Plamenatz, and Sanderson, all of which have a bearing on matters discussed in *The German Ideology*.

15. Letter to Annenkov

In December 1846 Marx received Proudhon's work *System of Economic Contradictions*. He summarized very clearly the differences between Proudhon's method and his own in a lengthy letter to Annenkov, a well-to-do Russian whom Marx had got to know in Paris. (See also the Introduction to the next section.)

... What is society, whatever its form may be? The product of men's reciprocal action. Are men free to choose this or that form of society for themselves? By no means. Assume a particular state of development in the productive forces of man and you will get a particular form of commerce and consumption. Assume particular stages of development in production, commerce, and consumption and you will have a corresponding social constitution, a corresponding organization of the family, of orders or of classes, in a word, a corresponding civil society. Assume a particular civil society and you will get particular political conditions which are only the official expression of civil society. M. Proudhon will never understand this because he thinks he is doing something great by appealing from the state to society—that is to say, from the official résumé of society to official society.

It is superfluous to add that men are not free to choose their productive forces—which are the basis of all their history—for every productive force is an acquired force, the product of former activity. The productive forces are therefore the result of practical human energy; but this energy is itself conditioned by the circumstances in which men find themselves, by the productive forces already acquired, by the social form which exists before they do, which they do not create, which is the product of the preceding generation. Because of this simple fact that every succeeding generation finds itself in possession of the productive forces acquired by the previous generation, which serve it as the raw material for new production, a coherence arises in human history, a history of humanity which takes shape is all the more a history of humanity as the productive forces of man and therefore his social relations have been more developed. Hence it necessarily follows that the social history of men is never anything but the history of their individual development, whether they are conscious of it or not. Their material relations are the basis of all their relations. These material relations are only the necessary forms in which their material and individual activity is realized.

M. Proudhon mixes up ideas and things. Men never relinquish what they have won, but this does not mean that they never relinquish the social form in which they have acquired certain productive forces. On the contrary, in order that they may not be deprived of the result attained, and forfeit the fruits of civilization, they are obliged, from the moment when the form of their commerce no longer corresponds to the productive forces acquired, to change all their traditional social forms. I am using the word 'commerce' here in its widest sense, as we use *Verkehr* in German. For example: the privileges, the institution of guilds and corporations, the regulatory regime of the Middle Ages, were social relations that alone corresponded to the acquired productive forces and to the social condition which had previously existed and from which these institutions had arisen. Under the protection of the regime of corporations and regulations, capital was accumulated, overseas trade was developed, colonies were founded. But the fruits of this men would have forfeited if they had tried to retain the forms under whose shelter these fruits had ripened. Hence burst two thunderclaps—the Revolutions of 1640 and 1688. All the old economic forms, the social relations corresponding to them, the political conditions which were the official expression of the old civil society, were destroyed in England. Thus the economic forms in which men produce, consume, and exchange, are transitory and historical. With the acquisition of new productive faculties, men change their mode of production and with the mode of production all the economic relations which are merely the necessary relations of this particular mode of production.

This is what M. Proudhon has not understood and still less demonstrated. M. Proudhon, incapable of following the real movement of history, produces a phantasmagoria which presumptuously claims to be dialectical. He does not feel it necessary to speak of the seventeenth, the eighteenth, or the nineteenth centuries, for his history proceeds in the misty realm of imagination and rises far above space and time. In short, it is not history but old Hegelian junk, it is not profane history—a history of man—but sacred history—a history of ideas. From his point of view man is only the instrument of which the idea or the eternal reason makes use in order to unfold itself. The evolutions of which M. Proudhon speaks are understood to be evolutions such as are accomplished within the mystic womb of the absolute idea. If you tear the veil from this mystical language, what it comes to is that M. Proudhon is offering you the order in which economic categories arrange themselves inside his own head. It will not require great exertion on my part to prove to you that it is the order of a very disorderly mind.

Monsieur Proudhon has very well grasped the fact that men produce cloth, linen, silks, and it is a great merit on his part to have grasped this small amount! What he has not grasped is that these men, according to their abilities, also produce the social relations amid which they prepare cloth and linen. Still less has he understood that men, who produce their social relations in accordance with their material productivity, also produce ideas, categories, that is to say the abstract ideal expression of these same social relations. Thus the categories are no more eternal than the relations they express. They are historical and transitory products. For M. Proudhon, on the contrary, abstractions, categories are the primordial cause. According to him they, and not men, make history. The abstraction, the category taken as such, i.e. apart from men and their material activities, is of course immortal, unchangeable, unmoved; it is only one form of the being of pure reason; which is only another way of saying that the abstraction as such is abstract. An admirable tautology! . . .

BIBLIOGRAPHY

Original: K. Marx and F. Engels, *Correspondence*, Paris, 1971, Vol. 1, pp. 446 ff.

Present translation: K. Marx, *The Poverty of Philosophy*, Moscow, n.d., pp. 202 ff., 212. Reproduced by kind permission of Lawrence and Wishart Ltd.

Other translations: None.

Commentaries: See Bibliography to next section.

16. *The Poverty of Philosophy*

The following extracts were, like *The German Ideology*, written in an attempt to instil a minimal doctrinal cohesion into the League of the Just, which was in the process of being transformed into the more open Communist League. Marx had invited Proudhon in 1846 to become the Paris correspondent of the Brussels communists, but Proudhon had refused as he considered Marx's ideas too dogmatic. When Proudhon soon afterwards published his *System of Economic Contradictions*, subtitled *The Philosophy of Poverty*, Marx took the opportunity to launch a devastating attack on Proudhon under the title *The Poverty of Philosophy*.

In the first part of his book Marx attacks Proudhon's economic doctrines; in the second part, which forms the bulk of the extracts below, Marx criticizes Proudhon's attempt to use Hegel's dialectic and points out the difference between Proudhon's abstract speculation and his own account of the real movement of the productive relations. In the final passages, Marx refutes Proudhon's view of the uselessness of strikes on the grounds that the resulting higher wages only serve to increase prices. He finishes with a kind of anarchist manifesto which looks forward to the day when concerted working-class action would abolish class antagonisms and state power.

The Poverty of Philosophy, which was written in French, was regarded by Marx as the first scientific presentation of his theory.

Value and Labour time

... Everyone knows that when supply and demand are evenly balanced, the relative value of any product is accurately determined by the quantity of labour embodied in it, that is to say, that this relative value expresses the proportional relation precisely in the sense we have just attached to it. M. Proudhon inverts the order of things. Begin, he says, by measuring the relative value of a product by the quantity of labour embodied in it, and supply and demand will infallibly balance one another. Production will correspond to consumption, the product will always be exchangeable. Its current price will express exactly its true value. Instead of saying like everyone else: when the weather is fine, a lot of people are to be seen going out for a walk, M. Proudhon makes his people go out for a walk in order to be able to ensure them fine weather.

What M. Proudhon gives as the consequence of marketable value determined *a priori* by labour time could be justified only by a law couched more or less in the following terms:

Products will in future be exchanged in the exact ratio of the labour time they have cost. Whatever may be the proportion of supply to demand, the exchange of commodities will always be made as if they had been produced proportionately to the demand. Let M. Proudhon take it upon himself to formulate and lay down such a law, and we shall relieve him of the necessity of giving proofs. If, on the other hand, he insists on justifying his theory, not as a legislator, but as an economist, he will have to prove that the time needed to create a commodity indicates exactly the degree of its utility and marks its proportional relation to the demand, and in consequence, to the total amount of wealth. In this case, if a product is sold at a price equal to its cost of production, supply and demand will always be evenly balanced; for the cost of production is supposed to express the true relation between supply and demand.

Actually, M. Proudhon sets out to prove that the labour time needed to create a product indicates its true proportional relation to needs, so that the things whose production costs the least time are the most immediately useful, and so on, step by step. The mere production of a luxury object proves at once, according to this doctrine, that society has spare time which allows it to satisfy a need for luxury.

M. Proudhon finds the very proof of his thesis in the observation that the most useful things cost the least time to produce, that society always begins with the easiest industries and successively 'starts on the production of objects which cost more labour time and which correspond to a higher order of needs'.

M. Proudhon borrows from M. Dunoyer the example of extractive industry—fruit-gathering, pasturage, hunting, fishing, etc.—which is the simplest, the least costly of industries, and the one by which man began 'the first day of his second creation'. The first day of his first creation is recorded in Genesis, which shows us God as the world's first manufacturer.

Things happen in quite a different way from what M. Proudhon imagines. The very moment civilization begins, production begins to be founded on the antagonism of orders, estates, classes, and finally on the antagonism of ac-cumulated labour and actual labour. No antagonism, no progress. This is the law that civilization has followed up to our day. Till now the productive forces have been developed by virtue of this system of class antagonisms. To say now that, because all the needs of all the workers were satisfied, men could devote themselves to the creation of products of a higher order—to more complicated industries—would be to leave class antagonism out of account and turn all historical development upside down. It is like saying that because, under the Roman emperors, muræna were fattened in artificial fishponds, therefore there was enough to feed abundantly the whole Roman population. Actually, on the contrary, the Roman people had not enough to buy bread with, while the Roman aristocrats had slaves enough to throw as fodder to the muræna.

The price of food has almost continuously risen, while the price of manu-factured and luxury goods has almost continuously fallen. Take the agricul-tural industry itself: the most indispensable objects, like corn, meat, etc., rise in price, while cotton, sugar, coffee, etc., fall in a surprising proportion. And even among comestibles proper, the luxury articles, like artichokes, asparagus, etc., are today relatively cheaper than foodstuffs of prime necessity. In our age, the superfluous is easier to produce than the necessary. Finally, at different historical epochs the reciprocal price relations are not only different, but opposed to one another. In the whole of the Middle Ages agricultural products were relatively cheaper than manufactured products; in modern times they are in inverse ratio. Does this mean that the utility of agricultural products has diminished since the Middle Ages?

The use of products is determined by the social conditions in which the consumers find themselves placed, and these conditions themselves are based on class antagonism.

Cotton, potatoes, and spirits are objects of the most common use. Potatoes have engendered scrofula; cotton has to a great extent driven out flax and wool, although wool and flax are, in many cases, of greater utility, if only from the point of view of hygiene; finally, spirits have got the upper hand of beer and wine, although spirits used as an alimentary substance are everywhere recognized to be poison. For a whole century, governments struggled in vain against the European opium; economics prevailed, and dictated its orders to consumption.

Why are cotton, potatoes, and spirits the pivots of bourgeois society? Because the least amount of labour is needed to produce them, and, consequently, they have the lowest price. Why does the minimum price determine the maximum consumption? Is it by any chance because of the absolute utility of these objects, their intrinsic utility, their utility in as much as they correspond, in the most useful manner, to the needs of the worker as a man, and not to the man as a worker? No, it is because in a society founded on poverty the poorest products have the fatal prerogative of being used by the greatest number.

To say now that because the least costly things are in greater use, they must be of greater utility, is saying that the wide use of spirits, because of their low cost of production, is the most conclusive proof of their utility; it is telling the proletarian that potatoes are more wholesome for him than meat; it is accepting the present state of affairs; it is, in short, making an apology, with M. Proudhon, for a society without understanding it.

In a future society, in which class antagonism will have ceased, in which there will no longer be any classes, use will no longer be determined by the minimum time of production; but the time of production devoted to different articles will be determined by the degree of their social utility.

Class Antagonism

. . . In principle, there is no exchange of products—but there is the exchange of the labour which co-operated in production. The mode of exchange of products depends upon the mode of exchange of the productive forces. In general, the form of exchange of products corresponds to the form of production. Change the latter, and the former will change in consequence. Thus in the history of society we see that the mode of exchanging products is regulated by the mode of producing them. Individual exchange corresponds also to a definite mode of production which itself corresponds to class antagonism. There is thus no individual exchange without the antagonism of classes.

But the respectable conscience refuses to see this obvious fact. So long as one is a bourgeois, one cannot but see in this relation of antagonism a relation of harmony and eternal justice, which allows no one to gain at the expense of another. For the bourgeois, individual exchange can exist without any antagonism of classes. For him, these are two quite unconnected things. Individual exchange, as the bourgeois conceives it, is far from resembling individual exchange as it actually exists in practice.

Mr. Bray turns the illusion of the respectable bourgeois into an ideal he would like to attain. In a purified individual exchange, freed from all the elements of antagonism he finds in it, he sees an 'equalitarian' relation which he would like society to adopt generally.

Mr. Bray does not see that this equalitarian relation, this corrective ideal that he would like to apply to the world, is itself nothing but the reflection of the actual world; and that therefore it is totally impossible to reconstitute society on the basis of what is merely an embellished shadow of it. In proportion as this shadow takes on substance again, we perceive that this substance, far from being the transfiguration dreamt of, is the actual body of existing society. . . .

Method in Political Economy

Here we are, right in Germany! We shall now have to talk metaphysics while talking political economy. And in this again we shall but follow M. Proudhon's 'contradictions'. Just now he forced us to speak English, to become pretty well English ourselves. Now the scene is changing. M. Proudhon is transporting us to our dear fatherland and is forcing us, whether we like it or not, to become German again.

If the Englishman transforms men into hats, the German transforms hats into ideas. The Englishman is Ricardo, rich banker and distinguished economist; the German is Hegel, simple professor of philosophy at the University of Berlin.

Louis XV, the last absolute monarch and representative of the decadence of French royalty, had attached to his person a physician who was himself France's first economist. This doctor, this economist, represented the imminent and certain triumph of the French bourgeoisie. Doctor Quesnay made a science out of political economy; he summarized it in his famous *Tableau économique*. Besides the thousand and one commentaries on this table which have appeared, we possess one by the doctor himself. It is the 'analysis of the economic table', followed by 'seven important observations'.

M. Proudhon is another Dr. Quesnay. He is the Quesnay of the metaphysics of political economy.

Now metaphysics—indeed all philosophy—can be summed up, according to Hegel, in method. We must, therefore, try to elucidate the method of M. Proudhon, which is at least as foggy as the *Economic Table*. It is for this reason that we are making seven more or less important observations. If Dr. Proudhon is not pleased with our observations, well then, he will have to become an Abbé Baudeau and give the 'explanation of the economico-metaphysical method' himself.

First Observation

'We are not giving a history according to the order in time, but according to the sequence of ideas. Economic phases or categories are in their manifestation sometimes contemporary, sometimes inverted. . . . Economic theories have none the less their logical sequence and their serial relation in the understanding: it is this order that we flatter ourselves to have discovered.' (Proudhon, Vol. I, p. 146.)

M. Proudhon most certainly wanted to frighten the French by flinging quasi-Hegelian phrases at them. So we have to deal with two men; firstly with M. Proudhon, and then with Hegel. How does M. Proudhon distinguish himself from other economists? And what part does Hegel play in M. Proudhon's political economy?

Economists express the relations of bourgeois production, the division of labour, credit, money, etc., as fixed, immutable, eternal categories. M. Proudhon, who has these ready-made categories before him, wants to explain to us the act of formation, the genesis of these categories, principles, laws, ideas, thoughts.

Economists explain how production takes place in the above-mentioned relations, but what they do not explain is how these relations themselves are produced, that is, the historical movement which gave them birth. M. Proudhon, taking these relations for principles, categories, abstract thoughts, has merely to put into order these thoughts, which are to be found alphabetically arranged at the end of every treatise on political economy. The economists' material is the active, energetic life of man; M. Proudhon's material is the dogmas of the economists. But the moment we cease to pursue the historical movement of production relations, of which the categories are but the theoretical expression, the moment we want to see in these categories no more than ideas, spontaneous thoughts, independent of real relations, we are forced to attribute the origin of these thoughts to the movement of pure reason. How does pure, eternal, impersonal reason give rise to these thoughts? How does it proceed in order to produce them?

If we had M. Proudhon's intrepidity in the matter of Hegelianism we should say: it is distinguished in itself from itself. What does this mean? Impersonal reason, having outside itself neither a base on which it can pose itself, nor an object to which it can oppose itself, nor a subject with which it can compose itself, is forced to turn head over heels, in posing itself, opposing itself, and composing itself—position, opposition, composition. Or, to use Greek—we have thesis, antithesis, and synthesis. For those who do not know the Hegelian language, we shall give the consecrating formula:—affirmation, negation, and negation of the negation. That is what language means. It is certainly not Hebrew (with due apologies to M. Proudhon); but it is the language of this pure reason, separate from the individual. Instead of the ordinary individual with his ordinary manner of speaking and thinking we have nothing but this ordinary manner in itself—without the individual.

Is it surprising that everything, in the final abstraction—for we have here an abstraction, and not an analysis—presents itself as a logical category? Is it surprising that, if you let drop little by little all that constitutes the individuality of a house, leaving out first of all the materials of which it is composed, then the form that distinguishes it, you end up with nothing but a body; that, if you leave out of account the limits of this body, you soon have nothing but a space—that if, finally, you leave out of account the dimensions of this space, there is absolutely nothing left but pure quantity, the logical category? If we abstract thus from every subject all the alleged accidents, animate or inanimate, men or things, we are right in saying that in the final abstraction the only substance left is the logical categories. Thus the metaphysicians who, in making these abstractions, think they are making analyses, and who, the more they detach themselves from things, imagine themselves to be getting all the nearer to the point of penetrating to their core—these metaphysicians in turn are right in saying that things here below are embroideries of which the logical categories constitute the canvas. This is what distinguishes the philosopher from the Christian. The Christian, in spite of logic, has only one incarnation of the *Logos*; the philosopher has never finished with incarnations. If all that exists, all that lives on land and under water, can be reduced by abstraction to a logical category—if the whole real world can be drowned thus in a world of abstractions, in the world of logical categories—who need be astonished at it?

All that exists, all that lives on land and under water, exists and lives only by some kind of movement. Thus the movement of history produces social relations; industrial movement gives us industrial products, etc.

Just as by dint of abstraction we have transformed everything into a logical category, so one has only to make an abstraction of every characteristic distinctive of different movements to attain movement in its abstract condition—purely formal movement, the purely logical formula of movement. If one finds in logical categories the substance of all things, one imagines one has found in the logical formula of movement the absolute method, which not only explains all things, but also implies the movement of things.

It is of this absolute method that Hegel speaks in these terms: 'Method is the absolute, unique, supreme, infinite force, which no object can resist; it is the tendency of reason to find itself again, to recognize itself in every object.' (*Logic*, Vol. III.) All things being reduced to a logical category, and every movement, every act of production, to method, it follows naturally that every aggregate of products and production, of objects and of movement, can be reduced to a form of applied metaphysics. What Hegel has done for religion, law, etc., M. Proudhon seeks to do for political economy.

So what is this absolute method? The abstraction of movement. What is the abstraction of movement? Movement in abstract condition. What is movement in abstract condition? The purely logical formula of movement or the movement of pure reason. Wherein does the movement of pure reason consist? In posing itself, opposing itself, composing itself; in formulating itself as thesis, antithesis, synthesis; or, yet again, in affirming itself, negating itself, and negating its negation.

How does reason manage to affirm itself, to pose itself in a definite category? That is the business of reason itself and of its apologists.

But once it has managed to pose itself as a thesis, this thesis, this thought, opposed to itself, splits up into two contradictory thoughts—the positive and the negative, the yes and the no. The struggle between these two antagonistic elements comprised in the antithesis constitutes the dialectical movement. The yes becoming no, the no becoming yes, the yes becoming both yes and no, the no becoming both no and yes, the contraries balance, neutralize, paralyse each other. The fusion of these two contradictory thoughts constitutes a new thought, which is the synthesis of them. This thought splits up once again into two contradictory thoughts, which in turn fuse into a new synthesis. Of this travail is born a group of thoughts. This group of thoughts follows the same dialectic movement as the simple category, and has a contradictory group as antithesis. Of these two groups of thoughts is born a new group of thoughts, which is the synthesis of them.

Just as from the dialectic movement of the simple categories is born the group, so from the dialectic movement of the groups is born the series, and from the dialectic movement of the series is born the entire system.

Apply this method to the categories of political economy, and you have the logic and metaphysics of political economy, or, in other words, you have the economic categories that everybody knows, translated into a little-known language which makes them look as if they had newly blossomed forth in an intellect of pure reason; so much do these categories seem to engender one another, to be linked up and intertwined with one another by the very working of the dialectic movement. The reader must not get alarmed at these metaphysics with all their scaffolding of categories, groups, series, and systems. M. Proudhon, in spite of all the trouble he has taken to scale the heights of the system of contradictions, has never been able to raise himself above the first two rungs of simple thesis and antithesis; and even these he has mounted only twice, and on one of these two occasions he fell over backwards.

Up to now we have expounded only the dialectics of Hegel. We shall see later how M. Proudhon has succeeded in reducing it to the meanest proportions. Thus, for Hegel, all that has happened and is still happening is only just what is happening in his own mind. Thus the philosophy of history is nothing but the history of philosophy, of his own philosophy. There is no longer a 'history according to the order in time', there is only 'the sequence of ideas in the understanding'. He thinks he is constructing the world by the movement of thought, whereas he is merely reconstructing systematically and classifying by the absolute method the thoughts which are in the minds of all.

Second Observation

Economic categories are only the theoretical expressions, the abstractions of the social relations of production. M. Proudhon, holding things upside down like a true philosopher, sees in actual relations nothing but the incarnation of these principles, of these categories, which were slumbering—so M. Proudhon the philosopher tells us—in the bosom of the 'impersonal reason of humanity'.

M. Proudhon the economist understands very well that men make cloth, linen, or silk materials in definite relations of production. But what he has not understood is that these definite social relations are just as much produced by men as linen, flax, etc. Social relations are closely bound up with productive forces. In acquiring new productive forces men change their mode of production; and in changing their mode of production, in changing the way of earning their living, they change all their social relations. The hand-mill gives you society with the feudal lord; the steam-mill, society with the industrial capitalist.

The same men who establish their social relations in conformity with their material productivity, produce also principles, ideas, and categories in conformity with their social relations.

Thus these ideas, these categories, are as little eternal as the relations they express. They are historical and transitory products.

There is a continual movement of growth in productive forces, of destruction in social relations, of formation in ideas; the only immutable thing is the abstraction of movement—mors immortalis.

Third Observation

The production relations of every society form a whole. M. Proudhon considers economic relations as so many social phases, engendering one another, resulting one from the other like the antithesis from the thesis, and realizing in their logical sequence the impersonal reason of humanity.

The only drawback to this method is that when he comes to examine a single one of these phases, M. Proudhon cannot explain it without having recourse to all the other relations of society; which relations, however, he has not yet made his dialectic movement engender. When, after that, M. Proudhon, by means of pure reason, proceeds to give birth to these other phases, he treats them as if they were new-born babes. He forgets that they are of the same age as the first.

Thus, to arrive at the constitution of value, which for him is the basis of all economic evolutions, he could not do without division of labour, competition, etc. Yet in the series, in the understanding of M. Proudhon, in the logical sequence, these relations did not yet exist.

In constructing the edifice of an ideological system by means of the categories of political economy, the limbs of the social system are dislocated. The different limbs of society are converted into so many separate societies, following one upon the other. How, indeed, could the single logical formula of movement, of sequence, of time, explain the structure of society, in which all relations co-exist simultaneously and support one another?

Fourth Observation

Let us see now to what modifications M. Proudhon subjects Hegel's dialectics when he applies it to political economy.

For him, M. Proudhon, every economic category has two sides—one good, the other bad. He looks upon these categories as the petty bourgeois looks upon the great men of history: Napoleon was a great man; he did a lot of good; he also did a lot of harm.

The good side and the bad side, the advantages and the drawbacks, taken together form for M. Proudhon the contradiction in every economic category.

The problem to be solved: to keep the good side, while eliminating the bad.

Slavery is an economic category like any other. Thus it also has its two sides. Let us leave alone the bad side and talk about the good side of slavery. Needless to say we are dealing only with direct slavery, with Negro slavery in Surinam, in Brazil, in the Southern States of North America.

Direct slavery is just as much the pivot of bourgeois industry as machinery, credits, etc. Without slavery you have no cotton; without cotton you have no modern industry. It is slavery that gave the colonies their value; it is the colonies that created world trade, and it is world trade that is the precondition of large-scale industry. Thus slavery is an economic category of the greatest importance.

Without slavery North America, the most progressive of countries, would be transformed into a patriarchal country. Wipe North America off the map of the world, and you will have anarchy—the complete decay of modern commerce and civilization. Cause slavery to disappear and you will have wiped America off the map of nations.

Thus slavery, because it is an economic category, has always existed among the institutions of the peoples. Modern nations have been able only to disguise slavery in their own countries, but they have imposed it without disguise upon the New World.

What would M. Proudhon do to save slavery? He would formulate the problem thus: preserve the good side of this economic category, eliminate the bad.

Hegel has no problems to formulate. He has only dialectics. M. Proudhon has nothing of Hegel's dialectics but the language. For him the dialectic movement is the dogmatic distinction between good and bad. . . .

Let us for a moment consider M. Proudhon himself as a category. Let us examine his good and his bad side, his advantages and his drawbacks.

If he has the advantage over Hegel of setting problems which he reserves the right of solving for the greater good of humanity, he has the drawback of being stricken with sterility when it is a question of engendering a new category by dialectical birth-throes. What constitutes dialectical movement is the co-existence of two contradictory sides, their conflict and their fusion into a new category. The very setting of the problem of eliminating the bad side cuts short the dialectic movement. It is not the category which is posed and opposed to itself, by its contradictory nature, it is M. Proudhon who gets excited, perplexed, and frets and fumes between the two sides of the category.

Caught thus in a blind alley, from which it is difficult to escape by legal means, M. Proudhon takes a real flying leap which transports him at one bound into a new category. Then it is that to his astonished gaze is revealed the serial relation in the understanding.

He takes the first category that comes handy and attributes to it arbitrarily the quality of supplying a remedy for the drawbacks of the category to be purified. Thus, if we are to believe M. Proudhon, taxes remedy the drawbacks of monopoly; the balance of trade, the drawbacks of taxes; landed property, the drawbacks of credit.

By taking the economic categories thus successively, one by one, and making one the antidote to the other, M. Proudhon manages to make with this mixture of contradictions and antidotes to contradictions, two volumes of contradictions, which he rightly entitles *The System of Economic Contradictions*.

Fifth Observation

'In the absolute reason all these ideas . . . are equally simple and general. . . . In fact, we attain knowledge only by a sort of scaffolding of our ideas. But truth in itself is independent of these dialectical symbols and freed from the combinations of our minds.' (Proudhon, Vol. II, p. 97.)

Here all of a sudden, by a kind of switch-over of which we now know the secret, the metaphysics of political economy has become an illusion! Never has M. Proudhon spoken more truly. Indeed, from the moment the process of the dialectic movement is reduced to the simple process of opposing good to bad, of posing problems tending to eliminate the bad, and of administering one category as an antidote to another, the categories are deprived of all spontaneity; the idea 'ceases to function'; there is no life left in it. It is no longer posed or decomposed into categories. The sequence of categories has become a sort of scaffolding. Dialectics has ceased to be the movement of absolute reason. There is no longer any dialectics but only, at the most, absolutely pure morality.

When M. Proudhon spoke of the series in the understanding, of the logical sequence of categories, he declared positively that he did not want to give history according to the order in time, that is, in M. Proudhon's view, the historical sequence in which the categories have manifested themselves. Thus for him everything happened in the pure ether of reason. Everything was to be derived from this ether by means of dialectics. Now that he has to put this dialectics into practice, his reason is in default. M. Proudhon's dialectis runs counter to Hegel's dialectics, and now we have M. Proudhon reduced to saying that the order in which he gives the economic categories is no longer the order in which they engender one another. Economic evolutions are no longer the evolutions of reasons itself.

What then does M. Proudhon give us? Real history, which is, according to M. Proudhon's understanding, the sequence in which the categories have manifested themselves in order of time? No! History as it takes place in the idea itself? Still less! That is, neither the profane history of the categories, nor their sacred history! What history does he give us then? The history of his own contradictions. Let us see how they go, and how they drag M. Proudhon in their train.

Before entering upon this examination, which gives rise to the sixth important observation, we have yet another, less important observation to make.

Let us admit with M. Proudhon that real history, history according to the order in time, is the historical sequence in which ideas, categories, and principles have manifested themselves.

Each principle has had its own century in which to manifest itself. The principle of authority, for example, had the eleventh century, just as the principle of individualism had the eighteenth century. In logical sequence, it was the century that belonged to the principle, and not the principle that belonged to the century. In other words it was the principle that made the history, and not the history that made the principle. When, consequently, in order to save principles as much as to save history, we ask ourselves why a particular principle was manifested in the eleventh or in the eighteenth century rather than in any other, we are necessarily forced to examine minutely what men were like in the eleventh century, what they were like in the eighteenth, what were their respective needs, their productive forces, their mode of production, the raw materials of their production—in short, what were the relations between man and man which resulted from all these conditions of existence. To get to the bottom of all these questions—what is this but to draw up the real, profane history of men in every century and to present these men as both the authors and the actors of their own drama? But the moment you present men as the actors and authors of their own history, you arrive—by a detour—at the real starting-point, because you have abandoned those eternal principles of which you spoke at the outset.

M. Proudhon has not even gone far enough along the by-way which an ideologist takes to reach the main road of history.

Sixth Observation

Let us take the by-way with M. Proudhon.

We shall concede that economic relations, viewed as immutable laws, eternal principles, ideal categories, existed before active and energetic men did; we shall concede further that these laws, principles, and categories had, since the beginning of time, slumbered 'in the impersonal reason of humanity'. We have already seen that, with all these changeless and motionless eternities, there is no history left; there is at most history in the idea, that is, history reflected in the dialectic movement of pure reason. M. Proudhon, by saying that, in the dialectic movement, ideas are no longer differentiated, has done away with both the shadow of movement and the movement of shadows, by means of which one could still have created at least a semblance of history. Instead of that, he imputes to history his own impotence. He lays the blame on everything, even the French language. 'It is not correct then', says M. Proudhon, the philosopher, 'to say that something appears, that something is produced: in civilization as in the universe, everything has existed, has acted, from eternity. This applies to the whole of social economy.' (Vol. II, p. 102.)

So great is the productive force of the contradictions which function and which make M. Proudhon function, that, in trying to explain history, he is forced to deny it; in trying to explain the successive appearance of social relations, he denies that anything can appear: in trying to explain production, with all its phases, he questions whether anything can be produced!

Thus, for M. Proudhon, there is no longer any history: no longer any sequence of ideas. And yet his book still exists; and it is precisely that book which is, to use his own expression, 'history according to the sequence of ideas'. How shall we find a formula, for M. Proudhon is a man of formulas, to help him to clear all these contradictions in one leap?

To this end he has invented a new reason, which is neither the pure and virgin absolute reason, nor the common reason of men living and acting in different periods, but a reason quite apart—the reason of the person, Society—of the subject, Humanity—which under the pen of M. Proudhon figures at times also as social genius, general reason, or finally as human reason. This reason, decked out under so many names, betrays itself nevertheless, at every moment, as the individual reason of M. Proudhon, with its good and its bad side, its antidotes, and its problems.

'Human reason does not create truth,' hidden in the depths of absolute, eternal reason. It can only unveil it. But such truths as it has unveiled up to now are incomplete, insufficient, and consequently contradictory. Hence, economic categories, being themselves truths discovered, revealed by human reason, by social genius, are equally incomplete and contain within themselves the germ of contradiction. Before M. Proudhon, social genius saw only the antagonistic elements, and not the synthetic formula, both hidden simultaneously in absolute reason. Economic relations, which merely realize on earth these insufficient truths, these incomplete categories, these contradictory ideas, are consequently contradictory in themselves, and present two sides, one good, the other bad.

To find complete truth, the idea, in all its fullness, the synthetic formula that is to annihilate the contradiction, this is the problem of social genius. This again is why, in M. Proudhon's illusion, this same social genius has been harried from one category to another without ever having been able, despite all its battery of categories, to snatch from God or from absolute reason, a synthetic formula.

'At first, society (social genius) states a primary fact, puts forward a hypothesis ... a veritable antinomy, whose antagonistic results develop in the social economy in the same way as its consequences could have been deduced in the mind; so that industrial movement, following in all things the deduction of ideas, splits up into two currents, one of useful effects, the other of subversive results. To bring harmony into the constitution of this two-sided principle, and to solve this antinomy, society gives rise to a second, which will soon be followed by a third; and progress of social genius will take place in this manner, until, having exhausted all its contradictions—I suppose, but it is not proved that there is a limit to human contradictions—it returns in one leap to all its former positions and with a single formula solves all its problems.' (Vol. I, p. 133.)

Just as the antithesis was before turned into an antidote, so now the thesis becomes a hypothesis. This change of terms, coming from M. Proudhon, has no longer anything surprising for us! Human reason, which is anything but pure, having only incomplete vision, encounters at every step new problems to be solved. Every new thesis which it discovers in absolute reason and which is the negation of the first thesis, becomes for it a synthesis, which it accepts rather naïvely as the solution of the problem in question. It is thus that this reason frets and fumes in ever renewing contradictions until, coming to the end of the contradictions, it perceives that all its theses and syntheses are merely contradictory hypotheses. In its perplexity, 'human reason, social genius, returns in one leap to all its former positions, and in a single formula, solves all its problems'. This unique formula, by the way, constitutes M. Proudhon's true discovery. It is constituted value.

Hypotheses are made only in view of a certain aim. The aim that social genius, speaking through the mouth of M. Proudhon, set itself in the first place, was to eliminate the bad in every economic category, in order to have nothing left but the good. For it, the good, the supreme well-being, the real practical aim, is equality. And why did the social genius aim at equality rather than inequality, fraternity, catholicism, or any other principle? Because 'humanity has successively realized so many separate hypotheses only in view of a superior hypothesis', which precisely is equality. In other words: because equality is M. Proudhon's ideal. He imagines that the division of labour, credit, the workshop—all economic relations—were invented merely for the benefit of equality, and yet they always ended up by turning against it. Since history and the fiction of M. Proudhon contradict each other at every step, the latter concludes that there is a contradiction. If there is a contradiction, it exists only between his fixed idea and real movement.

Henceforth the good side of an economic relation is that which affirms equality; the bad side, that which negates it and affirms inequality. Every new category is a hypothesis of the social genius to eliminate the inequality engendered by the preceding hypothesis. In short, equality is the primordial intention, the mystical tendency, the providential aim that the social genius has constantly before its eyes as it whirls in the circle of economic contradictions. Thus Providence is the locomotive which makes the whole of M. Proudhon's economic baggage move better than his pure and volatilized reason. He has devoted to Providence a whole chapter, which follows the one on taxes.

Providence, providential aim, this is the great word used today to explain the movement of history. In fact, this word explains nothing. It is at most a rhetorical form, one of the various ways of paraphrasing facts.

It is a fact that in Scotland landed property acquired a new value by the development of English industry. This industry opened up new outlets for wool. In order to produce wool on a large scale, arable land had to be transformed into pasturage. To effect this transformation, the estates had to be concentrated. To concentrate the estates, small holdings had first to be abolished, thousands of tenants had to be driven from their native soil and a few shepherds in charge of millions of sheep to be installed in their place. Thus, by successive transformations, landed property in Scotland has resulted in the driving out of men by sheep. Now say that the providential aim of the institution of landed property in Scotland was to have men driven out by sheep, and you will have made providential history.

Of course, the tendency towards equality belongs to our century. To say now that all former centuries, with entirely different needs, means of production, etc., worked providentially for the realization of equality is, firstly, to substitute the means and the men of our century for the men and the means of earlier centuries and to misunderstand the historical movement by which the successive generations transformed the results acquired by the generations that preceded them. Economists know very well that the very thing that was for the one a finished product was for the other but the raw material for new production.

Suppose, as M. Proudhon does, that social genius produced, or rather improvised, the feudal lords with the providential aim of transforming the settlers into responsible and equally-placed workers: and you will have effected a substitution of aims and of persons worthy of the Providence that instituted landed property in Scotland, in order to give itself the malicious pleasure of driving out men by sheep.

But since M. Proudhon takes such a tender interest in Providence, we refer him to the *Histoire de l'économie politique* of M. de Villeneuve-Bargemont, who likewise goes in pursuit of a providential aim. This aim, however, is not equality, but catholicism.

Seventh and Last Observation

Economists have a singular method of procedure. There are only two kinds of institutions for them, artificial and natural. The institutions of feudalism are artificial institutions, those of the bourgeoisie are natural institutions. In this they resemble the theologians, who likewise establish two kinds of religion. Every religion which is not theirs is an invention of men, while their own is an emanation from God. When the economists say that present-day relations—the relations of bourgeois production—are natural, they imply that these are the relations in which wealth is created and productive forces developed in conformity with the laws of nature. These relations therefore are themselves natural laws independent of the influence of time. They are eternal laws which must always govern society. Thus there has been history, but there is no longer any. There has been history, since there were the institutions of feudalism, and in these institutions of feudalism we find quite different relations of production from those of bourgeois society, which the economists try to pass off as natural and as such, eternal.

Feudalism also had its proletariat—serfdom, which contained all the germs of the bourgeoisie. Feudal production also had two antagonistic elements which are likewise designated by the name of the good side and the bad side of feudalism, irrespective of the fact that it is always the bad side that in the end triumphs over the good side. It is the bad side that produces the movement which makes history, by providing a struggle. If, during the epoch of the domination of feudalism, the economists, enthusiastic over the knightly virtues, the beautiful harmony between rights and duties, the patriarchal life of the towns, the prosperous condition of domestic industry in the countryside, the development of industry organized into corporations, guilds, and fraternities, in short, everything that constitutes the good side of feudalism, had set themselves the problem of eliminating everything that cast a shadow on this picture—serfdom, privileges, anarchy—what would have happened? All the elements which called forth the struggle would have been destroyed, and the development of the bourgeoisie nipped in the bud. One would have set oneself the absurd problem of eliminating history.

After the triumph of the bourgeoisie there was no longer any question of the good or the bad side of feudalism. The bourgeoisie took possession of the productive forces it had developed under feudalism. All the old economic forms, the corresponding civil relations, the political state which was the official expression of the old civil society, were smashed.

Thus feudal production, to be judged properly, must be considered as a mode of production founded on antagonism. It must be shown how wealth was produced within this antagonism, how the productive forces were developed at the same time as class antagonisms, how one of the classes, the bad side, the drawback of society, went on growing until the material conditions for its emancipation had attained full maturity. Is not this as good as saying that the mode of production, the relations in which productive forces are developed, are anything but eternal laws, but that they correspond to a definite development of men and of their productive forces, and that a change in men's productive forces necessarily brings about a change in their relations of production? As the main thing is not to be deprived of the fruits of civilization, of the acquired productive forces, the traditional forms in which they were produced must be smashed. From this moment the revolutionary class becomes conservative.

The bourgeoisie begins with a proletariat which is itself a relic of the proletariat of feudal times. In the course of its historical development, the bourgeoisie necessarily develops its antagonistic character, which at first is more or less disguised, existing only in a latent state. As the bourgeoisie develops, there develops in its bosom a new proletariat, a modern proletariat; there develops a struggle between the proletarian class and the bourgeois class, a struggle which, before being felt, perceived, appreciated, understood, avowed, and proclaimed aloud by both sides, expresses itself, to start with, merely in partial and momentary conflicts, in subversive acts. On the other hand, if all the members of the modern bourgeoisie have the same interests inasmuch as they form a class as against another class, they have opposite, antagonistic interests inasmuch as they stand face to face with one another. This opposition of interests results from the economic conditions of their bourgeois life. From day to day it thus becomes clearer that the production relations in which the bourgeoisie moves have not a simple, uniform character, but a dual character; that in the selfsame relations in which wealth is produced, poverty is produced also; that in the selfsame relations in which there is a development of the productive forces, there is also a force producing repression; that these relations produce bourgeois wealth, i.e. the wealth of the bourgeois class, only by continually annihilating the wealth of the individual members of this class and by producing an ever-growing proletariat.

The more the antagonistic character comes to light, the more the economists, the scientific representatives of bourgeois production, find themselves in conflict with their own theory; and different schools arise.

We have the fatalist economists, who in their theory are as indifferent to what they call the drawbacks of bourgeois production as the bourgeois themselves are in practice to the sufferings of the proletarians who help them to acquire wealth. In this fatalist school there are Classics and Romantics. The Classics, like Adam Smith and Ricardo, represent a bourgeoisie which, while still struggling with the relics of feudal society, works only to purge economic relations of feudal taints, to increase the productive forces and to give a new upsurge to industry and commerce. The proletariat that takes part in this struggle and is absorbed in this feverish labour experiences only passing, accidental sufferings, and itself regards them as such. Economists like Adam Smith and Ricardo, who are the historians of this epoch, have no other mission than that of showing how wealth is acquired in bourgeois production relations, of formulating these relations into categories, into laws, and of showing how superior these laws, these categories, are for the production of wealth to the laws and categories of feudal society. Poverty is in their eyes merely the pang which accompanies every childbirth, in nature as in industry.

The Romantics belong to our own age, in which the bourgeoisie is in direct opposition to the proletariat; in which poverty is engendered in as great abundance as wealth. The economists now pose as blasé fatalists, who, from their elevated position, cast a proudly disdainful glance at the human machines who manufacture wealth. They copy all the developments given by their predecessors, and the indifference which in the latter was merely naïveté becomes in them coquetry.

Next comes the humanitarian school, which sympathizes with the bad side of present-day production relations. It seeks, by way of easing its conscience, to palliate even if slightly the real contrasts; it sincerely deplores the distress of the proletariat, the unbridled competition of the bourgeois among themselves; it counsels the workers to be sober, to work hard and to have few children; it advises the bourgeois to put a reasoned ardour into production. The whole theory of this school rests on interminable distinctions between theory and practice, between principles and results, between idea and application, between form and content, between essence and reality, between right and fact, between the good side and the bad side.

The philanthropic school is the humanitarian school carried to perfection. It denies the necessity of antagonism; it wants to turn all men into bourgeois; it wants to realize theory in so far as it is distinguished from practice and contains no antagonism. It goes without saying that, in theory, it is easy to make an abstraction of the contradictions that are met with at every moment in actual reality. This theory would therefore become idealized reality. The philanthropists, then, want to retain the categories which express bourgeois relations, without the antagonism which constitutes them and is inseparable from them. They think they are seriously fighting bourgeois practice, and they are more bourgeois than the others.

Just as the economists are the scientific representatives of the bourgeois class, so the Socialists and the Communists are the theoreticians of the proletarian class. So long as the proletariat is not yet sufficiently developed to constitute itself as a class, and consequently so long as the struggle itself of the proletariat with the bourgeoisie has not yet assumed a political character, and the productive forces are not yet sufficiently developed in the bosom of the bourgeoisie itself to enable us to catch a glimpse of the material conditions necessary for the emancipation of the proletariat and for the formation of a new society, these theoreticians are merely utopians who, to meet the wants of the oppressed classes, improvise systems and go in search of a regenerating science. But in the measure that history moves forward, and with it the struggle of the proletariat assumes clearer outlines, they no longer need to seek science in their minds; they have only to take note of what is happening before their eyes and to become its mouthpiece. So long as they look for science and merely make systems, so long as they are at the beginning of the struggle, they see in poverty nothing but poverty, without seeing in it the revolutionary, subversive side, which will overthrow the old society. From this moment, science, which is a product of the historical movement, has associated itself consciously with it, has ceased to be doctrinaire and has become revolutionary.

Let us return to M. Proudhon.

Every economic relation has a good and a bad side; it is the one point on which M. Proudhon does not give himself the lie. He sees the good side expounded by the economists; the bad side he sees denounced by the Socialists. He borrows from the economists the necessity of eternal relations; he borrows from the Socialists the illusion of seeing in poverty nothing but poverty. He is in agreement with both in wanting to fall back upon the authority of science. Science for him reduces itself to the slender proportions of a scientific formula; he is the man in search of formulas. Thus it is that M. Proudhon flatters himself on having given a criticism of both political economy and communism: he is beneath them both. Beneath the economists, since, as a philosopher who has at his elbow a magic formula, he thought he could dispense with going into purely economic details; beneath the Socialists, because he has neither courage enough nor insight enough to rise, be it even speculatively, above the bourgeois horizon.

He wants to be the synthesis—he is a composite error.

He wants to soar as the man of science above the bourgeois and the proletarians; he is merely the petty bourgeois, continually tossed back and forth between capital and labour, political economy and communism. . . .

On Strikes

Then again, a general rise in wages can never produce a more or less general rise in the price of goods. Actually, if every industry employed the same number of workers in relation to fixed capital or to the instruments used, a general rise in wages would produce a general fall in profits and the current price of goods would undergo no alteration.

But as the relation of manual labour to fixed capital is not the same in different industries, all the industries which employ a relatively greater mass of capital and fewer workers, will be forced sooner or later to lower the price of their goods. In the opposite case, in which the price of their goods is not lowered, their profit will rise above the common rate of profits. Machines are not wage-earners. Therefore, the general rise in wages will affect less those industries, which, compared with the others, employ more machines than workers. But as competition always tends to level the rate of profits, those profits which rise above the average rate cannot but be transitory. Thus, apart from a few fluctuations, a general rise in wages will lead, not as M. Proudhon says, to a general increase in prices, but to a partial fall, that is, a fall in the current price of the goods that are made chiefly with the help of machines.

The rise and fall of profits and wages expresses merely the proportion in which capitalists and workers share in the product of a day's work, without influencing in most instances the price of the product. But that 'strikes followed by an increase in wages culminate in a general rise in prices, in a dearth even'—these are notions which can blossom only in the brain of a poet who has not been understood.

In England strikes have regularly given rise to the invention and application of new machines. Machines were, it may be said, the weapon employed by the capitalists to quell the revolt of specialized labour. The self-acting mule, the greatest invention of modern industry, put out of action the spinners who were in revolt. If combinations and strikes had no other effect than that of making the efforts of mechanical genius react against them, they would still exercise an immense influence on the development of industry.
. . .

The first attempts of workers to associate among themselves always take place in the form of combinations.

Large-scale industry concentrates in one place a crowd of people unknown to one another. Competition divides their interests. But the maintenance of wages, this common interest which they have against their boss, unites them in a common thought of resistance—combination. Thus combination always has a double aim, that of stopping competition among the workers, so that they can carry on general competition with the capitalist. If the first aim of resistance was merely the maintenance of wages, combinations, at first isolated, constitute themselves into groups as the capitalists in their turn unite for the purpose of repression, and in the face of always united capital, the maintenance of the association becomes more necessary to them than that of wages. This is so true that English economists are amazed to see the workers sacrifice a good part of their wages in favour of associations, which, in the eyes of these economists, are established solely in favour of wages. In this struggle—a veritable civil war—all the elements necessary for a coming battle unite and develop. Once it has reached this point, association takes on a political character.

Economic conditions had first transformed the mass of the people of the country into workers. The combination of capital has created for this mass a common situation, common interests. This mass is thus already a class as against capital, but not yet for itself. In the struggle, of which we have noted only a few phases, this mass becomes united, and constitutes itself as a class for itself. The interests it defends become class interests. But the struggle of class against class is a political struggle.

In the bourgeoisie we have two phases to distinguish: that in which it constituted itself as a class under the regime of feudalism and absolute monarchy, and that in which, already constituted as a class, it overthrew feudalism and monarchy to make society into a bourgeois society. The first of these phases was the longer and necessitated the greater efforts. This too began by partial combinations against the feudal lords.

Much research has been carried out to trace the different historical phases that the bourgeoisie has passed through, from the commune up to its constitution as a class.

But when it is a question of making a precise study of strikes, combinations and other forms in which the proletarians carry out before our eyes their organization as a class, some are seized with real fear and others display a transcendental disdain.

An oppressed class is the vital condition for every society founded on the antagonism of classes. The emancipation of the oppressed class thus implies necessarily the creation of a new society. For the oppressed class to be able to emancipate itself it is necessary that the productive powers already acquired and the existing social relations should no longer be capable of existing side by side. Of all the instruments of production, the greatest productive power is the revolutionary class itself. The organization of revolutionary elements as a class supposes the existence of all the productive forces which could be engendered in the bosom of the old society.

Does this mean that after the fall of the old society there will be a new class domination culminating in a new political power? No.

The condition for the emancipation of the working class is the abolition of every class, just as the condition for the liberation of the third estate, of the bourgeois order, was the abolition of all estates and all orders.

The working class, in the course of its development, will substitute for the old civil society an association which will exclude classes and their antagonism, and there will be no more political power properly so called, since political power is precisely the official expression of antagonism in civil society.

Meanwhile, the antagonism between the proletariat and the bourgeoisie is a struggle of class against class, a struggle which carried to its highest expression is a total revolution. Indeed, is it at all surprising that a society founded on the opposition of classes should culminate in brutal contradiction, the shock of body against body, as its final *dénouement*?

Do not say that social movement excludes political movement. There is never a political movement which is not at the same time social.

It is only in an order of things in which there are no more classes and class antagonisms that social evolutions will cease to be political revolutions. Till then, on the eve of every general reshuffling of society, the last word of social science will always be:

'Struggle or death; bloody war or nothing. It is thus that the question is inevitably posed.'

George Sand

BIBLIOGRAPHY

Original: K. Marx, *Œuvres*, ed. M. Rubel, Paris, 1963 ff., Vol. 1, pp. 1 ff.

Present translation: K. Marx, *The Poverty of Philosophy*, Moscow, n.d., pp. 53 ff., 66 ff., 87 f., 115 ff., 188 f., 194 f. Reproduced by kind permission of Lawrence and Wishart Ltd.

Other translations: None.

Commentaries: H. Adams, *Karl Marx in his Earlier writings*, 2nd edn., London, 1965, Ch. 11.
L. Dupré, *The Philosophical Foundations of Marxism*, New York, 1966, Ch. 7.
J. Hoffman, *Revolutionary Justice*, Urbana, 1972, pp. 85 ff.
E. Mandel, *The Formation of Marx's Economic Thought*, London, 1971, Ch. 4.
N. Rothenstreich, *Basis Problems of Marx's Philosophy*, New York, 1965, pp. 100 ff.
G. Woodcock, *Proudhon*, London, 1956.

17. Moralizing Criticism and Critical Morality

This article was written in refutation of Karl Heinzen, a 'true socialist' who propagated the idea of a Republic based on 'humanity'. In the extracts below Marx points to the importance of classes in any social analysis and explains in what sense the proletariat must act as allies of the bourgeoisie in the next revolutionary upheaval.

. . . After vouchsafing such profound explanations about the 'connection of politics with social conditions' and the 'class relations' with the State power, Mr. Heinzen exclaims triumphantly: 'The "communistic narrow-mindedness" which divides men into classes, or antagonizes them according to their handicraft, has been avoided by me. I have left open the "possibility" that "humanity" is not always determined by "class" or the "length of one's purse".' Bluff common sense transforms the class distinction into the 'length of the purse' and the class antagonism into trade quarrels. The length of the purse is a purely quantitative distinction, which may perchance antagonize any two individuals of the same class. That the medieval guilds confronted each other on the basis of handicraft is well known. But it is likewise well known that the modern class distinction is by no means based on handicraft; rather the division of labour within the same class produces very different methods of work.

It is very 'possible' that particular individuals are not always influenced in their attitude by the class to which they belong, but this has as little effect upon the class struggle as the secession of a few nobles to the *tiers état* had on the French Revolution. And then these nobles at least joined a class, the revolutionary class, the bourgeoisie. But Mr. Heinzen sees all classes melt away before the solemn idea of 'humanity'.

If he believes that entire classes, which are based upon economic conditions independent of their will, and are set by these conditions in a relation of mutual antagonism, can break away from their real relations, by virtue of the quality of 'humanity' which is inherent in all men, how easy it should be for a prince to raise himself above his 'princedom', above his 'princely handicraft' by virtue of 'humanity'? Why does he take it amiss when Engels perceives a 'brave Emperor Joseph' behind his revolutionary phrases?

But if, on the one hand, Mr. Heinzen obliterates all distinctions, in addressing himself vaguely to the 'humanity' of Germans, so that he is obliged to include even the princes in his admonitions, on the other hand he finds himself obliged to set up at least one distinction among Germans, for without a distinction there can be no antagonism, and without an antagonism, no materials for political Capuchinian sermons.

Mr. Heinzen therefore divides Germans into princes and subjects.

The 'narrow-minded' communists see not only the political distinction of prince and subject, but also the social distinction of classes.

It is well known that, shortly after the July Revolution, the victorious bourgeoisie, in its September laws, made 'the incitement of class against class', probably also out of 'humanity', a criminal offence, to which imprisonment and fines were attached. It is further well known that the English bourgeois newspapers could not denounce the Chartist leaders and Chartist writers more effectively than by reproaching them with setting class against class. It is even notorious that, in consequence of inciting class against class, German writers are incarcerated in fortresses. Is not Mr. Heinzen this time talking the language of the French September laws, the English bourgeois newspapers, and the German penal code?

But no. The well-meaning Mr. Heinzen only fears that the communists 'are seeking to assure the princes a revolutionary Fontanelle'. Thus the Belgian liberals assure us that the radicals are in secret alliance with the catholics; the French liberals assure us that the democrats have an understanding with the legitimists. And the liberal Mr. Heinzen assures us that the communists have an understanding with the princes.

As I once pointed out in the Franco-German Annuals, Germany has her own Christian-Germanic plague. Her bourgeoisie was so retarded in its development that it is beginning its struggle with absolute monarchy and seeking to establish its political power at the moment when in all developed countries the bourgeoisie is already engaged in the most violent struggles with the working class, and when its political illusions are already obsolete so far as the intellect of Europe is concerned.

In this country, where the political poverty of absolute monarchy still exists with a whole appendage of decayed semi-feudal orders and conditions, there exist on the other hand, partly in consequence of the industrial development and Germany's dependence of the world market, the antagonisms between the bourgeoisie and the working class, and the struggle arising therefrom, an instance of which are the workers' revolts in Silesia and Bohemia. The German bourgeoisie therefore finds itself in a relation of antagonism to the proletariat before it has yet constituted itself politically as a class. The struggle among the subjects has broken out before ever princes and nobles have been got rid of, in spite of all Hambach songs.

Mr. Heinzen does not know how to explain these contradictory relations, which of course are also reflected in German literature, except by putting them on to his opponents' conscience and interpreting them as the consequence of the counter-revolutionary activities of the communists.

Meanwhile, the German workers are quite aware that the absolute monarchy does not and cannot hesitate one moment to greet them with a whiff of grapeshot in the service of the bourgeoisie. Why then should they prefer the direct rule of the bourgeoisie to the brutal oppression of absolute government, with its semi-feudal retinue? The workers know that the bourgeoisie must not only make them wider concessions than absolute monarchy, but that in the interests of its commerce and industry, the bourgeoisie must create against its will the conditions for the unity of the workers, and the unity of the workers is the first requisite for their victory. The workers know that the abolition of bourgeois property relations is not brought about by the maintenance of feudal property relations. They know their own revolutionary movement can only be accelerated through the revolutionary movement of the bourgeoisie against the feudal orders and the absolute monarchy. They know that their own struggle with the bourgeoisie can only break out on the day the bourgeoisie triumphs. In spite of all, they do not share Mr. Heinzen's middle-class illusions. They can and must take part in the middle-class revolution as a condition preliminary to the Labour revolution. But they cannot for a moment regard it as their objective. . . .

BIBLIOGRAPHY

Original: *MEW*, Vol. 4, pp. 349 ff.

Present translation: K. Marx, *Selected Essays*, ed. H. Stenning, London and New York, 1926, pp. 155 ff.

Other translations: None.

Commentaries: F. Mehring, *Karl Marx*, London, 1936, p. 134.
D. McLellan, *Karl Marx: His Life and Thought*, London and New York, 1973, pp. 174 ff.

III. 1848 and After

18. *The Communist Manifesto*

The Communist League, linking the main centres of communist activities in Paris, London, Brussels, and Cologne, was formed out of the League of the Just in June 1847, largely at the instigation of Marx and Engels. At a Congress in London in November 1847, the need was expressed for a clear formulation of the League's principles, and Marx and Engels were asked to draw up a statement. Engels had already composed a draft after the June Congress, and Marx incorporated some of this material when he wrote the *Manifesto* in Brussels in December and January. By the time it was published in February 1848, the series of revolutions that marked that year had already broken out.

The *Communist Manifesto* has four sections. The first gives a history of society as class society since the Middle Ages and ends with a prophecy of the victory of the proletariat over the present ruling class, the bourgeoisie. The second section describes the position of communists within the proletarian class, rejects bourgeois objections to communism, and then characterizes the communist revolution, the measures to be taken by the victorious proletariat, and the nature of the future communist society. The third section contains an extended criticism of other types of socialism—reactionary, bourgeois, and utopian. The final section provides a short description of communist tactics towards other opposition parties and finishes with an appeal for proletarian unity.

None of the ideas in the *Communist Manifesto* were new, and its ideas on revolution and history were obviously influenced by French socialists such as Babeuf, Saint-Simon, and Considérant; and the concept of class, with which the *Manifesto* begins, was first used by French bourgeois historians. What is new is the force of expression and the powerful synthesis afforded by the materialist conception of history. For many parts of the *Manifesto* are simply brilliant summaries of views put forward in *The German Ideology*. Marx and Engels continued to recognize this pamphlet as a classic expression of their views, though they would subsequently have wished to modify some of its ideas—particularly (in the light of the Paris Commune) those relating to the proletariat's appropriation of the state apparatus and the rather simplistic statements on immiserization and class polarization.

A spectre is haunting Europe—the spectre of Communism. All the Powers of old Europe have entered into a holy alliance to exorcise this spectre: Pope and Tsar, Metternich and Guizot, French Radicals and German police-spies.

Where is the party in opposition that has not been decried as Communistic by its opponents in power? Where the Opposition that has not hurled back the branding reproach of Communism, against the more advanced opposition parties, as well as against its reactionary adversaries?

Two things result from this fact.

I. Communism is already acknowledged by all European Powers to be itself a Power.

II. It is high time that Communists should openly, in the face of the whole world, publish their views, their aims, their tendencies, and meet this nursery tale of the Spectre of Communism with a Manifesto of the party itself.

To this end, Communists of various nationalities have assembled in London, and sketched the following Manifesto, to be published in the English, French, German, Italian, Flemish, and Danish languages.

I
Bourgeois and Proletarians

The history of all hitherto existing society is the history of class struggles.

Freeman and slave, patrician and plebeian, lord and serf, guild-master and journeyman—in a word, oppressor and oppressed, stood in constant opposition to one another, carried on an uninterrupted, now hidden, now open fight, a fight that each time ended either in a revolutionary re-constitution of society at large or in the common ruin of the contending classes.

In the earlier epochs of history, we find almost everywhere a complicated arrangement of society into various orders, a manifold gradation of social rank. In ancient Rome we have patricians, knights, plebeians, slaves; in the Middle Ages, feudal lords, vassals, guild-masters, journeymen, apprentices, serfs; in almost all of these classes, again, subordinate gradations.

The modern bourgeois society that has sprouted from the ruins of feudal society has not done away with class antagonisms. It has but established new classes, new conditions of oppression, new forms of struggle in place of the old ones.

Our epoch, the epoch of the bourgeoisie, possesses, however, this distinctive feature: it has simplified the class antagonisms. Society as a whole is more and more splitting up into two great hostile camps, into two great classes directly facing each other: Bourgeoisie and Proletariat.

From the serfs of the Middle Ages sprang the chartered burghers of the earliest towns. From these burgesses the first elements of the bourgeoisie were developed.

The discovery of America, the rounding of the Cape, opened up fresh ground for the rising bourgeoisie. The East Indian and Chinese markets, the colonization of America, trade with the colonies, the increase in the means of exchange and in commodities generally, gave to commerce, to navigation, to industry, an impulse never before known, and thereby, to the revolutionary element in the tottering feudal society, a rapid development.

The feudal system of industry, under which industrial production was monopolized by closed guilds, now no longer sufficed for the growing wants of the new markets. The manufacturing system took its place. The guild-masters were pushed on one side by the manufacturing middle class; division of labour between the different corporate guilds vanished in the face of division of labour in each single workshop.

Meantime the markets kept ever growing, the demand ever rising. Even manufacture no longer sufficed. Thereupon, steam and machinery revolutionized industrial production. The place of manufacture was taken by the giant, Modern Industry, the place of the industrial middle class, by industrial millionaires, the leaders of whole industrial armies, the modern bourgeois.

Modern industry has established the world-market, for which the discovery of America paved the way. This market has given an immense development to commerce, to navigation, to communication by land. This development has, in its turn, reacted on the extension of industry; and in proportion as industry, commerce, navigation, railways extended, in the same proportion the bourgeoisie developed, increased its capital, and pushed into the background every class handed down from the Middle Ages.

We see, therefore, how the modern bourgeoisie is itself the product of a long course of development, of a series of revolutions in the modes of production and of exchange.

Each step in the development of the bourgeoisie was accompanied by a corresponding political advance of that class. An oppressed class under the sway of the feudal nobility, an armed and self-governing association in the medieval commune; here independent urban republic (as in Italy and Germany), there taxable 'third estate' of the monarchy (as in France), afterwards, in the period of manufacture proper, serving either the semi-feudal or the absolute monarchy as a counterpoise against the nobility, and, in fact, corner-stone of the great monarchies in general, the bourgeoisie has at last, since the establishment of Modern Industry and of the world-market, conquered for itself, in the modern representative State, exclusive political sway. The executive of the modern State is but a committee for managing the common affairs of the whole bourgeoisie.

The bourgeoisie, historically, has played a most revolutionary part.

The bourgeoisie, wherever it has got the upper hand, has put an end to all feudal, patriarchal, idyllic relations. It has pitilessly torn asunder the motley feudal ties that bound man to his 'natural superiors', and has left remaining no other nexus between man and man than naked self-interest, than callous 'cash payment'. It has drowned the most heavenly ecstasies of religious fervour, of chivalrous enthusiasm, of philistine sentimentalism, in the icy water of egotistical calculation. It has resolved personal worth into exchange value, and in place of the numberless indefeasible chartered freedoms, has set up that single, unconscionable freedom—Free Trade. In one word, for exploitation, veiled by religious and political illusions, it has substituted naked, shameless, direct, brutal exploitation.

The bourgeoisie has stripped of its halo every occupation hitherto honoured and looked up to with reverent awe. It has converted the physician, the lawyer, the priest, the poet, the man of science into its paid wage-labourers.

The bourgeoisie has torn away from the family its sentimental veil, and has reduced the family relation to a mere money relation.

The bourgeoisie has disclosed how it came to pass that the brutal display of vigour in the Middle Ages, which Reactionists so much admire, found its fitting complement in the most slothful indolence. It has been the first to show what man's activity can bring about. It has accomplished wonders far surpassing Egyptian pyramids, Roman aqueducts, and Gothic cathedrals; it has conducted expeditions that put in the shade all former Exoduses of nations and crusades.

The bourgeoisie cannot exist without constantly revolutionizing the instruments of production, and thereby the relations of production, and with them the whole relations of society. Conservation of the old modes of production in unaltered form, was, on the contrary, the first condition of existence for all earlier industrial classes. Constant revolutionizing of production, uninterrupted disturbance of all social conditions, everlasting uncertainty and agitation distinguish the bourgeois epoch from all earlier ones. All fixed, fast-frozen relations, with their train of ancient and venerable prejudices and opinions, are swept away, all new-formed ones become antiquated before they can ossify. All that is solid melts into air, all that is holy is profaned, and man is at last compelled to face with sober senses, his real conditions of life, and his relations with his kind.

The need of a constantly expanding market for its products chases the bourgeoisie over the whole surface of the globe. It must nestle everywhere, settle everywhere, establish connections everywhere.

The bourgeoisie has through its exploitation of the world-market given a cosmopolitan character to production and consumption in every country. To the great chagrin of Reactionists, it has drawn from under the feet of industry the national ground on which it stood. All old-established national industries have been destroyed or are daily being destroyed. They are dislodged by new industries, whose introduction becomes a life-and-death question for all civilized nations, by industries that no longer work up indigenous raw material, but raw material drawn from the remotest zones; industries whose products are consumed, not only at home, but in every quarter of the globe. In place of the old wants, satisfied by the productions of the country, we find new wants, requiring for their satisfaction the products of distant lands and climes. In place of the old local and national seclusion and self-sufficiency, we have intercourse in every direction, universal interdependence of nations. And as in material, so also in intellectual production. The intellectual creations of individual nations become common property. National one-sidedness and narrow-mindedness become more and more impossible, and from the numerous national and local literatures, there arises a world literature.

The bourgeoisie, by the rapid improvement of all instruments of production, by the immensely facilitated means of communication, draws all, even the most barbarian, nations into civilization. The cheap prices of its commodities are the heavy artillery with which it batters down all Chinese walls, with which it forces the barbarians' intensely obstinate hatred of foreigners to capitulate. It compels all nations, on pain of extinction, to adopt the bourgeois mode of production; it compels them to introduce what it calls civilization into their midst, i.e., to become bourgeois themselves. In one word, it creates a world after its own image.

The bourgeoisie has subjected the country to the rule of the towns. It has created enormous cities, has greatly increased the urban population as compared with the rural, and has thus rescued a considerable part of the population from the idiocy of rural life. Just as it has made the country dependent on the towns, so it has made barbarian and semi-barbarian countries dependent on the civilized ones, nations of peasants on nations of bourgeois, the East on the West.

The bourgeoisie keeps more and more doing away with the scattered state of the population, of the means of production, and of property. It has agglomerated population, centralized means of production, and has concentrated property in a few hands. The necessary consequence of this was political centralization. Independent or but loosely connected provinces, with separate interests, laws, governments, and systems of taxation, became lumped together into one nation, with one government, one code of laws, one national class-interest, one frontier, and one customs-tariff.

The bourgeoisie, during its rule of scarcely one hundred years, has created more massive and more colossal productive forces than have all preceding generations together. Subjection of Nature's forces to man, machinery, application of chemistry to industry and agriculture, steam-navigation, railways, electric telegraphs, clearing of whole continents for cultivation, canalization of rivers, whole populations conjured out of the ground—what earlier century had even a presentiment that such productive forces slumbered in the lap of social labour?

We see then that the means of production and of exchange, on whose foundation the bourgeoisie built itself up, were generated in feudal society. At a certain stage in the development of these means of production and of exchange, the conditions under which feudal society produced and exchanged, the feudal organization of agriculture and manufacturing industry, in one word, the feudal relations of property became no longer compatible with the already developed productive forces; they became so many fetters. They had to be burst asunder; they were burst asunder.

Into their place stepped free competition, accompanied by a social and political constitution adapted to it, and by the economical and political sway of the bourgeois class.

A similar movement is going on before our own eyes. Modern bourgeois society with its relations of production, of exchange and of property, a society that has conjured up such gigantic means of production and of exchange, is like the sorcerer, who is no longer able to control the powers of the nether world which he has called up by his spells. The history of industry and commerce for many a decade past is but the history of the revolt of modern productive forces against modern conditions of production, against the property relations that are the conditions for the existence of the bourgeoisie and of its rule. It is enough to mention the commercial crises that by their periodical return put on trial, each time more threateningly, the existence of the entire bourgeois society. In these crises a great part not only of the existing products, but also of the previously created productive forces, are periodically destroyed. In these crises there breaks out an epidemic that, in all earlier epochs, would have seemed an absurdity—the epidemic of over-production. Society suddenly finds itself put back into a state of momentary barbarism; it appears as if a famine, a universal war of devastation, has cut off the supply of every means of subsistence; industry and commerce seem to be destroyed; and why? Because there is too much civilization, too much means of subsistence, too much industry, too much commerce. The produc-tive forces at the disposal of society no longer tend to further the develop-ment of the conditions of bourgeois property; on the contrary, they have become too powerful for these conditions, by which they are fettered, and so soon as they overcome these fetters, they bring disorder into the whole of bourgeois society, endanger the existence of bourgeois property. The condi-tions of bourgeois society are too narrow to comprise the wealth created by them. And how does the bourgeoisie get over these crises? On the one hand by enforced destruction of a mass of productive forces; on the other, by the conquest of new markets, and by the more thorough exploitation of the old ones. That is to say, by paving the way for more extensive and more destruc-tive crises, and by diminishing the means whereby crises are prevented.

The weapons with which the bourgeoisie felled feudalism to the ground are now turned against the bourgeoisie itself.

But not only has the bourgeoisie forged the weapons that bring death to itself; it has also called into existence the men who are to wield those weapons—the modern working class—the proletarians.

In proportion as the bourgeoisie, i.e., capital, is developed, in the same proportion is the proletariat, the modern working class, developed—a class of labourers, who live only so long as they find work, and who find work only so long as their labour increases capital. These labourers, who must sell themselves piecemeal, are a commodity, like every other article of commerce, and are consequently exposed to all the vicissitudes of competition, to all the fluctuations of the market.

Owing to the extensive use of machinery and to division of labour, the work of the proletarians has lost all individual character, and, consequently, all charm for the workman. He becomes an appendage of the machine, and it is only the most simple, most monotonous, and most easily acquired knack, that is required of him. Hence, the cost of production of a workman is restricted, almost entirely, to the means of subsistence that he requires for his maintenance, and for the propagation of his race. But the price of a commodity, and therefore also of labour, is equal to its cost of production. In proportion, therefore, as the repulsiveness of the work increases, the wage decreases. Nay more, in proportion as the use of machinery and division of labour increases, in the same proportion the burden of toil also increases, whether by prolongation of the working hours, by increase of the work exacted in a given time or by increased speed of the machinery, etc.

Modern industry has converted the little workshop of the patriarchal master into the great factory of the industrial capitalist. Masses of labourers, crowded into the factory, are organized like soldiers. As privates of the industrial army they are placed under the command of a perfect hierarchy of officers and sergeants. Not only are they slaves of the bourgeois class, and of the bourgeois State; they are daily and hourly enslaved by the machine, by the overlooker, and, above all, by the individual bourgeois manufacturer himself. The more openly this despotism proclaims gain to be its end and aim, the more petty, the more hateful, and the more embittering it is.

The less the skill and exertion of strength implied in manual labour, in other words, the more modern industry becomes developed, the more is the labour of men superseded by that of women. Differences of age and sex have no longer any distinctive social validity for the working class. All are instruments of labour, more or less expensive to use, according to their age and sex.

No sooner is the exploitation of the labourer by the manufacturer, so far, at an end, and he receives his wages in cash, than he is set upon by the other portions of the bourgeoisie, the landlord, the shopkeeper, the pawnbroker, etc.

The lower strata of the middle class—the small tradespeople, shop-keepers, and retired tradesmen generally, the handicraftsmen and peasants—all these sink gradually into the proletariat, partly because their diminutive capital does not suffice for the scale on which Modern Industry is carried on, and is swamped in the competition with the large capitalists, partly because their specialized skill is rendered worthless by new methods of production. Thus the proletariat is recruited from all classes of the population.

The proletariat goes through various stages of development. With its birth begins its struggle with the bourgeoisie. At first the contest is carried on by individual labourers, then by the workpeople of a factory, then by the operatives of one trade, in one locality, against the individual bourgeois who directly exploits them. They direct their attacks not against the bourgeois

conditions of production, but against the instruments of production themselves; they destroy imported wares that compete with their labour, they
smash to pieces machinery, they set factories ablaze, they seek to restore by
force the vanished status of the workman of the Middle Ages.

At this stage the labourers still form an incoherent mass scattered over the
whole country, and broken up by their mutual competition. If anywhere
they unite to form more compact bodies, this is not yet the consequence of
their own active union, but of the union of the bourgeoisie, which class, in
order to attain its own political ends, is compelled to set the whole proletariat
in motion, and is moreover yet, for a time, able to do so. At this stage,
therefore, the proletarians do not fight their enemies, but the enemies of
their enemies, the remnants of absolute monarchy, the landowners, the non-
industrial bourgeois, the petty bourgeoisie. Thus the whole historical
movement is concentrated in the hands of the bourgeoisie; every victory so
obtained is a victory for the bourgeoisie.

But with the development of industry the proletariat not only increases in
number; it becomes concentrated in greater masses, its strength grows, and
it feels that strength more. The various interests and conditions of life within
the ranks of the proletariat are more and more equalized, in proportion as
machinery obliterates all distinctions of labour, and nearly everywhere
reduces wages to the same low level. The growing competition among the
bourgeois, and the resulting commercial crises, make the wages of the workers ever more fluctuating. The unceasing improvement of machinery, ever
more rapidly developing, makes their livelihood more and more precarious;
the collisions between individual workmen and individual bourgeois take
more and more the character of collisions between two classes. Thereupon
the workers begin to form combinations (Trades' Unions) against the bourgeois; they club together in order to keep up the rate of wages; they found
permanent associations in order to make provision beforehand for these
occasional revolts. Here and there the contest breaks out into riots.

Now and then the workers are victorious, but only for a time. The real
fruit of their battles lies, not in the immediate result, but in the ever-
expanding union of the workers. This union is helped on by the improved
means of communication that are created by modern industry and that place
the workers of different localities in contact with one another. It was just this
contact that was needed to centralize the numerous local struggles, all of the
same character, into one national struggle between classes. But every class
struggle is a political struggle. And that union, to attain which the burghers
of the Middle Ages, with their miserable highways, required centuries, the
modern proletarians, thanks to railways, achieve in a few years.

This organization of the proletarians into a class, and consequently into a
political party, is continually being upset again by the competition between
the workers themselves. But it ever rises up again, stronger, firmer, mightier.
It compels legislative recognition of particular interests of the workers, by

taking advantage of the divisions among the bourgeoisie itself. Thus the ten-hours' bill in England was carried.

Altogether, collisions between the classes of the old society further in many ways the course of development of the proletariat. The bourgeoisie finds itself involved in a constant battle. At first with the aristocracy; later on, with those portions of the bourgeoisie itself whose interests have become antagonistic to the progress of industry; at all times, with the bourgeoisie of foreign countries. In all these battles it sees itself compelled to appeal to the proletariat, to ask for its help, and thus to drag it into the political arena. The bourgeoisie itself, therefore, supplies the proletariat with its own elements of political and general education, in other words, it furnishes the proletariat with weapons for fighting the bourgeoisie.

Further, as we have already seen, entire sections of the ruling classes are, by the advance of industry, precipitated into the proletariat, or are at least threatened in their conditions of existence. These also supply the proletariat with fresh elements of enlightenment and progress.

Finally, in times when the class struggle nears the decisive hour, the process of dissolution going on within the ruling class, in fact within the whole range of old society, assumes such a violent, glaring character, that a small section of the ruling class cuts itself adrift, and joins the revolutionary class, the class that holds the future in its hands. Just as, therefore, at an earlier period, a section of the nobility went over to the bourgeoisie, so now a portion of the bourgeoisie goes over to the proletariat, and in particular, a portion of the bourgeois ideologists, who have raised themselves to the level of comprehending theoretically the historical movement as a whole.

Of all the classes that stand face to face with the bourgeoisie today, the proletariat alone is a really revolutionary class. The other classes decay and finally disappear in the face of Modern Industry; the proletariat is its special and essential product.

The lower middle class, the small manufacturer, the shopkeeper, the artisan, the peasant, all these fight against the bourgeoisie, to save from extinction their existence as fractions of the middle class. They are therefore not revolutionary, but conservative. Nay more, they are reactionary, for they try to roll back the wheel of history. If by chance they are revolutionary, they are so only in view of their impending transfer into the proletariat; they thus defend not their present, but their future interests, they desert their own standpoint to place themselves at that of the proletariat.

The 'dangerous class', the social scum, that passively rotting mass thrown off by the lowest layers of old society, may, here and there, be swept into the movement by a proletarian revolution; its conditions of life, however, prepare it far more for the part of a bribed tool of reactionary intrigue.

In the conditions of the proletariat, those of old society at large are already virtually swamped. The proletarian is without property; his relation to his wife and children has no longer anything in common with the bourgeois family relations; modern industrial labour, modern subjection to capital, the same in England as in France, in America as in Germany, has stripped him of every trace of national character. Law, morality, religion are to him so many bourgeois prejudices, behind which lurk in ambush just as many bourgeois interests.

All the preceding classes that got the upper hand, sought to fortify their already acquired status by subjecting society at large to their conditions of appropriation. The proletarians cannot become masters of the productive forces of society, except by abolishing their own previous mode of appropriation, and thereby also every other previous mode of appropriation. They have nothing of their own to secure and to fortify; their mission is to destroy all previous securities for, and insurances of, individual property.

All previous historical movements were movements of minorities, or in the interests of minorities. The proletarian movement is the self-conscious, independent movement of the immense majority, in the interests of the immense majority. The proletariat, the lowest stratum of our present society, cannot stir, cannot raise itself up, without the whole superincumbent strata of official society being sprung into the air.

Though not in substance, yet in form, the struggle of the proletariat with the bourgeoisie is at first a national struggle. The proletariat of each country must, of course, first of all settle matters with its own bourgeoisie.

In depicting the most general phases of the development of the proletariat, we traced the more or less veiled civil war, raging within existing society, up to the point where that war breaks out into open revolution, and where the violent overthrow of the bourgeoisie lays the foundation for the sway of the proletariat.

Hitherto, every form of society has been based, as we have already seen, on the antagonism of oppressing and oppressed classes. But in order to oppress a class, certain conditions must be assured to it under which it can, at least, continue its slavish existence. The serf, in the period of serfdom, raised himself to membership in the commune, just as the petty bourgeois, under the yoke of feudal absolutism, managed to develop into a bourgeois. The modern labourer, on the contrary, instead of rising with the progress of industry, sinks deeper and deeper below the conditions of existence of his own class. He becomes a pauper, and pauperism develops more rapidly than population and wealth. And here it becomes evident, that the bourgeoisie is unfit any longer to be the ruling class in society, and to impose its conditions of existence upon society as an overriding law. It is unfit to rule because it is incompetent to assure an existence to its slave within his slavery, because it cannot help letting him sink into such a state, that it has to feed him, instead of being fed by him. Society can no longer live under this bourgeoisie, in other words, its existence is no longer compatible with society.

The essential condition for the existence, and for the sway of the bourgeois class, is the formation and augmentation of capital; the condition for capital is wage-labour. Wage-labour rests exclusively on competition between the labourers. The advance of industry, whose involuntary promoter is the bourgeoisie, replaces the isolation of the labourers, due to competition, by their revolutionary combination, due to association. The development of Modern Industry, therefore, cuts from under its feet the very foundation on which the bourgeoisie produces and appropriates products. What the bourgeoisie, therefore, produces, above all, is its own grave-diggers. Its fall and the victory of the proletariat are equally inevitable.

II
Proletarians and Communists

In what relation do the Communists stand to the proletarians as a whole?

The Communists do not form a separate party opposed to other working-class parties.

They have no interests separate and apart from those of the proletariat as a whole.

They do not set up any sectarian principles of their own, by which to shape and mould the proletarian movement.

The Communists are distinguished from the other working-class parties by this only: 1. In the national struggles of the proletarians of the different countries, they point out and bring to the front the common interests of the entire proletariat, independently of all nationality. 2. In the various stages of development which the struggle of the working class against the bourgeoisie has to pass through, they always and everywhere represent the interests of the movement as a whole.

The Communists, therefore, are on the one hand, practically, the most advanced and resolute section of the working-class parties of every country, that section which pushes forward all others; on the other hand, theoretically, they have over the great mass of the proletariat the advantage of clearly understanding the line of march, the conditions, and the ultimate general results of the proletarian movement.

The immediate aim of the Communists is the same as that of all the other proletarian parties: formation of the proletariat into a class, overthrow of the bourgeois supremacy, conquest of political power by the proletariat.

The theoretical conclusions of the Communists are in no way based on ideas or principles that have been invented, or discovered, by this or that would-be universal reformer.

They merely express, in general terms, actual relations springing from an existing class struggle, from a historical movement going on under our very eyes. The abolition of existing property relations is not at all a distinctive feature of Communism.

All property relations in the past have continually been subject to historical change consequent upon the change in historical conditions.

The French Revolution, for example, abolished feudal property in favour of bourgeois property.

The distinguishing feature of Communism is not the abolition of property generally, but the abolition of bourgeois property. But modern bourgeois private property is the final and most complete expression of the system of producing and appropriating products, that is based on class antagonisms, on the exploitation of the many by the few.

In this sense, the theory of the Communists may be summed up in the single sentence: Abolition of private property.

We Communists have been reproached with the desire of abolishing the right of personally acquiring property as the fruit of a man's own labour, which property is alleged to be the groundwork of all personal freedom, activity, and independence.

Hard-won, self-acquired, self-earned property! Do you mean the property of the petty artisan and of the small peasant, a form of property that preceded the bourgeois form? There is no need to abolish that; the development of industry has to a great extent already destroyed it, and is still destroying it daily.

Or do you mean modern bourgeois private property?

But does wage-labour create any property for the labourer? Not a bit. It creates capital, i.e., that kind of property which exploits wage-labour, and which cannot increase except upon condition of begetting a new supply of wage-labour for fresh exploitation. Property, in its present form, is based on the antagonism of capital and wage-labour. Let us examine both sides of this antagonism.

To be a capitalist, is to have not only a purely personal, but a social, status in production. Capital is a collective product, and only by the united action of many members, nay, in the last resort, only by the united action of all members of society, can it be set in motion.

Capital is, therefore, not a personal, it is a social power.

When, therefore, capital is converted into common property, into the property of all members of society, personal property is not thereby transformed into social property. It is only the social character of the property that is changed. It loses its class-character.

Let us now take wage-labour.

The average price of wage-labour is the minimum wage, i.e., that quantum of the means of subsistence which is absolutely requisite to keep the labourer in bare existence as a labourer. What, therefore, the wage-labourer appropriates by means of his labour merely suffices to prolong and reproduce a bare existence. We by no means intend to abolish this personal appropriation of the products of labour, an appropriation that is made for the maintenance and reproduction of human life, and that leaves no surplus wherewith to command the labour of others. All that we want to do away with is the miserable character of this appropriation, under which the labourer lives merely to increase capital, and is allowed to live only in so far as the interest of the ruling class requires it.

In bourgeois society, living labour is but a means to increase accumulated labour. In Communist society, accumulated labour is but a means to widen, to enrich, to promote the existence of the labourer.

In bourgeois society, therefore, the past dominates the present; in Communist society, the present dominates the past. In bourgeois society capital is independent and has individuality, while the living person is dependent and has no individuality.

And the abolition of this state of things is called by the bourgeois abolition of individuality and freedom! And rightly so. The abolition of bourgeois individuality, bourgeois independence, and bourgeois freedom is undoubtedly aimed at.

By freedom is meant, under the present bourgeois conditions of production, free trade, free selling and buying.

But if selling and buying disappears, free selling and buying disappears also. This talk about free selling and buying, and all the other 'brave words' of our bourgeoisie about freedom in general, have a meaning, if any, only in contrast with restricted selling and buying, with the fettered traders of the Middle Ages, but have no meaning when opposed to the Communistic abolition of buying and selling, of the bourgeois conditions of production, and of the bourgeoisie itself.

You are horrified at our intending to do away with private property. But in your existing society, private property is already done away with for nine-tenths of the population; its existence for the few is solely due to its non-existence in the hands of those nine-tenths. You reproach us, therefore, with intending to do away with a form of property, the necessary condition for whose existence is the non-existence of any property for the immense majority of society.

In one word, you reproach us with intending to do away with your property. Precisely so; that is just what we intend.

From the moment when labour can no longer be converted into capital, money, or rent, into a social power capable of being monopolized, i.e., from the moment when individual property can no longer be transformed into bourgeois property, into capital, from that moment, you say, individuality vanishes.

You must, therefore, confess that by 'individual' you mean no other person than the bourgeois, than the middle-class owner of property. This person must, indeed, be swept out of the way, and made impossible.

Communism deprives no man of the power to appropriate the products of society; all that it does is to deprive him of the power to subjugate the labour of others by means of such appropriation.

It has been objected that upon the abolition of private property all work will cease, and universal laziness will overtake us.

According to this, bourgeois society ought long ago to have gone to the dogs through sheer idleness; for those of its members who work acquire nothing, and those who acquire anything do not work. The whole of this objection is but another expression of the tautology: that there can no longer by any wage-labour when there is no longer any capital.

All objections urged against the Communistic mode of producing and appropriating material products have, in the same way, been urged against the Communistic modes of producing and appropriating intellectual products. Just as, to the bourgeois, the disappearance of class property is the disappearance of production itself, so the disappearance of class culture is to him identical with the disappearance of all culture.

That culture, the loss of which he laments, is, for the enormous majority, a mere training to act as a machine.

But don't wrangle with us so long as you apply, to our intended abolition of bourgeois property, the standard of your bourgeois notions of freedom, culture, law, etc. Your very ideas are but the outgrowth of the conditions of your bourgeois production and bourgeois property, just as your jurisprudence is but the will of your class made into a law for all, a will whose essential character and direction are determined by the economical conditions of existence of your class.

The selfish misconception that induces you to transform into eternal laws of nature and of reason the social forms springing from your present mode of production and form of property—historical relations that rise and disappear in the progress of production—this misconception you share with every ruling class that has preceded you. What you see clearly in the case of ancient property, what you admit in the case of feudal property, you are of course forbidden to admit in the case of your own bourgeois form of property.

Abolition of the family! Even the most radical flare up at this infamous proposal of the Communists.

On what foundation is the present family, the bourgeois family, based? On capital, on private gain. In its completely developed form this family exists only among the bourgeoisie. But this state of things finds its complement in the practical absence of the family among the proletarians, and in public prostitution.

The bourgeois family will vanish as a matter of course when its complement vanishes, and both will vanish with the vanishing of capital.

Do you charge us with wanting to stop the exploitation of children by their parents? To this crime we plead guilty.

But, you will say, we destroy the most hallowed of relations, when we replace home education by social.

And your education! Is not that also social, and determined by the social conditions under which you educate, by the intervention, direct or indirect, of society, by means of schools, etc.? The Communists have not invented the intervention of society in education; they do but seek to alter the character of that intervention, and to rescue education from the influence of the ruling class.

The bourgeois clap-trap about the family and education, about the hallowed co-relation of parent and child, becomes all the more disgusting, the more, by the action of Modern Industry, all family ties among the proletarians are torn asunder, and their children transformed into simple articles of commerce and instruments of labour.

But you Communists would introduce community of women, screams the whole bourgeoisie in chorus.

The bourgeois sees in his wife a mere instrument of production. He hears that the instruments of production are to be exploited in common, and, naturally, can come to no other conclusion than that the lot of being common to all will likewise fall to the women.

He has not even a suspicion that the real point aimed at is to do away with the status of women as mere instruments of production.

For the rest, nothing is more ridiculous than the virtuous indignation of our bourgeois at the community of women which, they pretend, is to be openly and officially established by the Communists. The Communists have no need to introduce community of women; it has existed almost from time immemorial.

Our bourgeois, not content with having the wives and daughters of their proletarians at their disposal, not to speak of common prostitutes, take the greatest pleasure in seducing each other's wives.

Bourgeois marriage is in reality a system of wives in common and thus, at the most, what the Communists might possibly be reproached with, is that they desire to introduce, in substitution for a hypocritically concealed, an openly legalized, community of women. For the rest, it is self-evident that the abolition of the present system of production must bring with it the abolition of the community of women springing from that system, i.e., of prostitution both public and private.

The Communists are further reproached with desiring to abolish countries and nationality.

The working men have no country. We cannot take from them what they have not got. Since the proletariat must first of all acquire political supremacy, must rise to be the leading class of the nation, must constitute itself *the* nation, it is, so far, itself national, though not in the bourgeois sense of the world.

National differences and antagonisms between peoples are daily more and more vanishing, owing to the development of the bourgeoisie, to freedom of commerce, to the world-market, to uniformity in the mode of production and in the conditions of life corresponding thereto.

The supremacy of the proletariat will cause them to vanish still faster. United action, of the leading civilized countries at least, is one of the first conditions for the emancipation of the proletariat.

In proportion as the exploitation of one individual by another is put an end to, the exploitation of one nation by another will also be put an end to. In proportion as the antagonism between classes within the nation vanishes, the hostility of one nation to another will come to an end.

The charges against Communism made from a religious, a philosophical, and, generally, from an ideological standpoint are not deserving of serious examination.

Does it require deep intuition to comprehend that man's ideas, views, and conceptions, in one word, man's consciousness, changes with every change in the conditions of his material existence, in his social relation, and in his social life?

What else does the history of ideas prove, than that intellectual production changes its character in proportion as material production is changed? The ruling ideas of each age have ever been the ideas of its ruling class.

When people speak of ideas that revolutionize society, they do but express the fact, that within the old society, the elements of a new one have been created, and that the dissolution of the old ideas keeps even pace with the dissolution of the old conditions of existence.

When the ancient world was in its last throes, the ancient religions were overcome by Christianity. When Christian ideas succumbed in the eighteenth century to rationalist ideas, feudal society fought its death battle with the then revolutionary bourgeoisie. The ideas of religious liberty and freedom of conscience merely gave expression to the sway of free competition within the domain of knowledge.

'Undoubtedly,' it will be said, 'religious, moral, philosophical, and juridical ideas have been modified in the course of historical development. But religion, morality, philosophy, political science, and law constantly survived this change.'

'There are, besides, eternal truths, such as Freedom, Justice, etc., that are common to all states of society. But Communism abolishes eternal truths, it abolishes all religion and all morality, instead of constituting them on a new basis; it therefore acts in contradiction to all past historical experience.'

What does this accusation reduce itself to? The history of all past society has consisted in the development of class antagonisms, antagonisms that assumed different forms at different epochs.

But whatever form they may have taken, one fact is common to all past ages, viz., the exploitation of one part of society by the other. No wonder, then, that the social consciousness of past ages, despite all the multiplicity and variety it displays, moves within certain common forms, or general ideas, which cannot completely vanish except with the total disappearance of class antagonisms.

The Communist revolution is the most radical rupture with traditional property relations; no wonder that its development involves the most radical rupture with traditional ideas.

But let us have done with the bourgeois objections to Communism.

We have seen above, that the first step in the revolution by the working class is to raise the proletariat to the position of ruling class, to win the battle of democracy.

The proletariat will use its political supremacy to wrest, by degrees, all capital from the bourgeoisie, to centralize all instruments of production in the hands of the State, i.e., of the proletariat organized as the ruling class; and to increase the total of productive forces as rapidly as possible.

Of course, in the beginning this cannot be effected except by means of despotic inroads on the rights of property, and on the conditions of bourgeois production; by means of measures, therefore, which appear economically insufficient and untenable, but which, in the course of the movement, outstrip themselves, necessitate further inroads upon the old social order, and are unavoidable as a means of entirely revolutionizing the mode of production.

These measures will of course be different in different countries.

Nevertheless, in the most advanced countries, the following will be pretty generally applicable.

1. Abolition of property in land and application of all rents of land to public purposes.

2. A heavy progressive or graduated income tax.

3. Abolition of all right of inheritance.

4. Confiscation of the property of all emigrants and rebels.

5. Centralization of credit in the hands of the State, by means of a national bank with State capital and an exclusive monopoly.

6. Centralization of the means of communication and transport in the hands of the State.

7. Extension of factories and instruments of production owned by the State; the bringing into cultivation of wastelands, and the improvement of the soil generally in accordance with a common plan.

8. Equal liability of all to labour. Establishment of industrial armies, especially for agriculture.

9. Combination of agriculture with manufacturing industries; gradual abolition of the distinction between town and country, by a more equable distribution of the population over the country.

10. Free education for all children in public schools. Abolition of children's factory labour in its present form. Combination of education with industrial production, etc., etc.

When, in the course of development, class distinctions have disappeared, and all production has been concentrated in the hands of associated individuals, the public power will lose its political character. Political power, properly so called, is merely the organized power of one class for oppressing another. If the proletariat during its contest with the bourgeoisie is compelled, by the force of circumstances, to organize itself as a class, if, by means of a revolution, it makes itself the ruling class, and, as such, sweeps away by force the old conditions of production, then it will, along with these conditions, have swept away the conditions for the existence of class antagonisms and of classes generally, and will thereby have abolished its own supremacy as a class.

In place of the old bourgeois society, with its classes and class antagonisms, we shall have an association, in which the free development of each is the condition for the free development of all.

III
Socialist and Communist Literature

1. Reactionary Socialism

(a) *Feudal Socialism.* Owing to their historical position, it became the vocation of the aristocracies of France and England to write pamphlets against modern bourgeois society. In the French revolution of July 1830, and in the English reform agitation, these aristocracies again succumbed to the hateful upstart. Thenceforth, a serious political contest was altogether out of question. A literary battle alone remained possible. But even in the domain of literature the old cries of the restoration period had become impossible.

In order to arouse sympathy, the aristocracy were obliged to lose sight, apparently, of their own interests, and to formulate their indictment against the bourgeoisie in the interest of the exploited working class alone. Thus the aristocracy took their revenge by singing lampoons on their new master, and whispering in his ears sinister prophecies of coming catastrophe.

In this way arose Feudal Socialism: half lamentation, half lampoon; half echo of the past, half menace of the future; at times, by its bitter, witty, and incisive criticism, striking the bourgeoisie to the very heart's core; but always ludicrous in its effect, through total incapacity to comprehend the march of modern history.

The aristocracy, in order to rally the people to them, waved the proletarian alms-bag in front for a banner. But the people, so often as it joined them, saw on their hindquarters the old feudal coats of arms, and deserted with loud and irreverent laughter.

One section of the French Legitimists and 'Young England' exhibited this spectacle.

In pointing out that their mode of exploitation was different to that of the bourgeoisie, the feudalists forget that they exploited under circumstances and conditions that were quite different, and that are now antiquated. In showing that, under their rule, the modern proletariat never existed, they forget that the modern bourgeoisie is the necessary offspring of their own form of society.

For the rest, so little do they conceal the reactionary character of their criticism that their chief accusation against the bourgeoisie amounts to this, that under the bourgeois regime a class is being developed, which is destined to cut up root and branch the old order of society.

What they upbraid the bourgeoisie with is not so much that it creates a proletariat, as that it creates a revolutionary proletariat.

In political practice, therefore, they join in all coercive measures against the working class; and in ordinary life, despite their highfalutin phrases, they stoop to pick up the golden apples dropped from the tree of industry, and to barter truth, love, and honour for traffic in wool, sugar-beet, and potato spirits.

As the parson has ever gone hand in hand with the landlord, so has Clerical Socialism with Feudal Socialism.

Nothing is easier than to give Christian asceticism a Socialist tinge. Has not Christianity declaimed against private property, against marriage, against the State? Has it not preached in the place of these charity and poverty, celibacy and mortification of the flesh, monastic life and Mother Church? Christian Socialism is but the holy water with which the priest consecrates the heart-burnings of the aristocrat.

(b) *Petty-Bourgeois Socialism.* The feudal aristocracy was not the only class that was ruined by the bourgeoisie, not the only class whose conditions of existence pined and perished in the atmosphere of modern bourgeois society. The medieval burgesses and the small peasant proprietors were the precursors of the modern bourgeoisie. In those countries which are but little developed, industrially and commercially, these two classes still vegetate side by side with the rising bourgeoisie.

In countries where modern civilization has become fully developed, a new class of petty bourgeois has been formed, fluctuating between proletariat and bourgeoisie and ever renewing itself as a supplementary part of bourgeois society. The individual members of this class, however, are being constantly hurled down into the proletariat by the action of competition, and, as modern industry develops, they even see the moment approaching when they will completely disappear as an independent section of modern society, to be replaced, in manufactures, agriculture, and commerce, by overlookers, bailiffs, and shopmen.

In countries like France, where the peasants constitute far more than half of the population, it was natural that writers who sided with the proletariat against the bourgeoisie should use, in their criticism of the bourgeois regime, the standard of the peasant and petty bourgeois, and from the standpoint of these intermediate classes should take up the cudgels for the working class.

Thus arose petty-bourgeois Socialism. Sismondi was the head of this school, not only in France but also in England.

This school of Socialism dissected with great acuteness the contradictions in the conditions of modern production. It laid bare the hypocritical apologies of economists. It proved, incontrovertibly, the disastrous effects of machinery and division of labour; the concentration of capital and land in a few hands; overproduction and crises; it pointed out the inevitable ruin of the petty bourgeois and peasant, the misery of the proletariat, the anarchy in production, the crying inequalities in the distribution of wealth, the industrial war of extermination between nations, the dissolution of old moral bonds, of the old family relations, of the old nationalities.

In its positive aims, however, this form of Socialism aspires either to restoring the old means of production and of exchange, and with them the old property relations, and the old society, or to cramping the modern means of production and of exchange, within the framework of the old property relations that have been, and were bound to be, exploded by those means. In either case, it is both reactionary and Utopian.

Its last words are: corporate guilds for manufacture, patriarchal relations in agriculture.

Ultimately, when stubborn historical facts had dispersed all intoxicating effects of self-deception, this form of Socialism ended in a miserable fit of the blues.

(c) *German, or 'True', Socialism*. The Socialist and Communist literature of France, a literature that originated under the pressure of a bourgeoisie in power, and that was the expression of the struggle against this power, was introduced into Germany at a time when the bourgeoisie, in that country, had just begun its contest with feudal absolutism.

German philosophers, would-be philosophers, and *beaux esprits*, eagerly seized on this literature, only forgetting that when these writings immigrated from France into Germany, French social conditions had not immigrated along with them. In contact with German social conditions, this French literature lost all its immediate practical significance, and assumed a purely literary aspect. Thus, to the German philosophers of the eighteenth century, the demands of the first French Revolution were nothing more than the demands of 'Practical Reason' in general, and the utterance of the will of the revolutionary French bourgeoisie signified in their eyes the laws of pure Will, of Will as it was bound to be, of true human Will generally.

The work of the German *literati* consisted solely in bringing the new French ideas into harmony with their ancient philosophical conscience, or rather, in annexing the French ideas without deserting their own philosophic point of view.

This annexation took place in the same way in which a foreign language is appropriated, namely, by translation.

It is well known how the monks wrote silly lives of Catholic Saints over the manuscripts on which the classical works of ancient heathendom had been written. The German *literati* reversed this process with the profane French literature. They wrote their philosophical nonsense beneath the French original. For instance, beneath the French criticism of the economic functions of money, they wrote 'Alienation of Humanity', and beneath the French criticism of the bourgeois State they wrote 'Dethronement of the Category of the General', and so forth.

The introduction of these philosophical phrases at the back of the French historical criticisms they dubbed 'Philosophy of Actions', 'True Socialism', 'German Science of Socialism', 'Philosophical Foundation of Socialism', and so on.

The French Socialist and Communist literature was thus completely emasculated. And, since it ceased in the hands of the German to express the struggle of one class with the other, he felt conscious of having overcome 'French one-sidedness' and of representing, not true requirements, but the requirements of Truth; not the interests of the proletariat, but the interests of Human Nature, of Man in general, who belongs to no class, has no reality, who exists only in the misty realm of philosophical fantasy.

This German Socialism, which took its schoolboy task so seriously and solemnly, and extolled its poor stock-in-trade in such mountebank fashion, meanwhile gradually lost its pedantic innocence.

The fight of the German, and, especially, of the Prussian bourgeoisie, against feudal aristocracy and absolute monarchy, in other words, the liberal movement, became more earnest.

By this, the long wished-for opportunity was offered to 'True' Socialism of confronting the political movement with the Socialist demands, of hurling the traditional anathemas against liberalism, against representative government, against bourgeois competition, bourgeois freedom of the press, bourgeois legislation, bourgeois liberty and equality, and of preaching to the masses that they had nothing to gain, and everything to lose, by this bourgeois movement. German Socialism forgot, in the nick of time, that the French criticism, whose silly echo it was, presupposed the existence of modern bourgeois society, with its corresponding economic conditions of existence, and the political constitution adapted thereto, the very things whose attainment was the object of the pending struggle in Germany.

To the absolute governments, with their following of parsons, professors, country squires, and officials, it served as a welcome scarecrow against the threatening bourgeoisie.

It was a sweet finish after the bitter pills of floggings and bullets with which these same governments, just at that time, dosed the German working-class risings.

While this 'True' Socialism thus served the governments as a weapon for fighting the German bourgeoisie, it at the same time, directly represented a reactionary interest, the interest of the German Philistines. In Germany the petty-bourgeois class, a relic of the sixteenth century, and since then constantly cropping up again under various forms, is the real social basis of the existing state of things.

To preserve this class is to preserve the existing state of things in Germany. The industrial and political supremacy of the bourgeoisie threatens it with certain destruction; on the one hand, from the concentration of capital; on the other, from the rise of a revolutionary proletariat. 'True' Socialism appeared to kill these two birds with one stone. It spread like an epidemic.

The robe of speculative cobwebs, embroidered with flowers of rhetoric, steeped in the dew of sickly sentiment, this transcendental robe in which the German Socialists wrapped their sorry 'eternal truths', all skin and bone, served wonderfully to increase the sale of their goods among such a public.

And on its part, German Socialism recognized, more and more, its own calling as the bombastic representative of the petty-bourgeois Philistine.

It proclaimed the German nation to be the model nation, and the German petty Philistine to be the typical man. To every villainous meanness of this model man it gave a hidden, higher, Socialistic interpretation, the exact contrary of its real character. It went to the extreme length of directly opposing the 'brutally destructive' tendency of Communism, and of proclaiming its supreme and impartial contempt of all class struggles. With very few exceptions, all the so-called Socialist and Communist publications that now (1847) circulate in Germany belong to the domain of this foul and enervating literature.

2. Conservative, or Bourgeois, Socialism

A part of the bourgeoisie is desirous of redressing social grievances, in order to secure the continued existence of bourgeois society.

To this section belong economists, philanthropists, humanitarians, improvers of the condition of the working class, organizers of charity, members of societies for the prevention of cruelty to animals, temperance fanatics, hole-and-corner reformers of every imaginable kind. This form of Socialism has, moreover, been worked out into complete systems.

We may cite Proudhon's *Philosophie de la misère* as an example of this form.

The Socialistic bourgeois want all the advantages of modern social conditions without the struggles and dangers necessarily resulting from them. They desire the existing state of society minus its revolutionary and disintegrating elements. They wish for a bourgeoisie without a proletariat. The bourgeoisie naturally conceives the world in which it is supreme to be the best; and bourgeois Socialism develops this comfortable conception into various more or less complete systems. In requiring the proletariat to carry out such a system, and thereby to march straightaway into the social New

Jerusalem, it but requires in reality that the proletariat should remain within the bounds of existing society, but should cast away all its hateful ideas concerning the bourgeoisie.

A second and more practical, but less systematic, form of this Socialism sought to depreciate every revolutionary movement in the eyes of the working class, by showing that no mere political reform, but only a change in the material conditions of existence, in economical relations, could be of any advantage to them. By changes in the material conditions of existence, this form of Socialism, however, by no means understands abolition of the bourgeois relations of production, an abolition that can be effected only by a revolution, but administrative reforms, based on the continued existence of these relations; reforms, therefore, that in no respect affect the relations between capital and labour, but, at the best, lessen the cost, and simplify the administrative work, of bourgeois government.

Bourgeois Socialism attains adequate expression when, and only when, it becomes a mere figure of speech.

Free trade: for the benefit of the working class. Protective duties: for the benefit of the working class. Prison Reform: for the benefit of the working class. This is the last word and the only seriously meant word of bourgeois Socialism.

It is summed up in the phrase: the bourgeois is a bourgeois—for the benefit of the working class.

3. *Critical-Utopian Socialism and Communism*

We do not here refer to that literature which, in every great modern revolution, has always given voice to the demands of the proletariat, such as the writings of Babeuf and others.

The first direct attempts of the proletariat to attain its own ends, made in times of universal excitement, when feudal society was being overthrown, these attempts necessarily failed, owing to the then undeveloped state of the proletariat, as well as to the absence of the economic conditions for its emancipation, conditions that had yet to be produced, and could be produced by the impending bourgeois epoch alone. The revolutionary literature that accompanied these first movements of the proletariat had necessarily a reactionary character. It inculcated universal asceticism and social levelling in its crudest form.

The Socialist and Communist systems properly so called, those of Saint-Simon, Fourier, Owen, and others, spring into existence in the early undeveloped period, described above, of the struggle between proletariat and bourgeoisie (see Section I. Bourgeoisie and Proletariat).

The founders of these systems see, indeed, the class antagonisms, as well as the action of the decomposing elements, in the prevailing form of society. But the proletariat, as yet in its infancy, offers to them the spectacle of a class without any historical initiative or any independent political movement.

Since the development of class antagonism keeps even pace with the development of industry, the economic situation, as they find it, does not as yet offer to them the material conditions for the emancipation of the proletariat. They therefore search after a new social science, after new social laws, that are to create these conditions.

Historical action is to yield to their personal inventive action, historically created conditions of emancipation to fantastic ones, and the gradual, spontaneous class-organization of the proletariat to an organization of society specially contrived by these inventors. Future history resolves itself, in their eyes, into the propaganda and the practical carrying out of their social plans.

In the formation of their plans they are conscious of caring chiefly for the interests of the working class, as being the most suffering class. Only from the point of view of being the most suffering class does the proletariat exist for them.

The undeveloped state of the class struggle, as well as their own surroundings, causes Socialists of this kind to consider themselves far superior to all class antagonisms. They want to improve the condition of every member of society, even that of the most favoured. Hence, they habitually appeal to society at large, without distinction of class; nay, by preference, to the ruling class. For how can people, when once they understand their system, fail to see in it the best possible plan of the best possible state of society?

Hence, they reject all political, and especially all revolutionary, action; they wish to attain their ends by peaceful means, and endeavour, by small experiments, necessarily doomed to failure, and by the force of example, to pave the way for the new social Gospel.

Such fantastic pictures of future society, painted at a time when the proletariat is still in a very undeveloped state and has but a fantastic conception of its own position, correspond with the first instinctive yearnings of that class for a general reconstruction of society.

But these Socialist and Communist publications contain also a critical element. They attack every principle of existing society. Hence they are full of the most valuable materials for the enlightenment of the working class. The practical measures proposed in them—such as the abolition of the distinction between town and country, of the family, of the carrying on of industries for the account of private individuals, and of the wage system, the proclamation of social harmony, the conversion of the functions of the State into a mere superintendence of production, all these proposals point solely to the disappearance of class antagonisms which were, at that time, only just cropping up, and which in these publications, are recognized in their earliest, indistinct, and undefined forms only. These proposals, therefore, are of a purely Utopian character.

The significance of Critical-Utopian Socialism and Communism bears an inverse relation to historical development. In proportion as the modern class struggle develops and takes definite shape, this fantastic standing apart from the contest, these fantastic attacks on it, lose all practical value and all theoretical justification. Therefore, although the originators of these systems were, in many respects, revolutionary, their disciples have, in every case, formed mere reactionary sects. They hold fast by the original views of their masters, in opposition to the progressive historical development of the proletariat. They, therefore, endeavour, and that consistently, to deaden the class struggle and to reconcile the class antagonisms. They still dream of experimental realization of their social Utopias, of founding isolated '*phalanstères*', of establishing 'Home Colonies', of setting up a 'Little Icaria'— duodecimo editions of the New Jerusalem—and to realize all these castles in the air they are compelled to appeal to the feelings and purses of the bourgeois. By degrees they sink into the category of the reactionary conservative Socialists depicted above, differing from these only by more systematic pedantry, and by their fanatical and superstitious belief in the miraculous effects of their social science.

They, therefore, violently oppose all political action on the part of the working class; such action, according to them, can only result from blind unbelief in the new Gospel.

The Owenites in England and the Fourierists in France, respectively, oppose the Chartists and the *Réformistes*.

IV

Position of the Communists in Relation to the Various Existing Opposition Parties

Section II has made clear the relations of the Communists to the existing working-class parties, such as the Chartists in England and the Agrarian Reformers in America.

The Communists fight for the attainment of the immediate aims for the enforcement of the momentary interests of the working class; but in the movement of the present, they also represent and take care of the future of that movement. In France the Communists ally themselves with the Social-Democrats, against the conservative and radical bourgeoisie, reserving, however, the right to take up a critical position in regard to phrases and illusions traditionally handed down from the great Revolution.

In Switzerland they support the Radicals, without losing sight of the fact that this party consists of antagonistic elements, partly of Democratic Socialists, in the French sense, partly of radical bourgeois.

In Poland they support the party that insists on an agrarian revolution as the prime condition for national emancipation, that party which fomented the insurrection of Cracow in 1846.

In Germany they fight with the bourgeoisie whenever it acts in a revolutionary way, against the absolute monarchy, the feudal squirearchy, and the petty bourgeoisie.

But they never cease, for a single instant, to instil into the working class the clearest possible recognition of the hostile antagonism between bourgeoisie and proletariat, in order that the German workers may straightaway use, as so many weapons against the bourgeoisie, the social and political conditions that the bourgeoisie must necessarily introduce along with its supremacy, and in order that, after the fall of the reactionary classes in Germany, the fight against the bourgeoisie itself may immediately begin.

The Communists turn their attention chiefly to Germany, because that country is on the eve of a bourgeois revolution that is bound to be carried out under more advanced conditions of European civilization, and with a much more developed proletariat, than that of England was in the seventeenth, and of France in the eighteenth century, and because the bourgeois revolution in Germany will be but the prelude to an immediately following proletarian revolution.

In short, the Communists everywhere support every revolutionary movement against the existing social and political order of things.

In all these movements they bring to the front, as the leading question in each, the property question, no matter what its degree of development at the time.

Finally, they labour everywhere for the union and agreement of the democratic parties of all countries.

The Communists disdain to conceal their views and aims. They openly declare that their ends can be attained only by the forcible overthrow of all existing social conditions. Let the ruling classes tremble at a Communistic revolution. The proletarians have nothing to lose but their chains. They have a world to win.

WORKING MEN OF ALL COUNTRIES, UNITE!

BIBLIOGRAPHY

Original: MEW, Vol. 4, pp. 461 ff.

Present translation: MESW, 1, pp. 33 ff. Reproduced by kind permission of Lawrence and Wishart Ltd.

Other translations: The first translation was by Helen Macfarlane, published in 1850 in Harney's Red Republican. The well-known translation, that by Samuel Moore reproduced here, dates from 1888. There is a further translation by Eden and Cedar Paul published in 1930.

Commentaries: D. Fernbach, Introduction to K. Marx, The 1848 Revolutions, New York and London, 1973.

R. Hunt, The Political Ideas of Marx and Engels, Pittsburg and London, 1975, Vol. 1, pp. 176 ff.

H. Laski, Introduction to The Communist Manifesto, Socialist Landmark, London, 1948.

D. Ryazanov, *The Communist Manifesto of Marx and Engels*, New York and London, 1930.

D. Struik, *The Birth of the Communist Manifesto*, New York, 1971.

A. J. P. Taylor, Introduction to: *The Communist Manifesto*, Harmondsworth, 1967.

Y. Wagner and M. Strauss, 'The Programme of the *Communist Manifesto* and its Theoretical Implications', *Political Studies*, Dec. 1969.

19. Wage-Labour and Capital

The following text was first published in the form of articles in the *Neue Rheinische Zeitung* during April 1849. Originally, however, they were delivered as lectures to the Workingmen's Club in Brussels at the beginning of December 1847. Marx had hoped to publish them in Brussels in February 1848, but the outbreak of revolution precluded this.

These articles contain Marx's first systematic exposition of his economic theories —in particular on the subject of relative and absolute immiserization. In the third edition of 1891 Engels considerably modified the text (for example, by consistently substituting 'labour power' for 'labour'). The present translation follows the original.

From various quarters we have been reproached with not having presented the economic relations which constitute the material foundation of the present class struggles and national struggles. We have designedly touched upon these relations only where they directly forced themselves to the front in political conflicts.

The point was, above all, to trace the class struggle in current history, and to prove empirically by means of the historical material already at hand and which is being newly created daily, that, with the subjugation of the working class that February and March had wrought, its opponents were simultaneously defeated—the bourgeois republicans in France and the bourgeois and peasant classes which were fighting feudal absolutism throughout the continent of Europe; that the victory of the 'honest republic' in France was at the same time the downfall of the nations that had responded to the February Revolution by heroic wars of independence; finally, that Europe, with the defeat of the revolutionary workers, had relapsed into its old double slavery, the Anglo-Russian slavery. The June struggle in Paris, the fall of Vienna, the tragicomedy of Berlin's November 1848, the desperate exertions of Poland, Italy, and Hungary, the starving of Ireland into submission—these were the chief factors which characterized the European class struggle between bourgeoisie and working class and by means of which we proved that every revolutionary upheaval, however remote from the class struggle its goal may appear to be, must fail until the revolutionary working class is victorious, that every social reform remains a utopia until the proletarian revolution and the feudalistic counter-revolution measure swords in a world war. In our presentation, as in reality, Belgium and Switzerland were tragicomic genre-pictures akin to caricature in the great historical tableau, the one being the model state of the bourgeois monarchy, the other the model state of the bourgeois republic, both of them states which imagine themselves to be as independent of the class struggle as of the European revolution.

Now, after our readers have seen the class struggle develop in colossal political forms in 1848, the time has come to deal more closely with the economic relations themselves on which the existence of the bourgeoisie and its class rule, as well as the slavery of the workers, are founded.

We shall present in three large sections: 1) the relation of wage labour to capital, the slavery of the worker, the domination of the capitalist; 2) the inevitable destruction of the middle bourgeois classes and of the so-called peasant estate under the present system; 3) the commercial subjugation and exploitation of the bourgeois classes of the various European nations by the despot of the world market—England.

We shall try to make our presentation as simple and popular as possible and shall not presuppose even the most elementary notions of political economy. We wish to be understood by the workers. Moreover, the most remarkable ignorance and confusion of ideas prevails in Germany in regard to the simplest economic relations, from the accredited defenders of the existing state of things down to the socialist miracle workers and the unrecognized political geniuses in which fragmented Germany is even richer than in sovereign princes.

Now, therefore, for the first question: What are wages? How are they determined?

If workers were asked: 'How much are your wages?' one would reply: 'I get a mark a day from my employer'; another, 'I get two marks,' and so on. According to the different trades to which they belong, they would mention different sums of money which they receive from their respective employers for the performance of a particular piece of work, for example, weaving a yard of linen or type-setting a printed sheet. In spite of the variety of their statements, they would all agree on one point: wages are the sum of money paid by the capitalist for a particular labour time or for a particular output of labour.

The capitalist, it seems, therefore, buys their labour with money. They sell him their labour for money. For the same sum with which the capitalist has bought their labour, for example, two marks, he could have bought two pounds of sugar or a definite amount of any other commodity. The two marks, which he bought two pounds of sugar, are the price of the two pounds of sugar. The two marks, with which he bought twelve hours' use of labour, are the price of twelve hours' labour. Labour, therefore, is a commodity, neither more nor less than sugar. The former is measured by the clock, the latter by the scales.

The workers exchange their commodity, labour, for the commodity of the capitalist, for money, and this exchange takes place in a definite ratio. So much money for so much labour. For twelve hours' weaving, two marks. And do not the two marks represent all the other commodities which I can buy for two marks? In fact, therefore, the worker has exchanged his commodity, labour, for other commodities of all kinds and that in a definite ratio.

By giving him two marks, the capitalist has given him so much meat, so much clothing, so much fuel, light, etc., in exchange for his day's labour. Accordingly, the two marks express the ratio in which labour is exchanged for other commodities, the exchange value of his labour. The exchange value of a commodity, reckoned in money, is what is called its price. Wages are only a special name for the price of labour, for the price of this peculiar commodity which has no other repository than human flesh and blood.

Let us take any worker, say, a weaver. The capitalist supplies him with the loom and yarn. The weaver sets to work and the yarn is converted into linen. The capitalist takes possession of the linen and sells it, say, for twenty marks. Now are the wages of the weaver a share in the linen, in the twenty marks, in the product of his labour? By no means. Long before the linen is sold, perhaps long before its weaving is finished, the weaver has received his wages. The capitalist, therefore, does not pay these wages with the money which he will obtain from the linen, but with money already in reserve. Just as the loom and the yarn are not the product of the weaver to whom they are supplied by his employer, so likewise with the commodities which the weaver receives in exchange for his commodity, labour. It was possible that his employer found no purchaser at all for his linen. It was possible that he did not get even the amount of the wages by its sale. It is possible that he sells it very profitably in comparison with the weaver's wages. All that has nothing to do with the weaver. The capitalist buys the labour of the weaver with a part of his available wealth, of his capital, just as he has bought the raw material—the yarn—and the instrument of labour—the loom—with another part of his wealth. After he has made these purchases, and these purchases include the labour necessary for the production of linen, he produces only with the raw materials and instruments of labour belonging to him. For the latter include now, true enough, our good weaver as well, who has as little share in the product or the price of the product as the loom has.

Wages are, therefore, not the worker's share in the commodity produced by him. Wages are the part of already existing commodities with which the capitalist buys for himself a definite amount of productive labour.

Labour is, therefore, a commodity which its possessor, the wage-worker, sells to capital. Why does he sell it? In order to live.

But the exercise of labour is the worker's own life-activity, the manifestation of his own life. And this life-activity he sells to another person in order to secure the necessary means of subsistence. Thus his life-activity is for him only a means to enable him to exist. He works in order to live. He does not even reckon labour as part of his life, it is rather a sacrifice of his life. It is a commodity which he has made over to another. Hence, also, the product of his activity is not the object of his activity. What he produces for himself is not the silk that he weaves, not the gold that he draws from the mine, not the palace that he builds. What he produces for himself is wages, and silk, gold, palace resolve themselves for him into a definite quantity of the means of subsistence, perhaps into a cotton jacket, some copper coins, and a lodging in a cellar. And the worker, who for twelve hours weaves, spins, drills, turns, builds, shovels, breaks stones, carries loads, etc.—does he consider this twelve hours' weaving, spinning, drilling, turning, building, shovelling, stone-breaking as a manifestation of his life, as life? On the contrary, life begins for him where this activity ceases, at table, in the public house, in bed. The twelve hours' labour, on the other hand, has no meaning for him as weaving, spinning, drilling, etc., but as earnings, which bring him to the table, to the public house, into bed. If the silk worm were to spin in order to continue its existence as a caterpillar, it would be a complete wage-worker. Labour was not always a commodity. Labour was not always wage labour, that is, free labour. The slave did not sell his labour to the slave-owner, any more than the ox sells its services to the peasant. The slave, together with his labour, is sold once and for all to his owner. He is a commodity which can pass from the hand of one owner to that of another. He is himself a commodity, but the labour is not his commodity. The serf sells only a part of his labour. He does not receive a wage from the owner of the land; rather the owner of the land receives a tribute from him.

The serf belongs to the land and turns over to the owner of the land the fruits thereof. The free labourer, on the other hand, sells himself and, indeed, sells himself piecemeal. He sells at auction eight, ten, twelve, fifteen hours of his life, day after day, to the highest bidder, to the owner of the raw materials, instruments of labour, and means of subsistence, that is, to the capitalist. The worker belongs neither to an owner nor to the land, but eight, ten, twelve, fifteen hours of his daily life belong to him who buys them. The worker leaves the capitalist to whom he hires himself whenever he likes, and the capitalist discharges him whenever he thinks fit, as soon as he no longer gets any profit out of him, or not the anticipated profit. But the worker, whose sole source of livelihood is the sale of his labour, cannot leave the whole class of purchasers, that is, the capitalist class, without renouncing his existence. He belongs not to this or that capitalist but to the capitalist class, and, moreover, it is his business to dispose of himself, that is, to find a purchaser within this capitalist class.

Now, before going more closely into the relation between capital and wage labour, we shall present briefly the most general relations which come into consideration in the determination of wages.

Wages, as we have seen, are the price of a definite commodity, of labour. Wages are, therefore, determined by the same laws that determine the price of every other commodity. The question, therefore, is, how is the price of a commodity determined?

By competition between buyers and sellers, by the relation of inquiry to delivery, of demand to supply. Competition, by which the price of a commodity is determined, is three-sided.

The same commodity is offered by various sellers. With goods of the same quality, the one who sells most cheaply is certain of driving the others out of the field and securing the greatest sale for himself. Thus, the sellers mutually contend among themselves for sales, for the market. Each of them desires to sell, to sell as much as possible and, if possible, to sell alone, to the exclusion of the other sellers. Hence, one sells cheaper than another. Consequently, competition takes place among the sellers, which depresses the price of the commodities offered by them.

But competition also takes place among the buyers, which in its turn causes the commodities offered to rise in price.

Finally competition occurs between buyers and sellers; the former desire to buy as cheaply as possible, the latter to sell as dearly as possible. The result of this competition between buyers and sellers will depend upon how the two above-mentioned sides of the competition are related, that is, whether the competition is stronger in the army of buyers or in the army of sellers. Industry leads two armies into the field against each other, each of which again carries on a battle within its own ranks, among its own troops. The army whose troops beat each other up the least gains the victory over the opposing host.

Let us suppose there are 100 bales of cotton on the market and at the same time buyers for 1,000 bales of cotton. In this case, therefore, the demand is ten times as great as the supply. Competition will be very strong among the buyers, each of whom desires to get one, and if possible all, of the 100 bales for himself. This example is no arbitrary assumption. We have experienced periods of cotton crop failure in the history of the trade when a few capitalists in alliance have tried to buy, not one hundred bales, but all the cotton stocks of the world. Hence, in the example mentioned, one buyer will seek to drive the other from the field by offering a relatively higher price per bale of cotton. The cotton sellers, who see that the troops of the enemy army are engaged in the most violent struggle among themselves and that the sale of all their hundred bales is absolutely certain, will take good care not to fall out among themselves and depress the price of cotton at the moment when their adversaries are competing with one another to force it up. Thus, peace suddenly descends on the army of the sellers. They stand facing the buyers as one man, fold their arms philosophically, and there would be no bounds to their demands were it not that the offers of even the most persistent and eager buyers have very definite limits.

If, therefore, the supply of a commodity is lower than the demand for it, then only slight competition, or none at all, takes place among the sellers. In the same proportion as this competition decreases, competition increases among the buyers. The result is a more or less considerable rise in commodity prices.

It is well known that the reverse case with a reverse result occurs more frequently. Considerable surplus of supply over demand; desperate competition among the sellers; lack of buyers; disposal of goods at ridiculously low prices.

But what is the meaning of a rise, a fall in prices; what is the meaning of high price, low price? A grain of sand is high when examined through a microscope, and a tower is low when compared with a mountain. And if price is determined by the relation between supply and demand, what determines the relation between supply and demand?

Let us turn to the first bourgeois we meet. He will not reflect for an instant but, like another Alexander the Great, will cut this metaphysical knot with the multiplication table. If the production of the goods which I sell has cost me 100 marks, he will tell us, and if I get 110 marks from the sale of these goods, within the year of course—then that is sound, honest, legitimate profit. But if I get in exchange 120 or 130 marks, that is a high profit; and if I get as much as 200 marks, that would be an extraordinary, an enormous profit. What, therefore, serves the bourgeois as his measure of profit? The cost of production of his commodity. If he receives in exchange for this commodity an amount of other commodities which it has cost less to produce, he has lost. If he receives in exchange for his commodity an amount of other commodities the production of which has cost more, he has gained. And he calculates the rise or fall of the profit according to the degree in which the exchange value of his commodity stands above or below zero— the cost of production.

We have thus seen how the changing relation of supply and demand causes now a rise and now a fall of prices, now high, now low prices. If the price of a commodity rises considerably because of inadequate supply or disproportionate increase of the demand, the price of some other commodity must necessarily have fallen proportionately, for the price of a commodity only expresses in money the ratio in which other commodities are given in exchange for it. If, for example, the price of a yard of silk material rises from five marks to six marks, the price of silver in relation to silk material has fallen and likewise the prices of all other commodities that have remained at their old prices have fallen in relation to the silk. One has to give a larger amount of them in exchange to get the same amount of silks. What will be the consequence of the rising price of a commodity? A mass of capital will be thrown into that flourishing branch of industry and this influx of capital into the domain of the favoured industry will continue until it yields the ordinary profits or, rather, until the price of its products, through overproduction, sinks below the cost of production.

Conversely, if the price of a commodity falls below its cost of production, capital will be withdrawn from the production of this commodity. Except in the case of a branch of industry which has become obsolete and must, therefore, perish, the production of such a commodity, that is, its supply, will go on decreasing owing to this flight of capital until it corresponds to the demand, and consequently its price is again on a level with its cost of production or, rather, until the supply has sunk below the demand, that is, until its price rises again above its cost of production, for the current price of a commodity is always either above or below its cost of production.

We see how capital continually migrates in and out, out of the domain of one industry into that of another. High prices bring too great an immigration and low prices too great an emigration.

We could show from another point of view how not only supply but also demand is determined by the cost of production. But this would take us too far away from our subject.

We have just seen how the fluctuations of supply and demand continually bring the price of a commodity back to the cost of production. The real price of a commodity, it is true, is always above or below its cost of production; but rise and fall reciprocally balance each other, so that within a certain period of time, taking the ebb and flow of the industry together, commodities are exchanged for one another in accordance with their cost of production, their price, therefore, being determined by their cost of production.

This determination of price by cost of production is not to be understood in the sense of the economists. The economists say that the average price of commodities is equal to the cost of production; that this is a law. The anarchical movement, in which rise is compensated by fall and fall by rise, is regarded by them as chance. With just as much right one could regard the fluctuations as the law and the determination by the cost of production as chance, as has actually been done by other economists. But it is solely these fluctuations, which, looked at more closely, bring with them the most fearful devastations and, like earthquakes, cause bourgeois society to tremble to its foundations—it is solely in the course of these fluctuations that prices are determined by the cost of production. The total movement of this disorder is its order. In the course of this industrial anarchy, in this movement in a circle, competition compensates, so to speak, for one excess by means of another.

We see, therefore, that the price of a commodity is determined by its cost of production in such manner that the periods in which the price of this commodity rises above its cost of production are compensated by the periods in which it sinks below the cost of production, and vice versa. This does not hold good, of course, for separate, particular industrial products but only for the whole branch of industry. Consequently, it also does not hold good for the individual industralist but only for the whole class of industrialists.

The determination of price by the cost of production is equivalent to the determination of price by the labour time necessary for the manufacture of a commodity, for the cost of production consists of 1) raw materials and depreciation of instruments, that is, of industrial products the production of which has cost a certain amount of labour time, and 2) of direct labour, the measure of which is, precisely, time.

Now, the same general laws that regulate the price of commodities in general of course also regulate wages, the price of labour.

Wages will rise and fall according to the relation of supply and demand, according to the turn taken by the competition between the buyers of labour, the capitalists, and the sellers of labour, the workers. The fluctuations in wages correspond in general to the fluctuations in prices of commodities. Within these fluctuations, however, the price of labour will be determined by the cost of production, by the labour time necessary to produce this commodity –labour power.

What, then, is the cost of production of labour?

It is the cost required for maintaining the worker as a worker and of developing him into a worker.

The less the period of training, therefore, that any work requires the smaller is the cost of production of the worker and the lower is the price of his labour, his wages. In those branches of industry in which hardly any period of apprenticeship is required and where the mere bodily existence of the worker suffices, the cost necessary for his production is almost confined to the commodities necessary for keeping him alive and capable of working. The price of his labour will, therefore, be determined by the price of the necessary means of subsistence.

Another consideration, however, also comes in. The manufacturer in calculating his cost of production and, accordingly, the price of the products takes into account the wear and tear of the instruments of labour. If, for example, a machine costs him 1,000 marks and wears out in ten years, he adds 100 marks annually to the price of the commodities so as to be able to replace the worn-out machine by a new one at the end of ten years. In the same way, in calculating the cost of production of simple labour, there must be included the cost of reproduction, whereby the race of workers is enabled to multiply and to replace worn-out workers by new ones. Thus the depreciation of the worker is taken into account in the same way as the depreciation of the machine.

The cost of production of simple labour, therefore, amounts to the cost of existence and reproduction of the worker. The price of this cost of existence and reproduction constitutes wages. Wages so determined are called the wage minimum. This wage minimum, like the determination of the price of commodities by the cost of production in general, does not hold good for the single individual but for the species. Individual workers, millions of workers, do not get enough to be able to exist and reproduce themselves; but the wages of the whole working class level down, within their fluctuations, to this minimum.

Now that we have arrived at an understanding of the most general laws which regulate wages like the price of any other commodity, we can go into our subject more specifically.

Capital consists of raw materials, instruments of labour, and means of subsistence of all kinds, which are utilized in order to produce new raw materials, new instruments of labour, and new means of subsistence. All these component parts of capital are creations of labour, products of labour, accumulated labour. Accumulated labour which serves as a means of new production is capital.

So say the economists.

What is a Negro slave? A man of the black race. The one explanation is as good as the other.

A Negro is a Negro. He only becomes a slave in certain relations. A cotton-spinning jenny is a machine for spinning cotton. It becomes capital only in certain relations. Torn from these relationships it is no more capital than gold in itself is money or sugar the price of sugar.

In production, men not only act on nature but also on one another. They produce only by co-operating in a certain way and mutually exchanging their activities. In order to produce, they enter into definite connections and relations with one another and only within these social connections and relations does their action on nature, does production, take place.

These social relations into which the producers enter with one another, the conditions under which they exchange their activities and participate in the whole act of production, will naturally vary according to the character of the means of production. With the invention of a new instrument of warfare, firearms, the whole internal organization of the army necessarily changed; the relationships within which individuals can constitute an army and act as an army were transformed and the relations of different armies to one another also changed.

Thus the social relations within which individuals produce, the social relations of production, change, are transformed, with the change and development of the material means of production, the productive forces. The relations of production in their totality constitute what are called the social relations, society, and, specifically, a society at a definite stage of historical development, a society with a peculiar, distinctive character. Ancient society, feudal society, bourgeois society are such totalities of production relations, each of which at the same time denotes a special stage of development in the history of mankind.

Capital, also, is a social relation of production. It is a bourgeois production relation, a production relation of bourgeois society. Are not the means of subsistence, the instruments of labour, the raw materials of which capital consists, produced and accumulated under given social conditions, in definite social relations? Are they not utilized for new production under given social conditions, in definite social relations? And is it not just this definite social character which turns the products serving for new production into capital?

Capital consists not only of means of subsistence, instruments of labour, and raw materials, not only of material products; it consists just as much of exchange values. All the products of which it consists are commodities. Capital is, therefore, not only a sum of material products; it is a sum of commodities, of exchange values, of social magnitudes.

Capital remains the same, whether we put cotton in place of wool, rice in place of wheat, or steamships in place of railways, provided only that the cotton, the rice, the steamships—the body of capital—have the same exchange value, the same price as the wool, the wheat, the railways in which it was previously incorporated. The body of capital can change continually without the capital suffering the slightest alteration.

But while all capital is a sum of commodities, that is, of exchange values, not every sum of commodities, of exchange values, is capital.

Every sum of exchange values is an exchange value. Every separate exchange value is a sum of exchange values. For instance, a house that is worth 1,000 marks is an exchange value of 1,000 marks. A piece of paper worth a pfennig is a sum of exchange values of one-hundredths of a pfennig. Products which are exchangeable for others are commodities. The particular ratio in which they are exchangeable constitutes their exchange value or, expressed in money, their price. The quantity of these products can change nothing in their quality of being commodities or representing an exchange value or having a definite price. Whether a tree is large or small it is a tree. Whether we exchange iron for other products in ounces or in hundredweights, does this make any difference in its character as commodity, as exchange value? It is a commodity of greater or lesser value, of higher or lower price, depending upon the quantity.

How, then, does any amount of commodities, of exchange value, become capital?

By maintaining and multiplying itself as an independent social power, that is, as the power of a portion of society, by means of its exchange for direct, living labour. The existence of a class which possesses nothing but its capacity to labour is a necessary prerequisite of capital.

It is only the domination of accumulated, past, materialized labour over direct, living labour that turns accumulated labour into capital.

Capital does not consist in accumulated labour serving living labour as a means for new production. It consists in living labour serving accumulated labour as a means for maintaining and multiplying the exchange value of the latter.

What takes place in the exchange between capitalist and wage-worker?

The worker receives means of subsistence in exchange for his labour, but the capitalist receives in exchange for his means of subsistence labour, the productive activity of the worker, the creative power whereby the worker not only replaces what he consumes but gives to the accumulated labour a greater value than it previously possessed. The worker receives a part of the available means of subsistence from the capitalist. For what purpose do these means of subsistence serve him? For immediate consumption. As soon, however, as I consume the means of subsistence, they are irretrievably lost to me unless I use the time during which I am kept alive by them in order to produce new means of subsistence, in order during consumption to create by my labour new values in place of the values which perish in being consumed. But it is just this noble reproductive power that the worker surrenders to the capitalist in exchange for means of subsistence received. He has, therefore, lost it for himself.

Let us take an example: a tenant farmer gives his day labourer five silver groschen a day. For these five silver groschen the labourer works all day on the farmer's field and thus secures him a return of ten silver groschen. The farmer not only gets the value replaced that he has to give the day labourer; he doubles it. He has therefore employed, consumed, the five silver groschen that he gave to the labourer in a fruitful, productive manner. He has bought with the five silver groschen just that labour and power of the labourer which produces agricultural products of double value and makes ten silver groschen out of five. The day labourer, on the other hand, receives in place of his productive power, the effect of which he has bargained away to the farmer, five silver groschen, which he exchanges for means of subsistence, and these he consumes with greater or less rapidity. The five silver groschen have, therefore, been consumed in a double way, reproductively for capital, for they have been exchanged for labour power which produced ten silver groschen, unproductively for the worker, for they have been exchanged for means of subsistence which have disappeared forever and the value of which he can only recover by repeating the same exchange with the farmer. Thus capital presupposes wage labour; wage labour presupposes capital. They reciprocally condition the existence of each other; they reciprocally bring forth each other.

Does a worker in a cotton factory produce merely cotton textiles? No, he produces capital. He produces values which serve afresh to command his labour and by means of it to create new values.

Capital can only increase by exchanging itself for labour power, by calling wage labour to life. The labour of the wage-worker can only be exchanged for capital by increasing capital, by strengthening the power whose slave it is. Hence, increase of capital is increase of the proletariat, that is, of the working class.

The interests of the capitalist and those of the worker are, therefore, one and the same, assert the bourgeois and their economists. Indeed! The worker perishes if capital does not employ him. Capital perishes if it does not exploit labour, and in order to exploit it, it must buy it. The faster capital intended for production, productive capital, increases, the more, therefore, industry prospers, the more the bourgeoisie enriches itself and the better business is, the more workers does the capitalist need, the more dearly does the worker sell himself.

The indispensable condition for a tolerable situation of the worker is, therefore, the fastest possible growth of productive capital.

But what is the growth of productive capital? Growth of the power of accumulated labour over living labour. Growth of the domination of the bourgeoisie over the working class. If wage labour produces the wealth of others that rules over it, the power that is hostile to it, capital, then the means of employment, that is, the means of subsistence, flow back to it from this hostile power, on condition that it makes itself afresh into a part of capital, into the lever which hurls capital anew into an accelerated movement of growth.

To say that the interests of capital and those of the workers are one and the same is only to say that capital and wage labour are two sides of one and the same relation. The one conditions the other, just as usurer and squanderer condition each other.

As long as the wage-worker is a wage-worker his lot depends upon capital. That is the much-vaunted community of interests between worker and capitalist.

If capital grows, the mass of wage labour grows, the number of wage-workers grows; in a word, the domination of capital extends over a greater number of individuals. Let us assume the most favourable case: when productive capital grows, the demand for labour grows; consequently, the price of labour, wages, goes up.

A house may be large or small; as long as the surrounding houses are equally small it satisfies all social demands for a dwelling. But let a palace arise beside the little house, and it shrinks from a little house to a hut. The little house shows now that its owner has only very slight or no demands to make; and however high it may shoot up in the course of civilization, if the neighbouring palace grows to an equal or even greater extent, the occupant of the relatively small house will feel more and more uncomfortable, dissatisfied, and cramped within its four walls.

A noticeable increase in wages presupposes a rapid growth of productive capital. The rapid growth of productive capital brings about an equally rapid growth of wealth, luxury, social wants, social enjoyments. Thus, although the enjoyments of the worker have risen, the social satisfaction that they give has fallen in comparison with the increased enjoyments of the capitalist, which are inaccessible to the worker, in comparison with the state of development of society in general. Our desires and pleasures spring from society; we measure them, therefore, by society and not by the objects which serve for their satisfaction. Because they are of a social nature, they are of a relative nature.

In general, wages are determined not only by the amount of commodities for which I can exchange them. They embody various relations.

What the workers receive for their labour is, in the first place, a definite sum of money. Are wages determined only by this money price?

In the sixteenth century, the gold and silver circulating in Europe increased as a result of the discovery of richer and more easily worked mines in America. Hence, the value of gold and silver fell in relation to other commodities. The workers received the same amount of coined silver for their labour as before. The money price of their labour remained the same, and yet their wages had fallen, for in exchange for the same quantity of silver they received a smaller amount of other commodities. This was one of the circumstances which furthered the growth of capital and the rise of the bourgeoisie in the sixteenth century.

Let us take another case. In the winter of 1847, as a result of a crop failure, the most indispensable means of subsistence, cereals, meat, butter, cheese, etc., rose considerably in price. Assume that the workers received the same sum of money for their labour power as before. Had not their wages fallen? Of course. For the same money they received less bread, meat, etc., in exchange. Their wages had fallen, not because the value of silver had diminished, but because the value of the means of subsistence had increased.

Assume, finally, that the money price of labour remains the same while all agricultural and manufactured goods have fallen in price owing to the employment of new machinery, a favourable season, etc. For the same money the workers can now buy more commodities of all kinds. Their wages, therefore, have risen, just because the money value of their wages has not changed.

Thus, the money price of labour, nominal wages, do not coincide with real wages, that is, with the sum of commodities which is actually given in exchange for the wages. If, therefore, we speak of a rise or fall of wages, we must keep in mind not only the money price of labour, the nominal wages.

But neither nominal wages, that is, the sum of money for which the worker sells himself to the capitalist, nor real wages, that is, the sum of commodities which he can buy for this money, exhaust the relations contained in wages.

Wages are, above all, also determined by their relation to the gain, to the profit of the capitalist—comparative, relative wages.

Real wages express the price of labour in relation to the price of other commodities; relative wages, on the other hand, express the share of direct labour, to capital.

Real wages may remain the same, they may even rise, and yet relative wages may fall. Let us suppose, for example, that all means of subsistence have gone down in price by two-thirds while wages per day have only fallen by one-third, that is to say, for example, from three marks to two marks. Although the worker can command a greater amount of commodities with these two marks than he previously could with three marks, yet his wages have gone down in relation to the profit of the capitalist. The profit of the capitalist (for example, the manufacturer) has increased by one mark; that is, for a smaller sum of exchange values which he pays to the worker, the latter must produce a greater amount of exchange values than before. The value of capital relative to the share of labour has risen. The division of social wealth between capital and labour has become still more unequal. With the same capital, the capitalist commands a greater quantity of labour. The power of the capitalist class over the working class has grown, the social position of the worker has deteriorated, has been depressed one step further below that of the capitalist.

What, then, is the general law which determines the rise and fall of wages and profit in their reciprocal relation?

They stand in inverse ratio to each other. Capital's exchange value, profit, rises in the same proportion as labour's share, wages, falls, and vice versa. Profit rises to the extent that wages fall; it falls to the extent that wages rise.

The objection will, perhaps, be made that the capitalist can profit by a favourable exchange of his products with other capitalists, by increase of the demand for his commodities, whether as a result of the opening of new markets, or as a result of a momentarily increased demand in the old markets, etc.; that the capitalist's profit can, therefore, increase by overreaching other capitalists, independently of the rise and fall of wages, of the exchange value of labour; or that the capitalist's profit may also rise owing to the improvement of the instruments of labour, a new application of natural forces, etc.

First of all, it will have to be admitted that the result remains the same, although it is brought about in reverse fashion. True, the profit has not risen because wages have fallen, but wages have fallen because the profit has risen. With the same amount of other people's labour, the capitalist has acquired a greater amount of exchange values, without having paid more for the labour on that account; that is, therefore, labour is paid less in proportion to the net profit which it yields the capitalist.

In addition, we recall that, in spite of the fluctuations in prices of commodities, the average price of every commodity, the ratio in which it is exchanged for other commodities, is determined by its cost of production. Hence the overreachings within the capitalist class necessarily balance one another. The improvement of machinery, new application of natural forces in the service of production, enable a larger amount of products to be created in a given period of time with the same amount of labour and capital, but not by any means a larger amount of exchange values. If, by the use of the spinning jenny, I can turn out twice as much yarn in an hour as before its invention, say, one hundred pounds instead of fifty, then in the long run I will receive for these hundred pounds no more commodities in exchange than formerly for the fifty pounds, because the cost of production has fallen by one-half, or because I can deliver double the product at the same cost.

Finally, in whatever proportion the capitalist class, the bourgeoisie, whether of one country or of the whole world market, shares the net profit of production within itself, the total amount of this net profit always consists only of the amount by which, on the whole, accumulated labour has been increased by direct labour. This total amount grows, therefore, in the proportion in which labour augments capital, that is, in the proportion in which profit rises in comparison with wages.

We see, therefore, that even if we remain within the relation of capital and wage labour, the interests of capital and the interests of wage labour are diametrically opposed.

A rapid increase of capital is equivalent to a rapid increase of profit. Profit can only increase rapidly if the exchange value of labour, if relative wages, decrease just as rapidly. Relative wages can fall although real wages rise simultaneously with nominal wages, with the money value of labour, if they do not rise, however, in the same proportion as profit. If, for instance, in times when business is good, wages rise by five per cent, profit on the other hand by thirty per cent, then the comparative, the relative wages, have not increased but decreased.

Thus if the income of the worker increases with the rapid growth of capital, the social gulf that separates the worker from the capitalist increases at the same time, and the power of capital over labour, the dependence of labour on capital, likewise increases at the same time.

To say that the worker has an interest in the rapid growth of capital is only to say that the more rapidly the worker increases the wealth of others, the richer will be the crumbs that fall to him, the greater is the number of workers that can be employed and called into existence, the more can the mass of slaves dependent on capital be increased.

We have thus seen that:

Even the most favourable situation for the working class, the most rapid possible growth of capital, however much it may improve the material existence of the worker, does not remove the antagonism between his interests and the interests of the bourgeoisie, the interests of the capitalists. Profit and wages remain as before in inverse proportion.

If capital is growing rapidly, wages may rise; the profit of capital rises incomparably more rapidly. The material position of the worker has improved, but at the cost of his social position. The social gulf that divides him from the capitalist has widened.

Finally:

To say that the most favourable condition for wage labour is the most rapid possible growth of productive capital is only to say that the more rapidly the working class increases and enlarges the power that is hostile to it, the wealth that does not belong to it and that rules over it, the more favourable will be the conditions under which it is allowed to labour anew at increasing bourgeois wealth, at enlarging the power of capital, content with forging for itself the golden chains by which the bourgeoisie drags it in its train.

Are growth of productive capital and rise of wages really so inseparably connected as the bourgeois economists maintain? We must not take their word for it. We must not even believe them when they say that the fatter capital is, the better will its slave be fed. The bourgeoisie is too enlightened, it calculates too well, to share the prejudices of the feudal lord who makes a display by the brilliance of his retinue. The conditions of existence of the bourgeoisie compel it to calculate.

We must, therefore, examine more closely:

How does the growth of productive capital affect wages?

If, on the whole, the productive capital of bourgeois society grows, a more manifold accumulation of labour takes place. The capitals increase in number and extent. The numerical increase of the capitals increases the competition between the capitalists. The increasing extent of the capitals provides the means for bringing more powerful labour armies with more gigantic instruments of war into the industrial battlefield.

One capitalist can drive another from the field and capture his capital only by selling more cheaply. In order to be able to sell more cheaply without ruining himself, he must produce more cheaply, that is, raise the productive power of labour as much as possible. But the productive power of labour is raised, above all, by a greater division of labour, by a more universal introduction and continual improvement of machinery. The greater the labour army among whom labour is divided, the more gigantic the scale on which machinery is introduced, the more does the cost of production proportionately decrease, the more fruitful is labour. Hence, a general rivalry arises among the capitalists to increase the division of labour and machinery and to exploit them on the greatest possible scale.

If, now, by a greater division of labour, by the utilization of new machines and their improvement, by more profitable and extensive exploitation of natural forces, one capitalist has found the means of producing with the same amount of labour or of accumulated labour a greater amount of products, of commodities, than his competitors, if he can, for example, produce a whole yard of linen in the same labour time in which his competitors weave half a yard, how will this capitalist operate?

He could continue to sell half a yard of linen at the old market price; this would, however, be no means of driving his opponents from the field and of enlarging his own sales. But in the same measure in which his production has expanded, his need to sell has also increased. The more powerful and costly means of production that he has called into life enable him, indeed, to sell his commodities more cheaply, they compel him, however, at the same time to sell more commodities, to conquer a much larger market for his commodities; consequently, our capitalist will sell his half yard of linen more cheaply than his competitors.

The capitalist will not, however, sell a whole yard as cheaply as his competitors sell half a yard, although the production of the whole yard does not cost him more than the half yard costs the others. Otherwise he would not gain anything extra but only get back the cost of production by the exchange. His possibly greater income would be derived from the fact of having set a larger capital into motion, but not from having made more of his capital than the others. Moreover, he attains the object he wishes to attain, if he puts the price of his goods only a small percentage lower than that of his competitors. He drives them from the field, he wrests from them at least a part of their sales, by underselling them. And, finally, it will be remembered that the current price always stands above or below the cost of production, according to whether the sale of the commodity occurs in a favourable or unfavourable industrial season. The percentage at which the capitalist who has employed new and more fruitful means of production sells above his real cost of production will vary, depending upon whether the market price of a yard of linen stands below or above its hitherto customary cost of production.

However, the privileged position of our capitalist is not of long duration; other competing capitalists introduce the same machines, the same division of labour, introduce them on the same or on a larger scale, and this introduction will become so general that the price of linen is reduced not only below its old, but below its new cost of production.

The capitalists find themselves, therefore, in the same position relative to one another as before the introduction of the new means of production, and if they are able to supply by these means double the production at the same price, they are now forced to supply the double product below the old price. On the basis of this new cost of production, the same game begins again. More division of labour, more machinery, enlarged scale of exploitation of machinery and division of labour. And again competition brings the same counteraction against this result.

We see how in this way the mode of production and the means of production are continually transformed, revolutionized, how the division of labour is necessarily followed by greater division of labour, the application of machinery by still greater application of machinery, work on a large scale by work on a still larger scale.

That is the law which again and again throws bourgeois production out of its old course and which compels capital to intensify the productive forces of labour. And because it has once intensified them, this law gives capital no rest and continually whispers in its ear: 'Go on! Go on!'

This law is none other than that which, within the fluctuations of trade periods, necessarily levels out the price of a commodity to its cost of production.

However powerful the means of production which a capitalist brings into the field, competition will make these means of production universal and from the moment when it has made them universal, the only result of the greater fruitfulness of his capital is that he must now supply for the same price ten, twenty, a hundred times as much as before. But, as he must sell perhaps a thousand times as much as before in order to outweigh the lower selling price by the greater amount of the product sold, because a more extensive sale is now necessary, not only in order to make more profit but in order to replace the cost of production—the instrument of production itself, as we have seen, becomes more and more expensive—and because this mass sale becomes a question of life and death not only for him but also for his rivals, the old struggle begins again all the more violently the more fruitful the already discovered means of production are. The division of labour and the application of machinery, therefore, will go on anew on an incomparably greater scale.

Whatever the power of the means of production employed may be, competition seeks to rob capital of the golden fruits of this power by bringing the price of the commodities back to the cost of production, by thus making cheaper production—the supply of ever greater amounts of products for the same total price—an imperative law to the same extent as production can be cheapened, that is, as more can be produced with the same amount of labour. Thus the capitalist would have won nothing by his own exertions but the obligation to supply more in the same labour time, in a word, more difficult conditions for the augmentation of the value of his capital. While, therefore, competition continually pursues him with its law of the cost of production and every weapon that he forges against his rivals recoils against himself, the capitalist continually tries to get the better of competition by incessantly introducing new machines, more expensive, it is true, but producing more cheaply, and new division of labour in place of the old, and by not waiting until competition has rendered the new ones obsolete.

If now we picture to ourselves this feverish simultaneous agitation on the whole world market, it will be comprehensible how the growth, accumulation, and concentration of capital results in an uninterrupted division of labour, and in the application of new and the perfecting of old machinery precipitately and on an ever more gigantic scale.

But how do these circumstances, which are inseparable from the growth of productive capital, affect the determination of wages?

The greater division of labour enables one worker to do the work of five, ten, or twenty; it therefore multiplies competition among the workers fivefold, tenfold, or twentyfold. The workers do not only compete by one selling himself cheaper than another; they compete by one doing the work of five, ten, twenty; and the division of labour, introduced by capital and continually increased, compels the workers to compete among themselves in this way.

Further, as the division of labour increases, labour is simplified. The special skill of the worker becomes worthless. He becomes transformed into a simple, monotonous productive force that does not have to use intense bodily or intellectual faculties. His labour becomes a labour that anyone can perform. Hence, competitors crowd upon him on all sides, and besides we remind the reader that the more simple and easily learned the labour is, the lower the cost of production needed to master it, the lower do wages sink, for, like the price of every other commodity, they are determined for by the cost of production.

Therefore, as labour becomes more unsatisfying, more repulsive, competition increases and wages decrease. The worker tries to keep up the amount of his wages by working more, whether by working longer hours or by producing more in one hour. Driven by want, therefore, he still further increases the evil effects of the division of labour. The result is that the more he works the less wages he receives, and for the simple reason that he competes to that extent with his fellow workers, hence makes them into so many competitors who offer themselves on just the same bad terms as he does himself, and that, therefore, in the last resort he competes with himself, with himself as a member of the working class.

Machinery brings about the same results on a much greater scale, by replacing skilled workers by unskilled, men by women, adults by children. It brings about the same results, where it is newly introduced, by throwing the hand workers on to the streets in masses, and, where it is developed, improved and replaced by more productive machinery, by discharging workers in smaller batches. We have portrayed above, in a hasty sketch, the industrial war of the capitalists among themselves; this war has the peculiarity that its battles are won less by recruiting than by discharging the army of labour. The generals, the capitalists, compete with one another as to who can discharge most soldiers of industry.

The economists tell us, it is true, that the workers rendered superfluous by machinery find new branches of employment.

They dare not assert directly that the same workers who are discharged find places in the new branches of labour. The facts cry out too loudly against this lie. They really only assert that new means of employment will open up for other component sections of the working class, for instance, for the portion of the young generation of workers that was ready to enter the branch of industry which has gone under. That is, of course, a great consolation for the disinherited workers. The worshipful capitalists will never want for fresh exploitable flesh and blood, and will let the dead bury their dead. This is a consolation which the bourgeois give themselves rather than one which they give the workers. If the whole class of wage-workers were to be abolished owing to machinery, how dreadful that would be for capital which, without wage labour, ceases to be capital!

Let us suppose, however, that those directly driven out of their jobs by machinery, and the entire section of the new generation that was already on the watch for this employment, find a new occupation. Does any one imagine that it will be as highly paid as that which has been lost? That would contradict all the laws of economics. We have seen how modern industry always brings with it the substitution of a more simple, subordinate occupation for the more complex and higher one.

How, then, could a mass of workers who have been thrown out of one branch of industry owing to machinery find refuge in another, unless the latter is lower, worse paid?

The workers who work in the manufacture of machinery itself have been cited as an exception. As soon as more machinery is demanded and used in industry, it is said, there must necessarily be an increase of machines, consequently of the manufacture of machines, and consequently of the employment of workers in the manufacture of machines; and the workers engaged in this branch of industry are claimed to be skilled, even educated workers.

Since the year 1840 this assertion, which even before was only half true, has lost all semblance of truth because ever more versatile machines have been employed in the manufacture of machinery, no more and no less than in the manufacture of cotton yarn, and the workers employed in the machine factories, confronted by highly elaborate machines, can only play the part of highly unelaborate machines.

But in place of the man who has been discharged owing to the machine, the factory employs maybe three children and one woman. And did not the man's wages have to suffice for the three children and a woman? Did not the minimum of wages have to suffice to maintain and to propagate the race? What, then, does this favourite bourgeois phrase prove? Nothing more than that now four times as many workers' lives are used up in order to gain a livelihood for one worker's family.

Let us sum up: The more productive capital grows, the more the division of labour and the application of machinery expands. The more the division of labour and the application of machinery expands, the more competition among the workers expands and the more their wages contract.

In addition, the working class gains recruits from the higher strata of society also; a mass of petty industrialists and small rentiers are hurled down into its ranks and have nothing better to do than urgently stretch out their arms alongside those of the workers. Thus the forest of uplifted arms demanding work becomes ever thicker, while the arms themselves become ever thinner.

That the small industrialist cannot survive in a contest one of the first conditions of which is to produce on an ever greater scale, that is, precisely to be a large and not a small industrialist, is self-evident.

That the interest on capital decreases in the same measure as the mass and number of capitals increase, as capital grows; that, therefore, the small rentier can no longer live on his interest but must throw himself into industry, and, consequently, help to swell the ranks of the small industrialists and thereby of candidates for the proletariat—all this surely requires no further explanation.

Finally, as the capitalists are compelled, by the movement described above, to exploit the already existing gigantic means of production on a larger scale and to set in motion all the mainsprings of credit to this end, there is a corresponding increase in industrial earthquakes, in which the trading world can only maintain itself by sacrificing a part of wealth, of products and even of productive forces to the gods of the nether world—in a word, crises increase. They become more frequent and more violent, if only because, as the mass of production, and consequently the need for extended markets, grows, the world market becomes more and more contracted, fewer and fewer new markets remain available for exploitation, since every preceding crisis has subjected to world trade a market hitherto unconquered or only superficially exploited. But capital does not live only on labour. A lord, at once aristocratic and barbarous, it drags with it into the grave the corpses of its slaves, whole hecatombs of workers who perish in the crises. Thus we see: if capital grows rapidly, competition among the workers grows incomparably more rapidly, that is, the means of employment, the means of subsistence, of the working class decrease proportionately so much the more, and, nevertheless, the rapid growth of capital is the most favourable condition for wage labour.

BIBLIOGRAPHY

Original: *MEW*, Vol. 6, pp. 397 ff.

Present translation: MESW, Vol. 1, pp. 79 ff. (considerably modified). Reproduced by kind permission of Lawrence and Wishart Ltd.

Other translations: None.

Commentaries: E. Mandel, *The Formation of Marx's Economic Thought*, London, 1971, Ch. 4.

20. Speech on Free Trade

The extracts below come from a speech that Marx delivered to the members of the Democratic Association in Brussels on 9 January 1848. Marx argues for Free Trade on the grounds that the economic supremacy of the bourgeoisie that is thereby implied is as necessary a stage in the class struggle as its political domination; but that both merely aid the unity of the proletariat and its ultimate emancipation.

... We are told that free trade would create an international division of labour, and thereby give to each country the production which is most in harmony with its natural advantages.

You believe perhaps, gentlemen, that the production of coffee and sugar is the natural destiny of the West Indies.

Two centuries ago, nature, which does not trouble herself about commerce, had planted neither sugar-cane nor coffee trees there.

And it may be that in less than half a century you will find there neither coffee nor sugar, for the East Indies, by means of cheaper production, have already successfully combated this alleged natural destiny of the West Indies. And the West Indies, with their natural wealth, are already as heavy a burden for England as the weavers of Dacca, who also were destined from the beginning of time to weave by hand.

One other thing must never be forgotten, namely, that, just as everything has become a monopoly, there are also nowadays some branches of industry which dominate all the others, and secure to the nations which most largely cultivate them the command of the world market. Thus in international commerce cotton alone has much greater commercial importance than all the other raw materials used in the manufacture of clothing put together. It is truly ridiculous to see the free-traders stress the few specialities in each branch of industry, throwing them into the balance against the products used in everyday consumption and produced most cheaply in those countries in which manufacture is most highly developed.

If the free-traders cannot understand how one nation can grow rich at the expense of another, we need not wonder, since these same gentlemen also refuse to understand how within one country one class can enrich itself at the expense of another.

Do not imagine, gentlemen, that in criticizing freedom of trade we have the least intention of defending the system of protection.

One may declare oneself an enemy of the constitutional regime without declaring oneself a friend of the ancient regime.

Moreover, the protectionist system is nothing but a means of establishing large-scale industry in any given country, that is to say, of making it dependent upon the world market, and from the moment that dependence upon the world market is established, there is already more or less dependence upon free trade. Besides this, the protective system helps to develop free competition within a country. Hence we see that in countries where the bourgeoisie is beginning to make itself felt as a class, in Germany for example, it makes great efforts to obtain protective duties. They serve the bourgeoisie as weapons against feudalism and absolute government, as a means for the concentration of its own powers and for the realization of free trade within the same country.

But, in general, the protective system of our day is conservative, while the free trade system is destructive. It breaks up old nationalities and pushes the antagonism of the proletariat and the bourgeoisie to the extreme point. In a word, the free trade system hastens the social revolution. It is in this revolutionary sense alone, gentlemen, that I vote in favour of free trade. . . .

BIBLIOGRAPHY

Original: K. Marx, *Œuvres*, ed. M. Rubel, Paris, 1963, Vol. 1, pp. 154 ff.

Present translation: K. Marx, *The Poverty of Philosophy*, Moscow, n.d., pp. 250 ff. Reproduced by kind permission of Lawrence and Wishart Ltd.

Other translations: None.

21. Articles for the Neue Rheinische Zeitung

On the outbreak of the 1848 revolutions, Marx went to Paris and then to Germany. His principal aim was the foundation of a newspaper, and the *Neue Rheinische Zeitung* started publication on 15 June with Marx as chief editor. The sub-title of the paper was 'organ of democracy', and it was more concerned at the outset to support the radical liberals than to put forward a directly proletarian point of view. Internally, the paper supported the left wing of the Frankfurt parliament against the King and in foreign policy urged a war against Russia in order to promote German unity. As the forces of reaction grew stronger towards the end of 1848, so the radicalism of the *Neue Rheinische Zeitung* increased. But it was not until April 1849 that Marx finally abandoned the policy of co-operation with the liberal democrats and urged the creation of an independent workers' party. The paper was suppressed in May 1849 and Marx returned to Paris.

The Revolution in Germany

... We do not make the utopian demand that at the outset a united indivisible German republic should be proclaimed, but we ask the so-called Radical-Democratic Party not to confuse the starting-point of the struggle and of the revolutionary movement with the goal. Both German unity and the German constitution can result only from a movement in which the internal conflicts and the war with the East will play an equally decisive role. The final act of constitution cannot be decreed, it coincides with the movement we have to go through. It is therefore not a question of putting into practice this or that view, this or that political idea, but of understanding the course of development. The National Assembly has to take only such steps as are practicable in the first instance. ...

Every provisional political set-up following a revolution calls for dictatorship, and an energetic dictatorship at that. From the very beginning we blamed Camphausen for not having acted in a dictatorial manner, for not having immediately smashed up and removed the remains of the old institutions. While thus Herr Camphausen indulged in constitutional fancies, the defeated party strengthened its positions within the bureaucracy and in the army, and occasionally even risked an open fight. The Assembly was convened for the purpose of agreeing on the terms of the constitution. It existed as an equal party alongside the Crown. Two equal powers under a provisional arrangement! It was this division of powers with the aid of which Herr Camphausen sought 'to save freedom'—it was this very division of powers under provisional arrangement that was bound to lead to conflicts. The Crown served as a cover for the counter-revolutionary aristocratic, military, and bureaucratic camarilla. The bourgeoisie stood behind the majority of the Assembly. The cabinet tried to mediate. Too weak to stand up for the bourgeoisie and the peasants and overthrow the power of the nobility, the bureaucracy, and the army chiefs at one blow, too unskilled to avoid always damaging the interests of the bourgeoisie by its financial measures, the cabinet merely succeeded in compromising itself in the eyes of all the parties and bringing about the very clash it sought to avoid. . . .

Assuming that arms will enable the counter-revolution to establish itself in the whole of Europe, money would then kill it in the whole of Europe. European bankruptcy, national bankruptcy, would be the fate nullifying the victory. Bayonets crumble like tinder when they come into contact with the salient 'economic' facts.

But developments will not wait for the bills of exchange drawn by the European states on European society to expire. The crushing counter-blow of the June revolution will be struck in Paris. With the victory of the 'red republic' in Paris, armies will be rushed from the interior of their countries to the frontiers and across them, and the real strength of the fighting parties will become evident. We shall then remember this June and this October and we too shall exclaim:

Vae victis!

The purposeless massacres perpetrated since the June and October events, the tedious offering of sacrifices since February and March, the very cannibalism of the counter-revolution will convince the nations that there is only one way in which the murderous death agonies of the old society and the bloody birth throes of the new society can be shortened, simplified, and concentrated, and that way is revolutionary terror. . . .

The history of the Prussian middle class, and that of the German middle class in general between March and December shows that a purely middle-class revolution and the establishment of bourgeois rule in the form of a constitutional monarchy is impossible in Germany, and that the only alternatives are either a feudal absolutist counter-revolution or a social republican revolution. . . .

England and the Revolution

. . . But England, the country that turns whole nations into her proletarians, that spans the whole world with her enormous arms, that has already once defrayed the cost of a European Restoration, the country in which class contradictions have reached their most acute and shameless form—England seems to be the rock which breaks the revolutionary waves, the country where the new society is stifled before it is born. England dominates the world market. Any upheaval in economic relations in any country of the European continent, in the whole European continent without England, is a storm in a teacup. Industrial and commercial relations within each nation are governed by its intercourse with other nations, and depend on its relations with the world market. But the world market is dominated by England and England is dominated by the bourgeoisie.

Thus, the liberation of Europe, whether brought about by the struggle of the oppressed nationalities for their independence or by overthrowing feudal absolutism, depends on the successful uprising of the French working class. Every social upheaval in France, however, is bound to be thwarted by the English bourgeoisie, by Great Britain's industrial and commercial domination of the world. Every partial social reform in France or on the European continent as a whole, if designed to be lasting, is merely a pious wish. Only a world war can break old England, as only this can provide the Chartists, the party of the organized English workers, with the conditions for a successful rising against their powerful oppressors. Only when the Chartists head the English government will the social revolution pass from the sphere of utopia to that of reality. But any European war in which England is involved is a world war, waged in Canada and Italy, in the East Indies and Prussia, in Africa and on the Danube. A European war will be the first result of a successful workers' revolution in France. England will head the counter-revolutionary armies, just as she did during the Napoleonic period, but the war itself will place her at the head of the revolutionary movement and she will repay the debt she owes to the revolution of the eighteenth century.

The table of contents for 1849 reads: 'Revolutionary rising of the French working class, world war.'

Taxes

... After God had created the world and kings by the grace of God, He left smaller-scale industry to men. Weapons and lieutenants' uniforms are made in a profane manner and the profane way of production cannot, like heavenly industry, create out of nothing. It needs raw materials, tools, and wages, weighty things that are categorized under the modest term of 'production costs'. These production costs are offset for the state through taxes and taxes are offset through the nation's work. From the economic point of view, therefore, it remains an enigma how any king can give any people anything. The people must first make weapons and give them to the king in order to be able to receive them from the king. The king can only give what has already been given to him. This from the economic point of view. However, constitutional kings arise at precisely those moments when people are beginning to understand the economic mystery. Thus the first beginnings of the fall of kings by the grace of God have always been questions of taxes. So too in Prussia. ...

Marx's Defence Speech at his Trial

... Society is not founded upon the law; this is a legal fiction. On the contrary, the law must be founded upon society, it must express the common interests and needs of society—as distinct from the caprice of the individuals—which arise from the material mode of production prevailing at the given time. This Code Napoléon, which I am holding in my hand, has not created modern bourgeois society. On the contrary, bourgeois society, which emerged in the eighteenth century and developed further in the nineteenth, merely finds its legal expression in this Code. As soon as it ceases to fit the social conditions, it becomes simply a bundle of paper. You cannot make the old laws the foundation of the new social development, any more than these old laws created the old social conditions.

They were engendered by the old conditions of society and must perish with them. They are bound to change with the changing conditions of life. To maintain the old laws in face of the new needs and demands of social development is essentially the same as hypocritically upholding the out-of-date particular interests of a minority in face of the up-to-date interests of the community. This maintenance of the legal basis aims at asserting minority interests as if they were the predominant interests, when they are no longer dominant; it aims at imposing on society laws which have been condemned by the conditions of life in this society, by the way the members of this society earn their living, by their commerce and their material production; it aims at retaining legislators who are concerned only with their particular interests; it seeks to misuse political power in order forcibly to place the interests of a minority above the interests of the majority. The maintenance of the legal basis is therefore in constant conflict with the existing needs, it hampers commerce and industry, it prepares the way for social crises, which erupt in political revolutions. . . .

The Crown did not want and could not want reconciliation. Gentlemen of the jury, let us not deceive ourselves concerning the nature of the struggle which began in March and was later waged between the National Assembly and the Crown. It was not an ordinary conflict between a cabinet and a parliamentary opposition, it was not a conflict between men who were ministers and men who wanted to become ministers, it was not a struggle between two political parties in a legislative chamber. It is quite possible that members of the National Assembly belonging to the minority or the majority believed that this was so. The decisive factor however, is not the opinion of the deputies, but the real historical position of the National Assembly as it emerged both from the European revolution and the March revolution it engendered. What took place here was not a political conflict between two parties within the framework of one society, but a conflict between two societies, a social conflict, which assumed a political form; it was the struggle of the old feudal bureaucratic society with modern bourgeois society, a struggle between the society of free competition and the society of the guilds, between the society of landownership and the industrial society, between a religious society and a scientific society. The political expression corresponding to the old society was the Crown by the grace of God, the bullying bureaucracy and the independent army. The social foundation corresponding to this old political power consisted of privileged aristocratic landownership with its enthralled or partially enthralled peasants, the small patriarchal or guild industries, the strictly separated estates, the sharp contradiction between town and country and, above all, the domination of the countryside over the town. The old political power—the Crown by the grace of God, the bullying bureaucracy, the independent army—realized that its essential material basis would disappear from under its feet, as soon as any change was made in the basis of the old society, privileged aristocratic landownership, the aristocracy itself, the domination of the countryside over the town, the dependent position of the rural population, and the laws corresponding to these conditions of life, such as the parish regulations, the criminal law. The National Assembly made such an attempt. On the other hand that old society realized that political power would be wrenched from its hands, as soon as the Crown, the bureaucracy, and the army lost their feudal privileges. The National Assembly wanted to abolish these privileges. It is not surprising, therefore, that the army, the bureaucracy, and the nobility joined forces in urging the Crown to effect a *coup de main*, and it is not surprising that the Crown, knowing that its own interests were closely interlinked with those of the old feudal bureaucratic society, allowed itself to be impelled to a *coup d'état*. For the Crown represented feudal aristocratic society, just as the National Assembly represented modern bourgeois society. The conditions of existence in modern bourgeois society require that the bureaucracy and the army, which controlled commerce and industry, should become their tools, be reduced to mere organs of bourgeois intercourse. This society cannot tolerate that restrictions are placed on agriculture by feudal privileges and on industry by bureaucratic tutelage. This is contrary to free competition, the vital principle of this society. It cannot tolerate that foreign trade relations should be determined by considerations of the palace's international policies instead of by the interests of national production. It must subordinate fiscal policy to the needs of production, whereas the old state has to subordinate production to the needs of the Crown by the grace of God and the patching up of the monarchical walls, the social pillars of this Crown. Just as a modern industry is indeed a leveller, so modern society must break down all legal and political barriers between town and country. Modern society still has classes, but no longer estates. Its development lies in the struggle between these classes, but the latter stand united against the estates and their monarchy by the grace of God. . . .

BIBLIOGRAPHY

Originals: *MEW*, Vols. 5 and 6.

Present translation: K. Marx and F. Engels, *Articles from the Neue Rheinische Zeitung 1848–49*, Moscow, 1972, pp. 33, 124 f., 148 f., 202, 205, 232 f., 241 f. Reproduced by kind permission of Lawrence and Wishart Ltd.

Other translations: K. Marx, *The Revolutions of 1848*, ed. D. Fernbach, London, 1973.

Commentaries: D. Fernbach, Introduction to the edition cited above.

R. Hunt, *The Political Ideas of Marx and Engels*, Pittsburgh and London, 1975, Vol. 1, pp. 191 ff.

D. McLellan, *Karl Marx: His Life and Thought*, London, 1973, pp. 197 ff.

F. Mehring, *Karl Marx: The Story of his Life*, London, 1966, pp. 152 ff.

B. Nicolaievsky and O. Maenchen-Helfen, *Karl Marx, Man and Fighter*, 3rd edn., London, 1973, pp. 165 ff.

22. Address to the Communist League

The Communist League had been dissolved during the 1848 revolutions on Marx's authority. It was reconstituted in London during 1849 and when Marx arrived there in August 1849 he soon became active in refugee politics and a dominant figure in the League. The Central Committee in London decided in March 1850 to issue an Address aimed at galvanizing the groups of the League in Germany in the face of growing reaction there. This Address was drafted by Marx and Engels. Marx was still extremely optimistic about the imminence of the next revolutionary wave and the Address reflects this optimism to such an extent that some have even called it Blanquist. The Address asserts the necessity of an independent workers' party and describes what the attitude of the proletarian party should be to bourgeois parties during a revolutionary period—an attitude that Marx sums up with the phrase 'permanent revolution'.

Brothers! In the two revolutionary years 1848–9 the League proved itself in double fashion: first, in that its members energetically took part in the movement in all places, that in the press, on the barricades, and on the battlefields they stood in the front ranks of the only decidedly revolutionary class, the proletariat. The League further proved itself in that its conception of the movement as laid down in the circulars of the congresses and of the Central Committee of 1847 as well as in the *Communist Manifesto* turned out to be the only correct one, that the expectations expressed in those documents were completely fulfilled and the conception of present-day social conditions, previously propagated only in secret by the League, is now on everyone's lips and is openly preached in the market places. At the same time the former firm organization of the League was considerably slackened. A large proportion of the members who directly participated in the revolutionary movement believed the time for secret societies to have gone by and public activities alone sufficient. The individual circles and communities allowed their connections with the Central Committee to become loose and gradually dormant. Consequently, while the democratic party, the party of the petty bourgeoisie, organized itself more and more in Germany, the workers' party lost its only firm foothold, remained organized at the most in separate localities for local purposes, and in the general movement thus came completely under the domination and leadership of the petty-bourgeois democrats. An end must be put to this state of affairs, the independence of the workers must be restored. The Central Committee realized this necessity and therefore already in the winter of 1848–9 it sent an emissary, Josef Moll, to Germany for the reorganization of the League. Moll's mission, however, was without lasting effect, partly because the German workers at that time had not acquired sufficient experience and partly because it was interrupted by the insurrection of the previous May. Moll himself took up the musket, entered the Baden-Palatinate army and fell on July 19 in the encounter at the Murg. The League lost in him one of its oldest, most active, and most trustworthy members, one who had been active in all the congresses and Central Committees and even prior to this had carried out a series of missions with great success. After the defeat of the revolutionary parties of Germany and France in July 1849, almost all the members of the Central Committee came together again in London, replenished their numbers with new revolutionary forces, and set about the reorganization of the League with renewed zeal.

Reorganization can only be carried out by an emissary, and the Central Committee considers it extremely important that the emissary should leave precisely at this moment when a new revolution is impending, when the workers' party, therefore, must act in the most organized, most unanimous, and most independent fashion possible if it is not to be exploited and taken in tow again by the bourgeoisie as in 1848.

Brothers! We told you as early as 1848 that the German liberal bourgeois would soon come to power and would immediately turn their newly acquired power against the workers. You have seen how this has been fulfilled. In fact it was the bourgeois who, immediately after the March movement of 1848, took possession of the state power and used this power to force back at once the workers, their allies in the struggle, into their former oppressed position. Though the bourgeoisie was not able to accomplish this without uniting with the feudal party, which had been disposed of in March, without finally even surrendering power once again to this feudal absolutist party, still it has secured conditions for itself which, in the long run, owing to the financial embarrassment of the government, would place power in its hands and would safeguard all its interests, if it were possible for the revolutionary movement to assume already now a so-called peaceful development. The bourgeoisie, in order to safeguard its rule, would not even need to make itself obnoxious by violent measures against the people, since all such violent steps have already been taken by the feudal counter-revolution. Developments, however, will not take this peaceful course. On the contrary, the revolution, which will accelerate this development, is near at hand, whether it will be called forth by an independent uprising of the French proletariat or by an invasion of the Holy Alliance against the revolutionary Babylon [i.e. Paris].

And the role, this so treacherous role which the German liberal bourgeois played in 1848 against the people, will in the impending revolution be taken over by the democratic petty bourgeois, who at present occupy the same position in the opposition as the liberal bourgeois before 1848. This party, the democratic party, which is far more dangerous to the workers than the previous liberal one, consists of three elements:

I. Of the most advanced sections of the big bourgeoisie, which pursue the aim of the immediate complete overthrow of feudalism and absolutism. This faction is represented by the one-time Berlin compromisers, by the tax resisters.

II. Of the democratic-constitutional petty bourgeois, whose main aim during the previous movement was the establishment of a more or less democratic federal state as striven for by their representatives, the Lefts in the Frankfurt Assembly, and later by the Stuttgart parliament, and by themselves in the campaign for the Reich Constitution.

III. Of the republican petty bourgeois, whose ideal is a German federative republic after the manner of Switzerland, and who now call themselves Red and social-democratic because they cherish the pious wish of abolishing the pressure of big capital on small capital, of the big bourgeois on the small bourgeois. The representatives of this faction were the members of the democratic congresses and committees, the leaders of the democratic associations, the editors of the democratic newspapers.

Now, after their defeat, all these factions call themselves Republicans or Reds, just as the republican petty bourgeois in France now call themselves Socialists. Where, as in Württemberg, Bavaria, etc., they still find opportunity to pursue their aims constitutionally, they seize the occasion to retain their old phrases and to prove by deeds that they have not changed in the least. It is evident, moreover, that the altered name of this party does not make the slightest difference in its attitude to the workers, but merely proves that they are now obliged to turn against the bourgeoisie, which is united with absolutism, and to seek support in the proletariat.

The petty-bourgeois democratic party in Germany is very powerful; it comprises not only the great majority of the bourgeois inhabitants of the towns, the small people in industry and trade, and the guild-masters; it numbers among its followers also the peasants and the rural proletariat, in so far as the latter has not yet found a support in the independent urban proletariat.

The relation of the revolutionary workers' party to the petty-bourgeois democrats is this: it marches together with them against the faction which it aims at overthrowing, it opposes them in everything whereby they seek to consolidate their position in their own interests.

Far from desiring to revolutionize all society for the revolutionary proletarians, the democratic petty bourgeois strive for a change in social conditions by means of which existing society will be made as tolerable and comfortable as possible for them. Hence they demand above all diminution of state expenditure by a curtailment of the bureaucracy and shifting the chief taxes on to the big landowners and bourgeois. Further, they demand the abolition of the pressure of big capital on small, through public credit institutions and laws against usury, by which means it will be possible for them and the peasants to obtain advances, on favourable conditions, from the state instead of from the capitalists; they also demand the establishment of bourgeois property relations in the countryside by the complete abolition of feudalism. To accomplish all this they need a democratic state structure, either constitutional or republican, that will give them and their allies, the peasants, a majority; also a democratic communal structure that will give them direct control over property and over a series of functions now performed by the bureaucrats.

The domination and speedy increase of capital is further to be counteracted partly by restricting the right of inheritance and partly by transferring as many jobs of work as possible to the state. As far as the workers are concerned, it remains certain above all that they are to remain wage-workers as before; the democratic petty bourgeois only desire better wages and a more secure existence for the workers and hope to achieve this through partial employment by the state and through charity measures; in short, they hope to bribe the workers by more or less concealed alms and to break their revolutionary potency by making their position tolerable for the moment. The demands of the petty-bourgeois democracy here summarized are not put forward by all of its factions at the same time and only a very few members of them consider that these demands constitute definite aims in their entirety. The further separate individuals or factions among them go, the more of these demands will they make their own, and those few who see their own programme in what has been outlined above might believe that thereby they have put forward the utmost that can be demanded from the revolution. But these demands can in nowise suffice for the party of the proletariat. While the democratic petty bourgeois wish to bring the revolution to a conclusion as quickly as possible, and with the achievement, at most, of the above demands, it is our interest and our task to make the revolution permanent, until all more or less possessing classes have been forced out of their position of dominance, until the proletariat has conquered state power, and the association of proletarians, not only in one country but in all the dominant countries of the world, has advanced so far that competition among the proletarians of these countries has ceased and that at least the decisive productive forces are concentrated in the hands of the proletarians. For us the issue cannot be the alteration of private property but only its annihilation, not the smoothing over of class antagonisms but the abolition of classes, not the improvement of existing society but the foundation of a new one. That during the further development of the revolution, the petty-bourgeois democracy will for a moment obtain predominating influence in Germany is not open to doubt. The question, therefore, arises as to what the attitude of the proletariat and in particular of the League will be in relation to it;

1. during the continuance of the present conditions where the petty-bourgeois democrats are likewise oppressed;

2. in the next revolutionary struggle, which will give them the upper hand;

3. after this struggle, during the period of preponderance over the over-thrown classes and the proletariat.

1. At the present moment, when the democratic petty bourgeois are every-where oppressed, they preach in general unity and reconciliation to the proletariat, they offer it their hand and strive for the establishment of a large opposition party which will embrace all shades of opinion in the democratic party, that is, they strive to entangle the workers in a party organization in which general social-democratic phrases predominate, behind which their special interests are concealed and in which the particular demands of the proletariat may not be brought forward for the sake of beloved peace. Such a union would turn out solely to their advantage and altogether to the disad-vantage of the proletariat. The proletariat would lose its whole independent, laboriously achieved position and once more sink down to being an appen-dage of official bourgeois democracy. This union must, therefore, be most decisively rejected. Instead of once again stooping to serve as the applauding chorus of the bourgeois democrats, the workers, and above all the League, must exert themselves to establish an independent, secret, and public organ-ization of the workers' party alongside the official democrats and make each section the central point and nucleus of workers' societies, in which the attitude and interests of the proletariat will be discussed independently of bourgeois influences. How far the bourgeois democrats are from seriously considering an alliance in which the proletarians would stand side by side with them with equal power and equal rights is shown, for example, by the Breslau democrats who, in their organ, the *Neue Oder-Zeitung*, most fur-iously attack the independently organized workers, whom they style Socialists. In the case of a struggle against a common adversary no special union is required. As soon as such an adversary has to be fought directly, the interests of both parties, for the moment, coincide, and, as previously, so also in the future, this connection, calculated to last only for the moment, will arise of itself. It is self-evident that in the impending bloody conflicts, as in all earlier ones, it is the workers who, in the main, will have to win the victory by their courage, determination, and self-sacrifice. As previously, so also in this struggle, the mass of the petty bourgeois will as long as possible remain hesitant, undecided, and inactive, and then, as soon as the issue has been decided, will seize the victory for themselves, will call upon the workers to maintain tranquillity and return to their work, will guard against so-called excesses, and bar the proletariat from the fruits of victory. It is not in the power of the workers to prevent the petty-bourgeois democrats from doing this, but it is in their power to make it difficult for them to gain the upper hand as against the armed proletariat, and to dictate such conditions to them that the rule of the bourgeois democrats will from the outset bear within it the seeds of their downfall, and that their subsequent extrusion by the rule of the proletariat will be considerably facilitated. Above all things, the workers must counteract, as much as is at all possible, during the conflict and im-

mediately after the struggle, the bourgeois endeavours to allay the storm, and must compel the democrats to carry out their present terrorist phrases. Their actions must be so aimed as to prevent the direct revolutionary excitement from being suppressed again immediately after the victory. On the contrary, they must keep it alive as long as possible. Far from opposing so-called excesses, instances of popular revenge against hated individuals or public buildings that are associated only with hateful recollections, such instances must not only be tolerated but the leadership of them taken in hand. During the struggle and after the struggle, the workers must, at every opportunity, put forward their own demands alongside the demands of the bourgeois democrats. They must demand guarantees for the workers as soon as the democratic bourgeois set about taking over the government. If necessary they must obtain these guarantees by force, and in general they must see to it that the new rulers pledge themselves to all possible concessions and promises—the surest way to compromise them. In general, they must in every way restrain as far as possible the intoxication of victory and the enthusiasm for the new state of things, which make their appearance after every victorious street battle, by a calm and dispassionate estimate of the situation and by unconcealed mistrust in the new government. Alongside the new official governments they must establish simultaneously their own revolutionary workers' governments, whether in the form of municipal committees and municipal councils or in the form of workers' clubs or workers' committees, so that the bourgeois-democratic governments not only immediately lose the support of the workers but from the outset see themselves supervised and threatened by authorities which are backed by the whole mass of the workers. In a word, from the first moment of victory, mistrust must be directed no longer against the conquered reactionary party, but against the workers' previous allies, against the party that wishes to exploit the common victory for itself alone.

2. But in order to be able energetically and threateningly to oppose this party, whose treachery to the workers will begin from the first hour of victory, the workers must be armed and organized. The arming of the whole proletariat with rifles, muskets, cannon, and munitions must be put through at once, the revival of the old Citizens' Guard directed against the workers must be resisted. However, where the latter is not feasible the workers must attempt to organize themselves independently as a proletarian guard with commanders elected by themselves and with a general staff of their own choosing, and to put themselves at the command not of the state authority but of the revolutionary community councils which the workers will have managed to get adopted. Where workers are employed at the expense of the state they must see that they are armed and organized in a separate corps with commanders of their own choosing or as part of the proletarian guard. Arms and ammunition must not be surrendered on any pretext; any attempt

at disarming must be frustrated, if necessary by force. Destruction of the influence of the bourgeois democrats upon the workers, immediate independent and armed organization of the workers, and the enforcement of conditions as difficult and compromising as possible upon the inevitable momentary rule of the bourgeois democracy—these are the main points which the proletariat and hence the League must keep in view during and after the impending insurrection.

3. As soon as the new governments have consolidated their positions to some extent, their struggle against the workers will begin. Here, in order to be able to offer energetic opposition to the democratic petty bourgeois, it is above all necessary that the workers shall be independently organized and centralized in clubs. After the overthrow of the existing governments, the Central Committee will, as soon as it is at all possible, betake itself to Germany, immediately convene a congress, and put before the latter the necessary proposals for the centralization of the workers' clubs under a leadership established in the chief seat of the movement. The speedy organization of at least a provincial interlinking of the workers' clubs is one of the most important points for the strengthening and development of the workers' party; the immediate consequence of the overthrow of the existing governments will be the election of a national representative assembly. Here the proletariat must see to it:

I. That no groups of workers are barred on any pretext or by any kind of trickery on the part of local authorities or government commissioners.

II. That everywhere workers' candidates are put up alongside the bourgeois-democratic candidates, that they should consist as far as possible of members of the League, and that their election is promoted by all possible means. Even where there is no prospect whatsoever of their being elected, the workers must put up their own candidates in order to preserve their independence, to count their forces, and to bring before the public their revolutionary attitude and party standpoint. In this connection they must not allow themselves to be seduced by such arguments of the democrats as, for example, that by so doing they are splitting the democratic party and making it possible for the reactionaries to win. The ultimate intention of all such phrases is to dupe the proletariat. The advance which the proletarian party is bound to make by such independent action is infinitely more important than the disadvantage that might be incurred by the presence of a few reactionaries in the representative body. If the democracy from the outset comes out resolutely and terroristically against the reaction, the influence of the latter in the elections will be destroyed in advance.

The first point on which the bourgeois democrats will come into conflict with the workers will be the abolition of feudalism. As in the first French Revolution, the petty bourgeois will give the feudal lands to the peasants as free property, that is to say, try to leave the rural proletariat in existence and form a petty-bourgeois peasant class which will go through the same cycle of impoverishment and indebtedness which the French peasant is now still going through.

The workers must oppose this plan in the interest of the rural proletariat and in their own interest. They must demand that the confiscated feudal property remain state property and be converted into workers' colonies cultivated by the associated rural proletariat with all the advantages of large-scale agriculture, through which the principle of common property immediately obtains a firm basis in the midst of the tottering bourgeois property relations. Just as the democrats combine with the peasants, so must the workers combine with the rural proletariat. Further, the democrats will work either directly for a federative republic or, if they cannot avoid a single and indivisible republic, they will at least attempt to cripple the central government by the utmost possible autonomy and independence for the communities and provinces. The workers, in opposition to this plan, must not only strive for a single and indivisible German republic, but also within this republic for the most determined centralization of power in the hands of the state authority. They must not allow themselves to be misguided by the democratic talk of freedom for the communities, of self-government, etc. In a country like Germany where there are still so many relics of the Middle Ages to be abolished, where there is so much local and provincial obstinacy to be broken, it must under no circumstances be permitted that every village, every town, and every province should put a new obstacle in the path of revolutionary activity, which can proceed with full force only from the centre. It is not to be tolerated that the present state of affairs should be renewed, that Germans must fight separately in every town and in every province for one and the same advance. Least of all is it to be tolerated that a form of property, namely, communal property, which still lags behind modern private property and which everywhere is necessarily passing into the latter, together with the quarrels resulting from it between poor and rich communities, as well as communal civil law, with its trickery against the workers, that exists alongside of state civil law, should be perpetuated by a so-called free communal constitution. As in France in 1793 so today in Germany it is the task of the really revolutionary party to carry through the strictest centralization.

We have seen how the democrats will come to power with the next movement, how they will be compelled to propose more or less socialistic measures. It will be asked what measures the workers ought to propose in reply. At the beginning of the movement, of course, the workers cannot yet propose any directly communistic measures. But they can:

1. Compel the democrats to interfere in as many spheres as possible of the hitherto existing social order, to disturb its regular course, and to compromise themselves as well as to concentrate the utmost possible productive forces, means of transport, factories, railways, etc., in the hands of the state;

2. They must drive the proposals of the democrats, who in any case will not act in a revolutionary but in a merely reformist manner, to the extreme and transform them into direct attacks upon private property; thus, for example, if the petty bourgeois propose purchase of the railways and factories, the workers must demand that these railways and factories shall be simply confiscated by the state without compensation as being the property of reactionaries. If the democrats propose proportional taxes, the workers must demand progressive taxes; if the democrats themselves put forward a moderately progressive tax, the workers must insist on a tax with rates that rise so steeply that big capital will be ruined by it; if the democrats demand the regulation of state debts, the workers must demand state bankruptcy. Thus, the demands of the workers must everywhere be governed by the concessions and measures of the democrats.

If the German workers are not able to attain power and achieve their own class interests without completely going through a lengthy revolutionary development, they at least know for a certainty this time that the first act of this approaching revolutionary drama will coincide with the direct victory of their own class in France and will be very much accelerated by it.

But they themselves must do the utmost for their final victory by clarifying their minds as to what their class interests are, by taking up their position as an independent party as soon as possible and by not allowing themselves to be seduced for a single moment by the hypocritical phrases of the democratic petty bourgeois into refraining from the independent organization of the party of the proletariat. Their battle-cry must be: The Revolution in Permanence.

BIBLIOGRAPHY

Original: *MEW*, Vol. 7, pp. 244 ff.

Present translation: *MESW*, Vol. 1, pp. 106 ff. Reproduced by kind permission of Lawrence and Wishart Ltd.

Other translations: None.

Commentaries: S. Avineri, *The Social and Political Thought of Karl Marx*, Cambridge, 1968, pp. 196 ff.

D. Fernbach, Introduction to: K. Marx, *The Revolutions of 1848*, New York and London, 1974.

R. Hunt, *The Political Ideas of Marx and Engels*, Pittsburgh and London, 1975, Vol. 1, pp. 235 ff.

G. Lichtheim, *Marxism: An Historical and Critical Study*, New York and London, 1971, pp. 122 ff.

B. Nicolaievsky and O. Maenchen-Helfen, *Karl Marx: Man and Fighter*, 3rd edn., London, 1973, pp. 219 ff.

B. Wolfe, Marxism: *100 Years in the Life of a Doctrine*, London, 1967, pp. 151 ff.

23. The Class Struggles in France

The following texts come from a series of articles that Marx composed during 1850 as his main contribution to the *Neue Rheinische Zeitung-Revue*, the monthly he had founded in London to continue the propaganda of the *Neue Rheinische Zeitung*. The articles were originally entitled '1848–49' and were republished by Engels in 1895 under the title *The Class Struggles in France*.

The articles are an analysis of the turning-points of the 1848 revolution in France viewed against a background of class and economic interest. Engels described them as 'Marx's first attempt to explain a section of contemporary history by means of his materialist conception'. All but the last section were written during the first months of 1850. The final section was written in the autumn of that year and shows the effect of Marx's reading in the British Museum during the summer: his belief in the imminence of a fresh revolutionary outbreak and his enthusiasm for Blanqui had weakened and he now considered an economic crisis as the necessary precondition for any revolution.

With the exception of only a few chapters, every more important part of the annals of the revolution from 1848 to 1849 carries the heading: Defeat of the revolution!

What succumbed in these defeats was not the revolution. It was the pre-revolutionary traditional appendages, results of social relationships which had not yet come to the point of sharp class antagonisms—persons, illusions, conceptions, projects from which the revolutionary party before the February Revolution was not free, from which it could be freed not by the victory of February, but only by a series of defeats.

In a word: the revolution made progress, forged ahead, not by its immediate tragicomic achievements, but on the contrary by the creation of a powerful, united counter-revolution, by the creation of an opponent in combat with whom, only, the party of overthrow ripened into a really revolutionary party.

To prove this is the task of the following pages.

After the July Revolution, when the liberal banker Laffitte led his companion, the Duke of Orleans, in triumph to the *Hôtel de Ville*, he let fall the words: 'From now on the bankers will rule.' Laffitte had betrayed the secret of the revolution.

It was not the French bourgeoisie that ruled under Louis Philippe, but one faction of it: bankers, stock-exchange kings, railway kings, owners of coal and iron mines and forests, a part of the landed proprietors associated with them—the so-called finance aristocracy. It sat on the throne, it dictated laws in the Chambers, it distributed public offices, from cabinet portfolios to tobacco bureau posts.

The industrial bourgeoisie proper formed part of the official opposition, that is, it was represented only as a minority in the Chambers. Its opposition was expressed all the more resolutely, the more unalloyed the autocracy of the finance aristocracy became, and the more it itself imagined that its domination over the working class was ensured after the mutinies of 1832, 1834, and 1839, which had been drowned in blood. Grandin, Rouen manufacturer and the most fanatical instrument of bourgeois reaction in the Constituent as well as in the Legislative National Assembly, was the most violent opponent of Guizot in the Chamber of Deputies. Léon Faucher, later known for his impotent efforts to climb into prominence as the Guizot of the French counter-revolution, in the last days of Louis Philippe waged a war of the pen for industry against speculation and its train-bearer, the government. Bastiat agitated in the name of Bordeaux and the whole of wine-producing France against the ruling system.

The petty bourgeoisie of all gradations, and the peasantry also were completely excluded from political power. Finally, in the official opposition or entirely outside the *pays légal* [enfranchised citizens], there were the ideological representatives and spokesmen of the above classes, their savants, lawyers, doctors, etc., in a word: their so-called men of talent.

Owing to its financial straits, the July monarchy was dependent from the beginning on the big bourgeoisie, and its dependence on the big bourgeoisie was the inexhaustible source of increasing financial straits. It was impossible to subordinate the administration of the state to the interests of national production without balancing the budget, without establishing a balance between state expenditures and revenues. And how was this balance to be established without limiting state expenditures, that is, without encroaching on interests which were so many props of the ruling system, and without redistributing taxes, that is, without shifting a considerable share of the burden of taxation on to the shoulders of the big bourgeoisie itself?

On the contrary, the faction of the bourgeoisie that ruled and legislated through the Chambers had a direct interest in the indebtedness of the state. The state deficit was really the main object of its speculation and the chief source of its enrichment. At the end of each year a new deficit. After the lapse of four or five years a new loan. And every new loan offered new opportunities to the finance aristocracy for defrauding the state, which was kept artificially on the verge of bankruptcy—it had to negotiate with the bankers under the most unfavourable conditions. Each new loan gave a further opportunity, that of plundering the public which invested its capital in state bonds by means of stock-exchange manipulations, into the secrets of which the government and the majority in the Chambers were initiated. In general, the instability of state credit and the possession of state secrets gave the bankers and their associates in the Chambers and on the throne the possibility of evoking sudden, extraordinary fluctuations in the quotations of government securities, the result of which was always bound to be the ruin of a mass of smaller capitalists and the fabulously rapid enrichment of the big gamblers. As the state deficit was in the direct interest of the ruling faction of the bourgeoisie, it is clear why the extraordinary state expenditure in the last years of Louis Philippe's reign was far more than double the extraordinary state expenditure under Napoleon, indeed reached a yearly sum of nearly 400,000,000 francs, whereas the whole average annual export of France seldom attained a volume amounting to 750,000,000 francs. The enormous sums which, in this way, flowed through the hands of the state facilitated, moreover, swindling contracts for deliveries, bribery, defalcations, and all kinds of roguery. The defrauding of the state, practised wholesale in connection with loans, was repeated retail in public works. What occurred in the relations between Chamber and Government became multiplied in the relations between individual departments and individual entrepreneurs.

The ruling class exploited the building of railways in the same way as it exploited state expenditures in general and state loans. The Chambers piled the main burdens on the state, and secured the golden fruits to the speculating finance aristocracy. One recalls the scandals in the Chamber of Deputies, when by chance it leaked out that all the members of the majority, including a number of ministers, had been interested as shareholders in the very railway constructions which as legislators they caused to be carried out afterwards at the cost of the state.

On the other hand, the smallest financial reform was wrecked due to the influence of the bankers. For example, the postal reform. Rothschild protested. Was it permissible for the state to curtail sources of revenue out of which interest was to be paid on its ever-increasing debt?

The July monarchy was nothing other than a joint-stock company for the exploitation of France's national wealth, the dividends of which were divided among ministers, Chambers, 240,000 voters, and their adherents. Louis Philippe was the director of this company—Robert Macaire [a stage character noted for swindling] on the throne. Trade, industry, agriculture, shipping, the interests of the industrial bourgeoisie, were bound to be continually endangered and prejudiced under this system. Cheap government, *gouvernement à bon marché*, was what it had inscribed in the July days on its banner.

Since the finance aristocracy made the laws, was at the head of the administration of the state, had command of all the organized public authorities, dominated public opinion through the actual state of affairs and through the press, the same prostitution, the same shameless cheating, the same mania to get rich was repeated in every sphere, from the Court to the Café Borgne [a general name for cafés of dubious reputation], to get rich not by production, but by pocketing the already available wealth of others. Clashing every moment with the bourgeois laws themselves, an unbridled assertion of unhealthy and dissolute appetites manifested itself, particularly at the top of bourgeois society—lusts wherein wealth derived from gambling naturally seeks its satisfaction, where pleasure becomes debauched, where money, filth, and blood commingle. The finance aristocracy, in its mode of acquisition as well as in its pleasures, is nothing but the rebirth of the *lumpenproletariat* on the heights of bourgeois society.

And the non-ruling factions of the French bourgeoisie cried: corruption! The people cried: *à bas les grands voleurs! à bas les assassins!* [down with the big robbers! down with the assassins!] when in 1847, on the most prominent stages of bourgeois society, the same scenes were publicly enacted that regularly lead the *lumpenproletariat* to brothels, to workhouses, and lunatic asylums, to the bar of justice, to the dungeon, and to the scaffold. The industrial bourgeoisie saw its interest endangered, the petty bourgeoisie was filled with moral indignation, the imagination of the people was offended, Paris was flooded with pamphlets—'The Rothschild Dynasty', 'Usurers Kings of the Epoch', etc.—in which the rule of the finance aristocracy was denounced and stigmatized with greater or less wit.

Rien pour la gloire! Glory brings no profit! *La paix partout et toujours!* [peace everywhere and always]. War depresses the quotations of the three and four per cents! the France of the Bourse jobbers had inscribed on her banner. Her foreign policy was therefore lost in a series of mortifications to French national sentiment, which reacted all the more vigorously when the rape of Poland was brought to its conclusion with the incorporation of Cracow by Austria, and when Guizot came out actively on the side of the Holy Alliance in the Swiss *Sonderbund* war. The victory of the Swiss liberals in this mimic war raised the self-respect of the bourgeois opposition in France; the bloody uprising of the people in Palermo worked like an electric shock on the paralysed masses of the people and awoke their great revolutionary memories and passions.

The eruption of the general discontent was finally accelerated and the mood for revolt ripened by two economic world events.

The potato blight and the crop failures of 1845 and 1846 increased the general ferment among the people. The dearth of 1847 called forth bloody conflicts in France as well as on the rest of the Continent. As against the shameless orgies of the finance aristocracy, the struggle of the people for the prime necessities of life! At Buzançais, hunger rioters executed; in Paris, oversatiated *escrocs* [crooks] snatched from the courts by the royal family!

The second great economic event which hastened the outbreak of the revolution was a general commercial and industrial crisis in England. Already heralded in the autumn of 1845 by the wholesale reverses of the speculators in railway shares, staved off during 1846 by a number of incidents such as the impending abolition of the corn duties, the crisis finally burst in the autumn of 1847 with the bankruptcy of the London wholesale grocers, on the heels of which followed the insolvencies of the land banks and the closing of the factories in the English industrial districts. The after-effect of this crisis on the Continent had not yet spent itself when the February Revolution broke out.

The devastation of trade and industry caused by the economic epidemic made the autocracy of the finance aristocracy still more unbearable. Throughout the whole of France the bourgeois opposition agitated at banquets for an electoral reform which should win for it the majority in the Chambers and overthrow the Ministry of the Bourse. In Paris the industrial crisis had, moreover, the particular result of throwing a multitude of manufacturers and big traders, who under the existing circumstances could no longer do any business in the foreign market, on to the home market. They set up large establishments, the competition of which ruined the small *épiciers* [grocers], and *boutiquiers* [shopkeepers] *en masse*. Hence the innumerable bankruptcies among this section of the Paris bourgeoisie, and hence their revolutionary action in February. It is well known how Guizot and the Chambers answered the reform proposals with an unambiguous challenge, how Louis Philippe too late resolved on a Ministry led by Barrot, how things went as far as hand-to-hand fighting between the people and the army, how the army was disarmed by the passive conduct of the National Guard, how the July monarchy had to give way to a Provisional Government.

The Provisional Government which emerged from the February barricades necessarily mirrored in its composition the different parties which shared in the victory. It could not be anything but a compromise between the different classes which together had overturned the July throne, but whose interests were mutually antagonistic. The great majority of its members consisted of representatives of the bourgeoisie. The republican petty bourgeoisie was represented by Ledru-Rollin and Flocon, the republican bourgeoisie by the people from the National, the dynastic opposition by Crémieux, Dupont de l'Eure, etc. The working class had only two representatives, Louis Blanc and Albert. Finally, Lamartine's presence in the Provisional Government did not really represent any real interest, any definite class, it represented the February Revolution itself, the common uprising with its illusions, its poetry, its visionary content, and its phrases. For the rest, the spokesman of the February Revolution, by his position and his views, belonged to the bourgeoisie.

If Paris, as a result of political centralization, rules France, the workers, in moments of revolutionary earthquakes, rule Paris. The first act in the life of the Provisional Government was an attempt to escape from this overpowering influence by an appeal from intoxicated Paris to sober France. Lamartine disputed the right of the barricade fighters to proclaim a republic on the ground that only the majority of Frenchmen had that right; they must await their votes, the Paris proletariat must not besmirch its victory by a usurpation. The bourgeoisie allows the proletariat only one usurpation—that of fighting.

Up to noon of 25 February the republic had not yet been proclaimed; on

the other hand, all the ministries had already been divided among the bourgeois elements of the Provisional Government and among the generals,
bankers, and lawyers of the *National*. But the workers were determined this
time not to put up with any bamboozlement like that of July 1830. They
were ready to take up the fight anew and to get a republic by force of arms.
With this message, Raspail betook himself to the *Hôtel de Ville*. In the name
of the Paris proletariat he commanded the Provisional Government to proclaim a republic; if this order of the people were not fulfilled within two
hours, he would return at the head of 200,000 men. The bodies of the fallen
were scarcely cold, the barricades were not yet cleared away, the workers not
yet disarmed, and the only force which could be opposed to them was the
National Guard. Under these circumstances the doubts born of considerations of state policy and the juristic scruples of conscience entertained by the
Provisional Government suddenly vanished. The time limit of two hours
had not yet expired when all the walls of Paris were resplendent with the
gigantesque historical words:

République française! Liberté, Égalité, Fraternité!

. . . Woe to June! re-echoes Europe.

The Paris proletariat was forced into the June insurrection by the bourgeoisie. This sufficed to mark its doom. Its immediate, avowed needs did not
drive it to engage in a fight for the forcible overthrow of the bourgeoisie, nor
was it equal to this task. The *Moniteur* had to inform it officially that the
time was past when the republic saw any occasion to bow and scrape to its
illusions, and only its defeat convinced it of the truth that the slightest
improvement in its position remains a utopia within the bourgeois republic,
a utopia that becomes a crime as soon as it wants to become a reality. In
place of its demands, exuberant in form, but petty and even bourgeois still in
content, the concession of which it wanted to wring from the February
republic, there appeared the bold slogan of revolutionary struggle:
Overthrow of the bourgeoisie! Dictatorship of the working class!

By making its burial-place the birthplace of the bourgeois republic, the
proletariat compelled the latter to come out forthwith in its pure form as the
state whose admitted object it is to perpetuate the rule of capital, the slavery
of labour. Having constantly before its eyes the scarred, irreconcilable,
invincible enemy—invincible because his existence is the condition of its
own life—bourgeois rule, freed from all fetters, was bound to turn immediately into bourgeois terrorism. With the proletariat removed for the
time being from the stage and bourgeois dictatorship recognized officially,
the middle strata of bourgeois society, the petty bourgeoisie and the peasant
class, had to adhere more and more closely to the proletariat as their position
became more unbearable and their antagonism to the bourgeoisie more
acute. Just as earlier they had to find the cause of their distress in its
upsurge, so now in its defeat.

If the June insurrection raised the self-assurance of the bourgeoisie all over the Continent, and caused it to league itself openly with the feudal monarchy against the people, who was the first victim of this alliance? The Continental bourgeoisie itself. The June defeat prevented it from consolidating its rule and from bringing the people, half satisfied and half out of humour, to a standstill at the lowest stage of the bourgeois revolution.

Finally, the defeat of June divulged to the despotic powers of Europe the secret that France must maintain peace abroad at any price in order to be able to wage civil war at home. Thus the peoples who had begun the fight for their national independence were abandoned to the superior power of Russia, Austria, and Prussia, but, at the same time, the fate of these national revolutions was made subject to the fate of the proletarian revolution, and they were robbed of their apparent autonomy, their independence of the great social revolution. The Hungarian shall not be free, nor the Pole, nor the Italian, as long as the worker remains a slave!

Finally, with the victories of the Holy Alliance, Europe has taken on a form that makes every fresh proletarian upheaval in France directly coincide with a world war. The new French revolution is forced to leave its national soil forthwith and conquer the European terrain, on which alone the social revolution of the nineteenth century can be accomplished.

Thus only the June defeat has created all the conditions under which France can seize the initiative of the European revolution. Only after being dipped in the blood of the June insurgents did the tricolour become the flag of the European revolution—the red flag!

And we exclaim: The revolution is dead!—Long live the revolution!

... The classes whose social slavery the constitution is to perpetuate, proletariat, peasantry, petty bourgeoisie, it puts in possession of political power through universal suffrage. And from the class whose old social power it sanctions, the bourgeoisie, it withdraws the political guarantees of this power. It forces the political rule of the bourgeoisie into democratic conditions, which at every moment help the hostile classes to victory and jeopardize the very foundations of bourgeois society. From the ones it demands that they should not go forward from political to social emancipation; from the others that they should not go back from social to political restoration.

These contradictions perturbed the bourgeois republicans little. To the extent that they ceased to be indispensable—and they were indispensable only as the protagonists of the old society against the revolutionary proletariat—they fell, a few weeks after their victory, from the position of a party to that of a coterie. And they treated the constitution as a big intrigue. What was to be constituted in it was, above all, the rule of the coterie. The President was to be a protracted Cavaignac; the Legislative Assembly a protracted Constituent Assembly. They hoped to reduce the political power of the masses of the people to a semblance of power, and to be able to make sufficient play with this sham power itself to keep continually hanging over the majority of the bourgeoisie the dilemma of the June days: realm of the National or realm of anarchy.

The work on the constitution, which was begun on 4 September, was finished on 23 October. On 2 September the Constituent Assembly had decided not to dissolve until the organic laws supplementing the constitution were enacted. None the less, it now decided to bring to life the creation that was most peculiarly its own, the President, already on 10 December, long before the circle of its own activity was closed. So sure was it of hailing, in the *homunculus* [little man] of the constitution, the son of his mother. As a precaution it was provided that if none of the candidates received two million votes, the election should pass over from the nation to the Constituent Assembly.

Futile provisions! The first day of the realization of the constitution was the last day of the rule of the Constituent Assembly. In the abyss of the ballot box lay its sentence of death. It sought the 'son of his mother' and found 'the nephew of his uncle'. Saul Cavaignac slew one million votes, but David Napoleon slew six million. Saul Cavaignac was beaten six times over.

10 December 1848 was the day of the peasant insurrection. Only from this day does the February of the French peasants date. The symbol that expressed their entry into the revolutionary movement, clumsily cunning, knavishly naïve, doltishly sublime, a calculated supersitition, a pathetic burlesque, a cleverly stupid anachronism, a world-historic piece of buffoonery, and an undecipherable hieroglyphic for the understanding of the civilized— this symbol bore the unmistakable physiognomy of the class that represents barbarism within civilization. The republic had announced itself to this class with the tax collector; it announced itself to the republic with the emperor. Napoleon was the only man who had exhaustively represented the interests and the imagination of the peasant class, newly created in 1789. By writing his name of the frontispiece of the republic, it declared war abroad and the enforcing of its class interests at home. Napoleon was to the peasants not a person but a programme. With banners, with beat of drums and blare of trumpets, they marched to the polling booths shouting: *plus d'impôts, à bas les riches, à bas la république, vive l'Empereur!* No more taxes, down with the rich, down with the republic, long live the emperor! Behind the emperor was hidden the peasant war. The republic that they voted down was the republic of the rich.

10 December was the *coup d'état* of the peasants, which overthrew the existing government. And from that day on, when they had taken a government from France and given a government to her, their eyes were fixed steadily on Paris. For a moment active heroes of the revolutionary drama, they could no longer be forced back into the inactive and spineless role of the chorus.

The other classes helped to complete the election victory of the peasants. To the proletariat, the election of Napoleon meant the deposition of

Cavaignac, the overthrow of the Constituent Assembly, the dismissal of bourgeois republicanism, the cassation of the June victory. To the petty bourgeoisie, Napoleon meant the rule of the debtor over the creditor. For the majority of the big bourgeoisie, the election of Napoleon meant an open breach with the faction of which it had had to make use, for a moment, against the revolution, but which became intolerable to it as soon as this faction sought to consolidate the position of the moment into a constitutional position. Napoleon in place of Cavaignac meant to this majority the monarchy in place of the republic, the beginning of the royalist restoration, a shy hint at Orleans, the lily [emblem of the Bourbons] hidden beneath the violet. Lastly, the army voted for Napoleon against the Mobile Guard, against the peace idyll, for war.

Thus it happened, as the *Neue Rheinische Zeitung* stated, that the most simple-minded man in France acquired the most multifarious significance. Just because he was nothing, he could signify everything save himself. . . .

In England—and the largest French manufacturers are petty bourgeois compared with their English rivals—we really find the manufacturers, a Cobden, a Bright, at the head of the crusade against the bank and the stock-exchange aristocracy. Why not in France? In England industry predominates; in France, agriculture. In England industry requires free trade; in France protective tariffs, national monopoly alongside the other monopolies. French industry does not dominate French production; the French industrialists, therefore, do not dominate the French bourgeoisie. In order to secure the advancement of their interests as against the remaining factions of the bourgeoisie, they cannot, like the English, take the lead of the movement and simultaneously push their class interests to the fore; they must follow in the train of the revolution, and serve interests which are opposed to the collective interests of their class. In February they had misunderstood their position; February sharpened their wits. And who is more directly threatened by the workers than the employer, the industrial capitalist? The manufacturer, therefore, of necessity became in France the most fanatical member of the party of Order. The reduction of his profit by finance, what is that compared with the abolition of profit by the proletariat? . . .

Little by little we have seen peasants, petty bourgeois, the middle classes in general, stepping alongside the proletariat, driven into open antagonism to the official republic and treated by it as antagonists. Revolt against bourgeois dictatorship, need of a change of society, adherence to democratic-republican institutions as organs of their movement, grouping round the proletariat as the decisive revolutionary power—these are the common characteristics of the so-called party of social-democracy, the party of the Red republic. This party of Anarchy, as its opponents christened it, is no less a coalition of different interests than the party of Order. From the smallest reform of the old social disorder to the overthrow of the old social order, from bourgeois liberalism to revolutionary terrorism—as far apart as this lie the extremes that form the starting-point and the finishing-point of the party of 'Anarchy'.

Abolition of the protective tariff—socialism! For it strikes at the monopoly of the industrial faction of the party of Order. Regulation of the state budget—socialism! For it strikes at the monopoly of the financial faction of the party of Order. Free admission of foreign meat and corn—socialism! For it strikes at the monopoly of the third faction of the party of Order, large landed property. The demands of the free-trade party, that is, of the most advanced English bourgeois party, appear in France as so many socialist demands. Voltairianism—socialism! For it strikes at a fourth faction of the party of Order, the Catholic. Freedom of the press, right of association, universal public education—socialism, socialism! They strike at the general monopoly of the party of Order.

So swiftly had the march of the revolution ripened conditions that the friends of reform of all shades, the most moderate claims of the middle classes, were compelled to group themselves round the banner of the most extreme party of revolution, round the red flag.

Yet, manifold as the socialism of the different large sections of the party of Anarchy was, according to the economic conditions and the total revolutionary requirements of their class or faction of a class arising out of these, in one point it is in harmony: in proclaiming itself to be the means of emancipating the proletariat and the emancipation of the latter, its object. Deliberate deception on the part of some; self-deception on the part of the others, who give out the world transformed according to their own needs as the best world for all, as the realization of all revolutionary claims and the elimination of all revolutionary collisions.

Behind the general socialist phrases of the 'party of Anarchy', which sound rather alike, there is concealed the socialism of the *National*, of the *Presse* and the *Siècle*, which more or less consistently wants to overthrow the rule of the finance aristocracy and to free industry and trade from their hitherto existing fetters. This is the socialism of industry, of trade and of agriculture, whose bosses in the party of Order deny these interests, in so far as they no longer coincide with their private monopolies. Socialism proper, petty-bourgeois socialism, socialism par excellence, is distinct from this bourgeois socialism, to which, as to every variety of socialism, a section of the workers and petty bourgeois naturally rallies. Capital hounds this class chiefly as its creditor, so it demands credit institutions; capital crushes it by competition, so it demands associations supported by the state; capital overwhelms it by concentration, so it demands progressive taxes, limitations on inheritance, taking over of large construction projects by the state, and other measures that forcibly stem the growth of capital. Since it dreams of the peaceful achievement of its socialism—allowing, perhaps, for a second February Revolution lasting a brief day or so—the coming historical process naturally appears to it as an application of systems, which the thinkers of society, whether in companies or as individual inventors, devise or have devised. Thus they become the eclectics or adepts of the existing socialist systems, of doctrinaire socialism, which was the theoretical expression of the proletariat only as long as it had not yet developed further into a free historical movement of its own.

While this utopia, doctrinaire socialism, which subordinates the total movement to one of its moments, which puts in place of common, social production the brainwork of individual pedants and, above all, in fantasy does away with the revolutionary struggle of the classes and its requirements by small conjurers' tricks or great sentimentality; while this doctrinaire socialism, which at bottom only idealizes present society, takes a picture of it without shadows, and wants to achieve its ideal athwart the realities of present society; while the proletariat surrenders this socialism to the petty bourgeoisie; while the struggle of the different socialist leaders among themselves sets forth each of the so-called systems as a pretentious adherence to one of the transit points of the social revolution as against another—the proletariat rallies more and more round revolutionary socialism, round communism, for which the bourgeoisie has itself invented the name of Blanqui. This socialism is the declaration of the permanence of the revolution, the class dictatorship of the proletariat as the necessary transit point to the abolition of class distinctions generally, to the abolition of all the relations of production on which they rest, to the abolition of all the social relations that correspond to these relations of production, to the revolutionizing of all the ideas that result from these social relations. . . .

In spite of the industrial and commercial prosperity that France momentarily enjoys, the mass of the people, the twenty-five million peasants, suffer from a great depression. The good harvests of the last few years have forced the prices of corn much lower even than in England, and the position of the peasants under such circumstances, in debt, sucked dry by usury, and crushed by taxes, must be anything but splendid. The history of the last three years has, however, provided sufficient proof that this class of the population is absolutely incapable of any revolutionary initiative.

Just as the period of crisis occurs later on the Continent than in England, so does that of prosperity. The original process always takes place in England; it is the demiurge of the bourgeois cosmos. On the Continent, the different phases of the cycle through which bourgeois society is ever speeding anew occur in secondary and tertiary form. First, the Continent exported incomparably more to England than to any other country. This export to England, however, in turn depends on the position of England, particularly with regard to the overseas market. Then England exports to the overseas lands incomparably more than the entire Continent, so that the quantity of Continental exports to these lands is always dependent on England's overseas exports at the time. While, therefore, the crises first produce revolutions on the Continent, the foundation for these is, nevertheless, always laid in

England. Violent outbreaks must naturally occur rather in the extremities of the bourgeois body than in its heart, since the possibility of adjustment is greater here than there. On the other hand, the degree to which the Continental revolutions react on England is at the same time the barometer which indicates how far these revolutions really call into question the bourgeois conditions of life, or how far they only hit their political formations.

With this general prosperity, in which the productive forces of bourgeois society develop as luxuriantly as is at all possible within bourgeois relationships, there can be no talk of a real revolution. Such a revolution is only possible in the periods when both these factors, the modern productive forces and the bourgeois productive forms come in collision with each other. The various quarrels in which the representatives of the individual factions of the Continental party of Order now indulge and mutually compromise themselves, far from providing the occasion for new revolutions are, on the contrary, possible only because the basis of the relationships is momentarily so secure and, what the reaction does not know, so bourgeois. From it all attempts of the reaction to hold up bourgeois development will rebound just as certainly as all moral indignation and all enthusiastic proclamations of the democrats. A new revolution is possible only in consequence of a new crisis. It is, however, just as certain as this crisis. . . .

BIBLIOGRAPHY

Original: *MEW*, Vol. 7, pp. 9 ff.

Present translation: *MESW*, Vol. 1, pp. 139 ff. Reproduced by kind permission of Lawrence and Wishart Ltd.

Other translations: K. Marx, *Surveys from Exile*, ed. D. Fernbach, New York and London, 1973.

Commentaries: D. Fernbach, Introduction to above volume.
N. Poulantzas, *Political Power and Social Classes*, London, 1971.
R. Price, *The Second French Republic*, London, 1972.
I. Zeitlin, *Marxism: A Re-examination*, Princeton, 1967, pp. 128 ff.

24. Speech to the Central Committee of the Communist League

In September 1850 the differences in the Central Committee of the League came to a head: Willich, Schapper, and their supporters wished to organize for an immediate revolution; Marx, who had spent the previous months reading economics in the British Museum, considered that a successful revolution would have to wait on economic events and that this could be a long process. Marx's solution was to establish two autonomous branches within the League. The following are the most important extracts from Marx's speech recommending this solution and his views on the prospects for revolution. In the event, the Willich–Schapper group seceded, and the League continued under Marx's leadership for two more years.

... It is necessary to form two branches here for the very reason that the unity of the League must at all cost be preserved. Quite apart from personal disagreements we have witnessed also differences of principle even in the Society. In the last debate 'the position of the German proletariat in the next revolution' was discussed and views were expressed by members of the minority on the Central Committee which directly oppose those in the last circular but one and even the 'Manifesto'. A German nationalist point of view was substituted for the universal outlook of the 'Manifesto', and the patriotic feelings of the German artisans were pandered to. The materialist standpoint of the 'Manifesto' has given way to idealism. The revolution is seen not as the product of realities of the situation but as the result of a mere effort of will. What we say to the workers is: You have 15, 20, 50 years of civil war to go through in order to change society and to train yourselves for the exercise of political power, whereas they say, we must take over at once, or else we may as well take to our beds. Just as the Democrats abused the word 'people' so now the word 'proletariat' has been degraded to a mere phrase. To make this phrase effective it was necessary to describe the petty bourgeois as proletarians so that in practice it was the petty bourgeois and not the proletarians who were represented. The actual revolutionary process had to be replaced by revolutionary catchwords. This debate has finally laid bare the differences in principle which lay behind the clash of personalities, and the time for action has now arrived. For it is personal antagonism that has furnished both parties with their battle-cries, and some members of the League have called the defenders of the 'Manifesto' reactionaries, hoping thereby to make them unpopular, a vain endeavour, as they do not seek popularity. The majority would be justified in dissolving the London branch and expelling the minority as being in conflict with the principles of the

League. I do not wish to put a motion to that effect as it would cause a pointless scandal and because these people are still communists in their own view even though the opinions they are now expressing are anti-communist and could at best be described as social-democratic. It is obvious, however, that it would be a mere waste of time, and a dangerous one at that, for us to remain together any longer. Schapper has often spoken of separation—very well, then, let us go ahead with it. I think that I have found the way to do so without destroying the party. . . . I have always defied the momentary opinions of the proletariat. If the best a party can do is to just fail to seize power, then we repudiate it. If the proletariat could gain control of the government the measures it would introduce would be those of the petty bourgeoisie and not those appropriate to the proletariat. Our party can only gain power when the situation allows it to put its own measures into practice. Louis Blanc is the best instance of what happens when you come to power prematurely. In France, moreover, it wasn't just the proletariat that gained power but the peasants and the petty bourgeois as well, and it is their demands that will necessarily prevail. The Paris Commune shows what can be accomplished without being in the government. And incidentally why do we hear nothing from Willich and the other members of the minority who approved the circular unanimously at the time? We cannot and will not split the League; we wish merely to divide the London Region into two branches. . . .

BIBLIOGRAPHY

Original: B. Nicolaievsky, 'Towards a History of the Communist League', *International Review of Social History*, Vol. 1, 1956, pp. 248 ff.

Present translation: K. Marx, *The Cologne Communist Trial*, ed. R. Livingstone, London, 1971, pp. 251 ff. Reproduced by kind permission of Lawrence and Wishart Ltd.

Other translations: K. Marx, *The Revolutions of 1848*, ed. D. Fernbach, New York and London, 1973.

Commentary: S. Avineri, *The Social and Political Thought of Karl Marx*, Cambridge, 1968, pp. 195 ff.

D. Fernbach, Introduction to edition cited above.

R. Hunt, *The Political Ideas of Marx and Engels*, Pittsburgh and London, 1975, Vol. 1, pp. 254 ff.

R. Livingstone, Introduction to edition cited above.

B. Nicolaievsky, article cited above.

B. Wolfe, *Marxism: 100 Years in the Life of a Doctrine*, London, 1967, pp. 151 ff.

25. *The Eighteenth Brumaire of Louis Bonaparte*

In December 1851 Louis Napoleon seized power in France and proclaimed himself Emperor, thereby consolidating the reaction that followed the revolution of 1848. Marx immediately composed a series of articles that were published by his friend, Weydemeyer, in a short-lived New York journal under the title *The Eighteenth Brumaire of Louis Bonaparte*. The title is an allusion to the date of Napoleon Bonaparte's *coup d'état* in 1799. It is Marx's most brilliant political pamphlet, whose intention, he wrote later, was to 'demonstrate how the class struggle in France created circumstances and relationships that made it possible for a grotesque mediocrity to play a hero's part'. The pamphlet is particularly interesting for Marx's views on class and the state.

Hegel remarks somewhere that all facts and personages of great importance in world history occur, as it were, twice. He forgot to add: the first time as tragedy, the second as farce. Caussidière for Danton, Louis Blanc for Robespierre, the *Montagne* of 1848 to 1851 for the *Montagne* of 1793 to 1795, the Nephew for the Uncle. And the same caricature occurs in the circumstances attending the second edition of the eighteenth Brumaire!

Men make their own history, but they do not make it just as they please; they do not make it under circumstances chosen by themselves, but under circumstances directly encountered, given, and transmitted from the past. The tradition of all the dead generations weighs like a nightmare on the brain of the living. And just when they seem engaged in revolutionizing themselves and things, in creating something that has never yet existed, precisely in such periods of revolutionary crisis they anxiously conjure up the spirits of the past to their service and borrow from them names, battle-cries, and costumes in order to present the new scene of world history in this time-honoured disguise and this borrowed language. Thus Luther donned the mask of the Apostle Paul, the Revolution of 1789 to 1814 draped itself alternately as the Roman republic and the Roman empire, and the Revolution of 1848 knew nothing better to do than to parody, now 1789, now the revolutionary tradition of 1793 to 1795. In like manner a beginner who has learnt a new language always translates it back into his mother tongue, but he has assimilated the spirit of the new language and can freely express himself in it only when he finds his way in it without recalling the old and forgets his native tongue in the use of the new.

Consideration of this conjuring up of the dead of world history reveals at once a salient difference. Camille Desmoulins, Danton, Robespierre, Saint-

Just, Napoleon, the heroes as well as the parties and the masses of the old French Revolution, performed the task of their time in Roman costume and with Roman phrases, the task of unchaining and setting up modern bourgeois society. The first ones knocked the feudal basis to pieces and mowed off the feudal heads which had grown on it. The other created inside France the conditions under which alone free competition could be developed, parcelled landed property exploited, and the unchained industrial productive power of the nation employed; and beyond the French borders he every-where swept the feudal institutions away, so far as was necessary to furnish bourgeois society in France with a suitable up-to-date environment on the European Continent. The new social formation once established, the antediluvian Colossi disappeared and with them resurrected Romanity—the Brutuses, Gracchi, Publicolas, the tribunes, the senators, and Caesar himself. Bourgeois society in its sober reality had begotten its true interpreters and mouthpieces in the Says, Cousins, Royer-Collards, Benjamin Constants, and Guizots; its real military leaders sat behind the office desks, and the pigheaded Louis XVIII was its political chief. Wholly absorbed in the production of wealth and in peaceful competitive struggle, it no longer comprehended that ghosts from the days of Rome had watched over its cradle. But unheroic as bourgeois society is, it nevertheless took heroism, sacrifice, terror, civil war, and battles of peoples to bring it into being. And in the classically austere traditions of the Roman republic its gladiators found the ideals and the art forms, the self-deceptions that they needed in order to conceal from themselves the bourgeois limitations of the content of their struggles and to keep their enthusiasm on the high plane of the great historical tragedy. Similarly, at another stage of development a century earlier, Cromwell and the English people had borrowed speech, passions, and illusions from the Old Testament for their bourgeois revolution. When the real aim had been achieved, when the bourgeois transformation of English society had been accomplished, Locke supplanted Habakkuk.

Thus the awakening of the dead in those revolutions served the purpose of glorifying the new struggles, not of parodying the old; of magnifying the given task in imagination, not of fleeing from its solution in reality; of finding once more the spirit of revolution, not of making its ghost walk about again.

From 1848 to 1851 only the ghost of the old revolution walked about, from Marrast, the *républicain en gants jaunes* [republican in yellow gloves], who disguised himself as the old Bailly, down to the adventurer who hides his commonplace repulsive features under the iron death mask of Napoleon. An entire people, which had imagined that by means of a revolution it had imparted to itself an accelerated power of motion, suddenly finds itself set back into a defunct epoch and, in order that no doubt as to the relapse may be possible, the old dates arise again, the old chronology, the old names, the old edicts, which had long become a subject of antiquarian erudition, and the old minions of the law, who had seemed long decayed. The nation feels like that mad Englishman in Bedlam who fancies that he lives in the times of the ancient Pharaohs and daily bemoans the hard labour that he must perform in the Ethiopian mines as a gold digger, immured in this subterranean prison, a dimly burning lamp fastened to his head, the overseer of the slaves behind him with a long whip, and at the exits a confused welter of barbarian mercenaries, who understand neither the forced labourers in the mines nor one another, since they speak no common language. 'And all this is expected of me,' sighs the mad Englishman, 'of me, a freeborn Briton, in order to make gold for the old Pharaohs.' 'In order to pay the debts of the Bonaparte family', sighs the French nation. The Englishman, so long as he was in his right mind, could not get rid of the fixed idea of making gold. The French, so long as they were engaged in revolution, could not get rid of the memory of Napoleon, as the election of 10 December proved. They hankered to return from the perils of revolution to the flesh-pots of Egypt, and 2 December 1851 was the answer. They have not only a caricature of the old Napoleon, they have the old Napoleon himself, caricatured as he must appear in the middle of the nineteenth century.

The social revolution of the nineteenth century cannot draw its poetry from the past, but only from the future. It cannot begin with itself before it has stripped off all superstition in regard to the past. Earlier revolutions required recollections of past world history in order to drug themselves concerning their own content. In order to arrive at its own content, the revolution of the nineteenth century must let the dead bury their dead. There the phrase went beyond the content; here the content goes beyond the phrase.

The February Revolution was a surprise attack, a taking of the old society unawares, and the people proclaimed this unexpected stroke as a deed of world importance, ushering in a new epoch. On 2 December the February Revolution is conjured away by a cardsharper's trick, and what seems overthrown is no longer the monarchy but the liberal concessions that were wrung from it by centuries of struggle. Instead of society having conquered a new content for itself, it seems that the state only returned to its oldest form, to the shamelessly simple domination of the sabre and the cowl. This is the answer to the *coup de main* [forceful act] of February 1848, given by the *coup de tête* [risky act] of December 1851. Easy come, easy go. Meanwhile the interval of time has not passed by unused. During the years 1848 to 1851 French society has made up, and that by an abbreviated because revolutionary method, for the studies and experiences which, in a regular, so to speak, textbook course of development, would have had to precede the February Revolution, if it was to be more than a ruffling of the surface. Society now seems to have fallen back behind its point of departure; it has in truth first to create for itself the revolutionary point of departure, the situation, the relations, the conditions under which alone modern revolution becomes serious.

Bourgeois revolutions, like those of the eighteenth century, storm swiftly from success to success; their dramatic effects outdo each other; men and things seem set in sparkling brilliants; ecstasy is the everyday spirit; but they are short-lived; soon they have attained their zenith, and a long crapulent depression lays hold of society before it learns soberly to assimilate the results of its storm-and-stress period. On the other hand, proletarian revolutions, like those of the nineteenth century, criticize themselves constantly, interrupt themselves continually in their own course, come back to the apparently accomplished in order to begin it afresh, deride with unmerciful thoroughness the inadequacies, weaknesses, and paltrinesses of their first attempts, seem to throw down their adversary only in order that he may draw new strength from the earth and rise again, more gigantic, before them, recoil ever and anon from the indefinite prodigiousness of their own aims, until a situation has been created which makes all turning back impossible, and the conditions themselves cry out:

Hic Rhodus, hic salta! [Here is Rhodes, jump here!]

For the rest, every fairly competent observer, even if he had not followed the course of French development step by step, must have had a presentiment that an unheard-of fiasco was in store for the revolution. It was enough to hear the self-complacent howl of victory with which Messieurs the Democrats congratulated each other on the expected gracious consequences of the second Sunday in May 1852 [Date on which term of office of President expired]. In their minds the second Sunday in May 1852 had become a fixed idea, a dogma, like the day on which Christ should reappear and the millennium begin, in the minds of the Chiliasts. As ever, weakness had taken refuge in a belief in miracles, fancied the enemy overcome when he was only conjured away in imagination, and it lost all understanding of the present in a passive glorification of the future that was in store for it and of the deeds it had *in petto* [hidden] but which it merely did not want to carry out as yet. Those heroes who seek to disprove their demonstrated incapacity by mutually offering each other their sympathy and getting together in a crowd had tied up their bundles, collected their laurel wreaths in advance and were just then engaged in discounting on the exchange market the republics *in partibus* [in exile] for which they had already providently organized the government personnel with all the calm of their unassuming disposition. 2 December struck them like a thunderbolt from a clear sky, and the peoples that in periods of pusillanimous depression gladly let their inward apprehension be drowned out by the loudest bawlers will perchance have convinced themselves that the times are past when the cackle of geese could save the Capitol.

The Constitution, the National Assembly, the dynastic parties, the blue and the red republicans, the heroes of Africa, the thunder from the platform, the sheet lightning of the daily press, the entire literature, the political names and the intellectual reputations, the civil law and the penal code, the *liberté*, *égalite*, *fraternité* and the second Sunday in May 1852—all has vanished like a phantasmagoria before the spell of a man whom even his enemies do not make out to be a magician. Universal suffrage seems to have survived only for a moment, in order that with its own hand it may make its last will and testament before the eyes of all the world and declare in the name of the people itself: All that exists deserves to perish.

It is not enough to say, as the French do, that their nation was taken unawares. A nation and a woman are not forgiven the unguarded hour in which the first adventurer that came along could violate them. The riddle is not solved by such turns of speech, but merely formulated differently. It remains to be explained how a nation of thirty-six millions can be surprised and delivered unresisting into captivity by three swindlers. . . .

The parliamentary party was not only dissolved into its two great factions, each of these factions was not only split up within itself, but the party of Order in parliament had fallen out with the party of Order outside parliament. The spokesmen and scribes of the bourgeoisie, its platform and its press, in short, the ideologists of the bourgeoisie and the bourgeoisie itself, the representatives and the represented, faced one another in estrangement and no longer understood one another.

The Legitimists in the provinces, with their limited horizon and their unlimited enthusiasm, accused their parliamentary leaders, Berryer and Falloux, of deserting to the Bonapartist camp and of defection from Henry V. Their fleur-de-lis minds believed in the fall of man, but not in diplomacy.

Far more fateful and decisive was the breach of the commercial bourgeoisie with its politicians. It reproached them, not as the Legitimists reproached theirs, with having abandoned their principles, but, on the contrary, with clinging to principles that had become useless.

I have already indicated above that since Fould's entry into the ministry the section of the commercial bourgeoisie which had held the lion's share of power under Louis Philippe, namely, the aristocracy of finance, had become Bonapartist. Fould represented not only Bonaparte's interest in the *bourse*, he represented at the same time the interests of the *bourse* before Bonaparte. The position of the aristocracy of finance is most strikingly depicted in a passage from its European organ, the London *Economist*. In its number of 1 February 1851, its Paris correspondent writes:

Now we have it stated from numerous quarters that above all things France demands tranquillity. The President declares it in his message to the Legislative Assembly; it is echoed from the tribune; it is asserted in the journals; it is announced from the pulpit; it is demonstrated by the sensitiveness of the public funds at the least prospect of disturbance, and their firmness the instant it is made manifest that the executive is victorious.

In its issue of 29 November 1851, *The Economist* declares in its own name:

'The President is the guardian of order, and is now recognized as such on every Stock Exchange of Europe.'

The aristocracy of finance, therefore, condemned the parliamentary struggle of the party of Order with the executive power as a disturbance of order, and celebrated every victory of the President over its ostensible representatives as a victory of order. By the aristocracy of finance must here be understood not merely the great loan promoters and speculators in public funds, in regard to whom it is immediately obvious that their interests coincide with the interests of the state power. All modern finance, the whole of the banking business, is interwoven in the closest fashion with public credit. A part of their business capital is necessarily invested and put out at interest in quickly convertible public funds. Their deposits, the capital placed at their disposal and distributed by them among merchants and industrialists, are partly derived from the dividends of holders of government securities. If in every epoch the stability of the state power signified Moses and the prophets to the entire money market and to the priests of this money market, why not all the more so today, when every deluge threatens to sweep away the old states, and the old state debts with them?

The industrial bourgeoisie, too, in its fanaticism for order, was angered by the squabbles of the parliamentary party of Order with the executive power. After their vote of 18 January on the occasion of Changarnier's dismissal, Thiers, Anglas, Sainte-Beuve, etc., received from their constituents in precisely the industrial districts public reproofs in which particularly their coalition with the *Montagne* was scourged as high treason to order. If, as we have seen, the boastful taunts, the petty intrigues which marked the struggle of the party of Order with the President merited no better reception, then, on the other hand, this bourgeois party, which required its representatives to allow the military power to pass from its own parliament to an adventurous pretender without offering resistance, was not even worth the intrigues that were squandered in its interests. It proved that the struggle to maintain its public interests, its own class interests, its political power, only troubled and upset it, as it was a disturbance of private business.

With barely an exception, the bourgeois dignitaries of the Departmental towns, the municipal authorities, the judges of the Commercial Courts, etc., everywhere received Bonaparte on his tours in the most servile manner, even when, as in Dijon, he made an unrestrained attack on the National Assembly, and especially on the party of Order.

When trade was good, as it still was at the beginning of 1851, the commercial bourgeoisie raged against any parliamentary struggle, lest trade be put out of humour. When trade was bad, as it continually was from the end of February 1851, the commercial bourgeoisie accused the parliamentary struggles of being the cause of stagnation and cried out for them to stop in order that trade might start again. The revision debates came on just in this bad period. Since the question here was whether the existing form of state was to be or not to be, the bourgeoisie felt itself all the more justified in demanding from its Representatives the ending of this torturous provisional arrangement and at the same time the maintenance of the *status quo*. There was no contradiction in this. By the end of the provisional arrangement it understood precisely its continuation, the postponement to a distant future of the moment when a decision had to be reached. The *status quo* could be maintained in only two ways: prolongation of Bonaparte's authority or his constitutional retirement and the election of Cavaignac. A section of the bourgeoisie desired the latter solution and knew no better advice to give its Representatives than to keep silent and leave the burning question untouched. They were of the opinion that if their Representatives did not speak, Bonaparte would not act. They wanted an ostrich parliament that would hide its head in order to remain unseen. Another section of the bourgeoisie desired, because Bonaparte was already in the presidential chair, to leave him sitting in it, so that everything might remain in the same old rut. They were indignant because their parliament did not openly infringe the Constitution and abdicate without ceremony.

The General Councils of the Departments, those provincial representative bodies of the big bourgeoisie, which met from 25 August onwards during the recess of the National Assembly, declared almost unanimously for revision, and thus against parliament and in favour of Bonaparte.

Still more unequivocally than in its falling out with its parliamentary representatives the bourgeoisie displayed its wrath against its literary representatives, its own press. The sentences to ruinous fines and shameless terms of imprisonment, on the verdicts of bourgeois juries, for every attack of bourgeois journalists on Bonaparte's usurpationist desires, for every attempt of the press to defend the political rights of the bourgeoisie against the executive power, astonished not merely France, but all Europe.

While the parliamentary party of Order, by its clamour for tranquillity, as I have shown, committed itself to quiescence, while it declared the political rule of the bourgeoisie to be incompatible with the safety and existence of the bourgeoisie, by destroying with its own hands in the struggle against the other classes of society all the conditions for its own regime, the parliamentary regime, the extra-parliamentary mass of the bourgeoisie, on the other hand, by its servility towards the President, by its vilification of parliament, by its brutal maltreatment of its own press, invited Bonaparte to suppress and annihilate its speaking and writing section, its politicians and its *literati*, its platform and its press, in order that it might then be able to pursue its private affairs with full confidence in the protection of a strong and unrestricted government. It declared unequivocally that it longed to get rid of its own political rule in order to get rid of the troubles and dangers of ruling.

And this extra-parliamentary bourgeoisie, which had already rebelled against the purely parliamentary and literary struggle for the rule of its own class and betrayed the leaders of this struggle, now dares after the event to indict the proletariat for not having risen in a bloody struggle, a life-and-death struggle on its behalf! This bourgeoisie, which every moment sacrificed its general class interests, that is, its political interests, to the narrowest and most sordid private interests, and demanded a similar sacrifice from its Representatives, now moans that the proletariat has sacrificed the bourgeoisie's ideal political interests to the proletariat's material interests. It poses as a lovely being that has been misunderstood and deserted in the decisive hour by the proletariat misled by Socialists. And it finds a general echo in the bourgeois world. Naturally, I do not speak here of German shyster politicians and riff-raff of the same persuasion. I refer, for example, to the already quoted *Economist*, which as late as 29 November 1851, that is, four days prior to the *coup d'état*, had declared Bonaparte to be the 'guardian of order', but the Thiers and Berryers to be 'anarchists', and on 27 December 1851, after Bonaparte had quieted these anarchists, is already vociferous concerning the treason to 'the skill, knowledge, discipline, mental influence, intellectual resources, and moral weight of the middle and upper ranks' committed by the masses of 'ignorant, untrained, and stupid *proletaires*'. The stupid, ignorant, and vulgar mass was none other than the bourgeois mass itself.

In the year 1851, France, to be sure, had passed through a kind of minor trade crisis. The end of February showed a decline in exports compared with 1850; in March trade suffered and factories closed down; in April the position of the industrial Departments appeared as desperate as after the February days; in May business had still not revived; as late as 28 June the holdings of the Bank of France showed, by the enormous growth of deposits and the equally great decrease in advances on bills of exchange, that production was at a standstill, and it was not until the middle of October that a progressive improvement of business again set in. The French bourgeoisie attributed this trade stagnation to purely political causes, to the struggle between parliament and the executive power, to the precariousness of a merely provisional form of state, to the terrifying prospect of the second Sunday in May 1852. I will not deny that all these circumstances had a depressing effect on some branches of industry in Paris and the Departments. But in any case this influence of the political conditions was only local and inconsiderable. Does this require further proof than the fact that the improvement of trade set in towards the middle of October, at the very moment when the political situation grew worse, the political horizon darkened and a thunderbolt from Elysium was expected at any moment? For the rest, the French bourgeois, whose 'skill, knowledge, spiritual insight, and intellectual resources' reach no further than his nose, could throughout the period of the Industrial Exhibition in London have found the cause of his commercial misery right under his nose. While in France factories were closed down, in England commercial bankruptcies broke out. While in April and May the industrial panic reached a climax in France, in April and May the commercial panic reached a climax in England. Like the French woollen industry, so the English woollen industry suffered, and as French silk manufacture, so did English silk manufacture. True, the English cotton mills continued working, but no longer at the same profits as in 1849 and 1850. The only difference was that the crisis in France was industrial, in England commercial; that while in France the factories stood idle, in England they extended operations, but under less favourable conditions than in preceding years; that in France it was exports, in England imports which were hardest hit. The common cause, which is naturally not to be sought within the bounds of the French political horizon, was obvious. The years 1849 and 1850 were years of the greatest material prosperity and of an over-production that appeared as such only in 1851. At the beginning of this year it was given a further special impetus by the prospect of the Industrial Exhibition. In addition there were the following special circumstances: first, the partial failure of the cotton crop in 1850 and 1851, then the certainty of a bigger cotton crop than had been expected; first the rise, then the sudden fall, in short, the fluctuations in the price of cotton. The crop of raw silk, in France at least, had turned out to be even below the average yield. Woollen manufacture, finally, had expanded so much since 1848 that the production of wool would not keep pace with it and the price of raw wool rose out of all proportion to the price of woollen manufactures. Here, then, in the raw material of three industries for the world market, we have already threefold material for a stagnation in trade. Apart from these special circumstances, the apparent crisis of 1851 was nothing else but the halt which overproduction and overspeculation invariably make in describing the industrial cycle, before they summon all their strength in order to rush feverishly through the final phase of this cycle and arrive once more at their starting-point, the general trade crisis. During such intervals in the history of trade commercial bankruptcies break out in England, while in France industry itself is reduced to idleness, being partly forced into retreat by the competition, just then becoming intolerable, of the English in all markets, and being partly singled out for attack as a luxury industry by every business stagnation. Thus, besides the general crisis, France goes through national trade crises of her own, which are nevertheless determined and conditioned far more by the general state of the world market than by French local influences. It will not be without interest to contrast the judgement of the English bourgeois with the prejudice of the French bourgeois. In its annual trade report for 1851, one of the largest Liverpool houses writes:

Few years have more thoroughly belied the anticipations formed at their commencement than the one just closed; instead of the great prosperity which was almost unanimously looked for, it has proved one of the most discouraging that has been seen for the last quarter of a century—this, of course, refers to the mercantile, not to the manufacturing, classes. And yet there certainly were grounds for anticipating the reverse at the beginning of the year—stocks of produce were moderate, money was abundant, and food was cheap, a plentiful harvest well secured, unbroken peace on the Continent, and no political or fiscal disturbances at home; indeed, the wings of commerce were never more unfettered. . . . To what source, then, is this disastrous result to be attributed? We believe to over-trading both in imports and exports. Unless our merchants will put more stringent limits to their freedom of action, nothing but a triennial panic can keep us in check.

Now picture to yourself the French bourgeois, how in the throes of this business panic his trade-crazy brain is tortured, set in a whirl, and stunned by rumours of *coups d'état* and the restoration of universal suffrage, by the struggle between parliament and the executive power, by the Fronde war between Orleanists and Legitimists, by the communist conspiracies in the south of France, by alleged *Jacqueries* [peasant uprisings] in the Departments of Nièvre and Cher, by the advertisements of the different candidates for the presidency, by the cheapjack slogans of the journals, by the threats of the republicans to uphold the Constitution and universal suffrage by force of arms, by the gospel-preaching *émigré* heroes *in partibus* [in exile], who announced that the world would come to an end on the second Sunday in May 1852—think of all this and you will comprehend why in this unspeakable, deafening chaos of fusion, revision, prorogation, constitution, conspiration, coalition, emigration, usurpation, and revolution, the bourgeois madly snorts at his parliamentary republic: '*Rather an end with terror than terror without end!*'

Bonaparte understood this cry. His power of comprehension was sharpened by the growing turbulence of creditors who, with each sunset which brought settling day, the second Sunday in May 1852, nearer, saw a movement of the stars protesting their earthly bills of exchange. They had become veritable astrologers. The National Assembly had blighted Bonaparte's hopes of a constitutional prorogation of his authority; the candidature of the Prince of Joinville forbade further vacillation.

If ever an event has, well in advance of its coming, cast its shadow before, it was Bonaparte's *coup d'état*. As early as 29 January 1849, barely a month after his election, he had made a proposal about it to Changarnier. In the summer of 1849 his own Prime Minister, Odilon Barrot, had covertly denounced the policy of *coups d'état*; in the winter of 1850 Thiers had openly done so. In May 1851, Persigny had sought once more to win Changarnier for the *coup*; the *Messager de l'Assemblée* had published an account of these negotiations. During every parliamentary storm, the Bonapartist journals threatened a *coup d'état*, and the nearer the crisis drew, the louder grew their tone. In the orgies that Bonaparte kept up every night with men and women of the 'swell mob', as soon as the hour of midnight approached and copious potations had loosened tongues and fired imaginations, the *coup d'état* was fixed for the following morning. Swords were drawn, glasses clinked, the Representatives were thrown out of the window, the imperial mantle fell upon Bonaparte's shoulders, until the following morning banished the spook once more and astonished Paris learned, from vestals of little reticence and from indiscreet paladins, of the danger it had once again escaped. During the months of September and October rumours of a *coup d'état* followed fast one after the other. Simultaneously, the shadow took on colour, like a variegated daguerreotype. Look up the September and October copies of the organs of the European daily press and you will find, word for word, intimations like the following: 'Paris is full of rumours of a *coup d'état*. The capital is to be filled with troops during the night, and the next morning is to bring decrees which will dissolve the National Assembly, declare the Department of the Seine in a state of siege, restore universal suffrage, and appeal to the people. Bonaparte is said to be seeking ministers for the execution of these illegal decrees.' The letters that bring these tidings always end with the fateful word 'postponed'. The *coup d'état* was ever the fixed idea of Bonaparte. With this idea he had again set foot on French soil. He was so obsessed by it that he continually betrayed it and blurted it out. He was so weak that, just as continually, he gave it up again. The shadow of the *coup d'état* had become so familiar to the Parisians as a spectre that they were not willing to believe in it when it finally appeared in the flesh. What allowed the *coup d'état* to succeed was, therefore, neither the reticent reserve of the chief of the Society of 10 December nor the fact that the National Assembly was caught unawares. If it succeeded, it succeeded despite his indiscretion and with its foreknowledge, a necessary, inevitable result of antecedent development.

On 10 October Bonaparte announced to his ministers his decision to restore universal suffrage; on the sixteenth they handed in their resignations, on the twenty-sixth Paris learned of the formation of the Thorigny ministry. Police Prefect Carlier was simultaneously replaced by Maupas; the head of the First Military Division, Magnan, concentrated the most reliable regiments in the capital. On 4 November, the National Assembly resumed its sittings. It had nothing better to do than to recapitulate in a short, succinct form the course it had gone through and to prove that it was buried only after it had died.

The first post that it forfeited in the struggle with the executive power was the ministry. It had solemnly to admit this loss by accepting at full value the Thorigny ministry, a mere shadow cabinet. The Permanent Commission had received M. Giraud with laughter when he presented himself in the name of the new ministers. Such a weak ministry for such strong measures as the restoration of universal suffrage! Yet the precise object was to get nothing through in parliament, but everything against parliament.

On the very first day of its re-opening, the National Assembly received the message from Bonaparte in which he demanded the restoration of universal suffrage and the abolition of the law of 31 May 1850. The same day his ministers introduced a decree to this effect. The National Assembly at once rejected the ministry's motion of urgency and rejected the law itself on 13 November by three hundred and fifty-five votes to three hundred and forty-eight. Thus, it tore up its mandate once more; it once more confirmed the fact that it had transformed itself from the freely elected representatives of the people into the usurpatory parliament of a class; it acknowledged once more that it had itself cut in two the muscles which connected the parliamentary head with the body of the nation.

If by its motion to restore universal suffrage the executive power appealed from the National Assembly to the people, the legislative power appealed by its Questors' Bill from the people to the army. This Questors' Bill was to establish its right of directly requisitioning troops, of forming a parliamentary army. While it thus designated the army as the arbitrator between itself and the people, between itself and Bonaparte, while it recognized the army as the decisive state power, it had to confirm, on the other hand, the fact that it had long given up its claim to dominate this power. By debating its right to requisition troops, instead of requisitioning them at once, it betrayed its doubts about its own powers. By rejecting the Questors' Bill, it made public confession of its impotence. This bill was defeated, its proponents lacking 108 votes of a majority. The *Montagne* thus decided the issue. It found itself in the position of Buridan's ass, not, indeed, between two bundles of hay with the problem of deciding which was the more attractive, but between two showers of blows with the problem of deciding which was the harder. On the one hand, there was the fear of Changarnier; on the other, the fear of Bonaparte. It must be confessed that the position was no heroic one.

On 18 November, an amendment was moved to the law on municipal elections introduced by the party of Order, to the effect that instead of three years', one year's domicile should suffice for municipal electors. The amendment was lost by a single vote, but this one vote immediately proved to be a mistake. By splitting up into its hostile factions, the party of Order had long ago forfeited its independent parliamentary majority. It showed now that there was no longer any majority at all in parliament. The National Assembly had become incapable of transacting business. Its atomic constituents were no longer held together by any force of cohesion; it had drawn its last breath; it was dead.

Finally, a few days before the catastrophe, the extra-parliamentary mass of the bourgeoisie was solemnly to confirm once more its breach with the bourgeoisie in parliament. Thiers, as a parliamentary hero infected more than the rest with the incurable disease of parliamentary cretinism, had, after the death of parliament, hatched out, together with the Council of State, a new parliamentary intrigue, a Responsibility Law by which the President was to be firmly held within the limits of the Constitution. Just as, on laying the foundation stone of the new market halls in Paris on 15 September, Bonaparte, like a second Masaniello, had enchanted the *dames des halles*, the fishwives—to be sure, one fishwife outweighed seventeen burgraves in real power; just as after the introduction of the Questors' Bill he enraptured the lieutenants he regaled in the Elysée, so now, on 25 November, he swept off their feet the industrial bourgeoisie, which had gathered at the circus to receive at his hands prize medals for the London Industrial Exhibition. I shall give the significant portion of his speech as reported in the *Journal des Débats*:

> With such unhoped-for successes, I am justified in reiterating how great the French republic would be if it were permitted to pursue its real interests and reform its institutions, instead of being constantly disturbed by demagogues, on the one hand, and by monarchist hallucinations, on the other. (Loud, stormy, and repeated applause from every part of the amphitheatre.) The monarchist hallucinations hinder all progress and all important branches of industry. In place of progress nothing but struggle. One sees men who were formerly the most zealous supporters of the royal authority and prerogative become partisans of a Convention merely in order to weaken the authority that has sprung from universal suffrage. (Loud and repeated applause.) We see men who have suffered most from the Revolution, and have deplored it most, provoke a new one, and merely in order to fetter the nation's will. . . . I promise you tranquillity for the future, etc., etc. (Bravo, bravo, a storm of bravos.)

Thus the industrial bourgeoisie applauds with servile bravos the *coup d'état* of 2 December, the annihilation of parliament, the downfall of its own rule, the dictatorship of Bonaparte. The thunder of applause on 25 November had its answer in the thunder of cannon on 4 December, and it was on the house of Monsieur Sallandrouze, who had clapped most, that they clapped most of the bombs.

Cromwell, when he dissolved the Long Parliament, went alone into its midst, drew out his watch in order that it should not continue to exist a minute after the time limit fixed by him, and drove out each one of the members of parliament with hilariously humorous taunts. Napoleon, smaller than his prototype, at least betook himself on the eighteenth Brumaire to the legislative body and read out to it, though in a faltering voice, its sentence of death. The second Bonaparte, who, moreover, found himself in possession of an executive power very different from that of Cromwell or Napoleon, sought his model not in the annals of world history, but in the annals of the Society of 10 December, in the annals of the criminal courts. He robs the Bank of France of twenty-five million francs, buys General Magnan with a million, the soldiers with fifteen francs apiece and liquor, comes together with his accomplices secretly like a thief in the night, has the houses of the most dangerous parliamentary leaders broken into and Cavaignac, Lamoricière, Le Flô, Changarnier, Charras, Thiers, Baze, etc., dragged from their beds, the chief squares of Paris and the parliamentary building occupied by troops, and cheapjack placards posted early in the morning on all the walls, proclaiming the dissolution of the National Assembly and the Council of State, the restoration of universal suffrage and the placing of the Seine Department in a state of siege. In like manner, he inserted a little later in the *Moniteur* a false document which asserted that influential parliamentarians had grouped themselves round him and formed a state *consulta* [council].

The rump parliament, assembled in the *mairie* building of the tenth *arrondissement* and consisting mainly of Legitimists and Orleanists, votes the deposition of Bonaparte amid repeated cries of 'Long live the Republic', unavailingly harangues the gaping crowds before the building, and is finally led off in the custody of African sharpshooters, first to the d'Orsay barracks, and later packed into prison vans and transported to the prisons of Mazas, Ham, and Vincennes. Thus ended the party of Order, the Legislative Assembly and the February Revolution. Before hastening to close, let us briefly summarize the latter's history:

I. First period. From 24 February to 4 May 1848. February period. Prologue. Universal brotherhood swindle.

II. Second period. Period of constituting the republic and of the Constituent National Assembly.

1. From 4 May to 25 June 1848. Struggle of all classes against the proletariat. Defeat of the proletariat in the June days.

2. From 25 June to 10 December 1848. Dictatorship of the pure bourgeois-republicans. Drafting of the Constitution. Proclamation of a state of siege in Paris. The bourgeois dictatorship set aside on 10 December by the election of Bonaparte as President.

3. From 20 December 1848 to 28 May 1849. Struggle of the Constituent Assembly with Bonaparte and with the party of Order in alliance with him. Passing of the Constituent Assembly. Fall of the republican bourgeoisie.

III. Third period. Period of the constitutional republic and of the Legislative National Assembly.

1. From 28 May 1849 to 13 June 1849. Struggle of the petty bourgeoisie with the bourgeoisie and with Bonaparte. Defeat of the petty-bourgeois democracy.

2. From 13 June 1849 to 31 May 1850. Parliamentary dictatorship of the party of Order. It completes its rule by abolishing universal suffrage, but loses the parliamentary ministry.

3. From 31 May 1850 to 2 December 1851. Struggle between the parliamentary bourgeoisie and Bonaparte.

(a) From 31 May 1850 to 12 January 1851. Parliament loses the supreme command of the army.

(b) From 12 January to 11 April 1851. It is worsted in its attempts to regain the administrative power. The party of Order loses its independent parliamentary majority. Its coalition with the republicans and the *Montagne*.

(c) From 11 April 1851 to 9 October 1851. Attempts at revision, fusion, prorogation. The party of Order decomposes into its separate constituents. The breach between the bourgeois parliament and press and the mass of the bourgeoisie becomes definite.

(d) From 9 October to 2 December 1851. Open breach between parliament and the executive power. Parliament performs its dying act and succumbs, left in the lurch by its own class, by the army, and by all the remaining classes. Passing of the parliamentary regime and of bourgeois rule. Victory of Bonaparte. Parody of restoration of empire.

On the threshold of the February Revolution, the social republic appeared as a phrase, as a prophecy. In the June days of 1848, it was drowned in the blood of the Paris proletariat, but it haunts the subsequent acts of the drama like a ghost. The democratic republic announces its arrival. On 13 June 1849, it is dissipated together with its petty bourgeois, who have taken to their heels, but in its flight it blows its own trumpet with redoubled boastfulness. The parliamentary republic, together with the bourgeoisie, takes possession of the entire stage; it enjoys its existence to the full, but 2 December 1851 buries it to the accompaniment of the anguished cry of the royalists in coalition: 'Long live the Republic!'

The French bourgeoisie balked at the domination of the working proletariat; it has brought the *lumpenproletariat* to domination, with the chief of the Society of 10 December at the head. The bourgeoisie kept France in breathless fear of the future terrors of red anarchy; Bonaparte discounted this future for it when, on 4 December, he had the eminent bourgeois of the Boulevard Montmartre and the Boulevard des Italiens shot down at their windows by the liquor-inspired army of order. It apotheosized the sword; the sword rules it. It destroyed the revolutionary press; its own press has been destroyed. It placed popular meetings under police supervision; its *salons* are under the supervision of the police. It disbanded the democratic National Guards; its own National Guard is disbanded. It imposed a state of siege; a state of siege is imposed upon it. It supplanted the juries by military commissions; its juries are supplanted by military commissions. It subjected public education to the sway of the priests; the priests subject it to their own education. It transported people without trial; it is being transported without trial. It repressed every stirring in society by means of the state power; every stirring in its society is suppressed by means of the state power. Out of enthusiasm for its purse, it rebelled against its own politicians and men of letters; its politicians and men of letters are swept aside, but its purse is being plundered now that its mouth has been gagged and its pen broken. The bourgeoisie never wearied of crying out to the revolution what Saint Arsenius cried out to the Christians: '*Fuge, tace, quiesce!* Flee, be silent, keep still!' Bonaparte cries to the bourgeoisie: '*Fuge, tace, quiesce!* Flee, be silent, keep still!'

The French bourgeoisie had long ago found the solution to Napoleon's dilemma: '*Dans cinquante ans l'Europe sera républicaine ou cosaque.*' [In fifty years Europe will be republican or Cossack]. It had found the solution to it in the '*république cosaque*'. No Circe, by means of black magic, has distorted that work of art, the bourgeois republic, into a monstrous shape. That republic has lost nothing but the semblance of respectability. Present-day France was contained in a finished state within the parliamentary republic. It only required a bayonet thrust for the bubble to burst and the monster to spring forth before our eyes.

Why did the Paris proletariat not rise in revolt after 2 December?

The overthrow of the bourgeoisie had as yet been only decreed; the decree had not been carried out. Any serious insurrection of the proletariat would at once have put fresh life into the bourgeoisie, would have reconciled it with the army and ensured a second June defeat for the workers.

On 4 December the proletariat was incited by bourgeois and *épicier* [grocer] to fight. On the evening of that day several legions of the National Guard promised to appear, armed and uniformed, on the scene of battle. For the bourgeois and the *épicier* had got wind of the fact that in one of his decrees of 2 December Bonaparte abolished the secret ballot and enjoined them to record their 'yes' or 'no' in the official registers after their names. The resistance of 4 December intimidated Bonaparte. During the night he caused placards to be posted on all the street corners of Paris, announcing the restoration of the secret ballot. The bourgeois and the *épicier* believed that they had gained their end. Those who failed to appear next morning were the bourgeois and the *épicier*.

By a *coup de main* during the night of 1 to 2 December, Bonaparte had robbed the Paris proletariat of its leaders, the barricade commanders. An army without officers, averse to fighting under the banner of the *Montagnards* because of the memories of June 1848 and 1849 and May 1850, it left to its vanguard, the secret societies, the task of saving the insurrectionary honour of Paris, which the bourgeoisie had so unresistingly surrendered to the soldiery that, later on, Bonaparte could sneeringly give as his motive for disarming the National Guard his fear that its arms would be turned against it itself by the anarchists!

'*C'est le triomphe complet et définitif du Socialisme!*' [It is the complete and definitive triumph of Socialism.] Thus Guizot characterized 2 December. But if the overthrow of the parliamentary republic contains within itself the germ of the triumph of the proletarian revolution, its immediate and palpable result was the victory of Bonaparte over parliament, of the executive power over the legislative power, of force without phrases over the force of phrases. In parliament the nation made its general will the law, that is, it made the law of the ruling class its general will. Before the executive power it renounces all will of its own and submits to the superior command of an alien will, to authority. The executive power, in contrast to the legislative power, expresses the heteronomy of a nation, in contrast to its autonomy. France, therefore, seems to have escaped the despotism of a class only to fall back beneath the despotism of an individual, and, what is more, beneath the authority of an individual without authority. The struggle seems to be settled in such a way that all classes, equally impotent and equally mute, fall on their knees before the rifle butt.

But the revolution is thoroughgoing. It is still journeying through purgatory. It does its work methodically. By 2 December 1851, it had completed one half of its preparatory work; it is now completing the other half. First it perfected the parliamentary power, in order to be able to overthrow it. Now that it has attained this, it perfects the executive power, reduces it to its purest expression, isolates it, sets it up against itself as the sole target, in order to concentrate all its forces of destruction against it. And when it has done this second half of its preliminary work, Europe will leap from its seat and exultantly exclaim: Well grubbed, old mole!

This executive power with its enormous bureaucratic and military organization, with its ingenious state machinery, embracing wide strata, with a host of officials numbering half a million, besides an army of another half million, this appalling parasitic body, which enmeshes the body of French society like a net and chokes all its pores, sprang up in the days of the absolute monarchy, with the decay of the feudal system, which it helped to hasten. The seigniorial privileges of the landowners and towns became transformed into so many attributes of the state power, the feudal dignitaries into paid officials and the motley pattern of conflicting medieval plenary powers into the regulated plan of a state authority whose work is divided and centralized as in a factory. The first French Revolution, with its task of breaking all separate local, territorial, urban and provincial powers in order to create the civil unity of the nation, was bound to develop what the absolute monarchy had begun: centralization, but at the same time the extent, the attributes, and the agents of governmental power. Napoleon perfected this state machinery. The Legitimist monarchy and the July monarchy added nothing but a greater division of labour, growing in the same measure as the division of labour within bourgeois society created new groups of interests, and, therefore, new material for state administration. Every common interest was straightway severed from society, counterposed to it as a higher, general interest, snatched from the activity of society's members themselves, and made an object of government activity, from bridge, a schoolhouse and the communal property of a village community to the railways, the national wealth and the national university of France. Finally, in its struggle against the revolution, the parliamentary republic found itself compelled to strengthen, along with the repressive measures, the resources and centralization of governmental power. All revolutions perfected this machine instead of smashing it. The parties that contended in turn for domination regarded the possession of this huge state edifice as the principal spoils of the victor.

But under the absolute monarchy, during the first Revolution, under Napoleon, bureaucracy was only the means of preparing the class rule of the bourgeoisie. Under the Restoration, under Louis Philippe, under the parliamentary republic, it was the instrument of the ruling class, however much it strove for power of its own.

Only under the second Bonaparte does the state seem to have made itself completely independent. As against civil society, the state machine has consolidated its position so thoroughly that the chief of the Society of 10 December suffices for its head, an adventurer blown in from abroad, raised on the shield by a drunken soldiery, which he has bought with liquor and sausages, and which he must continually ply with sausage anew. Hence the downcast despair, the feeling of most dreadful humiliation and degradation that oppresses the breast of France and makes her catch her breath. She feels dishonoured.

And yet the state power is not suspended in mid air. Bonaparte represents a class, and the most numerous class of French society at that, the small-holding peasants.

Just as the Bourbons were the dynasty of big landed property and just as the Orleans were the dynasty of money, so the Bonapartes are the dynasty of the peasants, that is, the mass of the French people. Not the Bonaparte who submitted to the bourgeois parliament, but the Bonaparte who dispersed the bourgeois parliament is the chosen of the peasantry. For three years the towns had succeeded in falsifying the meaning of the election of 10 December and in cheating the peasants out of the restoration of the empire. The election of 10 December 1848 has been consummated only by the *coup d'état* of 2 December 1851.

The small-holding peasants form a vast mass, the members of which live in similar conditions but without entering into manifold relations with one another. Their mode of production isolates them from one another instead of bringing them into mutual intercourse. The isolation is increased by France's bad means of communication and by the poverty of the peasants. Their field of production, the small holding, admits of no division of labour in its cultivation, no application of science, and, therefore, no diversity of development, no variety of talent, no wealth of social relationships. Each individual peasant family is almost self-sufficient; it itself directly produces the major part of its consumption and thus acquires its means of life more through exchange with nature than in intercourse with society. A small holding, a peasant, and his family; alongside them another small holding, another peasant and another family. A few score of these make up a village, and a few score of villages make up a Department. In this way, the great mass of the French nation is formed by simple addition of homologous magnitudes, much as potatoes in a sack form a sack of potatoes. In so far as millions of families live under economic conditions of existence that separate their mode of life, their interests, and their culture from those of the other classes, and put them in hostile opposition to the latter, they form a class. In so far as there is merely a local interconnection among these small-holding peasants, and the identity of their interests begets no community, no national bond, and no political organization among them, they do not form a class. They are consequently incapable of enforcing their class interests in their own name, whether through a parliament or through a convention. They cannot represent themselves, they must be represented. Their representative must at the same time appear as their master, as an authority over them, as an unlimited governmental power that protects them against the other classes and sends them rain and sunshine from above. The political influence of the small-holding peasants, therefore, finds its final expression in the executive power subordinating society to itself.

Historical tradition gave rise to the belief of the French peasants in the miracle that a man named Napoleon would bring all the glory back to them. And an individual turned up who gives himself out as the man because he bears the name of Napoleon, in consequence of the *Code Napoléon*, which lays down that *la recherche de la paternité est interdite* [inquiries concerning paternity are forbidden]. After a vagabondage of twenty years and after a series of grotesque adventures, the legend finds fulfilment and the man becomes Emperor of the French. The fixed idea of the Nephew was realized, because it coincided with the fixed idea of the most numerous class of the French people.

But, it may be objected, what about the peasant risings in half of France, the raids on the peasants by the army, the mass incarceration and transportation of peasants?

Since Louis XIV, France has experienced no similar persecution of the peasants 'on account of demagogic practices'.

But let there be no misunderstanding. The Bonaparte dynasty represents not the revolutionary, but the conservative peasant; not the peasant that strikes out beyond the condition of his social existence, the small holding, but rather the peasant who wants to consolidate this holding, not the country folk who, linked up with the towns, want to overthrow the old order through their own energies, but on the contrary those who, in stupefied seclusion within this old order, want to see themselves and their small holdings saved and favoured by the ghost of the empire. It represents not the enlightenment, but the superstition of the peasant; not his judgement, but his prejudice; not his future, but his past; not his modern Cevennes, but his modern Vendée.

The three years' rigorous rule of the parliamentary republic had freed a part of the French peasants from the Napoleonic illusion and had revolutionized them, even if only superficially; but the bourgeoisie violently repressed them, as often as they set themselves in motion. Under the parliamentary republic the modern and the traditional consciousness of the French peasant contended for mastery. This progress took the form of an incessant struggle between the schoolmasters and the priests. The bourgeoisie struck down the schoolmasters. For the first time the peasants made efforts to behave independently in the face of the activity of the government.

This was shown in the continual conflict between the *maires* and the prefects. The bourgeoisie deposed the *maires*. Finally, during the period of the parliamentary republic, the peasants of different localities rose against their own offspring, the army. The bourgeoisie punished them with states of siege and punitive expeditions. And this same bourgeoisie now cries out about the stupidity of the masses, the vile multitude, that has betrayed it to Bonaparte. It has itself forcibly strengthened the empire sentiments of the peasant class, it conserved the conditions that form the birthplace of this peasant religion. The bourgeoisie, to be sure, is bound to fear the stupidity of the masses as long as they remain conservative, and the insight of the masses as soon as they become revolutionary.

In the risings after the *coup d'état*, a part of the French peasants protested, arms in hand, against their own vote of 10 December 1848. The school they had gone through since 1848 had sharpened their wits. But they had made themselves over to the underworld of history; history held them to their word, and the majority was still so prejudiced that in precisely the reddest Departments the peasant population voted openly for Bonaparte. In its view, the National Assembly had hindered his progress. He had now merely broken the fetters that the towns had imposed on the will of the countryside. In some parts the peasants even entertained the grotesque notion of a convention side by side with Napoleon.

After the first revolution had transformed the peasants from semi-villeins into freeholders, Napoleon confirmed and regulated the conditions on which they could exploit undisturbed the soil of France which had only just fallen to their lot and slake their youthful passion for property. But what is now causing the ruin of the French peasant is his small holding itself, the division of the land, the form of property which Napoleon consolidated in France. It is precisely the material conditions which made the feudal peasant a small-holding peasant and Napoleon an emperor. Two generations have sufficed to produce the inevitable result: progressive deterioration of agriculture, progressive indebtedness of the agriculturist. The 'Napoleonic' form of property, which at the beginning of the nineteenth century was the condition for the liberation and enrichment of the French country folk, has developed in the course of this century into the law of their enslavement and pauperization. And precisely this law is the first of the *idées napoléoniennes* which the second Bonaparte has to uphold. If he still shares with the peasants the illusion that the cause of their ruin is to be sought, not in this small-holding property itself, but outside it, in the influence of secondary circumstances, his experiments will burst like soap bubbles when they come in contact with the relations of production.

The economic development of small-holding property has radically changed the relation of the peasants to the other classes of society. Under Napoleon, the fragmentation of the land in the countryside supplemented free competition and the beginning of big industry in the towns. The peasant class was the ubiquitous protest against the landed aristocracy which had just been overthrown. The roots that small-holding property struck in French soil deprived feudalism of all nutriment. Its landmarks formed the natural fortifications of the bourgeoisie against any surprise attack on the part of its old overlords. But in the course of the nineteenth century the feudal lords were replaced by urban usurers; the feudal obligation that went with the land was replaced by the mortgage; aristocratic landed property was replaced by bourgeois capital. The small holding of the peasant is now only the pretext that allows the capitalist to draw profits, interest, and rent from the soil, while leaving it to the tiller of the soil himself to see how he can extract his wages. The mortgage debt burdening the soil of France imposes on the French peasantry payment of an amount of interest equal to the annual interest on the entire British national debt. Small-holding property, in this enslavement by capital to which its development inevitably pushes forward, has transformed the mass of the French nation into troglodytes. Sixteen million peasants (including women and children) dwell in hovels, a large number of which have but one opening, others only two, and the most favoured only three. And windows are to a house what the five senses are to the head. The bourgeois order, which at the beginning of the century set the state to stand guard over the newly arisen small holding and manured it with laurels, has become a vampire that sucks out its blood and brains and throws them into the alchemistic cauldron of capital. The *Code Napoléon* is now nothing but a *codex* of distraints, forced sales, and compulsory auctions. To the four million (including children, etc.) officially recognized paupers, vagabonds, criminals, and prostitutes in France must be added five million who hover on the margin of existence and either have their haunts in the countryside itself or, with their rags and their children, continually desert the countryside for the towns and the towns for the countryside. The interests of the peasants, therefore, are no longer, as under Napoleon, in accord with, but in opposition, to the interests of the bourgeoisie, to capital. Hence the peasants find their natural ally and leader in the urban proletariat, whose task is the overthrow of the bourgeois order. But strong and unlimited government—and this is the second *idée napoléonienne*, which the second Napoleon has to carry out—is called upon to defend this 'material' order by force. This *ordre matériel* also serves as the catchword in all of Bonaparte's proclamations against the rebellious peasants.

Besides the mortgage which capital imposes on it, the small holding is burdened by taxes. Taxes are the source of life for the bureaucracy, the army, the priests, and the court, in short, for the whole apparatus of the executive power. Strong government and heavy taxes are identical. By its very nature, small-holding property forms a suitable basis for an all-powerful and innumerable bureaucracy. It creates a uniform level of relationships and persons over the whole surface of the land. Hence it also permits of uniform action from a supreme centre on all points of this uniform mass. It annihilates the aristocratic intermediate grades between the mass of the people and the state power. On all sides, therefore, it calls forth the direct interference of this state power and the interposition of its immediate organs. Finally, it produces an unemployed surplus population for which there is no place either on the land or in the towns, and which accordingly reaches out for state offices as a sort of respectable alms, and provokes the creation of state posts. By the new markets which he opened at the point of the bayonet, by the plundering of the Continent, Napoleon repaid the compulsory taxes with interest. These taxes were a spur to the industry of the peasant, whereas now they rob his industry of its last resources and complete his inability to resist pauperism. And an enormous bureaucracy, well-gallooned and well-fed, is the *idée napoléonienne* which is most congenial of all to the second Bonaparte. How could it be otherwise, seeing that alongside the actual classes of society he is forced to create an artificial caste, for which the maintenance of his regime becomes a bread-and-butter question? Accordingly, one of his first financial operations was the raising of officials' salaries to their old level and the creation of new sinecures.

Another *idée napoléonienne* is the domination of the priests as an instrument of government. But while in its accord with society, in its dependence on natural forces and its submission to the authority which protected it from above, the small holding that had newly come into being was naturally religious, the small holding that is ruined by debts, at odds with society and authority, and driven beyond its own limitations naturally becomes irreligious. Heaven was quite a pleasing accession to the narrow strip of land just won, more particularly as it makes the weather; it becomes an insult as soon as it is thrust forward as substitute for the small holding. The priest then appears as only the anointed bloodhound of the earthly police—another *idée napoléonienne*. On the next occasion, the expedition against Rome will take place in France itself, but in a sense opposite to that of M. de Montalembert.

Lastly, the culminating point of the *idées napoléoniennes* is the preponderance of the army. The army was the *point d'honneur* of the small-holding peasants, it was they themselves transformed into heroes, defending their new possessions against the outer world, glorifying their recently won nationhood, plundering and revolutionizing the world. The uniform was their own state dress; war was their poetry; the small holding, extended and rounded off in imagination, was their fatherland, and patriotism the ideal form of the sense of property. But the enemies against whom the French peasant has now to defend his property are not the Cossacks; they are the *huissiers* [bailiffs] and the tax collectors. The small holding lies no longer in the so-called fatherland, but in the register of mortgages. The army itself is no longer the flower of the peasant youth; it is the swamp-flower of the peasant *lumpenproletariat*. It consists in large measure of *remplaçants*, of substitutes, just as the second Bonaparte is himself only a *remplaçant*, the substitute for Napoleon. It now performs its deeds of valour by hounding the peasants in masses like chamois, by doing *gendarme* duty, and, if the internal contradictions of his system chase the chief of the Society of 10 December over the French border, his army, after some acts of brigandage, will reap, not laurels, but thrashings.

One sees that all *idées napoléoniennes* are ideas of the undeveloped small holding in the freshness of its youth; for the small holding that has outlived its day they are an absurdity. They are only the hallucinations of its death struggle, words that are transformed into phrases, spirits transformed into ghosts. But the parody of the empire was necessary to free the mass of the French nation from the weight of tradition and to work out in pure form the opposition between the state power and society. With the progressive undermining of small-holding property, the state structure erected upon it collapses. The centralization of the state that modern society requires arises only on the ruins of the military-bureaucratic government machinery which was forged in opposition to feudalism.

The condition of the French peasants provides us with the answer to the riddle of the general elections of 20 and 21 December, which bore the second Bonaparte up Mount Sinai, not to receive laws, but to give them.

Manifestly, the bourgeoisie had now no choice but to elect Bonaparte. When the puritans at the Council of Constance complained of the dissolute lives of the popes and wailed about the necessity of moral reform, Cardinal Pierre d'Ailly thundered at them: 'Only the devil in person can still save the Catholic Church, and you ask for angels.' In like manner, after the *coup d'état*, the French bourgeoisie cried: Only the chief of the Society of 10 December can still save bourgeois society! Only theft can still save property; only perjury, religion; bastardy, the family; disorder, order!

As the executive authority which has made itself an independent power, Bonaparte feels it to be his mission to safeguard 'bourgeois order'. But the strength of this bourgeois order lies in the middle class. He looks on himself, therefore, as the representative of the middle class and issues decrees in this sense. Nevertheless, he is somebody solely due to the fact that he has broken the political power of this middle class and daily breaks it anew. Consequently, he looks on himself as the adversary of the political and literary power of the middle class. But by protecting its material power, he generates its political power anew. The cause must accordingly be kept alive; but the effect, where it manifests itself, must be done away with. But this cannot pass off without slight confusions of cause and effect, since in their interaction both lose their distinguishing features. New decrees that obliterate the border line. As against the bourgeoisie, Bonaparte looks on himself, at the same time, as the representative of the peasants and of the people in general, who wants to make the lower classes of the people happy within the frame of bourgeois society. New decrees that cheat the 'True Socialists' of their statecraft in advance. But, above all, Bonaparte looks on himself as the chief of the Society of 10 December, as the representative of the *lumpenproletariat* to which he himself, his entourage, his government, and his army belong, and whose prime consideration is to benefit itself and draw California lottery prizes from the state treasury. And he vindicates his position as chief of the Society of 10 December with decrees, without decrees, and despite decrees.

This contradictory task of the man explains the contradictions of his government, the confused groping about which seeks now to win, now to humiliate, first one class and then another and arrays all of them uniformly against him, whose practical uncertainty forms a highly comical contrast to the imperious, categorical style of the government decrees, a style which is faithfully copied from the Uncle.

Industry and trade, hence the business affairs of the middle class, are to prosper in hothouse fashion under the strong government. The grant of innumerable railway concessions. But the Bonapartist *lumpenproletariat* is to enrich itself. The initiated play *tripotage* [speculation] on the *bourse* with the railway concessions. But no capital is forthcoming for the railways. Obligation of the Bank to make advances on railway shares. But, at the same time, the Bank is to be exploited for personal ends and therefore must be cajoled. Release of the Bank from the obligation to publish its report weekly. Leonine agreement of the Bank with the government. The people are to be given employment. Initiation of public works. But the public works increase the obligations of the people in respect of taxes. Hence reduction of the taxes by an onslaught on the *rentiers* [stockholders] by conversion of the five per cent bonds to four-and-a-half per cent. But, once more, the middle class must receive a *douceur* [sweetening]. Therefore doubling of the wine tax for the people, who buy it *en détail*, and halving of the wine tax for the middle class, who drink it *en gros*. Dissolution of the actual workers' associations, but promises of miracles of association in the future. The peasants are to be helped. Mortgage banks that expedite their getting into debt and accelerate the concentration of property. But these banks are to be used to make money out of the confiscated estates of the House of Orleans. No capitalist wants to agree to this condition, which is not in the decree, and the mortgage bank remains a mere decree, etc., etc.

Bonaparte would like to appear as the patriarchal benefactor of all classes. But he cannot give to one class without taking from another. Just as at the time of the Fronde it was said of the Duke of Guise that he was the most *obligeant* man in France because he had turned all his estates into his partisans' obligations to him, so Bonaparte would fain be the most *obligeant* man in France and turn all the property, all the labour of France into a personal obligation to himself. He would like to steal the whole of France in order to be able to make a present of her to France or, rather, in order to be able to buy France anew with French money, for as the chief of the Society of 10 December he must needs buy what ought to belong to him. And all the state institutions, the Senate, the Council of State, the legislative body, the Legion of Honour, the soldiers' medals, the washhouses, the public works, the railways, the *état-major* of the National Guard to the exclusion of privates, and the confiscated estates of the House of Orleans— all become parts of the institution of purchase. Every place in the army and in the government machine becomes a means of purchase. But the most important feature of this process, whereby France is taken in order to give to her, is the percentages that find their way into the pockets of the head and the members of the Society of 10 December during the turnover. The witticism with which Countess L., the mistress of M. de Morny, characterized the confiscation of the Orleans estates: '*C'est le premier vol de l'aigle*' [It is the first flight/theft of the eagle] is applicable to every flight of this eagle, which is more like a raven. He himself and his adherents call out to one another daily like that Italian Carthusian admonishing the miser who, with boastful display, counted up the goods on which he could yet live for years to come. '*Tu fai conto sopra i beni, bisogna prima far il conto sopra gli anni.*' [You would do better to count the years than count the goods.] Lest they make a mistake in the years, they count the minutes. A bunch of blokes push their way forward to the court, into the ministries to the head of the administration and the army, a crowd of the best of whom it must be said that no one knows whence he comes, a noisy, disreputable, rapacious bohème that crawls into gallooned coats with the same grotesque dignity as the high dignitaries of Soulouque. One can visualize clearly this upper stratum of the Society of 10 December, if one reflects that Véron-Crevel is its preacher of morals and Granier de Cassagnac its thinker. When Guizot, at the time of his ministry, utilized this Granier on a hole-and-corner newspaper against the dynastic opposition, he used to boast of him with the quip: '*C'est le roi des drôles*', 'he is the king of buffoons'. One would do wrong to recall the Regency or Louis XV in connection with Louis Bonaparte's court and clique. For 'often already, France has experienced a government of mistresses; but never before a government of *hommes entretenus* [kept men].'

Driven by the contradictory demands of his situation and being at the same time, like a conjurer, under the necessity of keeping the public gaze fixed on himself, as Napoleon's substitute, by springing constant surprises, that is to say, under the necessity of executing a *coup d'état en miniature* every day, Bonaparte throws the entire bourgeois economy into confusion, violates everything that seemed inviolable to the Revolution of 1848, makes some tolerant of revolution, others desirous of revolution, and produces actual anarchy in the name of order, while at the same time stripping its halo from the entire state machine, profanes it, and makes it at once loathsome and ridiculous. The cult of the Holy Tunic of Treves he duplicates at Paris in the cult of the Napoleonic imperial mantle. But when the imperial mantle finally falls on the shoulders of Louis Bonaparte, the bronze statue of Napoleon will crash from the top of the Vendôme Column.

BIBLIOGRAPHY

Original: MEW, Vol. 8, pp. 111 ff.

Present translation: MESW, Vol. 1, pp. 247 ff. Reproduced by kind permission of Lawrence and Wishart Ltd.

Other translations: K. Marx, *Surveys from Exile,* ed. D. Fernbach, New York and London, 1973.

Commentaries: D. Fernbach, Introduction to edition cited above.
F. Mehring, *Karl Marx,* London, 1936, pp. 213 ff.
N. Poulantzas, *Political Power and Social Classes,* London, 1973.
R. Price, *The Second French Republic,* London, 1972.
I. Zeitlin, *Marxism: A Re-examination,* Princeton, 1967, pp. 137 ff.

26. Journalism of the 1850s

Marx took to journalism in the early 1850s in order to make a living. The main paper for which he wrote was the *New York Daily Tribune*, a paper with a wide circulation and Fourierist leanings, for which he began to write in 1852. Marx was hired partly to attract the increasing number of potential German refugee readers in America, and his articles met with a wide response during the ten years he wrote for the paper. Although Marx himself had no high opinion of his work, he did incorporate into his articles a lot of the material on economics, technology, and agriculture that was involved in his more 'serious' studies. This gave his journalism a striking depth and long-term perspective.

British Political Parties

The political parties of Great Britain are sufficiently known in the United States. It will be sufficient to bring to mind, in a few strokes of the pen, the distinctive characteristics of each of them.

Up to 1846 the Tories passed as the guardians of the traditions of Old England. They were suspected of admiring in the British Constitution the eighth wonder of the world; to be *laudatores temporis acti* [praisers of time past], enthusiasts for the throne, the High Church, the privileges and liberties of the British subject. The fatal year, 1846, with its repeal of the Corn Laws, and the shout of distress which this repeal forced from the Tories, proved that they were enthusiasts for nothing but the rent of land, and at the same time disclosed the secret of their attachment to the political and religious institutions of Old England. These institutions are the very best institutions, with the help of which large landed property – the landed interest – has hitherto ruled England, and even now seeks to maintain its rule. The year 1846 brought to light in its nakedness the substantial class interest which forms the real base of the Tory party. The year 1846 tore down the traditionally venerable lion's hide, under which Tory class interest had hitherto hidden itself. The year 1846 transformed the Tories into Protectionists. Tory was the sacred name, Protectionist is the profane one; Tory was the political battle-cry, Protectionist is the economical shout of distress; Tory seemed an idea, a principle, Protectionist is an interest. Protectionists of what? Of their own revenues, of the rent of their own land. Then the Tories, in the end, are bourgeois as much as the remainder, for where is the bourgeois who is not a protectionist of his own purse? They are distinguished from the other bourgeois in the same way as rent of land is distinguished from commercial and industrial profit. Rent of land is conservative, profit is progressive; rent of land is national, profit is cosmopolitical;

rent of land believes in the State Church, profit is a dissenter by birth. The repeal of the Corn Laws in 1846 merely recognized an already accomplished fact, a change long since enacted in the elements of British civil society, viz., the subordination of the landed interest to the moneyed interest, of property to commerce, of agriculture to manufacturing industry, of the country to the city. Could this fact be doubted since the country population stands, in England, to the towns' population in the proportion of one to three? The substantial foundation of the power of the Tories was the rent of land. The rent of land is regulated by the price of food. The price of food, then, was artificially maintained at a high rate by the Corn Laws. The repeal of the Corn Laws brought down the price of food, which in its turn brought down the rent of land, and with sinking rent broke down the real strength upon which the political power of the Tories reposed.

What, then, are they trying to do now? To maintain a political power, the social foundation of which has ceased to exist. And how can this be attained? By nothing short of a counter-revolution, that is to say, by a reaction of the state against society. They strive to retain forcibly institutions and a political power which were condemned from the very moment at which the rural population found itself outnumbered three times by the population of the towns. And such an attempt must necessarily end with their destruction; it must accelerate and make more acute the social development of England; it must bring on a crisis.

The Tories recruit their army from the farmers, who have either not yet lost the habit of following their landlords as their natural superiors, or who are economically dependent upon them, or who do not yet see that the interest of the farmer and the interest of the landlord are no more identical than the respective interests of the borrower and of the usurer. They are followed and supported by the Colonial Interest, the Shipping Interest, the State Church party, in short, by all those elements which consider it necessary to safeguard their interests against the necessary results of modern manufacturing industry, and against the social revolution prepared by it.

Opposed to the Tories, as their hereditary enemies, stand the Whigs, a party with whom the American Whigs have nothing in common but the name.

The British Whig, in the natural history of politics, forms a species which, like all those of the amphibious class, exists very easily, but is difficult to describe. Shall we call them, with their opponents, Tories out of office, or, as continental writers love it, take them for the representatives of certain popular principles? In the latter case we should get embarrassed in the same difficulty as the historian of the Whigs, Mr Cooke, who, with great naïveté, confesses in his *History of Parties* that it is indeed a certain number of 'liberal, moral, and enlightened principles' which constitutes the Whig party, but that it was greatly to be regretted that during the more than a century and a half that the Whigs have existed, they have been, when in office, always prevented from carrying out these principles. So that in reality, according to the confession of their own historian, the Whigs represent something quite different from their professed 'liberal and enlightened principles'. Thus they are in the same position as the drunkard brought up before the Lord Mayor who declared that he represented the temperance principle but from some accident or other always got drunk on Sundays.

But never mind their principles; we can better make out what they are in historical fact; what they carry out, not what they once believed, and what they now want other people to believe with respect to their character.

The Whigs, as well as the Tories, form a fraction of the large landed proprietors of Great Britain. Nay, the oldest, richest, and most arrogant portion of English landed property is the very nucleus of the Whig party.

What, then, distinguishes them from the Tories? The Whigs are the aristocratic representatives of the bourgeoisie, of the industrial and commercial middle class. Under the condition that the bourgeoisie should abandon to them, to an oligarchy of aristocratic families, the monopoly of government and the exclusive possession of office, they make to the middle class, and assist it in conquering, all those concessions which in the course of social and political development have shown themselves to have become unavoidable and undelayable. Neither more nor less. And as often as such an unavoidable measure has been passed, they declare loudly that herewith the end of historical progress has been obtained; that the whole social movement has carried its ultimate purpose, and then they 'cling to finality'. They can support more easily than the Tories a decrease of their rental revenues, because they consider themselves as the heaven-born farmers of the revenues of the British Empire. They can renounce the monopoly of the Corn Laws, as long as they maintain the monopoly of government as their family property. Ever since the 'Glorious Revolution' of 1688 the Whigs, with short intervals, caused principally by the first French revolution and the consequent reaction, have found themselves in the enjoyment of the public offices. Whoever recalls to his mind this period of English history will find no other distinctive mark of Whigdom but the maintenance of their family oligarchy. The interests and principles which they represent besides, from time to time, do not belong to the Whigs; they are forced upon them by the development of the industrial and commercial class, the bourgeoisie. After 1688 we find them united with the Bankocracy, just then rising into importance, as we find them in 1846 united with the Millocracy. The Whigs as little carried the Reform Bill of 1831 as they carried the Free Trade Bill of 1846. Both reform movements, the political as well as the commercial, were movements of the bourgeoisie. As soon as either of these movements had ripened into irresistibility, as soon as, at the same time, it had become the safest means of turning the Tories out of office, the Whigs stepped forward, took up the direction of the government, and secured to themselves the governmental part of the victory. In 1831 they extended the political portion of reform as far as was necessary in order not to leave the middle class entirely dissatisfied; after 1846 they confined their free-trade measures so far as was necessary in order to save to the landed aristocracy the greatest possible amount of privileges. Each time they took the movement in hand in order to prevent its forward march, and to recover their own posts at the same time.

It is clear that from the moment when the landed aristocracy is no longer able to maintain its position as an independent power, to fight, as an independent party, for the government position, in short, that from the moment when the Tories are definitively overthrown, British history has no longer any room for the Whigs. The aristocracy once destroyed, what is the use of an aristocratic representation of the bourgeoisie against this aristocracy?

It is well known that in the Middle Ages the German emperors put the just then arising towns under imperial governors, '*advocati*', to protect these towns against the surrounding nobility. As soon as growing population and wealth gave them sufficient strength and independence to resist, and even to attack the nobility, the towns also drove out the noble governors, the *advocati*.

The Whigs have been these *advocati* of the British middle class, and their governmental monopoly must break down as soon as the landed monopoly of the Tories is broken down. In the same measure as the middle class has developed its independent strength, they have shrunk down from a party to a coterie.

It is evident what a distastefully heterogeneous mixture the character of the British Whigs must turn out to be: feudalists, who are at the same time Malthusians, money-mongers with feudal prejudices, aristocrats without point of honour, bourgeois without industrial activity, finality-men with progressive phrases, progressists with fanatical conservatism, traffickers in homeopathical fractions of reforms, fosterers of family-nepotism, grand masters of corruption, hypocrites of religion, Tartuffes of politics. The mass of the English people have a sound aesthetical common sense. They have an instinctive hatred against everything motley and ambiguous, against bats and Russellites. And then, with the Tories, the mass of the English people, the urban and rural proletariat, has in common the hatred against the 'money-monger'. With the bourgeoisie it has in common the hatred against aristocrats. In the Whigs it hates the one and the other, aristocrats and bourgeois, the landlord who oppresses, and the money lord who exploits it. In the Whig it hates the oligarchy which has ruled over England for more than a century, and by which the people is excluded from the direction of its own affairs.

The Peelites (Liberal Conservatives) are no party; they are merely the souvenir of a partyman, of the late Sir Robert Peel. But Englishmen are too prosaical for a souvenir to form, with them, the foundations for anything but elegies. And now that the people have erected brass and marble monuments to the late Sir Robert Peel in all parts of the country, they believe they are able so much the more to do without those perambulant Peel monuments, the Grahams, the Gladstones, the Cardwells, etc. The so-called Peelites are nothing but this staff of bureaucrats which Robert Peel had schooled for himself. And because they form a pretty complete staff, they forget for a moment that there is no army behind them. The Peelites, then, are old supporters of Sir Robert Peel, who have not yet come to a conclusion as to what party to attach themselves to. It is evident that a similar scruple is not a sufficient means for them to constitute an independent power. . . .

While the Tories, the Whigs, the Peelites—in fact, all the parties we have hitherto commented upon—belong more or less to the past, the Free Traders (the men of the Manchester School, the Parliamentary and Financial Reformers) are the official representatives of modern English society, the representatives of that England which rules the market of the world. They represent the party of the self-conscious bourgeoisie, of industrial capital striving to make available its social power as a political power as well, and to eradicate the last arrogant remnants of feudal society. This party is led on by the most active and most energetic portion of the English bourgeoisie—the manufacturers. What they demand is the complete and undisguised ascendancy of the bourgeoisie, the open, official subjection of society at large to the laws of modern, bourgeois production, and to the rule of those men who are the directors of that production. By free trade they mean the unfettered movement of capital; freed from all political, national, and religious shackles. The soil is to be a marketable commodity, and the exploitation of the soil is to be carried on according to the common commercial laws. There are to be manufacturers of food as well as manufacturers of twist and cottons, but no longer any lords of the land. There are, in short, not to be tolerated any political or social restrictions, regulations or monopolies, unless they proceed from 'the eternal laws of political economy', that is, from the conditions under which capital produces and distributes. The struggle of this party against the old English institutions, products of a superannuated, an evanescent stage of social development, is resumed in the watchword: Produce as cheap as you can, and do away with all the *faux frais* [unnecessary expenses] of production. And this watchword is addressed not only to the private individual, but to the nation at large principally.

Royalty, with its 'barbarous splendours', its court, its civil list and its flunkeys—what else does it belong to but to the *faux frais* of production? The nation can produce and exchange without royalty; away with the crown. The sinecures of the nobility, the House of Lords? *Faux frais* of production. The large standing army? *Faux frais* of production. The colonies? *Faux frais* of production. The State Church, with its riches, the spoils of plunder or of mendicity? *Faux frais* of production. Let parsons compete freely with each other, and everyone pay them according to his own wants. The whole circumstantial routine of English law, with its Court of Chancery? *Faux frais* of production. National wars? *Faux frais* of production. England can exploit foreign nations more cheaply while at peace with them.

You see, to these champions of the British bourgeoisie, to the men of the Manchester School, every institution of Old England appears in the light of a piece of machinery as costly as it is useless, and which fulfils no other purpose but to prevent the nation from producing the greatest possible quantity at the least possible expense, and to exchange its products in freedom. Necessarily, their last word is the bourgeois republic, in which free competition rules supreme in all spheres of life; in which there remains altogether that minimum only of government which is indispensable for the administration, internally and externally, of the common class interest and business of the bourgeoisie; and where this minimum of government is as soberly, as economically organized as possible. Such a party, in other countries, would be called democratic. But it is necessarily revolutionary, and the complete annihilation of Old England as an aristocratic country is the end which it follows up with more or less consciousness. Its nearest object, however, is the attainment of a parliamentary reform which should transfer to its hands the legislative power necessary for such a revolution.

But the British bourgeois are not excitable Frenchmen. When they intend to carry a parliamentary reform they will not make a February revolution. On the contrary. Having obtained, in 1846, a grand victory over the landed aristocracy by the repeal of the Corn Laws, they were satisfied with following up the material advantages of this victory, while they neglected to draw the necessary political and economic conclusions from it, and thus enabled the Whigs to reinstate themselves into their hereditary monopoly of government. During all the time from 1846 to 1852, they exposed themselves to ridicule by their battle-cry: Broad principles and practical (read *small*) measures. And why all this? Because in every violent movement they are obliged to appeal to the working class. And if the aristocracy is their vanishing opponent, the working class is their arising enemy. They prefer to compromise with the vanishing opponent rather than to strengthen the arising enemy, to whom the future belongs, by concessions of a more than apparent importance. Therefore, they strive to avoid every forcible collision with the aristocracy; but historical necessity and the Tories press them onwards. They cannot avoid fulfilling their mission, battering to pieces Old England, the England of the past; and the very moment when they will have conquered exclusive political dominion, when political dominion and economic supremacy will be united in the same hands, when, therefore, the struggle against capital will no longer be distinct from the struggle against the existing government—from that very moment will date the social revolution of England.

We now come to the Chartists, the politically active portion of the British working class. The six points of the Charter which they contend for contain nothing but the demand of universal suffrage, and of the conditions without which universal suffrage would be illusory for the working class, such as the ballot, payment of members, annual general elections. But universal suffrage is the equivalent for political power for the working class of England, where the proletariat forms the large majority of the population, where, in a long, though underground, civil war, it has gained a clear consciousness of its position as a class, and where even the rural districts know no longer any peasants, but only landlords, industrial capitalists (farmers), and hired labourers. The carrying of universal suffrage in England would, therefore, be a far more socialistic measure than anything which has been honoured with that name on the Continent.

Its inevitable result, here, is the political supremacy of the working class.

The Future Results of British Rule in India

... How came it that English supremacy was established in India? The paramount power of the Great Mogul was broken by the Mogul Viceroys. The power of the Viceroys was broken by the Mahrattas. The power of the Mahrattas was broken by the Afghans, and while all were struggling against all, the Briton rushed in and was enabled to subdue them all. A country not only divided between Mohammedan and Hindu, but between tribe and tribe, between caste and caste; a society whose framework was based on a sort of equilibrium, resulting from a general repulsion and constitutional exclusiveness between all its members. Such a country and such a society, were they not the predestined prey of conquest? If we knew nothing of the past history of Hindustan, would there not be the one great and incontestable fact, that even at this moment India is held in English thraldom by an Indian army maintained at the cost of India? India, then, could not escape the fate of being conquered, and the whole of her past history, if it be anything, is the history of the successive conquests she has undergone. Indian society has no history at all, at least no known history. What we call its history is but the history of the successive intruders who founded their empires on the passive basis of that unresisting and unchanging society. The question, therefore, is not whether the English had a right to conquer India, but whether we are to prefer India conquered by the Turk, by the Persian, by the Russian, to India conquered by the Briton.

England has to fulfil a double mission in India: one destructive, the other regenerating—the annihilation of old Asiatic society, and the laying of the material foundations of Western society in Asia.

Arabs, Turks, Tartars, Moguls, who had successively overrun India, soon became Hinduized, the barbarian conquerors being, by an eternal law of history, conquered themselves by the superior civilization of their subjects. The British were the first conquerors superior, and therefore, inaccessible to Hindu civilization. They destroyed it by breaking up the native communities, by uprooting the native industry, and by levelling all that was great and elevated in the native society. The historic pages of their rule in India report hardly anything beyond that destruction. The work of regeneration hardly transpires through a heap of ruins. Nevertheless it has begun.

The political unity of India, more consolidated, and extending farther than it ever did under the Great Moguls, was the first condition of its regeneration. That unity, imposed by the British sword, will now be strengthened and perpetuated by the electric telegraph. The native army, organized and trained by the British drill-sergeant, was the *sine qua non* of Indian self-emancipation, and of India ceasing to be the prey of the first foreign intruder. The free press, introduced for the first time into Asiatic society, and managed principally by the common offspring of Hindu and Europeans, is a new and powerful agent of reconstruction. The *Zemindars* [big landowners] and *Ryotwar* [peasant tenants] themselves, abominable as they are, involve two distinct forms of private property in land—the great desideratum of Asiatic society. From the Indian natives, reluctantly and sparingly educated at Calcutta, under English superintendence, a fresh class is springing up, endowed with the requirements for government and imbued with European science. Steam has brought India into regular and rapid communication with Europe, has connected its chief ports with those of the whole south-eastern ocean, and has revindicated it from the isolated position which was the prime law of its stagnation. The day is not far distant when, by a combination of railways and steam vessels, the distance between England and India, measured by time, will be shortened to eight days, and when that once fabulous country will thus be actually annexed to the Western world.

The ruling classes of Great Britain have had, till now, but an accidental, transitory and exceptional interest in the progress of India. The aristocracy wanted to conquer it, the moneyocracy to plunder it, and the millocracy to undersell it. But now the tables are turned. The millocracy have discovered that the transformation of India into a reproductive country has become of vital importance to them, and that, to that end, it is necessary, above all, to gift her with means of irrigation and of internal communication. They intend now drawing a net of railways over India. And they will do it. The results must be inappreciable.

It is notorious that the productive powers of India are paralysed by the utter want of means for conveying and exchanging its various produce. Nowhere, more than in India, do we meet with social destitution in the midst of natural plenty, for want of the means of exchange. It was proved before a Committee of the British House of Commons, which sat in 1848, that 'when grain was selling from 6s. to 8s. a quarter at Kandeish, it was sold at 64s. to 70s. at Poona, where the people were dying in the streets of famine, without the possibility of gaining supplies from Kandeish, because the clay-roads were impracticable'.

The introduction of railways may be easily made to subserve agricultural purposes by the formation of tanks, where ground is required for embankment, and by the conveyance of water along the different lines. Thus irrigation, the *sine qua non* of farming in the East, might be greatly extended, and the frequently recurring local famines, arising from the want of water, would be averted. The general importance of railways, viewed under this head, must become evident, when we remember that irrigated lands, even in the district near Ghauts, pay three times as much in taxes, afford ten or twelve times as much employment, and yield twelve or fifteen times as much profit, as the same area without irrigation.

Railways will afford the means of diminishing the amount and the cost of the military establishments. Col. Warren, Town Major of the Fort St. William, stated before a Select Committee of the House of Commons:

'The practicability of receiving intelligence from distant parts of the country in as many hours as at present it requires days and even weeks, and of sending instructions with troops and stores, in the more brief period, are considerations which cannot be too highly estimated. Troops could be kept at more distant and healthier stations than at present, and much loss of life from sickness would by this means be spared. Stores could not to the same extent be required at the various depots, and the loss by decay, and the destruction incidental to the climate, would also be avoided. The number of troops might be diminished in direct proportion to their effectiveness.'

We know that the municipal organization and the economical basis of the village communities has been broken up, but their worst feature, the dissolution of society into stereotype and disconnected atoms, has survived their vitality. The village isolation produced the absence of roads in India, and the absence of roads perpetuated the village isolation. On this plan a community existed with a given scale of low conveniences, almost without intercourse with other villages, without the desires and efforts indispensable to social advance. The British having broken up this self-sufficient inertia of the villages, railways will provide the new want of communication and intercourse. Besides, 'one of the effects of the railway system will be to bring into every village effected by it such knowledge of the contrivances and appliances of other countries, and such means of obtaining them, as will first put the hereditary and stipendiary village artisanship of India to full proof of its capabilities, and then supply its defects.' (Chapman, *The Cotton and Commerce of India*.)

I know that the English millocracy intend to endow India with railways with the exclusive view of extracting at diminished expenses the cotton and other raw materials for their manufactures. But when you have once introduced machinery into the locomotion of a country, which possesses iron and coals, you are unable to withhold it from its fabrication. You cannot maintain a net of railways over an immense country without introducing all those industrial processes necessary to meet the immediate and current wants of railway locomotion, and out of which there must grow the application of machinery to those branches of industry not immediately connected with railways. The railway-system will therefore become in India, truly the forerunner of modern industry. This is the more certain as the Hindus are allowed by British authorities themselves to possess particular aptitude for accommodating themselves to entirely new labour, and acquiring the requisite knowledge of machinery. Ample proof of this fact is afforded by the capacities and expertness of the native engineers in the Calcutta mint, where they have been for years employed in working the steam machinery, by the natives attached to the several steam engines in the Hurdwar coal districts, and by other instances. Mr. Campbell himself, greatly influenced as he is by the prejudices of the East India Company, is obliged to avow 'that the great mass of the Indian people possesses a great industrial energy, is well fitted to accumulate capital, and remarkable for a mathematical clearness of head, and talent for figures and exact sciences'. 'Their intellects', he says, 'are excellent.' Modern industry, resulting from the railway-system, will dissolve the hereditary divisions of labour, upon which rest the Indian castes, those decisive impediments to Indian progress and Indian power.

All the English bourgeoisie may be forced to do will neither emancipate nor materially mend the social condition of the mass of the people, depending not only on the development of the productive powers, but on their appropriation by the people. But what they will not fail to do is to lay down the material premises for both. Has the bourgeoisie ever done more? Has it ever effected a progress without dragging individuals and peoples through blood and dirt, through misery and degradation?

The Indians will not reap the fruits of the new elements of society scattered among them by the British bourgeoisie, till in Great Britain itself the now ruling classes shall have been supplanted by the industrial proletariat, or till the Hindus themselves shall have grown strong enough to throw off the English yoke altogether. At all events, we may safely expect to see, at a more or less remote period, the regeneration of that great and interesting country, whose gentle natives are, to use the expression of Prince Soltykov, even in the most inferior classes, '*plus fins et plus adroits que les Italiens*' [more subtle and adroit than the Italians], whose submission even is counterbalanced by a certain calm nobility, who, notwithstanding their natural languor, have astonished the British officers by their bravery, whose country has been the source of our languages, our religions, and who represent the type of the ancient German in the *Jat* and the type of the ancient Greek in the Brahmin.

I cannot part with the subject of India without some concluding remarks.

The profound hypocrisy and inherent barbarism of bourgeois civilization lies unveiled before our eyes, turning from its home, where it assumes respectable forms, to the colonies, where it goes naked. They are the defenders of property, but did any revolutionary party ever originate agrarian revolutions like those in Bengal, in Madras, and in Bombay? Did they not, in India, to borrow an expression of that great robber, Lord Clive himself, resort to atrocious extortion, when simple corruption could not keep pace with their rapacity? While they prated in Europe about the inviolable sanctity of the national debt, did they not confiscate in India the dividends of the *rayahs*, who had invested their private savings in the Company's own funds? While they combatted the French revolution under the pretext of defending 'our holy religion', did they not forbid, at the same time, Christianity to be propagated in India, and did they not, in order to make money out of the pilgrims streaming to the temples of Orissa and Bengal, take up the trade in the murder and prostitution perpetrated in the temple of Juggernaut? These are the men of 'Property, Order, Family, and Religion'.

The devastating effects of English industry, when contemplated with regard to India, a country as vast as Europe, and containing 150 millions of acres, are palpable and confounding. But we must not forget that they are only the organic results of the whole system of production as it is now constituted. That production rests on the supreme rule of capital. The centralization of capital is essential to the existence of capital as an independent power. The destructive influence of that centralization upon the markets of the world does but reveal, in the most gigantic dimensions, the inherent organic laws of political economy now at work in every civilized town. The bourgeois period of history has to create the material basis of the new world—on the one hand the universal intercourse founded upon the mutual dependency of mankind, and the means of that intercourse; on the other hand the development of the productive powers of man and the transformation of material production into a scientific domination of natural agencies. Bourgeois industry and commerce create these material conditions of a new world in the same way as geological revolutions have created the surface of the earth. When a great social revolution shall have mastered the results of the bourgeois epoch, the market of the world and the modern powers of production, and subjected them to the common control of the most advanced peoples, then only will human progress cease to resemble that hideous pagan idol, who would not drink the nectar but from the skulls of the slain.

BIBLIOGRAPHY

Originals: *New York Daily Tribune*, 21 and 25 Aug. 1852 and 8 Aug. 1853.

Editions: H. Christman, ed., *The American Journalism of Marx and Engels*, New York, 1966.
S. Avineri, ed., *Karl Marx on Colonialism and Modernization*, New York, 1968.
D. Fernbach, ed., *K. Marx: Surveys from Exile*, London, 1973.
T. Ferguson and S. O'Neil, eds., *K. Marx and F. Engels: The Collected Writings in the New York Daily Tribune*, New York, 1974.
Collections of Marx, *On Britain, On the Civil War in the United States*, etc. etc.

Commentary: Introductions to all editions cited above. On the last extract, see further:

E. Hobsbawm, Introduction to K. Marx, *Precapitalist Economic Formations*, New York and London, 1964.

L. Krader, *The Asiatic Mode of Production*, Assen, 1975.

G. Lichtheim, *Marx and the Asiatic Mode of Production*, *St. Anthony's Papers*, Vol. 14, Oxford, 1963.

27. Speech on the Anniversary of the *People's Paper*

This short speech was delivered in London on 14 April 1856 at a meeting to mark the fourth anniversary of the foundation of the Chartist *People's Paper*. Its theme is Marx's unshakeable conviction in the inevitability of revolution.

. . . The so-called Revolutions of 1848 were but poor incidents—small fractures and fissures in the dry crust of European society. However, they denounced the abyss. Beneath the apparently solid surface, they betrayed oceans of liquid matter, only needing expansion to rend into fragments continents of hard rock. Noisily and confusedly they proclaimed the emancipation of the Proletarian, i.e., the secret of the nineteenth century, and of the revolution of that century. That social revolution, it is true, was no novelty invented in 1848. Steam, electricity, and the self-acting mule were revolutionists of a rather more dangerous character than even citizens Barbès, Raspail, and Blanqui. But, although the atmosphere in which we live weighs upon every one with a 20,000 lb force, do you feel it? No more than European society before 1848 felt the revolutionary atmosphere enveloping and pressing it from all sides. There is one great fact, characteristic of this our nineteenth century, a fact which no party dares deny. On the one hand, there have started into life industrial and scientific forces which no epoch of the former human history had ever suspected. On the other hand, there exist symptoms of decay far surpassing the horrors recorded of the latter times of the Roman empire. In our days everything seems pregnant with its contrary. Machinery, gifted with the wonderful power of shortening and fructifying human labour, we behold starving and overworking it. The new-fangled sources of wealth, by some strange weird spell, are turned into sources of want. The victories of art seem bought by the loss of character. At the same pace that mankind masters nature, man seems to become enslaved to other men or to his own infamy. Even the pure light of science seems unable to shine but on the dark background of ignorance. All our invention and progress seem to result in endowing material forces with intellectual life, and in stultifying human life into a material force. This antagonism between modern industry and science on the one hand, modern misery and dissolution on the other hand; this antagonism between the productive powers, and the social relations of our epoch is a fact, palpable, overwhelming, and not to be controverted. Some parties may wail over it; others may wish to get rid of modern arts, in order to get rid of modern conflicts. Or they may imagine that so signal a progress in industry wants to be completed by as signal a regress in politics. On our part, we do not mistake the shape of the shrewd spirit that continues to mark all these contradictions. We know that to work well, the new-fangled forces of society only want to be mastered by new-fangled men—and such are the working men. They are as much the invention of modern time as machinery itself. In the signs that bewilder the middle class, the aristocracy and the poor prophets of regression, we do recognize our brave friend, Robin Goodfellow, the old mole that can work in the earth so fast, that worthy pioneer—the Revolution. The English working men are the first-born sons of modern industry. They will, then, certainly not be the last in aiding the social revolution produced by that industry, a revolution which means the emancipation of their own class all over the world, which is as universal as capital-rule and wages-slavery. I know the heroic struggles the English working class have gone through since the middle of the last century—struggles less glorious because they are shrouded in obscurity and burked by the middle-class historian. To revenge the misdeeds of the ruling class, there existed in the middle ages in Germany a secret tribunal, called the '*Vehmgericht*'. If a red cross was seen marked on a house, people knew that its owner was doomed by the '*Vehm*'. All the houses of Europe are now marked with the mysterious red cross. History is the judge—its executioner, the proletarian.

BIBLIOGRAPHY

Original: *The People's Paper*, 19 Apr. 1856.

28. Letters 1848–1857

Prospects for Revolution in Europe

Marx to Weydemeyer, 19 December 1849

... At present the most important movement is probably taking place here in England. There is on the one hand the agitation of the protectionists, supported by the fanaticized rural population—the consequences of the free corn trade are now beginning to manifest themselves in a form I predicted years ago. On the other hand there are the free traders, who as financial and parliamentary reformers have drawn the political and economic conclusions from their system in domestic affairs and as the peace party have drawn them in the sphere of foreign relations; finally, there are the Chartists who have joined forces with the bourgeoisie against the aristocracy while at the same time they have energetically resumed their own struggle against the bourgeoisie. The conflict of these parties will be impressive, and agitation will assume a stormier revolutionary form if, as I hope and not without good reasons, a Tory government replaces that of the Whigs. Another event, which is not yet evident on the continent, is the approach of an enormous industrial, agricultural, and commercial crisis. If the continent postpones its revolution until after the start of this crisis, it is possible that from the outset Britain will have to be an ally, even though an unpopular one, of the revolutionary continent. An earlier outbreak of the revolution—unless it is brought about by direct Russian intervention—would, in my opinion, be a misfortune; for just now, when trade is continuously expanding, neither the working masses in France, Germany, etc., nor the whole strata of shop-keepers, etc., are really in a revolutionary frame of mind, although they may utter revolutionary phrases. ...

Marx to Engels, 5 March 1856

... I promised, of course, that if circumstances permitted we would come to see the Rhenish workers; that any rising on their own, without initiative in Paris, Vienna, or Berlin, would be senseless; that if Paris does give the signal, it would be well to risk everything in any event, for then even a momentary defeat could have bad consequences only for the moment; that I would seriously consult my friends on the question of what could be done directly by the working-class population of the Rhine province itself, and that after a while they should send someone to London again, but should do nothing without previous arrangement.

Marx to Engels, 16 April 1856

... I fully agree with you about the Rhine province. The fatal thing for us is that I see something looming in the future which will smack of 'treason to the fatherland'. It will depend very much on the turn of things in Berlin whether we are not forced into a position similar to that of the Mayence Clubbists [supporters of the invading French armies in the early 1790s] in the old revolution. That would be hard. We who are so enlightened about our worthy brothers on the other side of the Rhine! The whole thing in Germany will depend on the possibility of backing the proletarian revolution by some second edition of the Peasant War. Then the affair will be splendid. ...

Marx to Engels, 8 October 1858

... We cannot deny that bourgeois society has experienced its sixteenth century a second time—a sixteenth century which will, I hope, sound the death-knell of bourgeois society just as the first one thrust it into existence. The specific task of bourgeois society is the establishment of a world market, at least in outline, and of production based upon this world market. As the world is round, this seems to have been completed by the colonization of California and Australia and the opening up of China and Japan. The difficult question for us is this: on the Continent the revolution is imminent and will immediately assume a socialist character. Is it not bound to be crushed in this little corner, considering that in a far greater territory the movement of bourgeois society is still in the ascendant? ...

Class Struggle and the Dictatorship of the Proletariat

Marx to Weydemeyer, 5 March 1852

... And now as to myself, no credit is due to me for discovering the existence of classes in modern society or the struggle between them. Long before me bourgeois historians had described the historical development of this class struggle and bourgeois economists the economic anatomy of the classes. What I did that was new was to prove: (1) that the existence of classes is only bound up with particular historical phases in the development of production, (2) that the class struggle necessarily leads to the dictatorship of the proletariat, (3) that this dictatorship itself only constitutes the transition to the abolition of all classes and to a classless society. ...

The Army and Historical Materialism

Marx to Engels, 25 September 1857

... The history of the army brings out more clearly than anything else the correctness of our conception of the connection between the productive forces and social relations. In general, the army is important for economic development. For instance, it was in the army that the ancients first fully developed a wage system. Similarly among the Romans the peculium castrense [private property of a son acquired through military service] was the first legal form in which the right of others than fathers of families to moveable property was recognized. So also the guild system among the corporation of fabri [joiners]. Here too the first use of machinery on a large scale. Even the special value of metals and their use as money appears to have been originally based—as soon as Grimm's stone age was passed—on their military significance. The division of labour within one banch was also first carried out in the armies. The whole history of the forms of bourgeois society is very strikingly epitomized here. If some day you can find time you must work the thing out from this point of view. . . .

NOTE: all extracts taken from *MESC*. Reproduced by kind permission of Lawrence and Wishart Ltd.

IV. The 'Economics'
1857–1867

29. *Grundrisse*

The *Grundrisse* is a very long manuscript written by Marx in the years 1857–8, which remained unpublished until 1941 and even then was virtually inaccessible until 1953. It was written in a hurry and covers in more or less detail the six parts into which Marx intended to divide his 'Economics'. Of these six parts the only one to be finished was *Capital*, which itself eventually grew into four volumes. Thus the *Grundrisse* is much more wide-ranging than *Capital*. But it is also more difficult to read because there is no coherent thread running through the manuscript in which he wrote everything down 'topsy-turvy', as he himself said. It also contains speculative elements that Marx would probably have ironed out in a published version.

The *Grundrisse* remains in many ways the most central of Marx's works. It is of wider scope than any later writings and takes up themes from the earlier works, in particular the *1844 Manuscripts*. The ideas of alienation, man as a social being, the dialectical categories of Hegel, and 'communist man' as the aim of history reappear here, though mediated through a profounder study of history and economics than was available to Marx in 1844.

The selections below fall into two parts: the first reproduces in its entirety the 'General Introduction' to the *Grundrisse*. Here Marx justifies his starting-point as 'the socially determined production of individuals' as opposed to the extra-societal individual of the eighteenth century. Marx then discusses the interrelationship of production, consumption, distribution, and exchange. In the third section he seeks to establish that the correct method of discussing economics is to start from simple theoretical concepts like value and labour and then proceed from them to the more complex but observable entities such as population or classes. The Introduction finishes with an incomplete discussion of the application of historical materialism to art. The rest of the selections are taken from the main body of the *Grundrisse* and deal with such diverse topics as the difference between feudalism and capitalism, the preconditions for a revolutionary crisis, automation, the abolition of the division of labour, and the nature of communist society.

General Introduction

1. *Production*

The subject of our discussion is first of all *material* production. Individuals producing in society, thus the socially determined production of individuals, naturally constitutes the starting-point. The individual and isolated hunter or fisher who forms the starting-point with Smith and Ricardo belongs to the insipid illusions of the eighteenth century. They are Robinson Crusoe stories which do not by any means represent, as students of the history of civilization imagine, a reaction against over-refinement and a return to a misunderstood natural life. They are no more based on such a naturalism than is Rousseau's *contrat social* which makes naturally independent individuals come in contact and have mutual intercourse by contract. They are the fiction and only the aesthetic fiction of the small and great adventure stories. They are, rather, the anticipation of 'civil society', which had been in the course of development since the sixteenth century and made gigantic strides towards maturity in the eighteenth. In this society of free competition the individual appears free from the bonds of nature, etc., which in former epochs of history made him part of a definite, limited human conglomeration. To the prophets of the eighteenth century, on whose shoulders Smith and Ricardo are still standing, this eighteenth-century individual, constituting the joint product of the dissolution of the feudal form of society and of the new forces of production which had developed since the sixteenth century, appears as an ideal whose existence belongs to the past; not as a result of history, but as its starting-point. Since that individual appeared to be in conformity with nature and corresponded to their conception of human nature, he was regarded not as developing historically, but as posited by nature. This illusion has been characteristic of every new epoch in the past. Steuart, who, as an aristocrat, stood more firmly on historical ground and was in many respects opposed to the spirit of the eighteenth century, escaped this simplicity of view.

The farther back we go into history, the more the individual and, therefore, the producing individual seems to depend on and belong to a larger whole: at first it is, quite naturally, the family and the clan, which is but an enlarged family; later on, it is the community growing up in its different forms out of the clash and the amalgamation of clans. It is only in the eighteenth century, in 'civil society', that the different forms of social union confront the individual as a mere means to his private ends, as an external necessity. But the period in which this standpoint—that of the isolated individual—became prevalent is the very one in which the social relations of society (universal relations according to that standpoint) have reached the highest state of development. Man is in the most literal sense of the word a *zoon politikon*, not only a social animal, but an animal which can develop into an individual only in society. Production by isolated individuals outside society—something which might happen as an exception to a civilized man who by accident got into the wilderness and already potentially possessed within himself the forces of society—is as great an absurdity as the idea of the development of language without individuals living together and talking to one another. We need not dwell on this any longer. It would not be necessary to touch upon this point at all, had not this nonsense—which, however, was justified and made sense in the eighteenth century—been transplanted, in all seriousness, into the field of political economy by Bastiat, Carey, Proudhon, and others. Proudhon and others naturally find it very pleasant, when they do not know the historical origin of a certain economic phenomenon, to give it a quasi-historico-philosophical explanation by going into mythology. Adam or Prometheus hit upon the scheme cut and dried, whereupon it was adopted, etc. Nothing is more tediously dry than the dreaming platitude.

Whenever wer speak, therefore, of production, we always have in mind production at a certain stage of social development, or production by social individuals. Hence, it might seem that in order to speak of production at all, we must either trace the historical process of development through its various phases, or declare at the outset that we are dealing with a certain historical period, as, for example, with modern capitalist production, which, as a matter of fact, constitutes the proper subject of this work. But all stages of production have certain landmarks in common, common purposes. 'Production in general' is an abstraction, but it is a rational abstraction, in so far as it singles out and fixes the common features, thereby saving us repetition. Yet these general or common features discovered by comparison constitute something very complex, whose constituent elements have different destinations. Some of these elements belong to all epochs, others are common to a few. Some of them are common to the most modern as well as to the most ancient epochs. No production is conceivable without them; but while even the most completely developed languages have laws and conditions in common with the least developed ones, what is characteristic of their development are the points of departure from the general and common. The conditions which generally govern production must be differentiated in order that the essential points of difference should not be lost sight of in view of the general uniformity which is due to the fact that the subject, mankind, and the object, nature, remain the same. The failure to remember this one fact is the source of all the wisdom of modern economists who are trying to prove the eternal nature and harmony of existing social conditions. Thus they say, for example, that no production is possible without some instrument of production, let that instrument be only the hand; that none is possible without past accumulated labour, even if that labour should consist of mere skill which has been accumulated and concentrated in the hand of the savage by repeated exercise. Capital is, among other things, also an instrument of production, also past impersonal labour. Hence capital is a universal, eternal, natural phenomenon; which is true if we disregard the specific properties which turn an 'instrument of production' and 'stored up labour' into capital. The entire history of the relationships of production appears to a man like Carey, for example, as a malicious perversion on the part of governments.

If there is no production in general there is also no general production. Production is always either some special branch of production, as, for example, agriculture, stock-raising, manufactures, etc., or an aggregate. But political economy is not technology. The connection between the general determinations of productions at a given stage of social development and the particular forms of production is to be developed elsewhere (later on).

Finally, production is never only of a particular kind. It is always a certain social body or a social subject that is engaged on a larger or smaller aggregate of branches of production. The connection between the real process and its scientific presentation also falls outside of the scope of this treatise. Production in general. Special branches of production. Production as a whole.

It is the fashion with economists to open their works with a general introduction, which is entitled 'production' (see, for example, John Stuart Mill) and deals with the general 'requisites of production'. This general introductory part consists of (or is supposed to consist of):

1. The conditions without which production is impossible, i.e. the essential conditions of all production. As a matter of fact, however, it can be reduced, as we shall see, to a few very simple definitions, which flatten out into shallow tautologies.

2. Conditions which further production more or less, as, for example, Adam Smith's discussion of a progressive and stagnant state of society.

In order to give scientific value to what serves with him as a mere summary, it would be necessary to study the degree of productivity by periods in the development of individual nations; such a study falls outside the scope of the present subject, and in so far as it does belong here is to be brought out in connection with the discussion of competition, accumulation, etc. The commonly accepted view of the matter gives a general answer to the effect that an industrial nation is at the height of its production at the moment when it reaches its historical climax in all respects. As a matter of fact a nation is at its industrial height so long as its main object is not gain, but the process of gaining. In that respect the Yankees stand above the English. Or, that certain races, climates, natural conditions, such as distance from the sea, fertility of the soil, etc., are more favourable to production than others. That again comes down to the tautology that the facility of creating wealth depends on the extent to which its elements are present both subjectively and objectively.

But all that is not what the economists are really concerned with in this general part. Their object is rather to represent production in contradistinction to distribution—see Mill, for example—as subject to eternal laws independent of history, and then to substitute bourgeois relations, in an underhand way, as immutable natural laws of society *in abstracto*. This is the more or less conscious aim of the entire proceeding. When it comes to distribution, on the contrary, mankind is supposed to have indulged in all sorts of arbitrary action. Quite apart from the fact that they violently break the ties which bind production and distribution together, so much must be clear from the outset: that, no matter how greatly the systems of distribution may vary at different stages of society, it should be possible here, as in the case of production, to discover the common features and to confound and eliminate all historical differences in formulating general human laws. For example, the slave, the serf, the wage-labourer—all receive a quantity of food, which enables them to exist as slave, serf, and wage-labourer. The conqueror, the official, the landlord, the monk or the Levite, who respectively live on tribute, taxes, rent, alms, and the tithe—all receive a part of the social product which is determined by laws different from those which determine the part received by the slave, etc. The two main points which all economists place under this head are, first, property; secondly, the protection of the latter by the administration of justice, police, etc. The objections to these two points can be stated very briefly.

1. All production is appropriation of nature by the individual within and through a definite form of society. In that sense it is a tautology to say that property (appropriation) is a condition of production. But it becomes ridiculous, when from that one jumps at once to a definite form of property, e.g. private property (which implies, besides, as a prerequisite the existence of an opposite form, viz. absence of property). History points rather to common property (e.g. among the Hindus, Slavs, ancient Celts, etc.) as the primitive form, which still plays an important part at a much later period as communal property. The question as to whether wealth grows more rapidly under this or that form of property is not even raised here as yet. But that there can be no such thing as production, nor, consequently, society, where property does not exist in any form, is a tautology. Appropriation which does not appropriate is a contradictio in subjecto.

2. Protection of gain, etc. Reduced to their real meaning, these commonplaces express more than their preachers know, namely, that every form of production creates its own legal relations, forms of government, etc. The crudity and the shortcomings of the conception lie in the tendency to see only an accidental reflective connection in what constitutes an organic union. The bourgeois economists have a vague notion that production is better carried on under the modern police than it was, for example, under club law. They forget that club law is also law, and that the right of the stronger continues to exist in other forms even under their 'government of law'.

When the social conditions corresponding to a certain stage of production are in a state of formation or disappearance, disturbances of production naturally arise, although differing in extent and effect.

To sum up: all the stages of production have certain destinations in common, which we generalize in thought: but the so-called general conditions of all production are nothing but abstract conceptions which do not go to make up any real stage in the history of production.

2. *The General Relation of Production to Distribution, Exchange, and Consumption*

Before going into a further analysis of production, it is necessary to look at the various divisions which economists put side by side with it. The most shallow conception is as follows: by production, the members of society appropriate (produce and shape) the products of nature to human wants; distribution determines the proportion in which the individual participates in this production; exchange brings him the particular products into which he wishes to turn the quantity secured by him through distribution; finally, through consumption the products become objects of use and enjoyment, of individual appropriation. Production yields goods adapted to our needs; distribution distributes them according to social laws; exchange distributes further what has already been distributed, according to individual wants; finally, in consumption the product drops out of the social movement, becoming the direct object of the individual want which it serves and satisfies in use. Production thus appears as the starting-point; consumption as the final end; and distribution and exchange as the middle; the latter has a double aspect, distribution being defined as a process carried on by society, exchange as one proceeding from the individual. The person is objectified in production; the material thing is subjectified in the person. In distribution, society assumes the part of go-between for production and consumption in the form of generally prevailing rules; in exchange this is accomplished by the accidental make-up of the individual.

Distribution determines what proportion (quantity) of the products the individual is to receive; exchange determines the products in which the individual desires to receive his share allotted to him by distribution.

Production, distribution, exchange, and consumption thus form a perfect connection, production standing for the general, distribution and exchange for the special, and consumption for the individual, in which all are joined together. To be sure this is a connection, but it does not go very deep. Production is determined according to the economists by universal natural laws, while distribution depends on social chance: distribution can, there-fore, have a more or less stimulating effect on production: exchange lies between the two as a formal social movement, and the final act of consump-tion, which is considered not only as a final purpose but also as a final aim, falls properly outside the scope of economics, except in so far as it reacts on the starting-point and causes the entire process to begin all over again.

The opponents of the economists—whether economists themselves or not—who reproach them with tearing apart, like barbarians, what is an organic whole, either stand on common ground with them or are below them. Nothing is more common than the charge that the economists have been considering production as an end in itself, too much to the exclusion of everything else. The same has been said with regard to distribution. This accusation is itself based on the economic conception that distribution exists side by side with production as a self-contained, independent sphere. Or, it is said, the various factors are not grasped in their unity. As though it were the textbooks that impress this separation upon life and not life upon the textbooks; and as though the subject at issue were a dialectical balancing of conceptions and not an analysis of real conditions.

Exchange and Circulation. The result we arrive at is not that production, distribution, exchange, and consumption are identical, but that they are all members of one entity, different aspects of one unit. Production predominates not only over production itself in the opposite sense of that term, but over the other elements as well. With production the process constantly starts over again. That exchange and consumption cannot be the predominating elements is self-evident. The same is true of distribution in the narrow sense of distribution of products; as for distribution in the sense of distribution of the agents of production, it is itself but a factor of production. A definite form of production thus determines the forms of consumption, distribution, exchange, and also the mutual relations between these various elements. Of course, production in its one-sided form is in its turn influenced by other elements, i.e. with the expansion of the market, i.e. of the sphere of exchange, production grows in volume and is subdivided to a greater extent.

With a change in distribution, production undergoes a change; as for example in the case of concentration of capital, of a change in the distribution of population in city and country, etc. Finally the demands of consumption also influence production. A mutual interaction takes place between the various elements. Such is the case with every organic body.

3. *The Method of Political Economy*

When we consider a given country from a politico-economic stand-point, we begin with its population, its subdivision into classes, location in city, country, or by the sea, occupation in different branches of production; then we study its exports and imports, annual production and consumption, prices of commodities, etc. It seems to be the correct procedure to commence with the real and the concrete, the actual prerequisites; in the case of political economy, to commence with population, which is the basis and the author of the entire productive activity of society. Yet on closer consideration it proves to be wrong. Population is an abstraction, if we leave out for example the classes of which it consists. These classes, again, are but an empty word unless we know what are the elements on which they are based, such as wage-labour, capital, etc. These imply, in their turn, exchange, division of labour, prices, etc. Capital, for example, does not mean anything without wage-labour, value, money, price, etc. If we start out, therefore, with population, we do so with a chaotic conception of the whole, and by closer analysis we will gradually arrive at simpler ideas; thus we shall proceed from the imaginary concrete to less and less complex abstractions, until we arrive at the simplest determinations. This once attained, we might start on our return journey until we finally came back to population, but this time not as a chaotic notion of an integral whole, but as a rich aggregate of many determinations and relations. The former method is the one which political economy had adopted in the past as its inception. The economists of the seventeenth century, for example, always started out with the living aggregate: population, nation, state, several states, etc., but in the end they invariably arrived by means of analysis at certain leading abstract general principles such as division of labour, money, value, etc. As soon as these separate elements had been more or less established by abstract reasoning, there arose the systems of political economy which start from simple conceptions such as labour, division of labour, demand, exchange value, and conclude with state, international exchange, and world market. The latter is manifestly the scientifically correct method. The concrete is concrete because it is a combination of many determinations, i.e. a unity of diverse elements. In our thought it therefore appears as a process of synthesis, as a result, and not as a starting-point, although it is the real starting-point and, therefore, also the starting-point of observation and conception. By the former method the complete conception passes into an abstract definition; by the latter the abstract definitions lead to the reproduction of the concrete subject in the course of reasoning. Hegel fell into the error, therefore, of considering the real as the result of self-coordinating, self-absorbed, and spontaneously operating thought, while the method of advancing from the abstract to the concrete is but the way of thinking by which the concrete is grasped and is reproduced in our mind as concrete. It is by no means, however, the process which itself generates the concrete. The simplest economic category, say, exchange value, implies the existence of population, population that is engaged in production under certain conditions; it also implies the existence of certain types of family, clan, or state, etc. It can have no other existence except as an abstract one-sided relation of an already given concrete and living aggregate.

As a category, however, exchange value leads an antediluvian existence. Thus the consciousness for which comprehending thought is what is most real in man, for which the world is only real when comprehended (and philosophical consciousness is of this nature), mistakes the movement of categories for the real act of production (which unfortunately receives only its impetus from outside), whose result is the world; that is true—here we have, however, again a tautology—in so far as the concrete aggregate, as a thought aggregate, the concrete subject of our thought, is in fact a product of thought, of comprehension; not, however, in the sense of a product of a self-emanating conception which works outside of and stands above observation and imagination, but of a conceptual working-over of observation and imagination. The whole, as it appears in our heads as a thought-aggregate, is the product of a thinking mind which grasps the world in the only way open to it, a way which differs from the one employed by the artistic, religious, or practical mind. The concrete subject continues to lead an independent existence after it has been grasped, as it did before, outside the head, so long as the head contemplates it only speculatively, theoretically. So that in the employment of the theoretical method in political economy, the subject, society, must constantly be kept in mind as the premiss from which we start.

But have these simple categories no independent historical or natural existence antedating the more concrete ones? That depends. For instance, in his *Philosophy of Right* Hegel rightly starts out with possession, as the simplest legal relation of individuals. But there is no such thing as possession before the family or the relations of lord and serf, which relations are a great deal more concrete, have come into existence. On the other hand, one would be right in saying that there are families and clans which only *possess*, but do not *own* things. The simpler category thus appears as a relation of simple family and clan communities with respect to property. In society the category appears as a simple relation of a developed organization, but the concrete substratum from which the relation of possession springs is always implied. One can imagine an isolated savage in possession of things. But in that case possession is no legal relation. It is not true that the family came as the result of the historical evolution of possession. On the contrary, the latter always implies the existence of this 'more concrete category of law'. Yet this much may be said, that the simple categories are the expression of relations in which the less developed concrete entity may have been realized without entering into the manifold relations and bearings which are mentally expressed in the concrete category; but when the concrete entity attains fuller development it will retain the same category as a subordinate relation.

Money may exist and actually had existed in history before capital or banks or wage-labour came into existence. With that in mind, it may be said that the more simple category can serve as an expression of the predominant relations of an undeveloped whole or of the subordinate relations of a more developed whole, relations which had historically existed before the whole developed in the direction expressed in the more concrete category. To this extent, the course of abstract reasoning, which ascends from the most simple to the complex, corresponds to the actual process of history.

On the other hand, it may be said that there are highly developed but historically less mature forms of society in which the highest economic forms are to be found, such as co-operation, advanced division of labour, etc., and yet there is no money in existence, e.g. Peru. In Slav communities also, money, as well as exchange to which it owes its existence, does not appear at all or very little within the separate communities, but it appears on their boundaries in their intercommunal traffic; in general, it is erroneous to consider exchange as a constituent element originating within the community. It appears at first more in the mutual relations between different communities than in those between the members of the same community. Furthermore, although money begins to play its part everywhere at an early stage, it plays an antiquity the part of a predominant element only in unidirectionally developed nations, viz. trading nations, and even in the most cultured antiquity, in Greece and Rome, it attains its full development, which constitutes the prerequisite of modern bourgeois society, only in the period of their decay. Thus this quite simple category attained its culmination in the past only at the most advanced stages of society. Even then it did not pervade all economic relations; in Rome, for example, at the time of its highest development, taxes and payments in kind remained the basis. As a matter of fact, the money system was fully developed there only so far as the army was concerned; it never came to dominate the entire system of labour. Thus, although the simple category may have existed historically before the more concrete one, it can attain its complete internal and external development only in complex forms of society, while the more concrete category has reached its full development in a less advanced form of society.

Labour is quite a simple category. The idea of labour in that sense, as labour in general, is also very old. Yet 'labour' thus simply defined by political economy is as much a modern category as the conditions which have given rise to this simple abstraction. The monetary system, for example, defines wealth quite objectively, as a thing external to itself in money. Compared with this point of view, it was a great step forward when the industrial or commercial system came to see the source of wealth not in the object but in the activity of persons, viz. in commercial and industrial labour. But even the latter was thus considered only in the limited sense of a money-producing activity. The physiocratic system marks still further progress in that it considers a certain form of labour, viz. agriculture, as the source of wealth, and wealth itself not in the disguise of money, but as a product in general, as the general result of labour. But corresponding to the limitations of the activity, this product is still only a natural product. Agriculture is productive, land is the source of production *par excellence*. It was a tremendous advance on the part of Adam Smith to throw aside all the limitations which mark wealth-producing activity and to define it as labour in general, neither industrial nor commercial nor agricultural, or one any more than the other. Along with the universal character of wealth–creating activity we now have the universal character of the object defined as wealth, viz. product in general, or labour in general, but as past, objectified labour. How difficult and how great was the transition is evident from the way Adam Smith himself falls back from time to time into the physiocratic system. Now it might seem as though this amounted simply to finding an abstract expression for the simplest relation into which men have been mutually entering as producers from times of yore, no matter under what form of society. In one sense this is true. In another it is not.

The indifference as to the particular kind of labour implies the existence of a highly developed aggregate of different species of concrete labour, none of which is any longer the predominant one. So the most general abstractions commonly arise only where there is the highest concrete development, where one feature appears to be jointly possessed by many and to be common to all. Then it cannot be thought of any longer in one particular form. On the other hand, this abstraction of labour is only the result of a concrete aggregate of different kinds of labour. The indifference to the particular kind of labour corresponds to a form of society in which individuals pass with ease from one kind of work to another, which makes it immaterial to them what particular kind of work may fall to their share. Labour has become here, not only categorially but really, a means of creating wealth in general and has no longer coalesced with the individual in one particular manner. This state of affairs has found its highest development in the most modern of bourgeois societies, the United States. It is only here that the abstraction of the category 'labour', 'labour in general', labour *sans phrase*, the starting-point of modern political economy, becomes realized in practice. Thus the simplest abstraction which modern political economy sets up as its starting-point, and which expresses a relation dating back to antiquity and prevalent under all forms of society, appears truly realized in this abstraction only as a category of the most modern society. It might be said that what appears in the United States as a historical product—viz. the indifference as to the particular kind of labour—appears among the Russians, for example, as a spontaneously natural disposition. But it makes all the difference in the world whether barbarians have a natural predisposition which makes them capable of applying themselves alike to everything, or whether civilized people apply themselves to everything. And, besides, this indifference of the Russians as to the kind of work they do corresponds to their traditional practice of remaining in the rut of a quite definite occupation until they are thrown out of it by external influences.

This example of labour strikingly shows how even the most abstract categories, in spite of their applicability to all epochs—just because of their abstract character—are by the very definiteness of the abstraction a product of historical conditions as well, and are fully applicable only to and under those conditions.

Bourgeois society is the most highly developed and most highly differentiated historical organization of production. The categories which serve as the expression of its conditions and the comprehension of its own organization enable it at the same time to gain an insight into the organization and the relationships of production which have prevailed under all the past forms of society, on the ruins and constituent elements of which it has arisen, and of which it still drags along some unsurmounted remains, while what had formerly been mere intimation has now developed to complete significance. The anatomy of the human being is the key to the anatomy of the ape. But the intimations of a higher animal in lower ones can be understood only if the animal of the higher order is already known. The bourgeois economy furnishes a key to ancient economy, etc. This is, however, by no means true of the method of those economists who blot out all historical differences and see the bourgeois form in all forms of society. One can understand the nature of tribute, tithes, etc., after one has learned the nature of rent. But they must not be considered identical.

Since, furthermore, bourgeois society is only a form resulting from the development of antagonistic elements, some relations belonging to earlier forms of society are frequently to be found in it, though in a crippled state or as a travesty of their former self, as for example communal property. While it may be said, therefore, that the categories of bourgeois economy contain what is true of all other forms of society, the statement is to be taken *cum grano salis* [with a grain of salt]. They may contain these in a developed or crippled or caricatured form, but always essentially different. The so-called historical development amounts in the last analysis to this, that the last form considers its predecessors as stages leading up to itself and always perceives them from a single point of view, since it is very seldom and only under certain conditions that it is capable of self-criticism; of course, we do not speak here of such historical periods as appear to their own contemporaries to be periods of decay. The Christian religion became capable of assisting us to an objective view of past mythologies as soon as it was ready for self-criticism to a certain extent, *dynamei*, so to speak. In the same way bourgeois political economy first came to understand the feudal, the ancient, and the oriental societies as soon as the self-criticism of bourgeois society had commenced. In as far as bourgeois political economy has not gone into the mythology of identifying the bourgeois system purely with the past, its criticism of the feudal system against which it still had to wage war resembled Christian criticism of the heathen religions or Protestant criticism of Catholicism.

In the study of economic categories, as in the case of every historical and social science, it must be borne in mind that, as in reality so in our mind, the subject, in this case modern bourgeois society, is given, and that the categories are therefore only forms of being, manifestations of existence, and frequently only one-sided aspects of this subject, this definite society; and that, expressly for that reason, the origin of political economy *as a science* does not by any means date from the time to which it is referred to *as such*. This is to be firmly kept in mind because it has an immediate and important bearing on the matter of the subdivisions of the science.

For instance, nothing seems more natural than to start with rent, with landed property, since it is bound up with land, the source of all production and all existence, and with the first form of production in all more or less settled communities, viz. agriculture. But nothing would be more erroneous. Under all forms of society there is a certain industry which predominates over all the rest and whose condition therefore determines the rank and influence of all the rest.

It is the universal light with which all the other colours are tinged and by whose peculiarity they are modified. It is a special ether which determines the specific gravity of everything that appears in it.

Let us take for example pastoral nations (mere hunting and fishing tribes are not as yet at the point from which real development commences). They engage in a certain form of agriculture, sporadically. The nature of land ownership is determined thereby. It is held in common and retains this form more or less according to the extent to which these nations hold on to traditions; such, for example, is land ownership among the Slavs. Among nations whose agriculture is carried on by a settled population—the settled state constituting a great advance—where agriculture is the predominant industry, such as in ancient and feudal societies, even the manufacturing industry and its organizations, as well as the forms of property which pertain to it, have more or less the characteristic features of the prevailing system of land ownership; society is then either entirely dependent upon agriculture, as in the case of ancient Rome, or, as in the Middle Ages, it imitates in its civic relations the forms or organization prevailing in the country. Even capital, with the exception of pure money capital, has, in the form of the traditional working tool, the characteristics of land ownership in the Middle Ages.

The reverse is true of bourgeois society. Agriculture comes to be more and more merely a branch of industry and is completely dominated by capital. The same is true of rent. In all the forms of society in which land ownership is the prevalent form, the influence of the natural element is the predominant one. In those where capital predominates, the prevailing element is the one historically created by society. Rent cannot be understood without capital, whereas capital can be understood without rent. Capital is the all-dominating economic power of bourgeois society. It must form the starting-point as well as the end and be developed before land ownership. After each has been considered separately, their mutual relation must be analysed.

It would thus be impractical and wrong to arrange the economic categories in the order in which they were the determining factors in the course of history. Their order of sequence is rather determined by the relation which they bear to one another in modern bourgeois society, and which is the exact opposite of what seems to be their natural order or the order of their historical development. What we are interested in is not the place which economic relations occupy in the historical succession of different forms of society. Still less are we interested in the order of their succession 'in the idea' (*Proudhon*), which is but a hazy conception of the course of history. We are interested in their organic connection within modern bourgeois society.

The sharp line of demarcation (abstract precision) which so clearly distinguished the trading nations of antiquity, such as the Phoenicians and the Carthaginians, was due to that very predominance of agriculture. Capital as trading or money capital appears in that abstraction where capital does not constitute as yet the predominating element of society. The Lombards and the Jews occupied the same position among the agricultural societies of the Middle Ages.

As a further illustration of the fact that the same category plays different parts at different stages of society, we may mention the following: one of the latest forms of bourgeois society, viz. joint stock companies, appears also at its beginning in the form of the great chartered monopolistic trading companies.

The concept of national wealth which is imperceptibly formed in the minds of the economists of the seventeenth century, and which in part continues to be entertained by those of the eighteenth century, is that wealth is produced solely for the state, but that the power of the latter is proportional to that wealth. It was as yet an unconsciously hypocritical way in which wealth announced itself and its own production as the aim of modern states, considering the latter merely as a means to the production of wealth.

The order of treatment must manifestly be as follows: first, the general abstract definitions which are more or less applicable to all forms of society, but in the sense indicated above. Secondly, the categories which go to make up the inner organization of bourgeois society and constitute the foundations of the principal classes: capital, wage-labour, landed property; their mutual relations; city and country; the three great social classes, the exchange between them; circulation, credit (private). Thirdly, the organization of bourgeois society in the form of the state, considered in relation to itself; the 'unproductive' classes; taxes; public debts; public credit; population; colonies; emigration. Fourthly, the international organization of production; international division of labour; international exchange; import and export; rate of exchange. Fifthly, the world market and crises.

4. *Production, Means of Production, and Conditions of Production: the Relations of Production and Distribution; the Connection between Form of State and Consciousness on the One Hand and Relations of Production and Distribution on the Other: Legal Relations: Family Relations*
Notes on the points to be mentioned here and not to be omitted:
    1. *War* attains complete development before peace; how certain economic phenomena, such as wage-labour, machinery, etc., are developed at an earlier date through war and in armies than within bourgeois society. The connection between productive force and commercial relationships is made especially plain in the case of the army.
    2. The relation between the previous idealistic methods of writing history and the realistic method; namely, the so-called history of civilization, which is all a history of religion and states. In this connection something may be said of the different methods hitherto employed in writing history. The so-called objective method. The subjective (the moral and others). The philosophical.
    3. *Secondary and tertiary*. Conditions of production which have been taken over or transplanted; in general, those that are not original. Here the effect of international relations must be introduced.

4. Objections to the materialistic character of this view. Its relation to naturalistic materialism.

5. The dialectic of the conception of productive force (means of production) and relation of production, a dialectic whose limits are to be determined and which does not do away with the concrete difference.

6. The unequal relation between the development of material production and art, for instance. In general, the conception of progress is not to be taken in the sense of the usual abstraction. In the case of art, etc., it is not so important and difficult to understand this disproportion as in that of practical social relations, e.g. the relation between education in the United States and Europe. The really difficulty point, however, that is to be discussed here is that of the unequal development of relations of production as legal relations. As, for example, the connection between Roman civil law (this is less true of criminal and public law) and modern production.

7. This conception of development appears to imply necessity. On the other hand, justification of accident. How. (Freedom and other points.) (The effect of means of communication.) World history has not always existed; history as world history is a result.

8. The starting-point is to be found in certain facts of nature embodied subjectively and objectively in clans, races, etc.

It is well known that certain periods of the highest development of art stand in no direct connection to the general development of society, or to the material basis and skeleton structure of its organization. Witness the example of the Greeks as compared with the modern nations, or even Shakespeare. As regards certain forms of art, e.g. the epos, it is admitted that they can never be produced in the universal epoch-making form as soon as art as such has come into existence; in other words, that in the domain of art certain important forms of it are possible only at a low stage of its development. If that be true of the mutual relations of different forms of art within the domain of art itself, it is far less surprising that the same is true of the relation of art as a whole to the general development of society. The difficulty lies only in the general formulation of these contradictions. No sooner are they specified than they are explained.

Let us take for instance the relation of Greek art, and that of Shakespeare's time, to our own. It is a well-known fact that Greek mythology was not only the arsenal of Greek art, but also the very ground from which it had sprung. Is the view of nature and of social relations which shaped Greek imagination and Greek art possible in the age of automatic machinery and railways and locomotives and electric telegraphs? Where does Vulcan come in as against Roberts & Co.? Jupiter, as against the lightning conductor? and Hermes, as against the *Crédit Mobilier*? All mythology masters and dominates and shapes the forces of nature in and through the imagination; hence it disappears as soon as man gains mastery over the forces of nature. What becomes of the Goddess Fama side by side with

Printing House Square? Greek art presupposes the existence of Greek mythology, i.e. that nature and even the form of society are wrought up in popular fancy in an unconsciously artistic fashion. That is its material. Not, however, any mythology taken at random, nor any accidental unconsciously artistic elaboration of nature (including under the latter all objects, hence also society). Egyptian mythology could never be the soil or womb which would give birth to Greek art. But in any event there had to be a mythology. In no event could Greek art originate in a society which excludes any mythological explanation of nature, any mythological attitude towards it, or which requires of the artist an imagination free from mythology.

Looking at it from another side: is Achilles possible side by side with powder and lead? Or is the *Iliad* at all compatible with the printing press and even printing machines? Do not singing and reciting and the muses necessarily go out of existence with the appearance of the printer's bar, and do not, therefore, the prerequisites of epic poetry disappear?

But the difficulty is not in grasping the idea that Greek art and epos are bound up with certain forms of social development. It lies rather in understanding why they still constitute for us a source of aesthetic enjoyment and in certain respects prevail as the standard and model beyond attainment.

A man cannot become a child again unless he becomes childish. But does he not enjoy the artless ways of the child, and must he not strive to reproduce its truth on a higher plane? Is not the character of every epoch revived, perfectly true to nature, in the child's nature? Why should the childhood of human society, where it had obtained its most beautiful development, not exert an eternal charm as an age that will never return? There are ill-bred children and precocious children. Many of the ancient nations belong to the latter class. The Greeks were normal children. The charm their art has for us does not conflict with the primitive character of the social order from which it had sprung. It is rather the product of the latter, and is due rather to the fact that the immature social conditions under which the art arose and under which alone it could appear can never return.

The Social Character of Production

Considered in the act of production itself, the labour of the individual is used by him as money to buy the product directly, that is, the object of his own activity; but it is particular money, used to buy this particular product. In order to be money in general, it must originate from general and not special labour; that is, it must originally be established as an element of general production. But on this presupposition it is not basically exchange that gives it its general character, but its presupposed social character will determine its participation in the products. The social character of production would make the product from the start a collective and general product.

The exchange originally found in production—which is an exchange not of exchange values but of activities determined by communal needs and communal aims—would from the start imply the participation of individuals in the collective world of products.

On the basis of exchange values, it is exchange that first makes of labour something general. In the other system labour is established as such before the exchange; that is, the exchange of products is not at all the medium by which participation of the individual in general production is brought about. There must of course be mediation. In the first case, we start with the autonomous production of private individuals (however much it is determined and modified subsequently by complex relationships), and mediation is carried out by the exchange of goods, exchange value and money, which are all expressions of one and the same relationship. In the second case, the presupposition itself is mediated, i.e. the precondition is collective production; the community is the foundation of production. The labour of the individual is established from the start as collective labour. But whatever the particular form of the product which he creates or helps to create, what he has bought with his labour is not this or that product, but a definite participation in collective production. Therefore he has no special product to exchange. His product is not an exchange value. The product does not have to change into any special form in order to have a general character for the individual. Instead of a division of labour necessarily engendered by the exchange of values, there is an organization of labour, which has as its consequence the participation of the individual in collective consumption.

In the first case, the social character of producion is established subsequently by the elevation of products to exchange values, and the exchange of these values. In the second case, the social character of production is a precondition, and participation in the world of production and in consumption is not brought about by the exchange of labour or the products of labour which are independent of it. It is brought about by the social conditions of production, within which the individual acts.

Thus the desire to turn individual labour directly into money (which also includes the product of labour), i.e. into a realized exchange value, means that the worker's labour must be designated as general labour. In other words, this means that those conditions are denied in which he must necessarily become money and exchange value, and is dependent on private exchange. This requirement can be satisfied only in conditions in which it is no longer set. On the basis of exchange values, neither the labour of the individual nor his product are directly general; to obtain this character, an objective mediation is required, money distinct from the product.

If we suppose communal production, the determination of time remains, of course, essential. The less time society requires in order to produce wheat, cattle, etc., the more time it gains for other forms of production, material or intellectual. As with a single individual, the universality of its development, its enjoyment, and its activity depends on saving time. In the final analysis, all forms of economics can be reduced to an economics of time. Likewise, society must divide up its time purposefully in order to achieve a production suited to its general needs; just as the individual has to divide his time in order to acquire, in suitable proportions, the knowledge he needs or to fulfil the various requirements of his activity.

On the basis of community production, the first economic law thus remains the economy of time, and the methodical distribution of working time between the various branches of production; and this law becomes indeed of much greater importance. But all this differs basically from the measurement of exchange values (labour and the products of labour) by labour time. The work of individuals participating in the same branch of activity, and the different kinds of labour are not only quantitatively but also qualitatively different. What is the precondition of a merely quantitative difference between things? The fact that their quality is the same. Thus units of labour can be measured quantitatively only if they are of equal and identical quality.

The Rise and Downfall of Capitalism

It is necessary to produce a precise analysis of the concept of capital, since it is the basic concept of modern economics just as capital itself, which is its abstract reflection, is the basis of bourgeois society. From a clear perception of the basic premiss of the relationship all the contradictions of bourgeois production must emerge as also the limit at which it progresses beyond itself.

It is important to notice that wealth as such, i.e. bourgeois wealth, is always most powerfully expressed in exchange value where it is established as a mediator between the extremes itself and use value. This mid-point always appears as the completed economic relationship because it unites the opposites and always appears in the end as a unilateral and superior power over and against the extremes. The movement or relationship that originally appears as a mediator between the extremes progresses in a necessary dialectic to the following result: it appears as its own mediator, as the subject whose moments are merely the extremes, whose status as independent preconditions it abolishes in order thus to establish itself as the only autonomous factor. Similarly, in the religious sphere, Christ the mediator between God and man—simple means of circulation between one and the other—becomes their unity, the God-man and then, as such, becomes more important than God; the saints become more important than Christ; and the priests more important than the saints.

The total economic expression, itself unilateral in relation to the extremes, is always exchange value when it is established as an intermediate link; as, for example, gold in simple circulation, or capital itself as a mediator between production and circulation. Inside capital itself one of its forms adopts the position of exchange value against another which is use value. Thus, for example, industrial capital appears as a producer in relation to the trader who represents circulation. Thus industrial capital represents the material side, and the other the formal side, wealth as such. At the same time, mercantile capital is itself again mediator between production (industrial capital) and circulation (the consuming public) or between exchange value and use value. For both sides mutually establish each other: production as money and circulation as use value (consuming public), the first as use value (product) and the second as exchange value (money). It is the same inside trade itself: the wholesaler, as mediator between producer and retailer or between producer and agriculturalist or between different producers, plays the same role of superior centre. It is also the case of the middleman in relation to the wholesaler; of the banker in relation to industrialists and businessmen; of the joint stock company in relation to simple production, and of the financier as mediator between the state and bourgeois society at its highest stage. Wealth as such presents itself all the more distinctly and broadly the farther it is removed from direct production and itself acts as mediator between sides which, considered in themselves, are already forms of economic relationship. Note that money becomes an end instead of a means and that capital, as the superior form of mediation, everywhere establishes the inferior form, labour, simply as a source of surplus value. For example, the bill-broker, banker etc., in relation to producers and farmers who are established relative to the former as labour (use value), whereas the banker situates himself in relation to them as capital, creation of surplus value etc. This appears in its weirdest form with financiers.

Capital is the immediate unity of product and money or, better, of production and circulation. Thus it is itself something immediate and its development consists in establishing itself and going beyond itself on this unity which is a determined and therefore a simple relationship. At first this unity appears in capital as something simple. . . .

Thus on the one hand production which is founded on capital creates universal industry—i.e. surplus labour, value-producing labour; on the other hand it creates a system of general exploitation of natural human attributes, a system of general profitability, whose vehicles seem to be just as much science as all the physical and intellectual characteristics. There is nothing which can escape, by its own elevated nature or self-justifying characteristics, from this cycle of social production and exchange. Thus capital first creates bourgeois society and the universal appropriation of nature and of social relationships themselves by the members of society. Hence the great civilizing influence of capital, its production of a stage of society compared with which all earlier stages appear to be merely local progress and idolatory of nature. Nature becomes for the first time simply an object for mankind, purely a matter of utility; it ceases to be recognized as a power in its own right; and the theoretical knowledge of its independent laws appears only as a stratagem designed to subdue it to human requirements, whether as the object of consumption or as the means of production. Pursuing this tendency, capital has pushed beyond national boundaries and prejudices, beyond the deification of nature and the inherited, self-sufficient satisfaction of existing needs confined within well-defined bounds, and the reproduction of the traditional way of life. It is destructive of all this, and permanently revolutionary, tearing down all obstacles that impede the development of productive forces, the expansion of needs, the diversity of production and the exploitation and exchange of natural and intellectual forces.

But because capital sets up any such boundary as a limitation, and is thus ideally over and beyond it, it does not in any way follow that it has really surmounted it, and since any such limitation contradicts its vocation, capitalist production moves in contradictions which are constantly overcome, only to be, again, constantly re-established. Still more so. The universality towards which it is perpetually driving finds limitations in its own nature, which at a certain stage of its development will make it appear as itself the greatest barrier to this tendency, leading thus to its own self-destruction.

To come closer to the heart of the question: first, there is a limit not inherent to production generally, but to production founded on capital. This limit is dual, or rather revealed one and the same when considered from two angles. It is sufficient here, in order to have discovered the basis of over-production—the fundamental contradiction of developed capitalism—to show that capital contains a particular limitation on production, a limitation which contradicts its general tendency to push beyond any barrier on its production. This is the discovery in general that capital is not, as the economists think, the absolute form for the development of productive forces, that is, not the absolute form of wealth coinciding absolutely with the development of productive forces. Viewed from the standpoint of capital, the stages of production that preceded it appear as so many fetters on the productive forces. But, correctly understood, capital itself appears as a condition for the development of productive forces so long as they need an external stimulant which is at the same time a brake. Capital disciplines productive forces, but becomes a superfluous burden at a certain stage of their development, in exactly the same way as the corporations were. These immanent limits must be identical with the nature of capital, with the essential determinations of its concept. These necessary limitations are:

1. Necessary labour as a limit to the exchange value of living labour power or of the salary of the industrial population.

2. Surplus value as a limit to surplus working time; and, in relation to relative surplus working time as a barrier to the development of productive forces.

3. What amounts to the same thing, transformation into money—ex-

change value in general—as a limit on production; or exchange based on value, or value based on exchange, as a limit on production.

4. Still the same point viewed as a limitation of the production of use values through exchange value; in other words, in order to become in general an object of production real wealth must adopt a determinate form, different from itself and thus absolutely not identical with it.

Hence over-production, which is the sudden recall of all these necessary moments of production based on capital; their neglect therefore brings a general depreciation in value. At the same time capital has the task imposed on it of beginning from a higher stage of the development of productive forces, etc. and renewing its search with a consequent ever greater collapse as capital. It is therefore clear that the higher the development of capital, the more it appears as a barrier to production, and therefore also to consumption, quite apart from the other contradictions which make it appear as a burdensome barrier to production and commerce. . . .

Alienated Labour

The additional value is thus again established as capital, as objectified labour entering into the exchange process with living labour, and thence dividing itself into a constant part—the objective conditions of labour, the existence of living labour power, the necessaries, food for the worker. In this second appearance of capital in this form, some points are cleared up which in its first appearance—as money, which is changing from the form of value into that of capital—were completely obscure. They are now solved through the process of valorization and production. At their first occurrence, the prerequisites themselves seemed to be exterior and derived from circulation; thus they did not arise from its internal nature, nor were they explained by it. These external prerequisites will now appear as elements in the movement of capital itself, so that capital itself has presupposed them as its own elements, irrespective of how they arose historically.

Within the production process itself, surplus value—the surplus value solicited as a result of the constraint of capital—appeared as surplus labour and even as living labour, which, however, since it cannot produce anything from nothing, finds its own objective conditions in advance. Now this surplus labour appears objectified as surplus product, and this surplus product, in order to valorize itself as capital, divides itself into a double form: as objective labour conditions (material and instrument) and as subjective labour conditions (food) for the living labour now to be put to work. The general form of value—objectified labour—and objectified labour arising from circulation is, naturally, the general and self-evident presupposition. Further: the surplus product in its totality—which objectifies surplus labour in its totality—now appears as surplus capital (as compared with the original capital, before it had undertaken this circulation), i.e as autonomous exchange value, which is opposed to the living labour power as its specific use all value. All the factors which were opposed to the living labour power as forces which were alien, external, and which consumed and utilized the living labour power under definite conditions which were themselves independent of it, are now established as its own product and result.

1. The surplus value or surplus product is nothing but a definite amount of objectified living labour—the sum of the surplus labour. This new value, which is opposed to living labour as an independent value to be exchanged against it, in fact as capital, is the product of labour. It is itself nothing but the general superfluity of labour over necessary labour—in an objective form, and thus as a value.

2. The special shapes that are assumed by this value in order to revalorize itself, i.e. to establish itself as capital—on the one hand as raw material and instrument, on the other as means of subsistence for labour during the act of production—are likewise, therefore, only special forms of surplus labour itself. Raw material and instrument are produced from it in such circumstances—or it itself becomes objective as raw material and instrument in such a proportion—that a definite sum of necessary work (necessary in the sense that it is living, and produces the means of subsistence which are its value) can be objectified in the surplus labour, and indeed incessantly objectified, in other words it can again continue the division of the objective and subjective conditions of its self-maintenance and self-reproduction. Moreover, while living labour is executing the process that reproduces its objective conditions, it has at the same time established raw material and instrument in such proportions that—as surplus labour, as labour beyond what is necessary—it can realize itself in them, and thus make them into material to create new values. The objective conditions of surplus labour—which are limited to the proportion of raw material and instrument above the requirements of necessary labour, while the objective conditions of necessary labour are divided within their objectivity into objective and subjective, into material elements of labour and subjective elements (means of subsistence for living labour)—thus now appear and are thus established as the product, the result, the objective form, the external existence of surplus labour itself. Originally, on the other hand, this seemed alien to living labour itself, as though capital was responsible for the fact that instrument and means of subsistence were present to such an extent that it was possible for living labour to realize itself not only as necessary labour but as surplus labour.

3. The independent and autonomous existence of value as against living labour power—

    hence its existence as capital—

    the objective, self-centred indifference, the alien nature of objective conditions of labour as against living labour power, reaching the point that—

(1) these conditions face the worker, as a person, in the person of the capitalist (as personifications with their own will and interest), this absolute separation and divorce of ownership (i.e. of the material conditions of labour from living labour power); these conditions are opposed to the worker as alien property, as the reality of another legal person and the absolute domain of their will—

and that

(2) labour hence appears as alien labour as opposed to the value personified in the capitalist or to the conditions of labour—

this absolute divorce between property and labour, between living labour power and the conditions of its realization, between objectified and living labour, between the value and the activity that creates value—

hence also the alien nature of the content of the work *vis-à-vis* the worker himself—

this separation now also appears as the product of labour itself, as an objectification of its own elements.

For through the new act of production itself (which merely confirmed the exchange between capital and living labour that had preceded it), surplus labour and thus surplus value, surplus product, in brief, the total result of labour (that of surplus labour as well as of necessary labour) is established as capital, as exchange value which is independently and indifferently opposed both to living labour power and to its mere use value.

Labour power has only adopted the subjective conditions of necessary labour—subsistence indispensable for productive labour power, i.e. its reproduction merely as labour power divorced from the conditions of its realization—and it has itself set up these conditions as objects and values, which stand opposed to it in an alien and authoritarian personification.

It comes out of this process not only no richer but actually poorer than when it entered it. For not only do the conditions of necessary labour that it has produced belong to capital; but also the possibility of creating values which is potentially present in labour power now likewise exists as surplus value, surplus product, in a word, as capital, as dominion over living labour power, as valued endowed with its own strength and will as opposed to the abstract, purposeless, purely subjective poverty of labour power. Labour power has not only produced alien wealth and its own poverty, but also the relationship of this wealth (as wealth concerned exclusively with itself) to itself as poverty, through the consumption of which wealth puts new life into itself and again makes itself fruitful. This all arose from the exchange in which labour power exchanged its living power for a quantity of objectified labour, except that this objectified labour—these conditions of its existence which exist outside it, and the independent external nature of these material conditions—appears as its own product. These conditions appear as though set up by labour power itself, both as its own objectification, and as the objectification of its own power which has an existence independent of it and, even more, rules over it, rules over it by its own doing.

'Thou shalt labour by the sweat of thy brow!' was Jehovah's curse that he bestowed upon Adam. A. Smith conceives of labour as such a curse. 'Rest' appears to him to be the fitting state of things, and identical with 'liberty' and 'happiness'. It seems to be far from A. Smith's thoughts that the individual, 'in his normal state of health, strength, activity, skill, and efficiency', might also require a normal portion of work, and of cessation from rest. It is true that the quantity of labour to be provided seems to be conditioned by external circumstances, by the purpose to be achieved, and the obstacles to its achievement that have to be overcome by labour. But neither does it occur to A. Smith that the overcoming of such obstacles may itself constitute an exercise in liberty, and that these external purposes lose their character of mere natural necessities and are established as purposes which the individual himself fixes. The result is the self-realization and objectification of the subject, therefore real freedom, whose activity is precisely labour. Of course he is correct in saying that labour has always seemed to be repulsive, and forced upon the worker from outside, in its historical forms of slave-labour, bond-labour and wage-labour, and that in this sense non-labour could be opposed to it as 'liberty and happiness'. This is doubly true of this contradictory labour which has not yet created the subjective and objective conditions (which it lost when it abandoned pastoral conditions) which make it into attractive labour and individual self-realization. This does not mean that labour can be made merely a joke, or amusement, as Fourier naïvely expressed it in shop-girl terms. Really free labour, the composing of music for example, is at the same time damned serious and demands the greatest effort. The labour concerned with material production can only have this character if (1) it is of a social nature, (2) it has a scientific character and at the same time is general work, i.e. if it ceases to be human effort as a definite, trained natural force, gives up its purely natural, primitive aspects and becomes the activity of a subject controlling all the forces of nature in the production process. Moreover, A. Smith is thinking only of the slaves of capital. For example, even the semi-artistic worker of the Middle Ages cannot be included in his definition. However, my immediate concern is not to discuss his philosophic view of labour, but only its economic aspect. Labour considered purely as a sacrifice and therefore as establishing a value, labour as the price to be paid for things and thus giving them a price according as they cost more or less labour, is a purely negative definition. In this way Mr. Senior was able, for example, to make capital into a source of production *sui generis* in the same sense as labour is a source of production of value, since the capitalist too is making a sacrifice, the sacrifice of abstinence, for, instead of directly consuming his produce, he is enriching himself. A pure negative accomplishes nothing. When the worker takes pleasure in his work—as, certainly, Senior's miser takes pleasure in his abstinence—the product loses nothing of its value. It is labour alone that produces; it is the only substance of products considered as values.[1] This is why working time (supposing it is of the same intensity) is the measure of value. The qualitative differences among workers—in so far as they are not the natural ones of sex, age, physical strength, etc., and express, fundamentally, not the qualitative value of labour, but its division and differentiation—are the result of historical processes. For the great majority of workers, these differences disappear again, since the work that they perform is simple; work that is of a higher quality, however, can be measured by economics in terms of simple labour. To say that working time, or the quantity of labour, measures values, means only that labour and values are measured by the same standard. Two things can be measured by the same standard only when they are of the same nature. Therefore products can be measured only by the standard of labour (working time) because they are by nature made from labour. They are objectified labour. As objects they may assume forms that show they were produced by labour and that finality has been imposed on them from the outside. This does not always occur; it is not possible to see objectified labour in an ox, nor in the products of nature that man reproduces. These forms, however, have nothing in common with each other; they exist as something constant so long as they have an existence as an activity measured by time, which can thus also be used to measure objectified labour. We shall examine later how far this measurement is linked to exchange, and to labour that is not yet socially organized, as a definite state of the social productive process. Use value is not connected with human activity as the source and creation of the product, it aims at producing an object that is useful for man. In so far as the product has a measure of its own, it is measured in terms of its natural properties—size, weight, length, capacity, measures of usefulness, etc. But as an effect, or as the static form of the force that has created it, is measured only by the volume of this force itself. The measure of labour is time. Simply because products are labour, they can be measured by the measure of labour, by the working time, or the quantity of labour consumed in them. The negation of rest, as a pure negation, as an ascetic sacrifice, accomplishes nothing. An individual may mortify the flesh and make a martyr of himself from morning to night, like the monks, but the amount of sacrifice that he makes will get him nowhere. The natural price of things is not the sacrifice made to obtain them. This is reminiscent of the pre-industrial era, in which riches were to be obtained by

[1] Proudhon's axiom that all work leaves a surplus shows how little he understands the position. What he denies to capital, he allows to be a natural property of labour. The point is rather that the working time necessary for the satisfaction of absolute necessities leaves some free time (which varies at the various stages of the development of the productive forces), so that surplus produce can thus be created if surplus labour is done. The object is to terminate this relationship, so that surplus produce itself can become necessary, and finally material production can leave everyone surplus time for other activities. There is no longer anything mystical about this. Originally the spontaneously developing association (the family) existed at the beginning together with a corresponding division of labour and co-operation. But then needs were slight in the beginning, and only developed with the productive forces. [Marx's footnote.]

sacrifices to the gods. There must be something else besides the sacrifice. Instead of speaking of a sacrifice of rest, one might speak of a sacrifice of laziness, of lack of freedom, of unhappiness—in fact, the negation of a negative condition. A. Smith considers labour from the psychological point of view, in relation to the pleasure or opposite that it gives to the individual. But in addition to being a feeling concerning his activity, work is something else: in the first place, in relationship to other people, for the mere sacrifice of A would be no use to B. Secondly, there is the worker's own particular relationship towards the object that he is making, and towards his own talents for work. Work is a positive, creative activity. The standard by which work is measured—i.e. time—naturally does not depend on its productivity. The measure consists of a unity, whose aliquot parts express a certain quantity. It certainly does not follow from this that the value of labour is constant; it is so only in so far as equal quantities of labour have the same unity of measurement. Pursuing this analysis further, we find that the values of products are measured by labour, not the labour actually employed, but the labour that is necessary for their production. Thus the condition of production is not the sacrifice but the labour. The equation expresses the condition of its reproduction given in the exchange, in other words, the possibility of renewing productive activity created by its own product.

Moreover, if A. Smith's idea of sacrifice correctly expresses the subjective relationship of the wage-earner to his own work, it still will not yield what he wishes it to—namely, that value is determined by means of the time worked. From the worker's point of view, even one hour of work may represent a great sacrifice. But the value of his work does not in the slightest depend on his feelings; nor does the value of the hour he has worked. A. Smith admits that this sacrifice may sometimes be bought more cheaply, sometimes more dearly: in which case one is struck by the fact that it must always be sold at the same price. In this also he is illogical. Farther on, he declares wages to be the standard by which value is measured, not the quantity of labour. To go to the slaughter is always the same sacrifice for the ox; this is no reason for beef to have a constant value.

Machinery, Automation, Free Time, and Communism

Historically, competition meant the abolition of guild coercion, governmental regulations, and the abolition of frontiers, tolls, etc., within a state—and in the world market it meant the abolition of tariffs, protection, and prohibition. In short, it was historically a negation of the limits and obstacles peculiar to the levels of production that obtained before the development of capital. These were described quite correctly, historically speaking, by the physiocrats as *laissez faire, laissez passer*, and advocated by them as such. Competition, however, has never been considered from the purely negative and purely historical aspect; indeed, even more stupid interpretations have been put forward, for example that competition represents the clash between individuals released from their chains and acting only in their own interests; or that it represents the repulsion and attraction of free individuals in relation to one another, and thus is the absolute form of individual liberty in the sphere of production and exchange. Nothing could be more wrong.

Although free competition has abolished the obstacles created by the relationships and means of pre-capitalist production, it should first be remembered that what were restrictions for capital were inherent frontiers for earlier means of production, within which they developed and moved naturally. These frontiers became obstacles only after productive forces and commercial relationships had sufficiently developed for capital to be the ruling principle of production. The frontiers that it tore down were obstacles to its own movement, development, and realization. It did not abolish all frontiers by any means, or all obstacles; only those that did not correspond to its needs, those that were obstacles for it. Within its own limitations—however much these may seem, from a higher point of view, to be obstacles in production, and have been fixed as such by the historical development of capital—it feels itself free and unhampered, that is, bounded only by itself, but its own conditions of existence.

In the same way the industry of the guilds in its heyday found that the guild organization gave it the freedom it needed, i.e. the production relationships corresponding to it. Guild industry gave rise to these relationships, developing them as its own inherent conditions, and thus not at all as external, restricting obstacles. The historical aspect of the negation of guild industry, etc., by capital, by means of free competition, means nothing more than that capital, sufficiently strengthened by a means of circulation adequate for its nature, tore down the historical barriers which interfered with and restricted its movement. But competition is far removed from possessing merely this historical significance, or from playing merely this negative role. *Free competition* is the relation of capital to itself as another capital, i.e. it is the real behaviour of capital as such. It is only then that the internal laws of capital—which appear only as tendencies in the early historical stages of its evolution—can be established; production founded on capital only establishes itself in so far as free competition develops, since free competition is the free development of the conditions and means of production founded on capital and of the process which constantly reproduces these conditions. It is not individuals but capital that establishes itself freely in free competition. So long as production founded on capital is the necessary and therefore the most suitable form in which social productive forces can develop, the movement of individuals within the pure conditions of capital will seem to be free. This liberty is then assured dogmatically by constant reference to the barriers that have been torn down by free competition. Free competition expresses the real development of capital. Because of it, individual capital finds imposed upon itself an external necessity that corresponds to the nature of capital, to the means of production founded on it, to the concept of capital. The mutual constraint that different portions of capital impose on each other, on labour, etc. (the competition of workers between themselves is only another form of the competition of capital), is the free and at the same time the real development of wealth as capital. So much so, that the profoundest economic theorists, Ricardo for example, begin by presuming the absolute domination of free competition, in order to study and formulate the laws that are suitable to capital, laws which at the same time appear as the vital tendencies that dominate it. Free competition, however, is the form suitable to the productive process of capital. The more it develops, the more clearly the shape of its movement is seen. What Ricardo, for example, has thus recognized (despite himself) is the historical nature of capital, and the limited character of free competition, which is still only the free movement of portions of capital, i.e. their movement within conditions that have nothing in common with those of any dissolved preliminary stages, but are their own conditions. The domination of capital is the prerequisite of free competition, just as the despotism of the Roman emperors was the prerequisite of the free Roman civil law. So long as capital is weak, it will rely on crutches taken from past means of production or from means of production that are disappearing as it comes onto the scene. As soon as it feels strong, it throws the crutches away and moves according to its own laws. As soon as it begins to feel and to be aware that it is itself an obstacle to development, it takes refuge in forms that, although they appear to complete the mastery of capital, are at the same time, by curbing free competition, the heralds of its dissolution, and of the dissolution of the means of production which are based on it. What lies in the nature of capital is only expressed in reality as an external necessity through competition, which means no more than that the various portions of capital impose the inherent conditions of capital on one another and on themselves. No category of the bourgeois economy—not even the first one, the determination of value—can become real by means of free competition, i.e. through the real process of capital, which appears as the interaction of portions of capital on one another and of all the other relationships of production and circulation that are determined by capital.

Hence the absurdity of considering free competition as being the final development of human liberty, and the negation of free competition as being the negation of individual liberty and of social production founded on individual liberty. It is only free development on a limited foundation—that of the dominion of capital. This kind of individual liberty is thus at the same time the most complete suppression of all individual liberty and total subjugation of individuality to social conditions which take the form of material forces—and even of all-powerful objects that are independent of the individuals relating to them. The only rational answer to the deification of free competition by the middle-class prophets, or its diabolization by the socialists, lies in its own development. If it is said that, within the limits of free competition, individuals by following their pure self-interest realize their social, or rather their general, interest, this means merely that they exert pressure upon one another under the conditions of capitalist production and that this clash between them can only give rise to the conditions under which their interaction took place. Moreover, once the illusion that competition is the supposedly absolute form of free individuality disappears, this proves that the conditions of competition, i.e. of production founded on capital, are already felt and thought of as a barrier, that they indeed already are such and will increasingly become so. The assertion that free competition is the final form of the development of productive forces, and thus of human freedom, means only that the domination of the middle class is the end of the world's history—of course quite a pleasant thought for yesterday's parvenus!

So long as the means of labour remains a means of labour, in the proper sense of the word, as it has been directly and historically assimilated by capital into its valorization process, it only undergoes a formal change, in that it appears to be the means of labour not only from its material aspect, but at the same time as a special mode of existence of capital determined by the general process of capital—it has become fixed capital. But once absorbed into the production process of capital, the means of labour undergoes various metamorphoses, of which the last is the machine, or rather an automatic system of machinery ('automatic' meaning that this is only the most perfected and most fitting form of the machine, and is what transforms the machinery into a system).

This is set in motion by an automaton, a motive force that moves of its own accord. The automaton consists of a number of mechanical and intellectual organs, so that the workers themselves can be no more than the conscious limbs of the automaton. In the machine, and still more in machinery as an automatic system, the means of labour is transformed as regards its use value, i.e. as regards its material existence, into an existence suitable for fixed capital and capital in general; and the form in which it was assimilated as a direct means of labour into the production process of capital is transformed into one imposed by capital itself and in accordance with it. In no respect is the machine the means of labour of the individual worker. Its distinctive character is not at all, as with the means of labour, that of transmitting the activity of the worker to its object; rather this activity is so arranged that it now only transmits and supervises and protects from damage the work of the machine and its action on the raw material.

With the tool it was quite the contrary. The worker animated it with his own skill and activity; his manipulation of it depended on his dexterity. The machine, which possesses skill and force in the worker's place, is itself the virtuoso, with a spirit of its own in the mechanical laws that take effect in it; and, just as the worker consumes food, so the machine consumes coal, oil, etc. (instrumental material) for its own constant self-propulsion. The worker's activity, limited to a mere abstraction, is determined and regulated on all sides by the movement of the machinery, not the other way round. The knowledge that obliges the inanimate parts of the machine, through their construction, to work appropriately as an automaton, does not exist in the consciousness of the worker, but acts upon him through the machine as an alien force, as the power of the machine itself. The appropriation of living labour by objectified labour—of valorizing strength or activity by self-sufficient value—which is inherent in the concept of capital, is established as the character of the production process itself—when production is based on machinery—as a function of its material elements and material movement. The production process has ceased to be a labour process in the sense that labour is no longer the unity dominating and transcending it. Rather labour appears merely to be a conscious organ, composed of individual living workers at a number of points in the mechanical system; dispersed, subjected to the general process of the machinery itself, it is itself only a limb of the system, whose unity exists not in the living workers but in the living (active) machinery, which seems to be a powerful organism when compared to their individual, insignificant activities. With the stage of machinery, objectified labour appears in the labour process itself as the dominating force opposed to living labour, a force represented by capital in so far as it appropriates living labour.

That the labour process is no more than a simple element in the valorization process is confirmed by the transformation on the material plane of the working tool into machinery, and of the living worker into a mere living accessory of the machine; they become no more than the means whereby its action can take place.

As we have seen, capital necessarily tends towards an increase in the productivity of labour and as great a diminution as possible in necessary labour. This tendency is realized by means of the transformation of the instrument of labour into the machine. In machinery, objectified labour is materially opposed to living labour as its own dominating force; it subordinates living labour to itself not only by appropriating it, but in the real process of production itself. The character of capital as value that appropriates value-creating activity is established by fixed capital, existing as machinery, in its relationship as the use value of labour power. Further, the value objectified in machinery appears as a prerequisite, opposed to which the valorizing power of the individual worker disappears, since it has become infinitely small.

In the large-scale production created by machines, any relationship of the product to the direct requirements of the producer disappears, as does any immediate use value. The form of production and the circumstances in which production takes place are so arranged that it is only produced as a vehicle for value, its use value being only a condition for this.

In machinery, objectified labour appears not only in the form of a product, or of a product utilized as a means of labour, but also in the force of production itself. The development of the means of labour into machinery is not fortuitous for capital; it is the historical transformation of the traditional means of labour into means adequate for capitalism. The accumulation of knowledge and skill, of the general productive power of society's intelligence, is thus absorbed into capital in opposition to labour and appears as the property of capital, or more exactly of fixed capital, to the extent that it enters into the production process as an actual means of production. Thus machinery appears as the most adequate form of fixed capital; and the latter, in so far as capital can be considered as being related to itself, is the most adequate form of capital in general. On the other hand, in so far as fixed capital is firmly tied to its existence as a particular use value, it no longer corresponds to the concept of capital which, as a value, can take up or throw off any particular form of use value, and incarnate itself in any of them indifferently. Seen from this aspect of the external relationships of capital, circulating capital seems to be the most adequate form of capital as opposed to fixed capital.

In so far as machinery develops with the accumulation of social knowledge and productive power generally, it is not in labour but in capital that general social labour is represented. Society's productivity is measured in fixed capital, exists within it in an objectified form; and conversely, the productivity of capital evolves in step with this general progress that capital appropriates gratis. We shall not go into the development of machinery in detail here. We are considering it only from the general aspect, to the extent that the means of labour, in its material aspect, loses its immediate form and opposes the worker materially as capital. Science thus appears, in the machine, as something alien and exterior to the worker; and living labour is subsumed under objectified labour, which acts independently. The worker appears to be superfluous in so far as his action is not determined by the needs of capital.

Thus the full development of capital does not take place—in other words, capital has not set up the means of production corresponding to itself—until the means of labour is not only formally determined as fixed capital, but has been transcended in its direct form, and fixed capital in the shape of a machine is opposed to labour within the production process. The production process as a whole, however, is not subordinated to the direct skill of the worker; it has become a technological application of science.

The tendency of capital is thus to give a scientific character to production, reducing direct labour to a simple element in this process. As with the transformation of value into capital, we see when we examine the development of capital more closely that on the one hand it presupposes a definite historical development of the productive forces (science being included among these forces), and on the other hand accelerates and compels this development.

The quantitative volume, and the efficiency (intensity) with which capital develops as fixed capital, thus shows in general the degree to which capital has developed as capital, as domination over living labour, and the degree to which it dominates the production process in general. It also expressed the accumulation of objectified productive forces and likewise of objectified labour.

But if capital only adequately displays its nature as use value within the production process in the form of machinery and other material forms of fixed capital, railways, for example (we shall return to this later), this never means that this use value (machinery by itself) is capital, or that machinery can be regarded as synonymous with capital; any more than gold would cease to have usefulness as gold if it were no longer used as money. Machinery does not lose its use value when it ceases to be capital. From the fact that machinery is the most suitable form of the use value of fixed capital, it does not follow that its subordination to the social relations of capitalism is the most suitable and final social production relationship for the utilization of machinery.

To the same degree that working time—the mere quantity of labour—is established by capital as the sole determining element, direct labour and its quantity cease to be the determining element in production and thus in the creation of use value. It is reduced quantitatively to a smaller proportion, just as qualitatively it is reduced to an indispensable but subordinate role as compared with scientific labour in general, the technological application of the natural sciences, and the general productive forces arising from the social organization of production. This force appears to be a natural gift of community labour, although it is a historical product. In this way capital works for its own dissolution as the dominant form of production.

The transformation of the process of production from the simple labour process into a scientific process, which subjects the forces of nature and converts them to the service of human needs, appears to be a property of fixed capital as opposed to living labour. Individual labour ceases altogether to be productive as such; or rather, it is productive only in collective labour, which subjects the forces of nature. This promotion of immediate labour to the level of community labour shows that individual work is reduced to helplessness vis-à-vis the concentration of common interest represented in capital. On the other hand, a property of circulating capital is the retention of labour in one branch of production thanks to coexisting labour in another branch.

In small-scale circulation, capital advances the worker his wages, which he exchanges for products necessary for his own consumption. The money that he receives only has this power because, at the same time, work is being carried out alongside him. It is only because capital has appropriated his labour that it can give him, with the money, control over the labour of others. This exchange of worker's own labour for that of others does not seem to be determined and conditioned by the simultaneous coexistence of the others' labour, but by the advance that capital has made to him.

It appears to be a property of the part of the circulating capital that is assigned to the worker, and of circulating capital in general, that the worker can proceed to the assimilation of what he himself needs during the process of production. This exchange appears as the material exchange not of simultaneous forces, but of capital; because of the existence of circulating capital. Thus all the forces of labour are transposed into forces of capital; in its fixed part, the productive force of labour (which is placed outside it and exists materially independently of it); and in its circulating part, we find first of all that the worker has himself produced the conditions for the renewal of his work, and secondly that the exchange of his labour is mediated through the coexisting labour of others in such a way that capital appears to make him the advance and to ensure the existence of labour in other branches. (Both the latter statements really belong to the chapter on accumulation.) Capital sets itself up as a mediator between the various labourers in the form of circulating capital.

Fixed capital, considered as a means of production, whose most adequate form is machinery, only produces value (i.e. increases the value of the product) in two cases: (1) in so far as it has value, i.e. in so far as it is itself the product of labour, a definite quantity of labour in an objectified form; (2) in so far as it increases the proportion of surplus labour to necessary labour by making it possible for labour, by increasing its productivity, to create more quickly a larger amount of the products needed for the sustenance of living labour power. To say that the worker is in co-operation with the capitalist because the latter makes the worker's labour easier by means of fixed capital, or shortens his labour, is a bourgeois phrase of the greatest absurdity. Fixed capital is in any case the product of labour, and is merely alien labour that has been appropriated by capital; and the capitalist could be said, rather, to be robbing labour of all its independent and attractive character by means of the machine. On the contrary, capital only uses machinery in so far as it enables the worker to devote more of his working time to the capitalist, to work longer for others and experience a larger part of his time as not belonging to him. Through this process, in fact, the quantity of labour necessary for the production of a particular object is reduced to a minimum, so that a maximum of labour can be valorized into the maximum number of such objects. The first aspect is important, because capital in this instance has quite unintentionally reduced human labour, the expenditure of energy, to a minimum. This will be to the advantage of emancipated labour, and is the condition of its emancipation.

All this shows the absurdity of Lauderdale's statement that fixed capital is independent of working time, and a self-contained source of value. It is such a source only in so far as it is itself objectified labour time, and establishes surplus labour time. The introduction of machines historically presupposes superfluous hands. Machinery only replaces labour when there is a super-fluity of labour force. Only in the imagination of economists does it come to the aid of the individual worker. It can only take effect with masses of workers, whose concentration as opposed to capital is one of its historical prerequisites, as we have seen. It does not arise in order to replace deficient labour power, but to reduce the mass of labour available to the necessary quantity. Only when labour power is present *en masse* is machinery introduced. (We shall return to this later.)

Lauderdale thinks he has made a great discovery when he says that mach-inery does not increase the productive power of labour but replaces it, or does what labour is unable to do by its own power. It is inherent in the concept of capital that the greater productive power of labour is seen as an increase of a force external to labour and as the enfeeblement of labour itself. The tools of labour make the worker independent—establish him as a pro-prietor. Machinery—as fixed capital—makes him dependent, expropriates him. Machinery only produces this effect to the extent that it is fixed capital, and it only has this character so long as the worker is related to the machine as a wage-earner, and the active individual in general as a mere worker.

Up to this point, fixed and circulating capital have appeared to be merely different, transitory determinations of capital; but now they are crystallized into special forms of existence of capital and circulating capital occurs along-side fixed capital. There are now two different kinds of capital. If we con-sider capital in a particular branch of production we see that it is divided into two parts, or that it is divided in a determined proportion into these two kinds of capital.

The difference within the production process—originally between the means of labour and labour material, and finally the product of labour—now appears as that between circulating capital (the first two) and fixed capital. The division of capital in its purely material aspect is now assimilated into its own form, which appears as what makes the distinction.

The mistake of such writers as Lauderdale, who state that capital, as such and independently of labour, creates value, and thus also surplus value (or profit), arises from their superficial view of the matter. Fixed capital, whose material form or use value is machinery, gives most appearance of truth to their fallacies. But against this, e.g. in *Labour Defended*, we see that the constructor of a road is willing to share it with the road user, but that the road itself cannot do this.

Circulating capital—provided only that it proceeds through its various stages—may decrease or increase, shorten or lengthen the circulation time, make the various stages of capitalist circulation easier or more difficult. Consequently, the surplus value that may be produced within a given time may be diminished without these interruptions—either because the number of reproductions is smaller, or because the quantity of capital constantly engaged in the process of production has contracted. In both cases there is no diminution of the presupposed value, but a decrease in its rate of growth.

As we have noted, the extent to which capital has developed is a measure of the development of heavy industry in general, and as soon as it has developed to a certain point, and thus has increased relative to the development of its productive forces (it being itself the objectification of these productive forces and their presupposed product), from this point onwards any interruption of the process of production will cause a diminution of capital itself and its presupposed value. The value of fixed capital is only reproduced to the extent that it is used upon the production process. If it is not used, it loses its use value without its value being transferred to the product. Thus the more fixed capital develops on a large scale, in the sense in which we have been considering it, the more the continuity of the production process or the constant flow of reproduction will become a condition and an external form of coercion of the means of production founded on capitalism.

From this standpoint, too, the appropriation of living labour by capital is directly expressed in machinery. It is a scientifically based analysis, together with the application of mechanical and chemical laws, that enables the machine to carry out the work formerly done by the worker himself. The development of machinery only follows this path, however, once heavy industry has reached an advanced stage, and the various sciences have been pressed into the service of capital, and, on the other hand, when machinery itself has yielded very considerable resources. Invention then becomes a branch of business, and the application of science to immediate production aims at determining the inventions at the same time as it solicits them. But this is not the way in which machinery in general came into being, still less the way that it progresses in detail. This way is a process of analysis—by subdivisions of labour, which transforms the worker's operations more and more into mechanical operations, so that, at a certain point, the mechanism can step into his place.

Thus we can see directly here how a particular means of labour is transferred from the worker to capital in the form of the machine and his own labour power devalued as a result of this transposition. Hence we have the struggle of the worker against machinery. What used to be the activity of the living worker has become that of the machine.

Thus the appropriation of his labour by capital is bluntly and brutally presented to the worker: capital assimilates living labour into itself 'as though love possessed its body'.

Real wealth develops much more (as is disclosed by heavy industry) in the enormous disproportion between the labour time utilized and its products, and also in the qualitative disproportion between labour that has been reduced to a mere abstraction, and the power of the production process that it supervises. Labour does not seem any more to be an essential part of the process of production. The human factor is restricted to watching and supervising the production process. (This applies not only to machinery, but also to the combination of human activities and the development of human commerce.)

The worker no longer inserts transformed natural objects as intermediaries between the material and himself; he now inserts the natural process that he has transformed into an industrial one between himself and inorganic nature, over which he has achieved mastery. He is no longer the principal agent of the production process: he exists alongside it. In this transformation, what appears as the mainstay of production and wealth is neither the immediate labour performed by the worker, nor the time that he works—but the appropriation by man of his own general productive force, his understanding of nature and the mastery of it; in a word, the development of the social individual. The theft of others' labour time upon which wealth depends today seems to be a miserable basis compared with this newly developed foundation that has been created by heavy industry itself. As soon as labour, in its direct form, has ceased to be the main source of wealth, then labour time ceases, and must cease, to be its standard of measurement, and thus exchange value must cease to be the measurement of use value. The surplus labour of the masses has ceased to be a condition for the development of wealth in general; in the same way that the non-labour of the few has ceased to be a condition for the development of the general powers of the human mind. Production based on exchange value therefore falls apart, and the immediate process of material production finds itself stripped of its impoverished, antagonistic form. Individuals are then in a position to develop freely. It is no longer a question of reducing the necessary labour time in order to create surplus labour, but of reducing the necessary labour of society to a minimum. The counterpart of this reduction is that all members of society can develop their education in the arts, sciences, etc., thanks to the free time and means available to all.

Capital is itself contradiction in action, since it makes an effort to reduce labour time to the minimum, while at the same time establishing labour time as the sole measurement and source of wealth. Thus it diminishes labour time in its necessary form, in order to increase its superfluous form; therefore it increasingly establishes superfluous labour time as a condition (a question of life and death) for necessary labour time. On the one hand it calls into life all the forces of science and nature, as well as those of social co-operation and commerce, in order to create wealth which is relatively independent of the labour time utilized. On the other hand it attempts to measure, in terms of labour time, the vast social forces thus created and imprisons them within the narrow limits that are required in order to retain the value already created *as* value. Productive forces and social relationships—the two different sides of the development of the social individual —appear to be, and are, only a means for capital, to enable it to produce from its own cramped base. But in fact they are the material conditions that will shatter this foundation.

Nature does not construct machines, locomotives, railways, electric telegraphs, self-acting mules, etc. These are products of human industry; natural material transformed into organs of the human will to dominate nature or to realize itself therein. They are organs of the human brain, created by human hands; the power of knowledge made into an object.

The development of fixed capital shows to what extent general social knowledge had become an immediate productive force, and thus up to what point the conditions for the social life process are themselves subjected to the control of the general intellect, and are remodelled to suit it, and to what extent social productive forces are produced not only in the form of knowledge but also as the direct organs of social practice; of the real life process.

Capital creates a great deal of disposable time, apart from the labour time that is needed for society in general and for each sector of society (i.e. space for the development of the individual's full productive forces, and thus also for those of society). This creation of non-working time is, from the capitalist standpoint, and from that of all earlier stages of development, non-working time or free time for the few. What is new in capital is that it also increases the surplus labour time of the masses by all artistic and scientific means possible, since its wealth consists directly in the appropriation of surplus labour time, since its direct aim is value, not use value. Thus, despite itself, it is instrumental in creating the means of social disposable time, and so in reducing working time for the whole of society to a minimum and thus making everyone's time free for their own development. But although its tendency is always to create disposable time, it also converts it into surplus labour. If it succeeds too well with the former, it will suffer from surplus production, and then the necessary labour will be interrupted as soon as no surplus labour can be valorized from capital. The more this contradiction develops, the clearer it becomes that the growth of productive forces can no longer be limited by the appropriation of the surplus labour of others; the masses of the workers must appropriate their own surplus labour.

When this has been done, disposable time ceases to have a contradictory character. Thus firstly, the labour time necessary will be measured by the requirements of the social individual, and secondly, social productivity will grow so rapidly that, although production is reckoned with a view to the wealth of all, the disposable time of all will increase. For real wealth is the developed productive force of all individuals. It is no longer the labour time but the disposable time which is the measure of wealth. Labour time as the measurement of wealth implies that wealth is founded on poverty, and that disposable time exists in and through opposition to surplus labour time; it implies that all an individual's time is working time, and degrades him to the level of a mere worker, and an instrument of labour. This is why the most developed machinery forces the worker to work longer hours than the savage does, or than the labourer himself when he only had the simplest and most primitive tools to work with. . . .

The development of heavy industry means that the basis upon which it rests—the appropriation of the labour time of others—ceases to constitute or to create wealth; and at the same time direct labour as such ceases to be the basis of production, since it is transformed more and more into a supervisory and regulating activity; and also because the product ceases to be made by individual direct labour, and results more from the combination of social activity. 'As the division of labour develops, almost all the work of any individual is a part of the whole, having no value or utility of itself. There is nothing on which the labourer can seize: this is my produce, this I will keep to myself.' In direct exchange between producers, direct individual labour is found to be realized in a particular product, or part of a product, and its common social character—as the objectification of general labour and the satisfaction of general need—is only established through exchange. The opposite takes place in the production process of heavy industry: on the one hand, once the productive forces of the means of labour have reached the level of an automatic process, the prerequisite is the subordination of the natural forces to the intelligence of society, while on the other hand individual labour in its direct form is transformed into social labour. In this way the other basis of this mode of production vanishes.

The labour time used for the production of fixed capital is related, within the production process of capital, to the time used for the production of circulating capital, as surplus labour time to necessary time. To the extent that production directed towards the satisfaction of immediate needs becomes more productive, a larger part of production can be directed towards the satisfaction of the needs of production itself, or the manufacture of means of production. In so far as the production of fixed capital is not materially directed either towards the production of immediate use values, or towards the production of values indispensable for the immediate reproduction of capital—which would again be an indirect representation of use value—but towards the production of means that serve to create value and not towards value as an immediate object; in other words, in so far as fixed capital concentrates on the creation of values and the means of valorization as the immediate object of production (the production of value is established materially in the object of production itself as the aim of production, the objectification of productive force, and the value-producing force of capital)—to that extent it is in the production of fixed capital that capital is established as an end in itself to a more powerful degree than in the production of circulating capital, and becomes effective as capital. Thus, in this sense too, the volume already possessed by fixed capital and the part occupied by its production in general production indicate the standard of development of wealth based on the mode of production of capital.

'The number of workers depends so far on circulating capital, as it depends on the quantity of products of coexisting labour, which labourers are allowed to consume.'

The passages that we have quoted from various economists all refer to fixed capital as the part of capital which is included in the production process. 'Flotating capital is consumed; fixed capital is merely used in the great process of production' (Economist, VI. 1). This is wrong, since it applies only to the part of the circulating capital that is itself consumed by fixed capital, by the material instruments. Only fixed capital is consumed 'in the great process of production', considered as the immediate production process.

Consumption within the production process is, however, in fact use. Moreover, the greater durability of fixed capital cannot be understood as a purely material phenomenon. The iron and wood of which my bed is made, the bricks out of which my house is constructed, or the marble statue that adorns a palace, are as durable as the iron and wood that are transformed into machinery. But it is not only for the technical reason that metal, etc., is most often used in machinery that durability is a necessary quality of the instrument and of the means of production, but because the instrument is intended constantly to play the same role in repeated processes of production. The solidity or durability of the means of production is a direct part of its use value. The more often it has to be renewed, the more expensive it becomes, and the greater the part of capital that must be transformed into it unprofitably. Its durability is thus its existence as a means of production. Its durability implies an increase in its productivity. With circulating capital on the other hand, if it is not transformed into fixed capital, durability does not in any way depend on the productive act itself; and therefore it is not a conceptually established element of it. . . .

'Since the general introduction of soulless machines into British factories, men have been treated, with a few exceptions, as secondary and subordinate machines, and much more attention has been given to the perfecting of raw material made of wood and metals than that made of bodies and minds.' [Robert Owen, *Essays on the Formation of the Human Character*, London, 1840, p. 31.]

Real economy—savings—consists in the saving of working time (the minimum, and reduction to the minimum, of production costs); but this saving is identical with the development of productivity. Economizing, therefore, does not mean the giving up of pleasure, but the development of power and productive capacity, and thus both the capacity for and the means of enjoyment. The capacity for enjoyment is a condition of enjoyment and therefore its primary means; and this capacity is the development of an individual's talents, and thus of the productive force. To economize on labour time means to increase the amount of free time, i.e. time for the complete development of the individual, which again reacts as the greatest productive force on the productive force of labour. From the standpoint of the immediate production process it may be considered as production of fixed capital; this fixed capital being man himself. It is also self-evident that immediate labour time cannot remain in its abstract contradiction to free time—as in the bourgeois economy. Work cannot become a game, as Fourier would like it to be; his great merit was that he declared that the ultimate object must be to raise to a higher level not distribution but the mode of production. Free time—which includes leisure time as well as time for higher activities—naturally transforms anyone who enjoys it into a different person, and it is this different person who then enters the direct process of production. The man who is being formed finds discipline in this process, while for the man who is already formed it is practice, experimental science, materially creative and self-objectifying knowledge, and he contains within his own head the accumulated wisdom of society. Both of them find exercise in it, to the extent that labour requires practical manipulation and free movement, as in agriculture.

As, little by little, the system of bourgeois economy develops, there develops also its negation, which is its final result. We are now still concerned with the direct process of production. If we consider bourgeois society as a whole, society itself seems to be the final result of the social process of production, i.e. man himself in his relation to society. Everything (such as the product, etc.) which has a fixed form appears only as an element, a vanishing element, in this movement. Even the immediate production process appears here only as an element. The conditions and objectifications of the process are themselves likewise elements of it, the subjects of which are only the individuals, but individuals who are related to one another, in relations which are both reproduced and created anew. It is their own constant process of movement in which they renew both themselves and the world of wealth which they create.

It is a fact that as the productive forces of labour develop, the objective conditions of labour (objectified labour) must grow in proportion to living labour. This is actually a tautology, for the growth of the productive forces of labour means merely that less direct labour is required in order to make a larger product, so that social wealth expresses itself more and more in the labour conditions that have been created by labour itself. From the point of view of capital, it does not appear that one of the elements of social activity (objectified labour) has become the ever more powerful body of the other element (subjective, living labour); rather it appears (and this is important for wage-labour) that the objective conditions of labour become more and more colossally independent of living labour—which is shown by their very extent—and social wealth becomes, in ever greater and greater proportions, an alien and dominating force opposing the worker. Stress is placed not on the state of objectification but on the state of alienation, estrangement, and abandonment, on the fact that the enormous objectified power which social labour has opposed to itself as one of its elements belongs not to the worker but to the conditions of production that are personified in capital. So long as the creation of this material form of activity, objectified in contrast to immediate labour power, occurs on the basis of capital and wage-labour, and so long as this process of objectification in fact seems to be a process of alienation as far as the worker is concerned, or to be the appropriation of alien labour from the capitalist's point of view, so long will this distortion and this inversion really exist and not merely occur in the imagination of both workers and capitalists. But this process of inversion is obviously merely a historical necessity, a necessity for the development of productive forces from a definite historical starting-point, or basis, but in no way an absolute necessity of production; it is, rather, ephemeral. The result and the immanent aim of the process is to destroy and transform this basis itself, as well as this form of the process. Bourgeois economists are so bogged down in their traditional ideas of the historical development of society in a single stage that the necessity of the objectification of the social forces of labour seems to them inseparable from the necessity of its alienation in relation to living labour.

But as living labour loses its immediate, individual character, whether subjective or entirely external, as individual activity becomes directly general or social, the objective elements of production lose this form of alienation. They are then produced as property, as the organic social body in which individuals are reproduced as individuals, but as social individuals. The conditions for their being such in the reproduction of their life, in their productive life process, have only been established by the historical economic process; these conditions are both objective and subjective conditions, which are the only two different forms of the same conditions.

The fact that the workers possess no property and the fact that objectified labour has property in living labour (in other words, that capital appropriates the labour of others) constitute the two opposite poles of the same relationship, and are the fundamental conditions of the bourgeois means of production and are in no sense a matter of indifference or chance. These means of distribution are the relations of production themselves, but *sub specie distributionis* (from the point of view of distribution). Thus it is quite absurd to say, as J. S. Mill does for example (*Principles of Political Economy*, 2nd edn. (London, 1849) 1.240), that: 'The laws and conditions of the production of wealth partake of the character of physical truths ... It is not so with the distribution of wealth. This is a matter of human institutions solely.' The 'laws and conditions' of the production of wealth and the laws of 'distribution of wealth' are the same laws in a different form; they both change and undergo the same historical process; they are, in general, never more than elements in a historical process.

No special sagacity is required in order to understand that, beginning with free labour or wage-labour for example, which arose after the dissolution of serfdom, machines can only develop in opposition to living labour, as a hostile power and alien property, i.e. they must, as capital, oppose the worker. But it is equally easy to see that machines do not cease to be agents of social production, once they become, for example, the property of associated workers. But in the first case, their means of distribution (the fact that they do not belong to the workers) is itself a condition of the means of production that is founded on wage-labour. In the second case, an altered means of distribution will derive from a new, altered basis of production emerging from the historical process.

Of the laws of modern political economy, the law of the tendency of the rate of profit to fall is, in every respect, the most important and the most essential for the understanding of the most difficult problems. From the historical point of view, also, it is the most important law, one which, in spite of its simplicity, has never yet been understood and still less been consciously enunciated. This fall in the rate of profit is bound up with (1) the already existing forces of production and the material foundation that it creates for a new production which in turn presupposes an enormous development of scientific powers; (2) the diminution of that part of capital already produced that has to be exchanged for direct labour, in other words a diminution in the direct labour which is necessary to reproduce an enormous quantity of values, which is expressed in a greater mass of products at low prices, since the sum total of prices equals the capital reproduced plus the profit; (3) the dimension of capital in general, including the portion of it that is not fixed capital. This implies great development in commerce, in exchange operations, and the market; the universality of simultaneous labour; means of communication; the presence of the necessary consumer funds to undertake this gigantic process (food, lodging for the workers, etc.). This being the case, it is plain that the already existing force of production that has been acquired as fixed capital, as well as scientific power, population—in short all the conditions of wealth—the most important conditions for the reproduction of wealth, that is, the rich development of the social individual, the development of the productive forces produced by capital itself in its historical development—all these factors, when they arrive at a certain stage, abolish the self-valorization of capital instead of establishing it. Beyond a certain point the development of productive forces becomes a barrier for capital. Thus capitalist relationships become a barrier for the development of the productive force of labour. On arrival at the point, capital, i.e. wage-labour, enters into the same relationship to the development of social wealth and productive forces as the guild system, serfdom, slavery, and is necessarily rejected as a fetter. Thus the last form of enslavement taken on by human activity—wage-labour on one side and capital on the other—is sloughed off and this process is in itself the result of the capitalist mode of production. The material and spiritual conditions of the negation of wage-labour and capital (which themselves are already the negation of earlier forms of unfree social production) are themselves the results of its process of production. The growing incompatibility of the productive development of society with its established relationships of production is expressed in acute contradictions, crises, and convulsions. The violent annihilation of capital, not through external relationships, but as the condition of its own self-preservation, is the most striking form in which notice is given to it to be gone and give room to a higher state of social production.

BIBLIOGRAPHY

Original: K. Marx, *Grundrisse der Kritik der politischen Oekonomie*, Berlin, 1953.

Present translation: *Marx's Grundrisse*, ed. D. McLellan, London, 1973, pp. 26 ff., 58 ff., 86 ff., 102 f., 111 ff., 141 ff., 150 ff.

Other translations: K. Marx, *Grundrisse*, ed. M. Nicolaus, Harmondsworth, 1973.
K. Marx, *Precapitalist Economic Formations*, ed. E. Hobsbawm, London, 1964.

Commentaries: H. Acton and G. Cohen, 'On some criticisms of historical materialism', *Proceedings of the Aristotelian Society*, Supp. vol. 44, 1970.
T. Carver, Introduction to K. Marx, *Texts on Method*, Oxford, 1974.
E. Hobsbawm, Introduction to edition cited above.
G. Lichtheim, 'Marx and the "Asiatic mode of Production"', *St. Anthony's Papers*, 1963, reprinted in *Karl Marx*, ed. T. Bottomore, New York, 1973.
E. Mandel, *The Formation of Marx's Economic Thought*, London, 1971, Chs. 6–8.
D. McLellan, Introduction to *Marx's Grundrisse*, cited above.
S. Moore, 'Marx and the Origins of Dialectical Materialism', *Inquiry* (Autumn 1971).
M. Nicolaus, 'The Unknown Marx', *New Left Review* (April 1968), reprinted in *Ideology in Social Science*, ed. R. Blackburn (London, 1972).
M. Nicolaus, Introduction to edition cited above.

30. Preface to *A Critique of Political Economy*

The *Critique of Political Economy*, intended as a writing-up of the first section of the *Grundrisse*, is of little intrinsic interest as its ideas are largely taken up again in the first chapters of *Capital*. However, the Preface, reproduced below, is interesting on two counts: it begins with a short intellectual autobiography and also contains a summary statement of the materialist conception of history which has become—often too exclusively—the 'classical' exposition of this idea.

I examine the system of bourgeois economics in the following order: capital, landed property, wage labour; state, foreign trade, world market. Under the first three headings, I investigate the economic conditions of life of the three great classes into which modern bourgeois society is divided; the interconnection of the three other headings is obvious at a glance. The first section of the first book, which deals with capital, consists of the following chapters: 1. Commodities; 2. Money, or simple circulation; 3. Capital in general. The first two chapters for the contents of the present part. The total material lies before me in the form of monographs, which were written at widely separated periods, for self-clarification, not for publication, and whose coherent elaboration according to the plan indicated will be dependent on external circumstances.

I am omitting a general introduction which I had jotted down because on close reflection any anticipation of results still to be proved appears to me to be disturbing, and the reader who on the whole desires to follow me must be resolved to ascend from the particular to the general. A few indications concerning the course of my own politico-economic studies may, on the other hand, appear in place here.

I was taking up law, which discipline, however, I only pursued as a subordinate subject along with philosophy and history. In the year 1842–3 , as editor of the *Rheinische Zeitung*, I experienced for the first time the embarrassment of having to take part in discussions on so-called material interests. The proceedings of the Rhenish Landtag on thefts of wood and parcelling of landed property, the official polemic which Herr von Schaper, then *Oberpräsident* of the Rhine Province, opened against the *Rheinische Zeitung* on the conditions of the Moselle peasantry, and finally debates on free trade and protective tariffs provided the first occasions for occupying myself with economic questions. On the other hand, at that time when the good will 'to go further' greatly outweighed knowledge of the subject, a philosophically weak echo of French socialism and communism made itself audible in the *Rheinische Zeitung*. I declared myself against this amateurism, but frankly confessed at the same time in a controversy with the *Allgemeine Augsburger Zeitung* that my previous studies did not permit me even to venture any judgement on the content of the French tendencies. Instead, I eagerly seized on the illusion of the managers of the *Rheinische Zeitung*, who thought that by a weaker attitude on the part of the paper they could secure a remission of the death sentence passed upon it, to withdraw from the public stage into the study.

The first work which I undertook for a solution of the doubts which assailed me was a critical review of the Hegelian philosophy of right, a work whose introduction appeared in 1844 in the *Deutsch–französische Jahrbücher*, published in Paris. My investigation led to the result that legal relations as well as forms of state are to be grasped neither from themselves nor from the so-called general development of the human mind, but rather have their roots in the material conditions of life, the sum total of which Hegel, following the example of the Englishmen and Frenchmen of the eighteenth century, combines under the name of 'civil society', that, however, the anatomy of civil society is to be sought in political economy. The investigation of the latter, which I began in Paris, I continued in Brussels, whither I had emigrated in consequence of an expulsion order of M. Guizot. The general result at which I arrived and which, once won, served as a guiding thread for my studies, can be briefly formulated as follows: In the social production of their life, men enter into definite relations that are indispensable and independent of their will, relations of production which correspond to a definite stage of development of their material productive forces. The sum total of these relations of production constitutes the economic structure of society, the real foundation, on which rises a legal and political superstructure and to which correspond definite forms of social consciousness. The mode of production of material life conditions the social, political, and intellectual life process in general. It is not the consciousness of men that determines their being, but, on the contrary, their social being that determines their consciousness. At a certain stage of their development, the material productive forces of society come in conflict with the existing relations of production, or—what is but a legal expression for the same thing—with the property relations within which they have been at work hitherto. From forms of development of the productive forces these relations turn into their fetters. Then begins an epoch of social revolution. With the change of the economic foundations the entire immense superstructre is more or less rapidly transformed. In considering such transformations a distinction should always be made between the material transformation of the economic conditions of production, which can be determined with the precision of natural science, and the legal, political, religious, aesthetic, or philosophic

—in short, ideological forms in which men become conscious of this conflict and fight it out. Just as our opinion of an individual is not based on what he thinks of himself, so can we not judge of such a period of transformation by its own consciousness; on the contrary, this consciousness must be explained rather from the contradictions of material life, from the existing conflict between the social productive forces and the relations of production. No social order ever perishes before all the productive forces for which there is room in it have developed; and new, higher relations of production never appear before the material conditions of their existence have matured in the womb of the old society itself. Therefore mankind always sets itself only such tasks as it can solve; since, looking at the matter more closely, it will always be found that the task itself arises only when the material conditions for its solution already exist or are at least in the process of formation. In broad outlines Asiatic, ancient, feudal, and modern bourgeois modes of production can be designated as progressive epochs in the economic formation of society. The bourgeois relations of production are the last antagonistic form of the social process of production—antagonistic not in the sense of individual antagonism, but of one arising from the social conditions of life of the individuals; at the same time the productive forces developing in the womb of bourgeois society create the material conditions for the solution of that antagonism. This social formation brings, therefore, the prehistory of human society to a close.

Friedrich Engels, with whom, since the appearance of his brilliant sketch on the criticism of the economic categories (in the *Deutsch–französische Jahrbücher*), I maintained a constant exchange of ideas by correspondence, had by another road (compare his *The Condition of the Working Class in England in 1844*) arrived at the same result as I, and when in the spring of 1845 he also settled in Brussels, we resolved to work out in common the opposition of our view to the ideological view of German philosophy, in fact, to settle accounts with our erstwhile philosophical conscience. The resolve was carried out in the form of a criticism of post-Hegelian philosophy. The manuscript, two large octavo volumes, had long reached its place of publication in Westphalia when we received the news that altered circumstances did not allow of its being printed. We abandoned the manuscript to the gnawing criticism of the mice all the more willingly as we had achieved our main purpose—self-clarification. Of the scattered works in which we put our views before the public at that time, now from one aspect, now from another, I will mention only the *Manifesto of the Communist Party*, jointly written by Engels and myself, and *Speech on Free Trade*, published by me. The decisive points of our view were first scientifically, although only polemically, indicated in my work published in 1847 and directed against Proudhon: *The Poverty of Philosophy*, etc. A dissertation written in German on *Wage Labour*, in which I put together my lectures on this subject delivered in the Brussels German Workers' Society, was interrupted, while being printed, by the February Revolution and my consequent forcible removal from Belgium.

The editing of the *Neue Rheinische Zeitung* in 1848 and 1849, and the subsequent events, interrupted my economic studies which could only be resumed in the year 1850 in London. The enormous amount of material for the history of political economy which is accumulated in the British Museum, the favourable vantage point afforded by London for the observation of bourgeois society, and finally the new stage of development upon which the latter appeared to have entered with the discovery of gold in California and Australia, determined me to begin afresh from the beginning and to work through the new material critically. These studies led partly of themselves into apparently quite remote subjects on which I had to dwell for a shorter or longer period. Especially, however, was the time at my disposal curtailed by the imperative necessity of earning my living. My contributions, during eight years now, to the first English-American newspaper, the *New York Tribune*, compelled an extraordinary scattering of my studies, since I occupy myself with newspaper correspondence proper only in exceptional cases. However, articles on striking economic events in England and on the Continent constituted so considerable a part of my contributions that I was compelled to make myself familiar with practical details which lie outside the sphere of the actual science of political economy.

This sketch of the course of my studies in the sphere of political economy is intended only to show that my views, however they may be judged and however little they coincide with the interested prejudices of the ruling classes, are the result of conscientious investigation lasting many years. But at the entrance to science, as at the entrance to hell, the demand must be posted:

> *Qui si convien lasciare ogni sospetto;*
> *Ogni viltà convien che qui sia morta*

[Here all mistrust must be left behind;
Here all cowardice must perish.]

Dante, *Divine Comedy*

BIBLIOGRAPHY

Original: *MEW*, Vol. 13, pp. 7 ff.

Present translation: *MESW*, Vol. 1, pp. 361 ff. Reproduced by kind permission of Lawrence and Wishart Ltd.

Other translations: None.

Commentaries: H. Acton and G. Cohen, 'On some Criticisms of Historical Materialism', *Proceedings of the Aristotelian Society*, Supp. vol. 44, 1970.
M. Bober, *Karl Marx's Interpretation of History*, 2nd edn., New York, 1965.

R. Daniels, 'Fate and Will in the Marxian Philosophy of History', *Journal of the History of Ideas*, 1960.

M. Dobb, Introduction to K. Marx, *Critique of Political Economy*, London, 1971.

M. Evans, *Karl Marx*, New York and London, 1975.

H. Fleischer, *Marxism and History*, London, 1973.

Z. Jordan, *The Evolution of Dialectical Materialism*, London, 1967.

J. Plamenatz, *German Marxism and Russian Communism*, London, 1954.

A. Prinz, 'Background to Marx's Preface of 1859', *Journal of the History of Ideas*, 1969.

J. Sanderson, *An Interpretation of the Political Ideas of Marx and Engels*, London, 1969.

I. Zeitlin, *Marxism: A Re-examination*, New York, 1967.

31. Theories of Surplus Value

Since the early 1850s Marx had intended to append to his 'Economics' a critical review of previous theorists. And when, in 1862/3, he lacked the mental energy to write up the *Grundrisse* for publication, he spent his time in composing this review. Like most of his works it grew while being written and ran to three largish volumes. It was published after Marx's death as a kind of fourth volume of *Capital*.

Most of the first volume deals with Adam Smith and his distinction between productive and unproductive labour; the second volume deals with Ricardo's theories of profit and rent, and the third with the English socialist followers of Ricardo. Since the *Theories of Surplus Value* is only a working draft, it contains much that is of little interest and a large part of it consists merely of extracts from classical economists. But interspersed there are passages of abiding interest such as those excerpted below.

Alienated Labour in Capitalist Society

. . . Already in its simple form this relation is an inversion—personification of the thing and materialization of the person: for what distinguishes this form from all previous forms is that the capitalist does not rule over the labourer through any personal qualities he may have, but only in so far as he is 'capital'; his domination is only that of materialized labour over living labour, of the labourer's product over the labourer himself.

The relation grows still more complicated and apparently more mysterious because, with the development of the specifically capitalist mode of production, it is not only these directly material things (all products of labour; considered as use-values, they are both material conditions of labour and products of labour; considered as exchange-values, they are materialized general labour time or money) that get up on their hind legs to the labourer and confront him as 'capital', but also the forms of socially developed labour—co-operation, manufacture (as a form of division of labour), the factory (as a form of social labour organized on machinery as its material basis)—all these appear as forms of the development of capital, and therefore the productive powers of labour built up on these forms of social labour—consequently also science and the forces of nature—appear as productive powers of capital. In fact, the unity of labour in co-operation, the combination of labour through the division of labour, the use for productive purposes in machine industry of the forces of nature and science alongside the products of labour—all this confronts the individual labourers themselves as something extraneous and objective, as a mere form of existence of the means of labour that are independent of them and control them, just as the means of labour themselves confront them, in their simple visible form as materials, instruments, etc., as functions of capital and consequently of the capitalist.

The social forms of their own labour or the forms of their own social labour are relations that have been formed quite independently of the individual labourers; the labourers, as subsumed under capital, become elements of these social formations—but these social formations do not belong to them. They therefore confront them as forms of capital itself, as combinations belonging to capital, as distinct from their individual labour power, arising from capital and incorporated in it. And this takes on a form that is all the more real the more on the one hand their labour power itself becomes so modified by these forms that it is powerless as an independent force, that is to say, outside this capitalist relationship, and that its independent capacity to produce is destroyed. And on the other hand, with the development of machinery, the conditions of labour seem to dominate labour also technologically, while at the same time they replace labour, oppress it, and make it superfluous in its independent forms.

In this process, in which the social character of their labour confronts them to a certain degree as capitalized (as for example in machinery the visible products of labour appear as dominating labour), the same naturally takes place with the forces of nature and science, the product of general historical development in its abstract quintessence—they confront the labourers as powers of capital. They are separate in fact from the skill and knowledge of the individual labourer—and although, in their origin, they too are the product of labour—wherever they enter into the labour process they appear as embodied in capital. The capitalist who makes use of a machine need not understand it. But science realized in the machine appears as capital in relation to the labourers. And in fact all these applications of science, natural forces and products of labour on a large scale, these applications founded on social labour, themselves appear only as means for the exploitation of labour, as means of appropriating surplus labour, and hence confront labour as powers belonging to capital. Capital naturally uses all these means only to exploit labour; but in order to exploit it, it must apply them in production. And so the development of the social productive powers of labour and the conditions for this development appear as acts of capital, towards which the individual labourer not only maintains a passive attitude, but which take place in opposition to him. . . .

Unproductive Labour

. . . Certain services, or the use-values, resulting from certain forms of activity or labour are embodied in commodities; others, on the contrary, leave no tangible result existing apart from the persons themselves who perform them; in other words, their result is not a vendible commodity. For example, the service a singer renders to me satisfies my aesthetic need; but what I enjoy exists only in an activity inseparable from the singer himself, and as soon as his labour, the singing, is at an end, my enjoyment too is at an end. I enjoy the activity itself—its reverberation on my ear. These services themselves, like the commodities which I buy, may be necessary or may only seem necessary—for example, the service of a soldier or physician or lawyer; or they may be services which give me pleasure. But this makes no difference to their economic character. If I am healthy and do not need a doctor or am lucky enough not to have to be involved in a lawsuit, then I avoid paying out money for medical or legal services as I do the plague.

Services may also be forced on me—the services of officials, etc.

If I buy the service of a teacher not to develop my faculties but to acquire some skill with which I can earn money—or if others buy this teacher for me—and if I really learn something (which in itself is quite independent of the payment for the service), then these costs of education, just as the costs of my maintenance, belong to the costs of production of my labour power. But the particular utility of this service alters nothing in the economic relation; it is not a relation in which I transform money into capital, or by which the supplier of this service, the teacher, transforms me into his capitalist, his master. Consequently it also does not affect the economic character of this relation whether the physician cures me, the teacher is successful in teaching me, or the lawyer wins my lawsuit. What is paid for is the performance of the service as such, and by its very nature the result cannot be guaranteed by those rendering the service. A large proportion of services belongs to the costs of consumption of commodities, as in the case of a cook, a maid, etc.

It is characteristic of all unproductive labours that they are at my command—as in the case of the purchase of all other commodities for consumption—only to the same extent as I exploit productive labourers. Of all persons, therefore, the productive labourer has the least command over the services of unproductive labourers. On the other hand, however, my power to employ productive labourers by no means grows in the same proportion as I employ unproductive labourers, but on the contrary diminishes in the same proportion, although [one has] most to pay for the compulsory services (State, taxes).

Productive labourers may themselves in relation to me be unproductive labourers. For example, if I have my house re-papered and the paper-hangers are wage-workers of a master who sells me the job, it is just the same for me as if I had bought a house already papered; as if I had expended money for a commodity for my consumption. But for the master who gets these labourers to hang the paper, they are productive labourers, for they produce surplus value for him.

How very unproductive, from the standpoint of capitalist production, the labourer is who indeed produces vendible commodities, but only to the amount equivalent to his own labour power, and therefore produces no surplus value for capital—can be seen from the passages in Ricardo saying that the very existence of such people is a nuisance. This is the theory and practice of capital.

We have seen: This process of production is not only a process of the production of commodities, but a process of the production of surplus value, the absorption of surplus labour, and hence a process of production of capital. The first formal act of exchange between money and labour or capital and labour is only potentially the appropriation of someone else's living labour by materialized labour. The actual process of appropriation takes place only in the actual production process, behind which lies as a past stage that first formal transaction—in which capitalist and labourer confront each other as mere owners of commodities, as buyer and seller. For which reason all vulgar economists—like Bastiat—go no further than the first formal transaction, precisely in order by this trick to get rid of the specific capitalist relation. The distinction is shown in a striking way by the exchange of money for unproductive labour. Here money and labour exchange with each other only as commodities. So that instead of this exchange forming capital, it is expenditure of revenue. . . .

Peasants and Artisans in Capitalist Society

. . . What then is the position of independent handicraftsmen or peasants who employ no labourers and therefore do not produce as capitalists? Either, as always in the case of peasants but for example not in the case of a gardener whom I get to come to my house, they are producers of commodities, and I buy the commodity from them—in which case for example it makes no difference that the handicraftsman produces it to order while the peasant produces his supply according to his means. In this capacity they confront me as sellers of commodities, not as sellers of labour, and this relation therefore has nothing to do with the exchange of capital for labour; therefore also it has nothing to do with the distinction between productive and unproductive labour, which depends entirely on whether the labour is exchanged for money or for money as money as capital. They therefore belong neither to the category of productive nor of unproductive labourers, although they are producers of commodities. But their production does not fall under the capitalist mode of production.

It is possible that these producers, working with their own means of production, not only reproduce their labour power but create surplus value, while their position enables them to appropriate for themselves their own surplus labour or a part of it (since a part of it is taken away from them in the form of taxes, etc.). And here we come up against a peculiarity that is characteristic of a society in which one definite mode of production predominates, even though not all productive relations have been subordinated to it. In feudal society, for example (as we can best observe in England because the system of feudalism was introduced here from Normandy ready made, and its form was impressed on what was in many respects a different social foundation), relations which were far removed from the nature of feudalism were given a feudal form; for example, simple money relations in which there was no trace of mutual personal service as between lord and vassal. It is for instance a fiction that the small peasant held his land in fief.

It is exactly the same in the capitalist mode of production. The independent peasant or handicraftsman is cut up into two persons. As owner of the means of production he is capitalist; as labourer he is his own wage-labourer. As capitalist he therefore pays himself his wages and draws his profit on his capital; that is to say, he exploits himself as wage-labourer, and pays himself, in the surplus value, the tribute that labour owes to capital. Perhaps he also pays himself a third portion as landowner (rent), in exactly the same way, as we shall see later, that the industrial capitalist, when he works with his own capital, pays himself interest, regarding this as something which he owes to himself not as industrial capitalist but qua capitalist pure and simple.

The determinate social character of the means of production in capitalist production—expressing a particular production relation—has so grown together with, and in the mode of thought of bourgeois society is so inseparable from, the material existence of these means of production as means of production, that the same determinateness (categorical determinateness) is assumed even where the relation is in direct contradiction to it. The means of production become capital only in so far as they have become separated from labourer and confront labour as an independent power. But in the case referred to the producer—the labourer—is the possessor, the owner, of his means of production. They are therefore not capital, any more than in relation to them he is a wage-labourer. Nevertheless they are looked on as capital, and he himself is split in two, so that he, as capitalist, employs himself as wage-labourer.

In fact this way of presenting it, however irrational it may be on first view, is nevertheless so far correct, that in this case the producer in fact creates his own surplus value on the assumption that he sells his commodity at its value, in other words, only his own labour is materialized in the whole product. But that he is able to appropriate for himself the whole product of his own labour, and that the excess of the value of his product over the average price for instance of his day's labour is not appropriated by a third person, a master, he owes not to his labour—which does not distinguish him from other labourers—but to his ownership of the means of production. It is therefore only through his ownership of these that he takes possession of his own surplus labour, and thus bears to himself as wage-labourer the relation of being his own capitalist.

Separation appears as the normal relation in this society. Where therefore it does not in fact apply, it is presumed and, as has just been shown, so far correctly; for (as distinct for example from conditions in Ancient Rome or Norway or in the north-west of the United States) in this society unity appears as accidental, separation as normal; and consequently separation is maintained as the relation even when one person unites the separate functions. Here emerges in a very striking way the fact that the capitalist as such is only a function of capital, the labourer a function of labour power. For it is also a law that economic development distributes functions among different persons; and the handicraftsman or peasant who produces with his own means of production will either gradually be transformed into a small capitalist who also exploits the labour of others, or he will suffer the loss of his means of production (in the first instance the latter may happen although he remains their nominal owner, as in the case of mortgages) and be transformed into a wage-labourer. This is the tendency in the form of society in which the capitalist mode of production predominates. . . .

Ricardo and the Value of Labour

. . . The value of labour is therefore determined by the means of subsistence which, in a given society, are traditionally necessary for the maintenance and reproduction of the labourers.

But why? By what law is the value of labour determined in this way?

Ricardo has in fact no answer, other than that the law of supply and demand reduces the average price of labour to the means of subsistence that are necessary (physically or socially necessary in a given society) for the maintenance of the labourer. He determines value here, in one of the basic propositions of the whole system, by demand and supply—as Say notes with malicious pleasure.

Instead of labour, Ricardo should have discussed labour power. But had he done so, capital would also have been revealed as the material conditions of labour, confronting the labourer as power that had acquired an independent existence, and capital would at once have been revealed as a definite social relationship. . . .

Ricardo and Surplus Value

. . . Apart from the confusion between labour and labour power, Ricardo defines the average wages or the value of labour correctly. For he says that the value of labour is determined neither by the money nor by the means of subsistence which the labourer receives, but by the labour time which it costs to produce it; that is, by the quantity of labour materialized in the means of subsistence of the labourer. This he calls the real wages. (See later.)

This definition of the value of labour, moreover, necessarily follows from his theory. Since the value of labour is determined by the value of the necessary means of subsistence on which this value is to be expended, and the value of the means of subsistence, like that of all other commodities, is determined by the quantity of labour they contain, it naturally follows that the value of labour equals the value of the means of subsistence, which equals the quantity of labour expended upon them.

However correct this formula is (apart from the direct opposition of labour and capital), it is, nevertheless, inadequate. Although in replacement of his wages the individual labourer does not directly produce—or reproduce, taking into account the continuity of this process—products on which he lives he may produce products which do not enter into his consumption at all, and even if he produces necessary means of subsistence, he may, due to the division of labour, only produce a single part of the necessary means of subsistence, for instance corn—and even that only in one form (for example in that of corn, not bread), but he produces commodities to the value of his means of subsistence, that is, he produces the value of his means of subsistence. This means, therefore, if we consider his daily average consumption, that the labour time which is contained in his daily means of subsistence, forms one part of his working-day. He works one part of the day in order to reproduce the value of his means of subsistence; the commodities which he produces in this part of the working-day have the same value, or represent a quantity of labour time equal to that contained in his daily means of subsistence. It depends on the value of these means of subsistence—in other words on the social productivity of labour and not on the productivity of the individual branch of production in which he works—how great a part of his working-day is devoted to the reproduction or production of the value, i.e., the equivalent, of his means of subsistence.

Ricardo of course assumes that the labour time contained in the daily means of subsistence is equal to the labour time which the labourer must work daily in order to reproduce the value of these means of subsistence. But by not directly showing that one part of the labourer's working-day is assigned to the reproduction of the value of his own labour power, he introduces a difficulty and obscures the clear understanding of the relationship. A twofold confusion arises from this. The origin of surplus value does not become clear, and consequently Ricardo is reproached by his successors for having failed to grasp and expound the nature of surplus value. That is part of the reason for their scholastic attempts at explaining it. But because the origin and nature of surplus value is in this way not clearly comprehended, the surplus labour plus the necessary labour, in short, the total working-day, is regarded as a fixed magnitude, the differences in the amount of surplus value are overlooked, and the productivity of capital, the compulsion to perform surplus labour—on the one hand to perform absolute surplus labour, and on the other its innate urge to shorten the necessary labour time—are not recognized, and therefore the historical justification for capital is not set forth. Adam Smith, however, had already stated the correct formula. Important as it was, to resolve value into labour, it was equally important to resolve surplus value into surplus labour, and to do so in explicit terms.

Ricardo starts out from the actual fact of capitalist production. The value of labour is smaller than the value of the product which it creates. The value of the product is therefore greater than the value of the labour which produces it, or the value of the wages. The excess of the value of the product over the value of the wages is the surplus value. (Ricardo wrongly uses the word profit, but, as we noted earlier, he identifies profit with surplus value here and is really speaking of the latter.) For him it is a fact, that the value of the product is greater than the value of the wages. How this fact arises remains unclear. The total working-day is greater than that part of the working-day which is required for the production of the wages. Why? That does not emerge. The magnitude of the total working-day is therefore wrongly assumed to be fixed, and directly entails wrong conclusions. The increase or decrease in surplus value can therefore be explained only from the growing or diminishing productivity of social labour which produces the means of subsistence. That is to say, only relative surplus value is understood.

It is obvious that if the labourer needed his whole day to produce his own means of subsistence (i.e., commodities equal to the value of his own means of subsistence), there could be no surplus value, and therefore no capitalist production and no wage-labour. This can only exist when the productivity of social labour is sufficiently developed to make possible some sort of excess of the total working-day over the labour time required for the reproduction of the wage—i.e., surplus labour, whatever its magnitude. But it is equally obvious, that with a given labour time (a given length of the working-day) the productivity of labour may be very different; on the other hand, with a given productivity of labour, the labour time, the length of the working-day, may be very different. Furthermore, it is clear that though the existence of surplus labour presupposes that the productivity of labour has reached a certain level, the mere possibility of this surplus labour (i.e., the existence of that necessary minimum productivity of labour), does not in itself make it a reality. For this to occur, the labourer must first be compelled to work in excess of the necessary time, and this compulsion is exerted by capital. This is missing in Ricardo's work, and therefore also the whole struggle over the regulation of the normal working-day.

At a low stage of development of the social productivity of labour, that is to say, where the surplus labour is relatively small, the class of those who live on the labour of others will generally be small in relation to the number of labourers. It can considerably grow (proportionately) in the measure in which productivity and therefore relative surplus value develop.

It is moreover understood that the value of labour varies greatly in the same country at different periods and in different countries during the same period. The temperate zones are however the home of capitalist production. The social productive power of labour may be very undeveloped; yet this may be compensated precisely in the production of the means of subsistence, on the one hand, by the fertility of the natural agents, such as the land; on the other hand, by the limited requirements of the population, due to climate, etc.—this is, for instance, the case in India. Where conditions are primitive, the minimum wage may be very small (quantitatively in use-values) because the social needs are not yet developed though it may cost much labour. But even if an average amount of labour were required to produce this minimum wage, the surplus value created, although it would be high in proportion to the wage (to the necessary labour time), would, even with a high rate of surplus value, be just as meagre (proportionately)—when expressed in terms of use-values—as the wage itself.

Let the necessary labour time be 10 hours, the surplus labour 2 hours, and the total working-day 12 hours. If the necessary labour-time were 12 hours, the surplus labour $2\frac{2}{5}$ hours and the total working-day $14\frac{2}{5}$ hours, then the values produced would be very different. In the first case they would amount to 12 hours, in the second to $14\frac{2}{5}$ hours. Similarly, the absolute magnitude of the surplus value: in the former case it would be 2 hours, in the latter $2\frac{2}{5}$. And yet the rate of surplus value or of surplus labour would be the same, because $2:10 = 2\frac{2}{5}:12$. If, in the second case, the variable capital which is laid out were greater, then so also would be the surplus value or surplus labour appropriated by it. If in the latter case, the surplus labour were to rise by $\frac{3}{5}$ hours instead of by $\frac{2}{5}$ hours, so that it would amount to 3 hours and the total working-day to 15 hours, then, although the necessary labour time or the minimum wage had increased, the rate of surplus value would have risen, for $2:10 = \frac{1}{5}$; but $3:12 = \frac{1}{4}$. Both could occur if, as a result of the corn, etc., becoming dearer, the minimum wage had increased from 10 to 12 hours. Even in this case, therefore, not only might the rate of surplus value remain the same, but the amount and rate of surplus value might grow.

But let us suppose that the necessary wage amounted to 10 hours, as previously, the surplus labour to 2 hours and all other conditions remained the same (that is, leaving out of account here any lowering in the production costs of constant capital). Now let the labourer work $2\frac{2}{5}$ hours longer, and appropriate 2 hours, while the $\frac{2}{5}$ forms surplus labour. In this case wages and surplus value would increase in equal proportion, the former, however, representing more than the necessary wage or the necessary labour time.

If one takes a given magnitude and divides it into two parts, it is clear that one part can only increase in so far as the other decreases, and vice versa.

But this is by no means the case with expanding (elastic) magnitudes. And the working-day represents such an elastic magnitude, as long as no normal working-day has been won. With such magnitudes, both parts can grow, either to an equal or unequal extent. An increase in one is not brought about by a decrease in the other and vice versa. This is moreover the once case in which wages and surplus value, in terms of exchange value, can both increase and possibly even in equal proportions. That they can increase in terms of use-value is self-evident; this can increase even if, for example, the value of labour decreases. From 1797 to 1815, when the price of corn and also the nominal wage rose considerably in England, the daily hours of labour increased greatly in the principal industries, which were then in a phase of ruthless expansion; and I believe that this arrested the fall in the rate of profit, because it arrested the fall in the rate of surplus value. In this case, however, whatever the circumstances, the normal working-day is lengthened and the normal span of life of the labourer, hence the normal duration of his labour power, is correspondingly shortened. This applies where a permanent lengthening of the working-day occurs. If it is only temporary, in order to compensate for a temporary rise in wages, it may (except in the case of children and women) have no other result than to prevent a fall in the rate of profit in those enterprises where the nature of the work makes a prolongation of labour time possible. (This is least possible in agriculture.)

Ricardo did not consider this at all since he investigated neither the origin of surplus value nor absolute surplus value and therefore regarded the working-day as a given magnitude. For this case, therefore, *his law*— that surplus value and wages (he erroneously says profit and wages) in terms of exchange-value can rise or fall only in *inverse* proportion—*is incorrect*.

Firstly let us assume that the necessary labour time and the surplus labour remain constant. That is 10 hours + 2 hours; the working-day equals 12 hours, surplus value equals 2 hours; the rate of surplus value is $\frac{1}{5}$.

In the second example the necessary labour time remains the same; surplus labour increases from 2 to 4 hours. Hence $10 + 4 =$ a working-day of 14 hours; surplus value equals 4 hours; rate of surplus value is $4:10 = \frac{4}{10} = \frac{2}{5}$.

In both cases the necessary labour time is the same; but the surplus value in the one case is twice as great as in the other and the working-day in the second case is one-sixth longer than in the first. Furthermore, although the wage is the same, the values produced, corresponding to the quantities of labour, would be very different; in the first case it would be equal to 12 hours, in the second to $12 + \frac{12}{6} = 14$ hours. It is therefore wrong to say that, provided the wage is the same (in terms of value, of necessary labour time), the surplus value contained in two commodities is proportionate to the quantities of labour contained in them. This is only correct where the normal working-day is the same.

Let us further assume that as a result of the rise in the productive power of labour, the necessary wage (although it remains constant in terms of use-values) falls from 10 to 9 hours and similarly that the surplus labour time falls from 2 to $1\frac{4}{5}$ hours ($\frac{9}{5}$). In this case $10:9 = 2:1\frac{4}{5}$. Thus the surplus labour time would fall in the same proportion as the necessary labour time. The rate of surplus value would be the same in both cases, for $2 = \frac{10}{5}$ and $1\frac{4}{5} = \frac{9}{5}$. $1\frac{4}{5}:9 = 2:10$. The quantity of use values that could be bought with the surplus value, would—according to the assumption—also remain the same. (But this would apply only to those use values which are necessary means of subsistence.) The working-day would decrease from 12 to $10\frac{4}{5}$ hours. The amount of value produced in the second case would be smaller than that produced in the first. And despite these unequal quantities of labour, the rate of surplus value would be the same in both cases.

In discussing surplus value we have distinguished between surplus value and the rate of surplus value. Considered in relation to one working-day, the surplus value is equal to the absolute number of hours which it represents, 2, 3, etc. The rate is equal to the proportion of this number of hours to the number of hours which makes up the necessary labour time. This distinction is very important, because it indicates the varying length of the working-day. If the surplus value equals 2 hours, then the rate is $\frac{1}{5}$, if the necessary labour time is 10 hours; and $\frac{1}{6}$, if the necessary labour time is 12 hours. In the first case the working-day consists of 12 hours and in the second of 14. In the first case the rate of surplus value is greater, while at the same time the labourer works a smaller number of hours per day. In the second case the rate of surplus value is smaller, the value of the labour power is greater, while at the same time the labourer works a greater number of hours per day. This shows that, with a constant surplus value, but a working-day of unequal length, the rate of surplus value may be different. The earlier case, $10:2$ and $9:1\frac{4}{5}$, shows how with a constant rate of surplus value, but a working-day of unequal length, the surplus value itself may be different, in one case 2 hours and in the other $1\frac{4}{5}$ hours.

I have shown previously (Chapter II), that if the length of the working-day and the necessary labour time, and therefore the rate of surplus value are given, the amount of surplus value depends on the number of workers simultaneously employed by the same capital. This was a tautological statement. For if 1 working-day gives me 2 surplus hours, than 12 working-days give me 24 surplus hours or 2 surplus days. The statement, however, becomes very important in connection with the determination of profit, which is equal to the proportion of surplus value to the capital advanced, thus depending on the absolute amount of surplus value. It becomes important because capitals of equal size but different organic composition employ unequal numbers of labourers; they must thus produce unequal amounts of surplus value, and therefore unequal profits. With a falling rate of surplus value, the profit may rise and with a rising rate of surplus value, the profit may fall; or the profit may remain unchanged, if a rise or fall in the rate of surplus value is compensated by a counter movement affecting the number of workers employed. Here we see immediately, how extremely wrong it is to identify the laws relating to the rise and fall of surplus value with the laws relating to the rise and fall of profit. If one merely considers the simple law of surplus value, then it seems a tautology to say that with a given rate of surplus value (and a given length of the working-day), the absolute amount of surplus value depends on the amount of capital employed. For an increase in this amount of capital and an increase in the number of labourers simultaneously employed are, on the assumption made, identical, or merely different expressions of the same fact. But when one turns to an examination of profit, where the amount of the total capital employed and the number of workers employed vary greatly for capitals of equal size, then the importance of the law becomes clear.

Ricardo starts by considering commodities of a given value, that is to say, commodities which represent a given quantity of labour. And from this starting-point, absolute and relative surplus value appear to be always identical. (This at any rate explains the one-sidedness of his mode of procedure and corresponds with his whole method of investigation: to start with the value of the commodities as determined by the definite labour time they contain, and then to examine to what extent this is affected by wages, profits, etc.) This appearance is nevertheless false, since it is not a question of commodities here, but of capitalist production, of commodities as products of capital.

Assume that a capital employs a certain number of workers, for example 20, and that wages amount to £20. To simplify matters let us assume that the fixed capital is nil, i.e., we leave it out of account. Further, assume that these 20 workers spin £80 of cotton into yarn, if they work 12 hours per day. If 1 lb of cotton costs 1s. then 20 lb costs £1 and £80 represents 1600 lb. If 20 workers spin 1600 lb in 12 hours, then they spin $\frac{1600}{12}$ lb, which is $133\frac{1}{3}$ lb in one hour. Thus, if the necessary labour time is 10 hours, then the surplus labour time is 2 hours and this equals $266\frac{2}{3}$ lb of yarn. The value of the 1600 lb would be £104. For if 10 hours of work equal £20, then 1 hour of work equals £2 and 2 hours of work £4, hence 12 hours of work are equal to £24. (Raw material £80 + £24 the newly-created value are equal to £104.)

But if each of the workers worked 4 hours of surplus labour, then their product would be equal to £8 (I mean the surplus value which he creates—his product is in fact equal to £28). The total product would be £121$\frac{1}{3}$. And this £121$\frac{1}{3}$ would be the equivalent of 1866$\frac{2}{3}$ lb of yarn. As before, since the conditions of production remained the same, 1 lb of yarn would have the same value; it would contain the same amount of labour time. Moreover, according to the assumption, the necessary wages—their value, the labour time they contained—would have remained unchanged.

Whether these 1866$\frac{2}{3}$ lb of yarn were being produced under the first set of conditions or under the second, i.e., with 2 or with 4 hours surplus labour, they would have the same value in both cases. The value therefore of the additional $266\frac{2}{3}$ lb of cotton that are spun, is $£13\frac{1}{3}$. This, added to the $£80$ for the 1600 lb, amounts to $£93\frac{1}{3}$ and in both cases 4 working-hours more for 20 men amount to $£8$. Altogether $£28$ for the labour, that is $£121\frac{1}{3}$. The wages are, in both cases, the same. The pound of yarn costs in both cases $1\frac{3}{10}s.$ Since the value of the pound of cotton is $1s.$, what remained for the newly-added labour in 1 lb of yarn would in both cases amount to $\frac{3}{10}s.$, equal to $3\frac{3}{5}d.$ (or $\frac{18}{5}d.$).

Nevertheless, under the conditions assumed, the relation between value and surplus value in each pound of yarn would be very different. In the first case, since the necessary labour was equal to $£20$ and the surplus labour to $£4$, or since the former amounted to 10 hours and the latter to 2 hours, the ratio of surplus labour to necessary labour would be $2:10 = \frac{2}{10} = \frac{1}{5}.$ (Similarly $£4:£20 = \frac{4}{20} = \frac{1}{5}.$) The $3\frac{3}{5}d.$ newly-added labour in a pound of yarn would in this case contain $\frac{1}{5}$ unpaid labour, that is $\frac{18}{25}d.$ or $\frac{72}{25}$ farthings equal to $2\frac{22}{25}$ farthings. In the second case, on the other hand, the necessary labour would be $£20$ (10 working-hours), the surplus labour $£8$ (4 working-hours). The ratio of surplus labour to necessary labour would be $8:20 = \frac{8}{20} = \frac{4}{10} = \frac{2}{5}.$ Thus the $3\frac{3}{5}d.$ of newly-added labour in a pound of yarn would contain $\frac{2}{5}$ unpaid labour, i.e., $5\frac{19}{25}$ farthings or $1d.$ $1\frac{19}{25}$ farthings. Although the yarn has the same value in both cases and although the same wages are paid in both cases, the surplus value in a pound of yarn is in one case twice as large as in the other. The ratio of value of labour to surplus value is of course the same in the individual commodity, that is, in a portion of the product, as in the whole product.

In the one case, the capital advanced is $£93\frac{1}{3}$ for cotton, and how much for wages? The wages for 1600 lb amount to $£20$ here, hence for the additional $266\frac{2}{3}$ lb a further $£3\frac{1}{3}$. This makes $£23\frac{1}{3}$. And the total capital outlay is $£93\frac{1}{3} + £23\frac{1}{3} = £116\frac{2}{3}$. The product comes to $£121\frac{1}{3}$. (The additional outlay in variable capital, of $£3\frac{1}{3}$ only yields $13\frac{1}{3}s.$ $£\frac{2}{3}$ surplus value. $£20:£4 = £3\frac{1}{3} + £\frac{2}{3}.$)

In the other case, however, the capital outlay would amount to only $£93\frac{1}{9} + £20 = £113\frac{1}{3}$, and $£4$ would have to be added to the $£4$ surplus value. The same number of pounds of yarn are produced in both cases and both have the same value, that is to say, they represent equal total quantities of labour, but these equal total quantities of labour are set in motion by capitals of unequal size, although the wages are the same; but the working-days are of unequal length and, therefore, unequal quantities of unpaid labour are produced. Taking the individual pound of yarn, the wages paid for it, or the amounts of paid labour a pound contains, are different. The same wages are spread over a larger volume of commodities here, not because labour is more productive in the one case than in the other, but because the total amount of unpaid labour which is set into motion in one case is greater than in the other. With the same quantity of paid labour, therefore, more pounds of yarn are produced in the one case than in the other, although in both cases the same quantities of yarn are produced, representing the same quantity of total labour (paid and unpaid). If, on the other hand, the productivity of labour had increased in the second case, then the value of the pound of yarn would at all events have fallen, whatever the ratio of surplus value to variable capital.

In such a case, therefore, it would be wrong to say that—because the value of the pound of yarn is fixed at $1s. 3\frac{3}{5}d.$, the value of the labour which is added is also fixed and amounts to $3\frac{3}{5}d.$, and the wages, i.e., the necessary labour time, remain, according to the assumption, unchanged—the surplus value must be the same and the two capitals under otherwise equal conditions would have produced the yarn with equal profits. This would be correct if we were concerned with one pound of yarn, but we are in fact concerned here with a capital which has produced $1866\frac{2}{3}$ lb of yarn. And in order to know the amount of profit (actually of surplus value) on one pound, we must know the length of the working-day, or the quantity of unpaid labour (when the productivity is given) that the capital sets in motion. But this information cannot be gathered by looking at the individual commodity.

Thus Ricardo deals only with what I have called the relative surplus value. From the outset he assumes, as Adam Smith and his predecessors seem to have done as well, that the length of the working-day is given. (At most, Adam Smith mentions differences in the length of the working-day in different branches of labour, which are levelled out or compensated by the relatively greater intensity of labour, difficulty, unpleasantness, etc.) On the basis of this postulate Ricardo, on the whole, explains relative surplus value correctly. Before we give the principal points of his theory, we shall cite a few more passages to illustrate Ricardo's point of view.

The labour of a million of men in manufactures, will always produce the same value, but will not always produce the same riches.

This means that the product of their daily labour will always be the product of a million working-days containing the same labour time; this is wrong, or is only true where the same normal working-day—taking into account the various difficulties etc. in different branches of labour—has been generally established.

Even then, however, the statement is wrong in the general form in which it is expressed here. If the normal working-day is 12 hours, and the annual product of one man is, in terms of money, £50 and the value of money remains unchanged, then, in this case, the product of 1 million men would always amount to £50 million per year. If the necessary labour is 6 hours, then the capital laid out for these million men would be £25,000,000 per annum. The surplus value would also be £25 million. The product would always be 50 million, whether the workers received 25 or 30 or 40 million. But in the first case the surplus value would be 25 million, in the second it would be 20 million and in the third 10 million. If the capital advanced consisted only of variable capital, i.e., only of the capital which is laid out in the wages of these 1 million men, then Ricardo would be right. He is, therefore, only right in the one case, where the total capital equals the variable capital; a presupposition which pervades all his, and Adam Smith's, observations regarding the capital of society as a whole, but in capitalist production this precondition does not exist in a single branch of industry, much less in the production of society as a whole.

That part of the constant capital which enters into the labour process without entering into the process of the creation of value, does not enter into the product, into the value of the product, and, therefore, important as it is in the determination of the general rate of profit, it does not concern us here, where we are considering the value of the annual product. But matters are quite different with that part of constant capital which enters into the annual product. We have seen that a portion of this part of constant capital, or what appears as constant capital in one sphere of production, appears as a direct product of labour within another sphere of production, during the same production period of one year; a large part of the capital laid out annually, which appears to be constant capital from the standpoint of the individual capitalist or the particular sphere of production, therefore, resolves itself into variable capital from the standpoint of society or of the capitalist class. This part is thus included in the 50 million, in that part of the 50 million which forms variable capital or is laid out in wages.

But the position is different with that part of constant capital which is used up in order to replace the constant capital consumed in industry and agriculture—with the consumed part of the constant capital employed in those branches of production which produce constant capital, raw material in its primary form, fixed capital, and auxiliary materials. The value of this part reappears, it is reproduced in the product. In what proportion it enters into the value of the whole product depends entirely on its actual magnitude—provided the productivity of labour does not change; but however the productivity may change, this part of the constant capital will always have a definite magnitude. (On average, apart from certain exceptions in agriculture, the amount of the product, i.e., the riches—which Ricardo distinguishes from the value—produced by one million men will, indeed, also depend on the magnitude of this constant capital which is antecedent to production.) This part of the value of the product would not exist without the new labour of the million men during the year. On the other hand, the labour of the million men would not yield the same amount of product without this constant capital which exists independently of their year's labour. It enters into the labour process as a condition of production but not a single additional hour is worked in order to reproduce the value of this part.

As value it is, therefore, not the result of the year's labour, although its value would not have been reproduced without this year's labour.

If the part of the constant capital which enters into the product were 25 million, then the value of the product of the one million men would be 75 million; if this part of the constant capital were 10 million, then the value of the product would only be 60 million, etc. And since the ratio of constant capital to variable capital increases in the course of capitalist development, the value of the annual product of a million men will tend to rise continuously, in proportion to the growth of the past labour which plays a part in their annual production. This alone shows that Ricardo was unable to understand either the essence of accumulation or the nature of profit.

With the growth in the proportion of constant to variable capital, grows also the productivity of labour, the productive forces brought into being, with which social labour operates. As a result of this increasing productivity of labour, however, a part of the existing constant capital is continuously depreciated in value, for its value depends not on the labour time that it cost originally, but on the labour time with which it can be reproduced, and this is continuously diminishing as the productivity of labour grows. Although, therefore, the value of the constant capital does not increase in proportion to its amount, it increases nevertheless, because its amount increases even more rapidly than its value falls. But we shall return later to Ricardo's views on accumulation.

It is evident, however, that if the length of the working-day is given, the value of the annual product of the labour of one million men will differ greatly according to the different amount of constant capital that enters into the product; and that, despite the growing productivity of labour, the value of this product will be greater where the constant capital forms a large part of the total capital, than under social conditions where it forms a relatively small part of the total capital. With the advance in the productivity of social labour, accompanied as it is by the growth of constant capital, a relatively ever increasing part of the annual product of labour will, therefore, fall to the share of capital as such, and thus property in the form of capital (apart from revenue) will be constantly increasing, and proportionately that part of value which the individual worker and even the working class creates will be steadily decreasing, compared with the product of their past labour that confronts them as capital. The alienation and the antagonism between labour power and the objective conditions of labour which have become independent in the form of capital, thereby grow continuously. (Not taking into account the variable capital, i.e., that part of the product of the annual labour which is required for the reproduction of the working class; even these means of subsistence, however, confront them as capital.)

Ricardo's view, that the working-day is given, limited, a fixed magnitude, is also expressed by him in other forms, for instance: 'They' (the wages of labour and profit of stock) are 'together always of the same value', in other words this only means that the (daily) labour time whose product is divided between the wages of labour and the profits of stock, is always the same, is constant.

Wages and profits together will be of the same value.

I hardly need to repeat here that in these passages one should always read 'surplus value' instead of 'profit'. 'Wages and profits taken together will continue always of the same value.' 'Wages are to be estimated by their real value, viz., by the quantity of labour and capital employed in producing them, and not by their nominal value either in coats, hats, money, or corn.'

The value of the means of subsistence which the worker obtains (buys with his wages), corn, clothes, etc., is determined by the total labour time required for their production, the quantity of immediate labour as well as the quantity of materialized labour necessary for their production. But Ricardo confuses the issue because he does not state it plainly, he does not say: 'their real value, viz., that quantity of the working-day required to reproduce the value of their the workers own necessaries, the equivalent of the necessaries paid to them, or exchanged for their labour.' Real wages have to be determined by the average time which the worker must work each day in order to produce or reproduce his own wages. . . .

Ricardo and the Middle Class

. . . There are two tendencies which constantly cut across one another; firstly, to employ as little labour as possible, in order to produce the same or a greater quantity of commodities, in order to produce the same or a greater net produce, surplus value, net revenue; secondly, to employ the largest possible number of workers (although as few as possible in proportion to the quantity of commodities produced by them), because—at a given level of productivity—the mass of surplus value and of surplus product grows with the amount of labour employed. The one tendency throws the labourers on to the streets and makes a part of the population redundant, the other absorbs them again and extends wage-slavery absolutely, so that the lot of the worker is always fluctuating but he never escapes from it. The worker, therefore, justifiably regards the development of the productive power of his own labour as hostile to himself; the capitalist, on the other hand, always treats him as an element to be eliminated from production. These are the contradictions with which Ricardo struggles in this chapter. What he forgets to emphasize is the constantly growing number of the middle classes, those who stand between the workman on the one hand and the capitalist and landlord on the other. The middle classes maintain themselves to an ever increasing extent directly out of revenue; they are a burden weighing heavily on the working base, and increase the social security and power of the upper ten thousand. . . .

Production and Consumption

... Once the distinction between constant capital and variable capital has been grasped, a distinction which arises simply out of the immediate process of production, out of the relationship of the different component parts of capital to living labour, it also becomes evident that in itself it has nothing to do with the absolute amount of the consumption goods produced, although plenty with the way in which these are realized. The way, however, of realizing the gross revenue in different commodities is not, as Ricardo has it, and Barton intimates it, the cause, but the effect of the immanent laws of capitalistic production, leading to a diminishing proportion, compared with the total amount of produce, of that part of it which forms the fund for the reproduction of the labouring class. If a large part of the capital consists of machinery, raw materials, auxiliary materials etc., then a smaller portion of the working class as a whole will be employed in the reproduction of the means of subsistence which enter into the consumption of the workers. This relative diminution in the reproduction of variable capital, however, is not the reason for the relative decrease in the demand for labour, but on the contrary, its effect. Similarly: a larger section of the workers employed in the production of articles of consumption which enter into revenue in general, will produce articles of consumption that are consumed by—are exchanged against the revenue of—capitalists, landlords, and their retainers (state, church etc.), and a smaller section will produce articles destined for the revenue of the workers. But this again is effect, not cause. A change in the social relation of workers and capitalists, a revolution in the conditions governing capitalist production, would change this at once. The revenue would be 'realized in different commodities', to use an expression of Ricardo's.

There is nothing in the, so to speak, physical conditions of production which forces the above to take place. The workmen, if they were dominant, if they were allowed to produce for themselves, would very soon, and without great exertion, bring the capital (to use a phrase of the vulgar economists) up to the standard of their needs. The very great difference is whether the available means of production confront the workers as capital and can therefore be employed by them only in so far as it is necessary for the increased production of surplus value and surplus produce for their employers, in other words whether the means of production employ the workers, or whether the workers, as subjects, employ the means of production—in the accusative case—in order to produce wealth for themselves. It is of course assumed here that capitalist production has already developed the productive forces of labour in general to a sufficiently high level for this revolution to take place. ...

The Unhistorical Outlook of Classical Economy

Ricardo championed bourgeois production in so far as it signified the most unrestricted development of the social productive forces, unconcerned for the fate of those who participate in production, be they capitalists or workers. He insisted upon the historical justification and necessity of this stage of development. His very lack of a historical sense of the past meant that he regarded everything from the historical standpoint of his time. Malthus also wishes to see the freest possible development of capitalist production, however only in so far as the condition of this development is the poverty of its main basis, the working classes, but at the same time he wants it to adapt itself to the 'consumption needs' of the aristocracy and its branches in State and Church, to serve as the material basis for the antiquated claims of the representatives of interests inherited from feudalism and the absolute monarchy. Malthus wants bourgeois production as long as it is not revolutionary, constitutes no historical factor of development, but merely creates a broader and more comfortable material basis for the 'old' society.

On the one hand, therefore, there is the working class, which, according to the population principle, is always redundant in relation to the means of life available to it, overpopulation arising from under-production; then there is the capitalist class, which, as a result of this population principle, is always able to sell the workers' own product back to them at such prices that they can only obtain enough to keep body and soul together; then there is an enormous section of society consisting of parasites and gluttonous drones, some of them masters and some servants, who appropriate, partly under the title of rent and partly under political titles, a considerable mass of wealth gratis from the capitalists, whose commodities they pay for above their value with money extracted from these same capitalists; the capitalist class, driven into production by the urge for accumulation, the economically unproductive sections representing prodigality, the mere urge for consumption. This is moreover advanced as the only way to avoid overproduction, which exists alongside overpopulation in relation to production. The best remedy for both is declared to be overconsumption by the classes standing outside production. The disproportion between the labouring population and production is eliminated by part of the product being devoured by non-producers and idlers. The disproportion arising from over-production by the capitalists is eliminated by means of overconsumption by the owners of wealth.

. . . At any rate nobody has better and more precisely than Ricardo elaborated the point that bourgeois production is not production of wealth for the producers (as he repeatedly calls the workers) and that therefore the production of bourgeois wealth is something quite different from the production of 'abundance', of 'necessaries and luxuries' for the men who produce them, as this would have to be the case if production were only a means for satisfying the needs of the producers through production dominated by use-value alone. Nevertheless, the same Ricardo says:

'If we lived in one of Mr. Owen's parallelograms, and enjoyed all our productions in common, then no one could suffer in consequence of abundance, but as long as society is constituted as it now is, abundance will often be injurious to producers, and scarcity beneficial to them' ([Ricardo], *On Protection to Agriculture*, fourth ed., London, 1822, p. 21).

Ricardo regards bourgeois, or more precisely, capitalist production as the absolute form of production, whose specific forms of production relations can therefore never enter into contradiction with, or enfetter, the aim of production—abundance—which includes both mass and variety of use-values, and which in turn implies a profuse development of man as producer, an all-round development of his productive capacities. And this is where he lands in an amusing contradiction: when we are speaking of value and riches, we should have only society as a whole in mind. But when we speak of capital and labour, then it is self-evident that 'gross revenue' only exists in order to create 'net revenue'. In actual fact, what he admires most about bourgeois production is that its definite forms—compared with previous forms of production—provide scope for the boundless development of the productive forces. When they cease to do this, or when contradictions appear within which they do this, he denies the contradictions, or rather, expresses the contradiction in another form by representing wealth as such—the mass of use-values in itself—without regard to the producers, as the *ultima Thule* [ultimate aim].

Sismondi is profoundly conscious of the contradictions in capitalist production; he is aware that, on the one hand, its forms—its production relations—stimulate unrestrained development of the productive forces and of wealth; and that, on the other hand, these relations are conditional, that their contradictions of use-value and exchange-value, commodity and money, purchase and sale, production and consumption, capital and wage-labour, etc., assume ever greater dimensions as productive power develops. He is particularly aware of the fundamental contradiction: on the one hand, unrestricted development of the productive forces and increase of wealth which, at the same time, consists of commodities and must be turned into cash; on the other hand, the system is based on the fact that the mass of producers is restricted to the necessaries. Hence, according to Sismondi, crises are not accidental, as Ricardo maintains, but essential outbreaks—occurring on a large scale and at definite periods—of the immanent contradictions. He wavers constantly: should the State curb the productive forces to make them adequate to the production relations, or should the production relations be made adequate to the productive forces? He often retreats into the past, becomes a *laudator temporis acti* [praiser of times past], or he seeks

to exorcise the contradictions by a different adjustment of revenue in relation to capital, or of distribution in relation to production, not realizing that
the relations of distribution are only the relations of production seen from a
different aspect. He forcefully criticizes the contradictions of bourgeois
production but does not understand them, and consequently does not understand the process whereby they can be resolved. However, at the bottom of
his argument is indeed the inkling that new forms of the appropriation of
wealth must correspond to productive forces and the material and social
conditions for the production of wealth which have developed within
capitalist society; that the bourgeois forms are only transitory and contradictory forms, in which wealth attains only an antithetical existence and appears
everywhere simultaneously as its opposite. It is wealth which always has
poverty as its prerequisite and only develops by developing poverty as well.
. . .

The Progress of Capitalist Production: A Summary

. . . The primitive accumulation of capital. Includes the centralization of the
conditions of labour. It means that the conditions of labour acquire an
independent existence in relation to the worker and to labour itself. This
historical act is the historical genesis of capital, the historical process of
separation which transforms the conditions of labour into capital and labour
into wage-labour. This provides the basis for capitalist production.

Accumulation of capital on the basis of capital itself, and therefore also on
the basis of the relationship of capital and wage-labour, reproduces the
separation and the independent existence of material wealth as against labour
on an ever-increasing scale.

Concentration of capital. Accumulation of large amounts of capital by the
destruction of the smaller capitals. Attraction. Decapitalization of the intermediate links between capital and labour. This is only the last degree and the
final form of the process which transforms the conditions of labour into
capital, then reproduces capital and the separate capitals on a larger scale
and finally separates from their owners the various capitals which have come
into existence at many points of society, and centralizes them in the hands of
big capitalists. It is in this extreme form of the contradiction and conflict
that production—even though in alienated form—is transformed into social
production. There is social labour, and in the real labour process the
instruments of production are used in common. As functionaries of the
process which at the same time accelerates this social production and thereby
also the development of the productive forces, the capitalists become superfluous in the measure that they, on behalf of society, enjoy the usufruct and
that they become overbearing as owners of this social wealth and commanders of social labour. Their position is similar to that of the feudal lords

whose exactions in the measure that their services became superfluous with the rise of bourgeois society, became mere outdated and inappropriate privileges and who therefore rushed headlong to destruction. . . .

BIBLIOGRAPHY

Original: *MEW*, Vol. 26, parts 1, 2, and 3.

Present translation: K. Marx, *Theories of Surplus Value*, Moscow, 1963–8: Vol. 1, pp. 390 ff., 405 ff.; Vol. 2, pp. 400, 404 ff., 573, 579 f.; Vol. 3, pp. 52 ff., 314 f. Reproduced by kind permission of Lawrence and Wishart Ltd.

Other translations: None.

Commentaries: I. Gough, 'Marx's Theory of Productive and Unproductive Labour', *New Left Review*, Vol. 76, 1972.
Introduction to Vol. 1 of Moscow edition cited above.
R. Meek, *Studies in the Labour Theory of Value*, 2nd edn., London, 1972.
M. Nicolaus, 'Proletariat and Middle Class in Marx', *Studies on the Left*, Jan.–Feb., 1967.
P. Walton and A. Gamble, *From Alienation to Surplus Value*, London 1972, Chs. 6 and 7.

32. Capital

Capital represents only a part of the work on 'Economics' that Marx had mapped out in the late 1850s. Thus it is an expansion of only a section—though the most important section—of the material contained in the *Grundrisse*. And of the four volumes of this 'section', Marx only managed to complete the first. The four were drafted out in reverse order: the manuscript for the *Theories of Surplus Value* was completed in 1862–3, the third volume of *Capital* in 1864, and the first in 1865–6. It was only this first volume that Marx saw through to publication.

Volume One of *Capital* consists of two very distinct parts: the first nine chapters contain a very abstract discussion of the central concepts of value, labour, surplus value, etc. It is not only this abstraction that makes them difficult; it is also the Hegelian mode of expression and the fact that, while the concepts used by Marx were familiar to mid-nineteenth-century economists, they abandoned by the later orthodoxy of the marginalist school. Modern economists have tended to discuss the functioning of the capitalist system as given and concentrate particularly on prices, whereas Marx wished to examine the mode of production which gave rise to the capitalist system and which would, he believed, bring about its own destruction.

Following the first nine chapters, there is a masterly account of the genesis of capitalism which makes pioneering use of the statistical material then becoming increasingly available. It is one of the best illustrations of applied historical materialism.

Volume Two of *Capital* is rather technical, and discusses the circulation of capital and the genesis of economic crises. Volume Three begins with a discussion of value and prices and the tendency of profits to fall, but trails off towards the end with the dramatically incomplete section on classes.

The following selections are intended to give the most important parts of the argument of the first volume, supplemented by a few key sections from the third volume.

From Volume One

From the Prefaces: i. 1867

'Every beginning is difficult' holds in all sciences. To understand the first chapter, especially the section that contains the analysis of commodities, will, therefore, present the greatest difficulty. That which concerns more especially the analysis of the substance of value and the magnitude of value, I have, as much as it was possible, popularized. The value-form, whose fully developed shape is the money-form, is very elementary and simple. Nevertheless, the human mind has for more than 2000 years sought in vain to get to the bottom of it, while on the other hand, to the successful analysis of much more composite and complex forms there has been at least an approximation. Why? Because the body, as an organic whole, is more easy of study than are the cells of that body. In the analysis of economic forms, moreover, neither microscopes nor chemical reagents are of use. The force of abstraction must replace both. But in bourgeois society the commodity-form of the product of labour—or the value-form of the commodity—is the economic cell-form. To the superficial observer, the analysis of these forms seems to turn upon minutiae. It does in fact deal with minutiae, but they are of the same order as those dealt with in microscopic anatomy.

With the exception of the section on value-form, therefore, this volume cannot stand accused on the score of difficulty. I presuppose, of course, a reader who is willing to learn something new and therefore to think for himself.

The physicist either observes physical phenomena where they occur in their most typical form and most free from disturbing influence, or, wherever possible, he makes experiments under conditions that assure the occurrence of the phenomenon in its normality. In this work I have to examine the capitalist mode of production, and the conditions of production and exchange corresponding to that mode. Up to the present time, their classic ground is England. That is the reason why England is used as the chief illustration in the development of my theoretical ideas. If, however, the German reader shrugs his shoulders at the condition of the English industrial and agricultural labourers, or in optimist fashion comforts himself with the thought that in Germany things are not nearly so bad; I must plainly tell him, 'De te fabula narratur!' [The story is about *you*!]

Intrinsically, it is not a question of the higher or lower degree of development of the social antagonisms that result from the natural laws of capitalist production. It is a question of these laws themselves, of these tendencies working with iron necessity towards inevitable results. The country that is more developed industrially only shows, to the less developed, the image of its own future.

But apart from this. Where capitalist production is fully naturalized among the Germans (for instance, in the factories proper), the condition of things is much worse than in England, because the counterpoise of the Factory Acts is wanting. In all other spheres, we, like all the rest of Continental Western Europe, suffer not only from the development of capitalist production, but also from the incompleteness of that development. Alongside modern evils, a whole series of inherited evils oppress us, arising from the passive survival of antiquated modes of production, with their inevitable train of social and political anachronisms. We suffer not only from the living, but from the dead. Le mort saisit le vif! [The dead seize the living!]

The social statistics of Germany and the rest of Continental Western Europe are, in comparison with those of England, wretchedly compiled. But they raise the veil just enough to let us catch a glimpse of the Medusa head behind it. We should be appalled at the state of things at home, if, as in England, our governments and parliaments appointed periodically commissions of inquiry into economic conditions; if these commissions were armed with the same plenary powers to get at the truth; if it was possible to find for this purpose men as competent, as free from partisanship and respect of persons as are the English factory inspectors, her medical reporters on public health, her commissioners of inquiry into the exploitation of women and children, into housing and food. Perseus wore a magic cap that the monsters he hunted down might not see him. We draw the magic cap down over eyes and ears as a make-believe that there are no monsters.

Let us not deceive ourselves on this. As in the eighteenth century, the American war of independence sounded the tocsin for the European middle class, so in the nineteenth century, the American Civil War sounded it for the European working class. In England the progress of social disintegration is palpable. When it has reached a certain point, it must react on the Continent. There it will take a form more brutal or more humane, according to the degree of development of the working class itself. Apart from higher motives, therefore, their own most important interests dictate to the classes that are for the nonce the ruling ones the removal of all legally removable hindrances to the free development of the working class. It is for this reason, as well as others, that I have given so large a space in this volume to the history, the details, and the results of English factory legislation. One nation can and should learn from others. And even when a society has got on the right track for the discovery of the natural laws of its movement—and it is the ultimate aim of this work to lay bare the economic law of motion of modern society—it can neither clear by bold leaps, nor remove by legal enactments, the obstacles offered by the successive phases of its normal development. But it can shorten and lessen the birth-pangs.

To prevent possible misunderstanding, a word. I paint the capitalist and the landlord in no sense *couleur de rose*. But here individuals are dealt with only in so far as they are the personifications of economic categories, embodiments of particular class relations and class interests. My standpoint, from which the evolution of the economic formation of society is viewed as a process of natural history, can less than any other make the individual responsible for relations whose creature he socially remains, however much he may subjectively raise himself above them.

In the domain of Political Economy, free scientific inquiry meets not merely the same enemies as in all other domains. The peculiar nature of the material it deals with summons as foes into the field of battle the most violent, mean, and malignant passions of the human breast, the Furies of private interest. The English Established Church, e.g., will more readily pardon an attack on 38 of its 39 articles than on $\frac{1}{39}$ of its income. Nowadays atheism itself is *culpa levis* [minor fault], as compared with criticism of existing property relations. Nevertheless, there is an unmistakeable advance. I refer, e.g., to the Blue book published within the last few weeks: 'Correspondence with Her Majesty's Missions Abroad, regarding Industrial Questions and Trades' Unions.' The representatives of the English Crown in foreign countries there declare in so many words that in Germany, in France, to be brief, in all the civilized states of the European continent, a radical change in the existing relations between capital and labour is as evident and inevitable as in England. At the same time, on the other side of the Atlantic Ocean, Mr. Wade, vice-president of the United States, declared in public meetings that, after the abolition of slavery, a radical change of the relations of capital and of property in land is next upon the order of the day. These are signs of the times, not to be hidden by purple mantles or black cassocks. They do not signify that tomorrow a miracle will happen. They show that, within the ruling classes themselves, a foreboding is dawning, that the present society is no solid crystal, but an organism capable of change, and is constantly changing.

The second volume of this work will treat of the process of the circulation of capital (Book II), and of the varied forms assumed by capital in the course of its development (Book III), the third and last volume (Book IV), the history of the theory.

Every opinion based on scientific criticism I welcome. As to the prejudices of so-called public opinion, to which I have never made concessions, now as aforetime the maximum of the great Florentine is mine:

Segui il tuo corso, e lascia dir le genti. [Follow your path and let people say what they will.]

ii. 1872

... The *European Messenger* of St. Petersburg, in an article dealing exclusively with the method of *Das Kapital*, finds my method of inquiry severely realistic, but my method of presentation, unfortunately, German-dialectical. It says: 'At first sight, if the judgement is based on the external form of the presentation of the subject, Marx is the most ideal of ideal philosophers, always in the German, i.e., the bad sense of the word. But in point of fact he is infinitely more realistic than all his forerunners in the work of economic criticism. He can in no sense be called an idealist.' I cannot answer the writer better than by aid of a few extracts from his own criticism, which may interest some of my readers to whom the Russian original is inaccessible.

After a quotation from the preface to my Criticism of Political Economy, where I discuss the materialistic basis of my method, the writer goes on: 'The one thing which is of moment to Marx is to find the law of the phenomena with whose investigation he is concerned; and not only is that law of moment to him, which governs these phenomena, in so far as they have a definite form and mutual connection within a given historical period.

Of still greater moment to him is the law of their variation, of their development, i.e., of their transition from one form into another, from one series of connections into a different one. This law once discovered, he investigates in detail the effects in which it manifests itself in social life. Consequently, Marx only troubles himself about one thing: to show, by rigid scientific investigation the necessity of successive determinate orders of social conditions, and to establish, as impartially as possible, the facts that serve him for fundamental starting-points. For this it is quite enough, if he proves, at the same time, both the necessity of the present order of things, and the necessity of another order into which the first must inevitably pass over; and this all the same, whether men believe or do not believe it, whether they are conscious or unconscious of it. Marx treats the social movement as a process of natural history, governed by laws not only independent of human will, consciousness, and intelligence, but rather, on the contrary, determining that will, consciousness, and intelligence. . . . If in the history of civilization the conscious element plays a part so subordinate, then it is self-evident that a critical inquiry whose subject-matter is civilization, can, less than anything else, have for its basis any form of, or any result of, consciousness. That is to say that not the idea, but the material phenomenon alone can serve as its starting-point. Such an inquiry will confine itself to the confrontation and the comparison of a fact, not with ideas, but with another fact. For this inquiry, the one thing of moment is that both facts be investigated as accurately as possible, and that they actually form, each with respect to the other, different momenta of an evolution; but most important of all is the rigid analysis of the series of successions, of the sequences and concatenations in which the different stages of such an evolution present themselves. But, it will be said, the general laws of economic life are one and the same, no matter whether they are applied to the present or the past. This Marx directly denies. According to him, such abstract laws do not exist. On the contrary, in his opinion every historical period has laws of its own. . . . As soon as society has outlived a given period of development, and is passing over from one given stage to another, it begins to be subject also to other laws. In a word, economic life offers us a phenomenon analogous to the history of evolution in other branches of biology. The old economists misunderstood the nature of economic laws when they likened them to the laws of physics and chemistry. A more thorough analysis of phenomena shows that social organisms differ among themselves as fundamentally as plants or animals. Nay, one and the same phenomenon falls under quite different laws in consequence of the different structure of these organisms as a whole, of the variations of their individual organs, of the different conditions in which those organs function, etc. Marx, e.g., denies that the law of population is the same at all times and in all places. He asserts, on the contrary, that every stage of development has its own law of population. . . . With the varying degree of development of productive power, social conditions and the laws governing them vary too. While Marx sets himself the task of following and explaining from this point of view the economic system established by the sway of capital, he is only formulating, in a strictly scientific manner, the aim that every accurate investigation into economic life must have. The scientific value of such an inquiry lies in the disclosing of the special laws that regulate the origin, existence, development, death of a given social organism and its replacement by another and higher one. And it is this value that, in point of fact, Marx's book has.'

While the writer pictures what he takes to be actually my method, in this striking and as far as concerns my own application of it generous way, what else is he picturing but the dialectic method?

Of course the method of presentation must differ in form from that of inquiry. The latter has to appropriate the material in detail, to analyse its different forms of development, to trace out their inner connection. Only after this work is done, can the actual movement be adequately described. If this is done successfully, if the life of the subject-matter is ideally reflected as in a mirror, then it may appear as if we had before us a mere *a priori* construction.

My dialectic method is not only different from the Hegelian, but is its direct opposite. To Hegel, the life-process of the human brain, i.e., the process of thinking, which, under the name of 'the Idea', he even transforms into an independent subject, is the demiurgos of the real world, and the real world is only the external, phenomenal form of 'the Idea'. With me, on the contrary, the ideal is nothing else than the material world reflected by the human mind, and translated into forms of thought.

The mystifying side of Hegelian dialectic I criticized nearly thirty years ago, at a time when it was still the fashion. But just as I was working at the first volume of *Das Kapital*, it was the good pleasure of the peevish, arrogant, mediocre epigoni who now talk big in cultured Germany, to treat Hegel in same way as the brave Moses Mendelssohn in Lessing's time treated Spinoza, i.e., as a 'dead dog'. I therefore openly avowed myself the pupil of that mighty thinker, and even here and there, in the chapter on the theory of value, coquetted with the modes of expression peculiar to him. The mystification which dialectic suffers in Hegel's hands by no means prevents him from being the first to present its general form of working in a comprehensive and conscious manner. With him it is standing on its head. It must be turned right side up again, if you should discover the rational kernel within the mystical shell.

In its mystified form, dialectic became the fashion in Germany, because it seemed to transfigure and to glorify the existing stage of things. In its rational form it is a scandal and abomination to bourgeoisdom and its doctrinaire professors, because it includes in its comprehension and affirmative recognition of the existing state of things at the same time also the recognition of the negation of that state, of its inevitable breaking-up; because it regards every historically developed social form as in fluid movement, and therefore takes into account its transient nature not less than its momentary existence; because it lets nothing impose upon it, and is in its essence critical and revolutionary.

The contradictions inherent in the movement of capitalist society impress themselves upon the practical bourgeois most strikingly in the changes of the periodic cycle, through which modern industry runs, and whose crowning point is the universal crisis. That crisis is once again approaching, although as yet but in its preliminary stage; and by the universality of its theatre and the intensity of its action it will drum dialectics even into the heads of the mushroom upstarts of the new, holy Prusso-German empire. . . .

Commodities: Use-Value and Exchange-Value

The wealth of those societies in which the capitalist mode of production prevails presents itself as 'an immense accumulation of commodities', its unit being a single commodity. Our investigation must therefore begin with the analysis of a commodity.

A commodity is, in the first place, an object outside us, a thing that by its properties satisfies human wants of some sort or another. The nature of such wants, whether, for instance, they spring from the stomach or from fancy, makes no difference. Neither are we here concerned to know how the object satisfies these wants, whether directly as means of subsistence, or indirectly as means of production.

Every useful thing, as iron, paper, etc., may be looked at from the two points of view: of quality and quantity. It is an assemblage of many properties, and may therefore be of use in various ways. To discover the various uses of things is the work of history. So also is the establishment of socially recognized standards of measure for the quantities of these useful objects. The diversity of these measures has its origin partly in the diverse nature of the objects to be measured, partly in convention.

The utility of a thing makes it a use-value. But this utility is not a thing of air. Being limited by the physical properties of the commodity, it has no existence apart from that commodity. A commodity, such as iron, corn, or a diamond, is therefore, so far as it is a material thing, a use-value, something useful. This property of a commodity is independent of the amount of labour required to appropriate its useful qualities. When treating of use-value, we always assume we are dealing with definite quantities, such as dozens of watches, yards of linen, or tons of iron. The use-values of commodities furnish the material for a special study, that of the commercial knowledge of commodities. Use-values become a reality only by use or consumption; they also constitute the substance of all wealth, whatever may be the social form of that wealth. In the form of society we are about to consider, they are, in addition, the material depositories of exchange-value.

Exchange-value, at first sight, presents itself as a quantitative relation, as the proportion in which values in use of one sort are exchanged for those of another sort, a relation constantly changing with time and place. Hence exchange-value appears to be something accidental and purely relative, and consequently an intrinsic value, i.e. an exchange-value that is inseparably connected with, inherent in, commodities, seems a contradiction in terms. Let us consider the matter a little more closely.

A given commodity, e.g., a quarter of wheat is exchanged for x blacking, y silk, or z gold, etc.—in short, for other commodities in the most different proportions. Instead of one exchange-value, the wheat has, therefore, a great many. But since x blacking, y silk, or z gold, etc., each represent the exchange-value of one quarter of wheat, x blacking, y silk, z gold, etc., must, as exchange-values, be replaceable by each other, or equal to each other. Therefore, first: the valid exchange-values of a given commodity express something equal; secondly, exchange-value, generally, is only the mode of expression, the phenomenal form, of something contained in it, yet distinguishable from it.

Let us take two commodities, e.g., corn and iron. The proportions in which they are exchangeable, whatever those proportions may be, can always be represented by an equation in which a given quantity of corn is equated to some quantity of iron: e.g., 1 quarter corn = x cwt. iron. What does this equation tell us? It tells us that in two different things—in 1 quarter of corn and x cwt. of iron, there exists in equal quantities something common to both. The two things must therefore be equal to a third, which in itself is neither the one nor the other. Each of them, so far as it is exchange-value, must therefore be reducible to this third.

A simple geometrical illustration will make this clear. In order to calculate and compare the areas of rectilinear figures, we decompose them into triangles. But the area of the triangle itself is expressed by something totally different from its visible figure, namely, by half the product of the base into the altitude. In the same way the exchange-values of commodities must be capable of being expressed in terms of something common to them all, of which thing they represent a greater or less quantity.

This common 'something' cannot be either a geometrical, a chemical, or any other natural property of commodities. Such properties claim our attention only in so far as they affect the utility of those commodities, make them use-values. But the exchange of commodities is evidently an act characterized by a total abstraction from use-value. Then one use-value is just as good as another, provided only it be present in sufficient quantity. Or, as old Barbon says, 'one sort of wares is as good as another, if the values be equal. There is no difference or distinction in things of equal value. . . . A hundred pounds' worth of lead or iron is of as great value as one hundred pounds' worth of silver or gold.' As use-values, commodities are, above all, of different qualities, but as exchange-values they are merely different quantities, and consequently do not contain an atom of use-value.

If then we leave out of consideration the use-value of commodities, they have only one common property left, that of being products of labour. But even the product of labour itself has undergone a change in our hands. If we make abstraction from its use-value, we make abstraction at the same time from the material elements and shapes that make the product a use-value; we see in it no longer a table, a house, yarn, or any other useful thing. Its existence as a material thing is put out of sight. Neither can it any longer be regarded as the product of the labour of the joiner, the mason, the spinner, or of any other definite kind of productive labour. Along with the useful qualities of the products themselves, we put out of sight both the useful character of the various kinds of labour embodied in them, and the concrete forms of that labour; there is nothing left but what is common to them all; all are reduced to one and the same sort of labour, human labour in the abstract.

Let us now consider the residue of each of these products; it consists of the same unsubstantial reality in each, a mere congelation of homogeneous human labour, of labour power expended without regard to the mode of its expenditure. All that these things now tell us is that human labour power has been expended in their production, that human labour is embodied in them. When looked at as crystals of this social substance, common to them all, they are—Values.

We have seen that when commodities are exchanged, their exchange-value manifests itself as something totally independent of their use-value. But if we abstract from their use-value, there remains their Value as defined above. Therefore, the common substance that manifests itself in the exchange-value of commodities, whenever they are exchanged, is their value. The progress of our investigation will show that exchange-value is the only form in which the value of commodities can manifest itself or be expressed. For the present, however, we have to consider the nature of value independently of this, its form.

A use-value, or useful article, therefore, has value only because human labour in the abstract has been embodied or materialized in it. How, then, is the magnitude of this value to be measured? Plainly, by the quantity of the value-creating substance, the labour, contained in the article. The quantity of labour, however, is measured by its duration, and labour time in its turn finds its standard in weeks, days, and hours.

Some people might think that if the value of a commodity is determined by the quantity of labour spent on it, the more idle and unskilful the labourer, the more valuable would his commodity be, because more time would be required in its production. The labour, however, that forms the substance of value, is homogeneous human labour, expenditure of one uniform labour power. The total labour power of society, which is embodied in the sum total of the values of all commodities produced by that society, counts here as one homogeneous mass of human labour power, composed though it be of innumerable individual units. Each of these units is the same as any other, so far as it has the character of the average labour power of society, and takes effect as such; that is, so far as it requires for producing a commodity no more time than is needed on average, no more than is socially necessary. The labour time socially necessary is that required to produce an article under the normal conditions of production, and with the average degree of skill and intensity prevalent at the time. The introduction of power-looms into England probably reduced by one-half the labour required to weave a given quantity of yarn into cloth. The hand-loom weavers, as a matter of fact, continued to require the same time as before; but for all that, the product of one hour of their labour represented after the change only half an hour's social labour, and consequently fell to one-half its former value.

We see then that that which determines the magnitude of the value of any article is the amount of labour socially necessary, or the labour time socially necessary for its production. Each individual commodity, in this connection, is to be considered as an average sample of its class. Commodities, therefore, in which equal quantities of labour are embodied, or which can be produced in the same time, have the same value. The value of one commodity is to the value of any other, as the labour time necessary for the production of the one is to that necessary for the production of the other. 'As values, all commodities are only definite masses of congealed labour time.'

The value of a commodity would therefore remain constant, if the labour time required for its production also remained constant. But the latter changes with every variation in the productiveness of labour. This productiveness is determined by various circumstances, among others, by the average amount of skill of the workmen, the state of science, and the degree of its practical application, the social organization of production, the extent and capabilities of the means of production, and by physical conditions. For example, the same amount of labour in favourable seasons is embodied in eight bushels of corn, and in unfavourable, only in four. The same labour extracts from rich mines more metal than from poor mines. Diamonds are of very rare occurrence on the earth's surface, and hence their discovery costs, on an average, a great deal of labour time. Consequently much labour is represented in a small compass. Jacob doubts whether gold has ever been paid for at its full value. This applies still more to diamonds. According to Eschwege, the total produce of the Brazilian diamond mines for the eighty years ending in 1823, had not realized the price of one-and-a-half years' average produce of the sugar and coffee plantations of the same country, although the diamonds cost much more labour, and therefore represented more value. With richer mines, the same quantity of labour would embody itself in more diamonds, and their value would fall. If we could succeed, at a small expenditure of labour, in converting carbon into diamonds, their value might fall below that of bricks. In general, the greater the productiveness of labour, the less is the labour time required for the production of an article, the less is the amount of labour crystallized in that article, and the less is its value; and vice versa, the less the productiveness of labour, the greater is the labour time required for the production of an article, and the greater is its value. The value of a commodity, therefore, varies directly as the quantity, and inversely as the productiveness, of the labour incorporated in it.

A thing can be a use-value, without having value. This is the case whenever its utility to man is not due to labour. Such are air, virgin soil, natural meadows, etc. A thing can be useful, and the product of human labour, without being a commodity. Whoever directly satisfies his wants with the produce of his own labour creates, indeed, use-values, but not commodities. In order to produce the latter, he must not only produce use-values, but use-values for others, social use-values. (And not only for others. The medieval peasant produced quit-rent-corn for his feudal lord and tithe-corn for his parson. But neither the quit-rent-corn nor the tithe-corn became commodities by reason of the fact that they had been produced for others. To become a commodity a product must be transferred to another, whom it will serve as a use-value, by means of an exchange.) Lastly, nothing can have value without being an object of utility. If the thing is useless, so is the labour contained in it; the labour does not count as labour, and therefore creates no value.

At first sight a commodity presented itself to us as a complex of two things —use-value and exchange-value. Later on, we saw also that labour, too, possesses the same twofold nature; for, so far as it finds expression in value, it does not possess the same characteristics that belong to it as a creator of use-values. I was the first to point out and to examine critically this twofold nature of the labour contained in commodities. As this point is the pivot on which a clear comprehension of Political Economy turns, we must go more into detail.

Let us take two commodities such as a coat and 10 yards of linen, and let the former be double the value of the latter, so that, if 10 yards of linen = W, the coat = 2W.

The coat is a use-value that satisfies a particular want. Its existence is the result of a special sort of productive activity, the nature of which is determined by its aim, mode of operation, subject, means, and result. The labour, whose utility is thus represented by the value in use of its product, or which manifests itself by making its product a use-value, we call useful labour. In this connection we consider only its useful effect.

As the coat and the linen are two qualitatively different use-values, so also are the two forms of labour that produce them, tailoring and weaving. Were these two objects not qualitatively different, not produced respectively by labour of different quality, they could not stand to each other in the relation of commodities. Coats are not exchanged for coats, one use-value is not exchanged for another of the same kind.

To all the different varieties of values in use there correspond as many different kinds of useful labour, classified according to the order, genus, species, and variety to which they belong in the social division of labour. This division of labour is a necessary condition for the production of commodities, but it does not follow, conversely, that the production of commodities is a necessary condition for the division of labour. In the primitive Indian community there is social division of labour, without production of commodities. Or, to take an example nearer home, in every factory the labour is divided according to a system, but this division is not brought about by the operatives mutually exchanging their individual products. Only such products can become commodities with regard to each other, as result from different kinds of labour, each kind being carried on independently and for the account of private individuals.

To resume, then: In the use-value of each commodity there is contained useful labour, i.e., productive activity of a definite kind and exercised with a definite aim. Use-values cannot confront each other as commodities, unless the useful labour embodied in them is qualitatively different in each of them. In a community, the produce of which in general takes the form of commodities, i.e., in a community of commodity producers, this qualitative difference between the useful forms of labour that are carried on independently by individual producers, each on their own account, develops into a complex system, a social division of labour.

Anyhow, whether the coat be worn by the tailor or by his customer, in either case it operates as a use-value. Nor is the relation between the coat and the labour that produced it altered by the circumstance that tailoring may have become a special trade, an independent branch of the social division of labour. Wherever the want of clothing forced them to it, the human race made clothes for thousands of years, without a single man becoming a tailor. But coats and linen, like every other element of material wealth that is not the spontaneous produce of Nature, must invariably owe their existence to a special productive activity, exercised with a definite aim, an activity that appropriates particular nature-given materials to particular human wants. So far therefore as labour is a creator of use-value, is useful labour, it is a necessary condition, independent of all forms of society for the existence of the human race; it is an eternal nature-imposed necessity, without which there can be no material exchanges between man and Nature, and therefore no life.

The use-values, coat, linen, etc., i.e., the bodies of commodities, are combinations of two elements—matter and labour. If we take away the useful labour expended upon them, a material substratum is always left, which is furnished by Nature without the help of man. The latter can work only as Nature does, that is by changing the form of matter. Nay more, in this work of changing the form he is constantly helped by natural forces. We see, then, that labour is not the only source of material wealth, of use-values produced by labour. As William Petty puts it, labour is its father and the earth its mother.

Let us now pass from the commodity considered as a use-value to the value of commodities.

By our assumption, the coat is worth twice as much as the linen. But this is a mere quantitative difference, which for the present does not concern us. We bear in mind, however, that if the value of the coat is double that of 10 yds of linen, 20 yds of linen must have the same value as one coat. So far as they are values, the coat and the linen are things of a like substance, objective expressions of essentially identical labour. But tailoring and weaving are, qualitatively, different kinds of labour. There are, however, states of society in which one and the same man does tailoring and weaving alternately, in which case these two forms of labour are mere modifications of the labour of the same individual, and not special and fixed functions of different persons; just as the coat which our tailor makes one day, and the trousers which he makes another day, imply only a variation in the labour of one and the same individual. Moreover, we see at a glance that, in our capitalist society, a given portion of human labour is, in accordance with the varying demand, at one time supplied in the form of tailoring, at another in the form of weaving. This change may possibly not take place without friction, but take place it must.

Productive activity, if we leave out of sight its special form, viz., the useful character of the labour, is nothing but the expenditure of human labour power. Tailoring and weaving, though qualitatively different productive activities, are each a productive expenditure of human brains, nerves, and muscles, and in this sense are human labour. They are but two different modes of expending human labour power. Of course, this labour power, which remains the same under all its modifications, must have attained a certain pitch of development before it can be expended in a multiplicity of modes. But the value of a commodity represents human labour in the abstract, the expenditure of human labour in general. And just as in society, a general or a banker plays a great part, but mere man, on the other hand, a very shabby part, so here with mere human labour. It is the expenditure of simple labour power, i.e., of the labour power which, on an average, apart from any special development, exists in the organism of every ordinary individual. Simple average labour, it is true, varies in character in different countries and at different times, but in a particular society it is given. Skilled labour counts only as simple labour intensified, or rather, as multiplied simple labour, a given quantity of skilled being considered equal to a greater quantity of simple labour. Experience shows that this reduction is constantly being made. A commodity may be the product of the most skilled labour, but its value, by equating it to the product of simple unskilled labour, represents a definite quantity of the latter labour alone. The different proportions in which different sorts of labour are reduced to unskilled labour as their standard, are established by a social process that goes on behind the backs of the producers, and, consequently, appear to be fixed by custom. For simplicity's sake we shall henceforth account every kind of labour to be unskilled, simple labour; by this we do no more than save ourselves the trouble of making the reduction.

Just as, therefore, in viewing the coat and linen as values, we abstract from their different use-values, so it is with the labour represented by those values: we disregard the difference between its useful forms, weaving and tailoring. As the use-values, coat and linen, are combinations of special productive activities with cloth and yarn, while the values, coat and linen, are, on the other hand, mere homogeneous congelations of indifferentiated labour, so the labour embodied in these latter values does not count by virtue of its productive relation to cloth and yarn, but only as being expenditure of human labour power. Tailoring and weaving are necessary factors in the creation of the use-values, coat and linen, precisely because these two kinds of labour are of different qualities; but only in so far as abstraction is made from their special qualities, only in so far as both possess the same quality of being human labour, do tailoring and weaving form the substance of the values of the same articles.

Coats and linen, however, are not merely values, but values of definite magnitude, and according to our assumption, the coat is worth twice as much as the ten yards of linen. Whence this difference in their values? It is owing to the fact that the linen contains only half as much labour as the coat, and consequently, that in the production of the latter, labour power must have been expended during twice the time necessary for the production of the former.

While, therefore, with reference to use-value, the labour contained in a commodity counts only qualitatively, with reference to value it counts only quantitatively, and must first be reduced to human labour pure and simple. In the former case, it is a question of How and What, in the latter of How much? How long a time? Since the magnitude of the value of a commodity represents only the quantity of labour embodied in it, it follows that all commodities, when taking in certain proportions, must be equal in value.

If the productive power of all the different sorts of useful labour required for the production of a coat remains unchanged, the sum of the values of the coats produced increases with their number. If one coat represents x days' labour, two coats represent 2x days' labour, and so on. But assume that the duration of the labour necessary for the production of a coat becomes doubled or halved. In the first case, one coat is worth as much as two coats were before; in the second case, two coats are only worth as much as one was before, although in both cases one coat renders the same service as before, and the useful labour embodied in it remains of the same quality. But the quantity of labour spent on its production has altered.

An increase in the quantity of use-values is an increase of material wealth. With two coats two men can be clothed, with one coat only one man. Nevertheless, an increased quantity of material wealth may correspond to a simultaneous fall in the magnitude of its value. This antagonistic movement has its origin in the twofold character of labour. Productive power has reference, of course, only to labour of some useful concrete form; the efficacy of any special productive activity during a given time being dependent on its productiveness. Useful labour becomes, therefore, a more or less abundant source of products, in proportion to the rise or fall of its productiveness. On the other hand, no change in this productiveness affects the labour represented by value. Since productive power is an attribute of the concrete useful forms of labour, of course it can no longer have any bearing on that labour, so soon as we make abstraction from those concrete useful forms. However then productive power may vary, the same labour, exercised during equal periods of time, always yields equal amounts of value. But it will yield, during equal periods of time, different quantities of values in use; more, if the productive power rise, fewer, if it fall. The same change in productive power, which increases the fruitfulness of labour, and, in consequence, the quantity of use-values produced by that labour, will diminish the total value of this increased quantity of use-values, provided such change shorten the total labour time necessary for their production; and vice versa.

On the one hand all labour is, speaking physiologically, an expenditure of human labour power, and in its character of identical abstract human labour, it creates and forms the value of commodities. On the other hand, all labour is the expenditure of human labour power in a special form and with a definite aim, and in this, its character of concrete useful labour, it produces use-values.

Commodities come into the world in the shape of use-values, articles, or goods, such as iron, linen, corn, etc. This is their plain, homely, bodily form. They are, however, commodities, only because they are something twofold, both objects of utility, and, at the same time, depositories of value. They manifest themselves therefore as commodities, or have the form of commodities, only in so far as they have two forms, a physical or natural form, and a value-form.

The reality of the value of commodities differs in this respect from Mistress Quickly, that we don't know 'where to have it'. The value of commodities is the very opposite of the coarse materiality of their substance, not an atom of matter enters into its composition. Turn and examine a single commodity, by itself, as we will, yet in so far as it remains an object of value, it seems impossible to grasp it. If, however, we bear in mind that the value of commodities has a purely social reality, and that they acquire this reality only in so far as they are expressions or embodiments of one identical social substance, viz., human labour, it follows as a matter of course, that value can only manifest itself in the social relation of commodity to commodity. In fact we started from exchange-value, or the exchange relation of commodities, in order to get at the value that lies hidden behind it. We must now return to this form under which value first appeared to us.

Everyone knows, if he knows nothing else, that commodities have a value-form common to them all, and presenting a marked contrast with the varied bodily forms of their use-values. I mean their money-form. Here, however, a task is set us, the performance of which has never yet even been attempted by bourgeois economy, the task of tracing the genesis of this money-form, of developing the expression of value implied in the value-relation of commodities, from its simplest, almost imperceptible outline, to the dazzling money-form. By doing this we shall, at the same time, solve the riddle presented by money.

The simplest value-relation is evidently that of one commodity to some one other commodity of a different kind. Hence the relation between the values of two commodities supplies us with the simplest expression of the value of a single commodity.

The whole mystery of the form of value lies hidden in this elementary form. Its analysis, therefore, is our real difficulty.

Here two different kinds of commodities (in our example the linen and the coat) evidently play two different parts. The linen expresses its value in the coat; the coat serves as the material in which that value is expressed. The former plays an active, the latter a passive, part. The value of the linen is represented as relative value, or appears in relative form. The coat officiates as equivalent, or appears in equivalent form.

The relative form and the equivalent form are two intimately connected, mutually dependent, and inseparable elements of the expression of value; but, at the same time, are mutually exclusive, antagonistic extremes—i.e., poles of the same expression. They are allotted respectively to the two different commodities brought into relation by that expression. It is not possible to express the value of linen in linen. 20 yards of linen = 20 yards of linen is no expression of value. On the contrary, such an equation merely says that 20 yards of linen are nothing else than 20 yards of linen, a definite quantity of the use-value linen. The value of the linen can therefore be expressed only relatively—i.e., in some other commodity. The relative form of the value of the linen presupposes, therefore, the presence of some other commodity—here the coat—under the form of an equivalent. On the other hand, the commodity that figures as the equivalent cannot at the same time assume the relative form. That second commodity is not the one whose value is expressed. Its function is merely to serve as the material in which the value of the first commodity is expressed.

No doubt, the expression 20 yards of linen = 1 coat, or 20 yards of linen are worth 1 coat, implies the opposite relation: 1 coat = 20 yards of linen, or 1 coat is worth 20 yards of linen. But, in that case, I must reverse the equation, in order to express the value of the coat relatively; and, so soon as I do that, the linen becomes the equivalent instead of the coat. A single commodity cannot, therefore, simultaneously assume, in the same expression of value, both forms. The very polarity of these forms makes them mutually exclusive.

Whether, then, a commodity assumes the relative form, or the opposite equivalent form, depends entirely upon its accidental position in the expression of value—that is, upon whether it is the commodity whose value is being expressed or the commodity in which value is being expressed.

In order to discover how the elementary expression of the value of a commodity lies hidden in the value-relation of two commodities, we must, in the first place, consider the latter entirely apart from its quantitative aspect. The usual mode of procedure is generally the reverse, and in the value-relation nothing is seen but the proportion between definite quantities of two different sorts of commodities that are considered equal to each other. It is apt to be forgotten that the magnitudes of different things can be compared quantitatively, only when those magnitudes are expressed in terms of the same unit. It is only as expressions of such a unit that they are of the same denomination, and therefore commensurable.

Whether 20 yards of linen = 1 coat or = 20 coats or = x coats—that is, whether a given quantity of linen is worth few or many coats, every such statement implies that the linen and coats, as magnitudes of value, are expressions of the same unit, things of the same kind. Linen = coat is the basis of the equation.

But the two commodities whose identity of quality is thus assumed, do not play the same part. It is only the value of the linen that is expressed. And how? By its reference to the coat as its equivalent, as something that can be exchanged for it. In this relation the coat is the mode of existence of value, is value embodied, for only as such is it the same as the linen. On the other hand, the linen's own value comes to the front, receives independent expression, for it is only as being value that it is comparable to the coat as a thing of equal value, or exchangeable with the coat. To borrow an illustration from chemistry, butyric acid is a different substance from propyl formate. Yet both are made up of the same chemical substances, carbon (C), hydrogen (H), and oxygen (O), and that, too, in like proportions—namely, $C_4H_8O_2$. If now we equate butyric acid to propyl formate, then, in the first place, propyl formate would be, in this relation, merely a form of existence of $C_4H_8O_2$; and in the second place, we should be stating that butyric acid also consists of $C_4H_8O_2$. Therefore, by thus equating the two substances, expression would be given to their chemical composition, while their different physical forms would be neglected.

If we say that, as values, commodities are mere congelations of human labour, we reduce them by our analysis, it is true, to the abstraction, value; but we ascribe to this value no form apart from their bodily form. It is otherwise in the value-relation of one commodity to another. Here, the one stands forth in its character of value by reason of its relation to the other.

By making the coat the equivalent of the linen, we equate the labour embodied in the former to that in the latter. Now, it is true that the tailoring, which makes the coat, is concrete labour of a different sort from the weaving which makes the linen. But the act of equating it to the weaving reduces the tailoring to that which is really equal in the two kinds of labour, to their common character of human labour. In this roundabout way, then, the fact is expressed, that weaving also, in so far as it weaves value, has nothing to distinguish it from tailoring, and, consequently, is abstract human labour. It is the expression of equivalence between different sorts of commodities that alone brings into relief the specific character of value-creating labour, and this it does by actually reducing the different varieties of labour embodied in the different kinds of commodities to their common quality of human labour in the abstract.

There is, however, something else required beyond the expression of the specific character of the labour of which the value of the linen consists. Human labour power in motion, or human labour, creates value, but is not itself value. It becomes value only in its congealed state, when embodied in the form of some object. In order to express the value of the linen as a congelation of human labour, that value must be expressed as having objective existence, as being a something materially different from the linen itself, and yet a something common to the linen and all other commodities. The problem is already solved.

When occupying the position of equivalent in the equation of value, the coat ranks qualitatively as the equal of the linen, as something of the same kind, because it is value. In this position it is a thing in which we see nothing but value, or whose palpable bodily form represents value. Yet the coat itself, the body of the commodity, coat, is a mere use-value. A coat as such no more tells us it is value, than does the first piece of linen we take hold of. This shows that when placed in value-relation to the linen, the coat signifies more than when out of that relation, just as many a man strutting about in a gorgeous uniform counts for more than when in mufti.

In the production of the coat, human labour power, in the shape of tailoring, must have been actually expended. Human labour is therefore accumulated in it. In this aspect the coat is a depository of value, but though worn to a thread, it does not let this fact show through. And as equivalent of the linen in the value equation, it exists under this aspect alone, counts therefore as embodied value, as a body that is value. A, for instance, cannot be 'your majesty' to B, unless at the same time majesty in B's eyes assumes the bodily form of A, and, what is more, with every new father of the people, changes its features, hair, and many other things besides.

Hence, in the value equation, in which the coat is the equivalent of the linen, the coat officiates as the form of value. The value of the commodity linen is expressed by the bodily form of the commodity coat, the value of one by the use-value of the other. As a use-value, the linen is something palpably different from the coat; as value, it is the same as the coat, and now has the appearance of a coat. Thus the linen acquires a value-form different from its physical form. The fact that it is value is made manifest by its equality with the coat, just as the sheep's nature of a Christian is shown in his resemblance to the Lamb of God.

We see, then, that all that our analysis of the value of commodities has already told us is told us by the linen itself, as soon as it comes into communication with another commodity, the coat. Only it betrays its thoughts in that language with which alone it is familiar, the language of commodities. In order to tell us that its own value is created by labour in its abstract character of human labour, it says that the coat, in so far as it is worth as much as the linen, and therefore is value, consists of the same labour as the linen. In order to inform us that its sublime reality as value is not the same as its buckram body, it says that value has the appearance of a coat, and consequently that so far as the linen is value, it and the coat are as like as two peas. We may here remark, that the language of commodities has, besides Hebrew, many other more or less correct dialects. The German 'Wertsein', to be worth, for instance, expresses in a less striking manner than the Romance verbs 'valere', 'valer', 'valoir', that the equating of commodity B to commodity A is commodity A's own mode of expressing its value. *Paris vaut bien une messe.* [Paris is easily worth a Mass.]

By means, therefore, of the value-relation expressed in our equation, the bodily form of commodity B becomes the value-form of commodity A, or the body of commodity B acts as a mirror to the value of commodity A. By putting itself in relation with commodity B, as value *in propria persona* [in its own person] as the matter of which human labour is made up, the commodity A converts the value in use, B, into the substance in which to express its, A's, own value. The value of A, thus expressed in the use-value of B, has taken the form of relative value.

Every commodity, whole value it is intended to express, is a useful object of given quantity, as 15 bushels of corn, or 100 lb of coffee. And a given quantity of any commodity contains a definite quantity of human labour. The value-form must therefore not only express value generally, but also value in definite quantity. Therefore, in the value-relation of commodity A to commodity B, of the linen to the coat, not only is the latter, as value in general, made the equal in quality of the linen, but a definite quantity of coat (1 coat) is made the equivalent of a definite quantity (20 yards) of linen.

The equation, 20 yards of linen = 1 coat, or 20 yards of linen are worth one coat, implies that the same quantity of value-substance (congealed labour) is embodied in both; that the two commodities have each cost the same amount of labour or the same quantity of labour time. But the labour time necessary for the production of 20 yards of linen or 1 coat varies with every change in the productiveness of weaving or tailoring. We have now to consider the influence of such changes on the quantitative aspect of the relative expression of value.

I. Let the value of the linen vary, that of the coat remaining constant. If, say in consequence of the exhaustion of flax-growing soil, the labour time necessary for the production of the linen be doubled, the value of the linen will also be doubled. Instead of the equation, 20 yards of linen = 1 coat, we should have 20 yards of linen = 2 coats, since 1 coat would now contain only half the labour time embodied in 20 yards of linen. If, on the other hand, in consequence, say, of improved looms, this labour time were reduced by one-half, the value of the linen would fall by one-half. Consequently, we should have 20 yards of linen = $\frac{1}{2}$ coat. The relative value of commodity A, i.e., its value expressed in commodity B, rises and falls directly as the value of A, the value of B being supposed constant.

II. Let the value of the linen remain constant, while the value of the coat varies. If, under these circumstances, in consequence, for instance, of a poor crop of wool, the labour time necessary for the production of a coat becomes doubled, we have instead of 20 yards of linen = 1 coat, 20 yards of linen = $\frac{1}{2}$ coat. If, on the other hand, the value of the coat sinks by one-half, then 20 yards of linen = 2 coats. Hence, if the value of commodity A remains constant, its relative value expressed in commodity B rises and falls inversely as the value of B.

If we compare the different cases in I and II, we see that the same change of magnitude in relative value may arise from totally opposite causes. Thus, the equation, 20 yards of linen = 1 coat, becomes 20 yards of linen = 2 coats, either, because, the value of the linen has doubled, or because the value of the coat has fallen by one-half; and it becomes 20 yards of linen = $\frac{1}{2}$ coat, either, because the value of the linen has fallen by one-half, or because the value of the coat has doubled.

III. Let the quantities of labour time respectively necessary for the production of the linen and the coat vary simultaneously in the same direction and in the same proportion. In this case 20 yards of linen continue equal to 1 coat, however much their values may have altered. Their change of value is seen as soon as they are compared with a third commodity, whose value has remained constant. If the values of all commodities rose or fell simultaneously, and in the same proportion, their relative values would remain unaltered. Their real change of value would appear from the diminished or increased quantity of commodities produced in a given time.

IV. The labour time respectively necessary for the production of the linen and the coat, and therefore the value of these commodities may simultaneously vary in the same direction, but at unequal rates, or in opposite directions, or in other ways. The effect of all these possible different variations, on the relative value of a commodity, may be deduced from the results of I, II, and III.

Thus real changes in the magnitude of value are neither unequivocally nor exhaustively reflected in their relative expression, that is, in the equation expressing the magnitude of relative value. The relative value of a commodity may vary, although its value remains constant. Its relative value may remain constant, although its value varies; and finally, simultaneous variations in the magnitude of value and in that of its relative expression by no means necessarily correspond in amount. . . .

The Fetishism of Commodities

A commodity appears, at first sight, a very trivial thing, and easily understood. Its analysis shows that it is, in reality, a very queer thing, abounding in metaphysical subtleties and theological niceties. So far as it is a value in use, there is nothing mysterious about it, whether we consider it from the point of view that by its properties it is capable of satisfying human wants, or from the point that those properties are the product of human labour. It is as clear as noonday, that man, by his industry, changes the forms of the materials furnished by Nature, in such a way as to make them useful to him. The form of wood, for instance, is altered, by making a table out of it. Yet, for all that, the table continues to be that common, everyday thing, wood. But, so soon as it steps forth as a commodity, it is changed into something transcendent. It not only stands with its feet on the ground, but, in relation to all other commodities, it stands on its head, and evolves out of its wooden brain grotesque ideas, far more wonderful than 'table-turning' ever was.

The mystical character of commodities does not originate, therefore, in their use-value. Just as little does it proceed from the nature of the determining factors of value. For, in the first place, however varied the useful kinds of labour, or productive activities, may be, it is a physiological fact, that they are functions of the human organism, and that each such function, whatever may be its nature or form, is essentially the expenditure of human brain, nerves, muscles, etc. Secondly, with regard to that which forms the groundwork for the quantitative determination of value, namely, the duration of that expenditure, or the quantity of labour, it is quite clear that there is a palpable difference between its quantity and the quality. In all states of society, the labour time that it costs to produce the means of subsistence must necessarily be an object of interest to mankind, though not of equal interest in different stages of development. And lastly, from the moment that men in any way work for one another, their labour assumes a social form.

Whence, then, arises the enigmatical character of the product of labour, so soon as it assumes the form of commodities? Clearly from this form itself. The equality of all sorts of human labour is expressed objectively by their products all being equally values; the measure of the expenditure of labour power by the duration of that expenditure takes the form of the quantity of value of the products of labour; and finally, the mutual relations of the producers, within which the social character of their labour affirms itself, take the form of a social relation between the products.

A commodity is therefore a mysterious thing, simply because in it the social character of men's labour appears to them as an objective character stamped upon the product of that labour; because the relation of the producers to the sum total of their own labour is presented to them as a social relation, existing not between themselves, but between the products of their labour. This is the reason why the products of labour become commodities, social things whose qualities are at the same time perceptible and imperceptible by the senses. In the same way the light from an object is perceived by us not as the subjective excitation of our optic nerve, but as the objective form of something outside the eye itself. But, in the act of seeing, there is at all events, an actual passage of light from one thing to another, from the external object to the eye. There is a physical relation between physical things. But it is different with commodities. There, the existence of the things *qua* commodities, and the value relation between the products of labour which stamps them as commodities, have absolutely no connection with their physical properties and with the material relations arising therefrom. There it is a definite social relation between men, that assumes, in their eyes, the fantastic form of a relation between things. In order, therefore, to find an analogy, we must have recourse to the mist-enveloped regions of the religious world. In that world the productions of the human brain appear as independent beings endowed with life, and entering into relation both with one another and the human race. So it is in the world of commodities with the products of men's hands. This I call the Fetishism which attaches itself to the products of labour, so soon as they are produced as commodities, and which is therefore inseparable from the production of commodities.

This Fetishism of commodities has its origin, as the foregoing analysis has already shown, in the peculiar social character of the labour that produces them.

As a general rule, articles of utility become commodities, only because they are products of the labour of private individuals or groups of individuals who carry on their work independently of each other. The sum total of the labour of all these private individuals forms the aggregate labour of society. Since the producers do not come into social contact with each other until they exchange their products, the specific social character of each producer's labour does not show itself except in the act of exchange. In other words, the labour of the individual asserts itself as a part of the labour of society only by means of the relations which the act of exchange establishes directly between the products, and indirectly, through them, between the producers. To the latter, therefore, the relations connecting the labour of one individual with that of the rest appear, not as direct social relations between individuals at work, but as what they really are, material relations between persons and social relations between things. It is only by being exchanged that the products of labour acquire, as values, one uniform social status, distinct from their varied forms of existence as objects of utility. This division of a product into a useful thing and a value becomes practically important, only when exchange has acquired such an extension that useful articles are produced for the purpose of being exchanged, and their character as values has therefore to be taken into account, beforehand, during production. From this moment the labour of the individual producer acquires socially a twofold character. On the one hand, it must, as a definite useful kind of labour, satisfy a definite social want, and thus hold its place as part and parcel of the collective labour of all, as a branch of a social division of labour that has sprung up spontaneously. On the other hand, it can satisfy the manifold wants of the individual producer himself, only in so far as the mutual exchangeability of all kinds of useful private labour is an established social fact, and therefore the private useful labour of each producer ranks on an equality with that of all others. The equalization of the most different kinds of labour can be the result only of an abstraction from their inequalities, or of reducing them to their common denominator, viz., expenditure of human labour power or human labour in the abstract. The twofold social character of the labour of the individual appears to him, when reflected in his brain, only under those forms which are impressed upon that labour in everyday practice by the exchange of products. In this way, the character that his own labour possesses of being socially useful takes the form of the condition that the product must be not only useful, but useful for others, and the social character that his particular labour has of being the equal of all other particular kinds of labour, takes the form that all the physically different articles that are the products of labour have one common quality, viz., that of having value.

Hence, when we bring the products of our labour into relation with each other as values, it is not because we see in these articles the material receptacles of homogeneous human labour. Quite the contrary: whenever, by an exchange, we equate as values our different products, by that very act, we also equate, as human labour, the different kinds of labour expended upon them. We are not aware of this, nevertheless we do it. Value, therefore, does not stalk about with a label describing what it is. It is value, rather, that converts every product into a social hieroglyphic. Later on, we try to decipher the hieroglyphic, to get behind the secret of our own social products; for to stamp an object of utility as a value, is just as much a social product as

language. The recent scientific discovery that the products of labour, so far
as they are values, are but material expressions of the human labour spent in
their production marks, indeed, an epoch in the history of the development
of the human race, but by no means dissipates the mist through which the
social character of labour appears to us to be an objective character of the
products themselves. The fact that in the particular form of production with
which we are dealing, viz., the production of commodities, the specific social
character of private labour carried on independently, consists in the equality
of every kind of that labour, by virtue of its being human labour, which
character, therefore, assumes in the product the form of value—this fact
appears to the producers, notwithstanding the discovery above referred to, to
be just as real and final, as the fact that, after the discovery by science of the
component gases of air, the atmosphere itself remained unaltered.

What, first of all, practically concerns producers when they make an
exchange, is the question, how much of some other product they get for their
own? in what proportions are the products exchangeable? When these
proportions have, by custom, attained a certain stability, they appear to
result from the nature of the products, so that, for instance, one ton of iron
and two ounces of gold appear as naturally to be of equal value as a pound of
gold and a pound of iron, in spite of their different physical and chemical
qualities, appear to be of equal weight. The character of having value, when
once impressed upon products, obtains fixity only by reason of their acting
and reacting upon each other as quantities of value. These quantities vary
continually, independently of the will, foresight, and action of the producers.
To them, their own social action takes the form of the action of objects,
which rule the producers instead of being ruled by them. It requires a fully
developed production of commodities before, from accumulated experience
alone, the scientific conviction springs up that all the different kinds of
private labour, which are carried on independently of each other, and yet as
spontaneously developed branches of the social division of labour, are continually being reduced to the quantitative proportions in which society
requires them. And why? Because, in the midst of all the accidental and ever
fluctuating exchange-relations between the products, the labour time socially
necessary for their production forcibly asserts itself like an overriding law of
Nature. The law of gravity thus asserts itself when a house falls about our
ears. The determination of the magnitude of value by labour time is therefore a secret, hidden under the apparent fluctuations in the relative values
of commodities. Its discovery, while removing all appearance of mere ac-
cidentality from the determination of the magnitude of the values of
products, yet in no way alters the mode in which that determination takes
place.

Man's reflections on the forms of social life, and consequently, also, his
scientific analysis of those forms, take a course directly opposite to that of
their actual historical development. He begins, *post festum* [after the event],

with the results of the process of development ready to hand before him. The characters that stamp products as commodities, and whose establishment is a necessary preliminary to the circulation of commodities, have already acquired the stability of natural, self-understood forms of social life, before man seeks to decipher, not their historical character, for in his eyes they are immutable, but their meaning. Consequently it was the analysis of the prices of commodities that alone led to the determination of the magnitude of value, and it was the common expression of all commodities in money that alone led to the establishment of their characters as values. It is, however, just this ultimate money-form of the world of commodities that actually conceals, instead of disclosing, the social character of private labour, and the social relations between the individual producers. When I state that coats or boots stand in a relation to linen, because it is the universal incarnation of abstract human labour, the absurdity of the statement is self-evident. Nevertheless, when the producers of coats and boots compare those articles with linen, or, what is the same thing, with gold or silver, as the universal equivalent, they express the relation between their own private labour and the collective labour of society in the same absurd form.

The categories of bourgeois economy consist of such like forms. They are forms of thought expressing with social validity the conditions and relations of a definite, historically determined mode of production, viz., the production of commodities. The whole mystery of commodities, all the magic and necromancy that surrounds the products of labour as long as they take the form of commodities, vanishes therefore, as soon as we come to other forms of production.

Since Robinson Crusoe's experiences are a favourite theme with political economists, let us take a look at him on his island. Moderate though he be, yet some few wants he has to satisfy, and must therefore do a little useful work of various sorts, such as making tools and furniture, taming goats, fishing and hunting. Of his prayers and the like we take no account, since they are a source of pleasure to him, and he looks upon them as so much recreation. In spite of the variety of his work, he knows that his labour, whatever its form, is but the activity of one and the same Robinson, and, consequently, that it consists of nothing but different modes of human labour. Necessity itself compels him to apportion his time accurately between his different kinds of work. Whether one kind occupies a greater space in his general activity than another depends on the difficulties, greater or less as the case may be, to be overcome in attaining the useful effect aimed at. This our friend Robinson soon learns by experience, and having rescued a watch, ledger, and pen and ink from the wreck, commences, like a true-born Briton, to keep a set of books. His stock-book contains a list of the objects of utility that belong to him, of the operations necessary for their production; and lastly, of the labour time that definite quantities of those objects have, on an average, cost him. All the relations between Robinson and the objects that form this wealth of his own creation, are here so simple and clear as to be intelligible without exertion, even to Mr. Sedley Taylor. And yet those relations contain all that is essential to the determination of value.

Let us now transport ourselves from Robinson's island bathed in light to the European middle ages shrouded in darkness. Here, instead of the independent man, we find everyone dependent, serfs and lords, vassals and suzerains, laymen and clergy. Personal dependence here characterizes the social relations of production just as much as it does the other spheres of life organized on the basis of that production. But for the very reason that personal dependence forms the groundwork of society, there is no necessity for labour and its products to assume a fantastic form different from their reality. They take the shape, in the transactions of society, of services in kind and payments in kind. Here the particular and natural form of labour, and not, as in a society based on production of commodities, its general abstract form is the immediate social form of labour. Compulsory labour is just as properly measured by time, as commodity-producing labour; but every serf knows that what he expends in the service of his lord is a definite quantity of his own personal labour power. The tithe to be rendered to the priest is more matter of fact than his blessing. No matter, then, what we may think of the parts played by the different classes of people themselves in this society, the social relations between individuals in the performance of their labour appear at all events as their own mutual personal relations, and are not disguised under the shape of social relations between the products of labour.

For an example of labour in common or directly associated labour, we have no occasion to go back to that spontaneously developed form which we find on the threshold of the history of all civilized races. We have one close at hand in the patriarchal industries of a peasant family, that produces corn, cattle, yarn, linen, and clothing for home use. These different articles are, as regards the family, so many products of its labour, but as between themselves, they are not commodities. The different kinds of labour, such as tillage, cattle tending, spinning, weaving, and making clothes, which result in the various products, are in themselves, and such as they are, direct social functions, because functions of the family, which, just as much as a society based on the production of commodities, possesses a spontaneously developed system of division of labour. The distribution of the work within the family, and the regulation of the labour time of the several members, depend as well upon differences of age and sex as upon natural conditions varying with the seasons. The labour power of each individual, by its very nature, operates in this case merely as a definite portion of the whole labour power of the family, and therefore the measure of the expenditure of individual labour power by its duration, appears here by its very nature as a social character of their labour.

Let us now picture to ourselves, by way of change, a community of free individuals, carrying on their work with the means of production in common, in which the labour power of all the different individuals is consciously applied as the combined labour power of the community. All the characteristics of Robinson's labour are here repeated, but with this difference, that they are social, instead of individual. Everything produced by him was exclusively the result of his own personal labour, and therefore simply an object of use for himself. The total product of our community is a social product. One portion serves as fresh means of production and remains social. But another portion is consumed by the members as means of subsistence. A distribution of this portion among them is consequently necessary. The mode of this distribution will vary with the productive organization of the community, and the degree of historical development attained by the producers. We will assume, but merely for the sake of a parallel with the production of commodities, that the share of each individual producer in the means of subsistence is determined by his labour time. Labour time would, in that case, play a double part. Its apportionment in accordance with a definite social plan maintains the proper proportion between the different kinds of work to be done and the various wants of the community. On the other hand, it also serves as a measure of the portion of the common labour borne by each individual, and of his share in the part of the total product destined for individual consumption. The social relations of the individual producers, with regard both to their labour and to its products, are in this case perfectly simple and intelligible, and that with regard not only to production but also to distribution.

The religious world is but the reflex of the real world. And for a society based upon the production of commodities, in which the producers in general enter into social relations with one another by treating their products as commodities and values, whereby they reduce their individual private labour to the standard of homogeneous human labour—for such a society, Christianity with its *cultus* of abstract man, more especially in its bourgeois developments, Protestantism, Deism, etc., is the most fitting form of religion. In the ancient Asiatic and other ancient modes of production, we find that the conversion of products into commodities, and therefore the conversion of men into producers of commodities, holds a subordinate place, which, however, increases in importance as the primitive communities approach nearer and nearer to their dissolution. Trading nations, properly so called, exist in the ancient world only in its interstices, like the gods of Epicurus in the Intermundia, or like Jews in the pores of Polish society. Those ancient social organisms of production are, as compared with bourgeois society, extremely simple and transparent. But they are founded either on the immature development of man individually, who has not yet severed the umbilical cord that unites him with his fellowmen in a primitive tribal community, or upon direct relations of subjection. They can arise and exist only when the development of the productive power of labour has not risen beyond a low stage, and when, therefore, the social relations within the sphere of material life, between man and man, and between man and Nature, are correspondingly narrow. This narrowness is reflected in the ancient worship of Nature, and in the other elements of the popular religions. The religious reflex of the real world can, in any case, only then finally vanish, when the practical relations of everyday life offer to man none but perfectly intelligible and reasonable relations with regard to his fellow-men and to Nature. . . .

The life-process of society, which is based on the process of material production, does not strip off its mystical veil until it is treated as production by freey associated men, and is consciously regulated by them in accordance with a settled plan. This, however, demands for society a certain material groundwork or set of conditions of existence which in their turn are the spontaneous product of a long and painful process of development.

Political Economy has indeed analysed, however incompletely, value and its magnitude, and has discovered what lies beneath these forms. But it has never once asked the question why labour is represented by the value of its product and labour-time by the magnitude of that value. These formulas, which bear stamped upon them in unmistakable letters, that they belong to a state of society in which the process of production has the mastery over man, instead of being controlled by him, such formulas appear to the bourgeois intellect to be as much a self-evident necessity imposed by Nature as productive labour itself. Hence forms of social production that preceded the bourgeois form are treated by the bourgeoisie in much the same way as the Fathers of the Church treated pre-Christian religions. . . .

To what extent some economists are misled by the Fetishism inherent in commodities, or by the objective appearance of the social characteristics of labour, is shown, among other ways, by the dull and tedious quarrel over the part played by Nature in the formation of exchange-value. Since exchange-value is a definite social manner of expressing the amount of labour bestowed upon an object, Nature has no more to do with it, than it has in fixing the course of exchange.

The mode of production in which the product takes the form of a commodity, or is produced directly for exchange, is the most general and most embryonic form of bourgeois production. It therefore makes its appearance at an early date in history, though not in the same predominating and characteristic manner as nowadays. Hence its Fetish character is comparatively easy to be seen through. But when we come to more concrete forms, even this appearance of simplicity vanishes. Whence arose the illusions of the monetary system? To it gold and silver, when serving as money, did not represent a social relation between producers, but were natural objects with strange social properties. And modern economy, which looks down with such disdain on the monetary system, does not its superstition come out as clear as noonday, whenever it treats of capital? How long is it since economy discarded the physiocratic illusion that rents grow out of the soil and not out of society?

But not to anticipate, we will content ourselves with yet another example relating to the commodity-form. Could commodities themselves speak, they would say: Our use-value may be a thing that interests men. It is no part of us as objects. What, however, does belong to us as objects is our value. Our natural intercourse as commodities proves it. In the eyes of each other we are nothing but exchange-values. Now listen how those commodities speak through the mouth of the economist. 'Value'—(i.e. exchange-value) 'is a property of things, riches'—(i.e. use-value) 'of man. Value, in this sense, necessarily implies exchanges, riches do not.' 'Riches' (use-value) 'are the attribute of men, value is the attribute of commodities. A man or a community is rich, a pearl or a diamond is valuable ... A pearl or a diamond is valuable' as a pearl or diamond. So far no chemist has ever discovered exchange-value either in a pearl or a diamond. The economical discoverers of this chemical element, who by the by lay special claim to critical acumen, find however that the use-value of objects belongs to them independently of their material properties, while their value, on the other hand, forms a part of them as objects. What confirms them in this view is the peculiar circumstance that the use-value of objects is realized without exchange, by means of a direct relation between the objects and man, while, on the other hand, their value is realized only by exchange, that is, by means of a social process. Who fails here to call to mind our good friend, Dogberry, who informs neighbour Seacoal, that, 'To be a well-favoured man is the gift of fortune; but to read and write comes by nature.' ...

Exchange and Money

Every owner of a commodity wishes to part with it in exchange only for those commodities whose use-value satisfies some want of his. Looked at in this way, exchange is for him simply a private transaction. On the other hand, he desires to realize the value of his commodity, to convert it into any other suitable commodity of equal value, irrespective of whether his own commodity has or has not any use-value for the owner of the other. From this point of view, exchange is for him a social transaction of a general character. But one and the same set of transactions cannot be simultaneously for all owners of commodities both exclusively private and exclusively social and general. ...

Up to this point, however, we are acquainted only with one function of money, namely, to serve as the form of manifestation of the value of commodities, or as the material in which the magnitudes of their values are socially expressed. An adequate form of manifestation of value, a fit embodiment of abstract, undifferentiated, and therefore equal human labour, that material alone can be whose every sample exhibits the same uniform qualities. On the other hand, since the difference between the magnitudes of value is purely quantitative, the money-commodity must be susceptible of merely quantitative differences, must therefore be divisible at will, and equally capable of being reunited. Gold and silver possess these properties by Nature.

The use-value of the money-commodity becomes twofold. In addition to its special use-value as a commodity (gold, for instance, serving to stop teeth, to form the raw material of articles of luxury, etc.), it acquires a formal use-value, originating in its specific social function. . . .

The first chief function of money is to supply commodities with the material for the expression of their values, or to represent their values as magnitudes of the same denomination, qualitatively equal, and quantitatively comparable. It thus serves as a universal measure of value. And only by virtue of this function does gold, the equivalent commodity par excellence, become money. . . .

A commodity strips off its original commodity-form on being alienated, i.e., on the instant its use-value actually attracts the gold, that before existed only ideally in its price. The realization of a commodity's price, or of its ideal value-form, is therefore at the same time the realization of the ideal use-value of money; the conversion of a commodity into money is the simultaneous conversion of money into a commodity. The apparently single process is in reality a double one. From the pole of the commodity-owner it is a sale, from the opposite pole of the money-owner, it is a purchase. In other words, a sale is a purchase, C—M is also M—C. . . .

We will assume that the two gold pieces, in consideration of which our weaver has parted with his linen, are the metamorphosed shape of a quarter of wheat. The sale of the linen, C—M, is at the same time its purchase, M—C. But the sale is the first act of a process that ends with a transaction of an opposite nature, namely, the purchase of a Bible; the purchase of the linen, on the other hand, ends a movement that began with a transaction of an opposite nature, namely, with the sale of the wheat. . . .

C—M (linen—money), which is the first phase of C—M—C (linen—money—Bible), is also M—C (money—linen), the last phase of another movement C—M—C (wheat—money—linen). The first metamorphosis of one commodity, its transformation from a commodity into money, is therefore also invariably the second metamorphosis of some other commodity, the retransformation of the latter from money into a commodity. . . .

The circulation of commodities differs from the direct exchange of products (barter), not only in form, but in substance. Only consider the course of events. The weaver has, as a matter of fact, exchanged his linen for a Bible, his own commodity for that of some one else. But this is true only so far as he himself is concerned. The seller of the Bible, who prefers something to warm his inside, no more thought of exchanging his Bible for linen than our weaver knew that wheat had been exchanged for his linen. B's commodity replaces that of A, but A and B do not mutually exchange those commodities. It may, of course, happen that A and B make simultaneous purchases, the one from the other; but such exceptional transactions are by no means the necessary result of the general conditions of the circulation of commodities. We see here, on the one hand, how the exchange of commodities breaks through all local and personal bounds inseparable from direct barter, and develops the circulation of the products of social labour; and on the other hand, how it develops a whole network of social relations spontaneous in their growth and entirely beyond the control of the actors. It is only because the farmer has sold his wheat that the weaver is enabled to sell his linen, only because the weaver has sold his linen that our Hotspur is enabled to sell his Bible, and only because the latter has sold the water of everlasting life that the distiller is enabled to sell his eau-de-vie, and so on.

. . .

The General Formula for Capital

The circulation of commodities is the starting-point of capital. The production of commodities, their circulation, and that more developed form of their circulation called commerce, these form the historical groundwork from which it rises. The modern history of capital dates from the creation in the sixteenth century of a world-embracing commerce and a world-embracing market.

If we abstract from the material substance of the circulation of commodities, that is, from the exchange of the various use-values, and consider only the economic forms produced by this process of circulation, we find its final result to be money: this final product of the circulation of commodities is the first form in which capital appears.

As a matter of history, capital, as opposed to landed property, invariably takes the form at first of money; it appears as moneyed wealth, as the capital of the merchant and of the usurer. But we have no need to refer to the origin of capital in order to discover that the first form of appearance of capital is money. We can see it daily under our very eyes. All new capital, to commence with, comes on the stage, that is, on the market, whether of commodities, labour, or money, even in our days, in the shape of money that by a definite process has to be transformed into capital.

The first distinction we notice between money that is money only, and money that is capital, is nothing more than a difference in their form of circulation.

The simplest form of the circulation of commodities is C—M—C, the transformation of commodities into money, and the change of the money back again into commodities; or selling in order to buy. But alongside of this form we find another specifically different form: M—C—M, the transformation of money into commodities, and the change of commodities back again into money; or buying in order to sell. Money that circulates in the latter manner is thereby transformed into, becomes capital, and is already potentially capital.

Now let us examine the circuit M—C—M a little closer. It consists, like the other, of two antithetical phases. In the first phase, M—C, or the purchase, the money is changed into a commodity. In the second phase, C—M, or the sale, the commodity is changed back again into money. The combination of these two phases constitutes the single movement whereby money is exchanged for a commodity, and the same commodity is again exchanged for money; whereby a commodity is bought in order to be sold, or, neglecting the distinction in form between buying and selling, whereby a commodity is bought with money, and then money is bought with a commodity. The result, in which the phases of the process vanish, is the exchange of money for money, M—M. If I purchase 2000 lb of cotton for £100, and resell the 2000 lb of cotton for £110, I have, in fact, exchanged £100 for £110, money for money.

Now it is evident that the circuit M—C—M would be absurd and without meaning if the intention were to exchange by this means two equal sums of money, £100 for £100. The miser's plan would be far simpler and surer; he sticks to his £100 instead of exposing it to the dangers of circulation. And yet, whether the merchant who has paid £100 for his cotton sells it for £110, or lets it go for £100, or even £50, his money has, at all events, gone through a characteristic and original movement, quite different in kind from that which it goes through in the hands of the peasant who sells corn, and with the money thus set free buys clothes. We have therefore to examine first the distinguishing characteristics of the forms of the circuits M—C—M and C—M—C, and in doing this the real difference that underlies the mere difference of form will reveal itself.

Let us see, in the first place, what the two forms have in common.

Both circuits are resolvable into the same two antithetical phases, C—M, a sale, and M—C, a purchase. In each of these phases the same material elements—a commodity, and money, and the same economical dramatis personae, a buyer and a seller—confront one another. Each circuit is the unity of the same two antithetical phases, and in each case this unity is brought about by the intervention of three contracting parties, of whom one only sells, another only buys, while the third both buys and sells.

What, however, first and foremost distinguishes the circuit C—M—C from the circuit M—C—M, is the inverted order of succession of the two phases. The simple circulation of commodities begins with a sale and ends with a purchase, while the circulation of money as capital begins with a purchase and ends with a sale. In the one case both the starting-point and the goal are commodities, in the other they are money. In the first form the movement is brought about by the intervention of money, in the second by that of a commodity.

In the circulation C—M—C, the money is in the end converted into a commodity, that serves as a use-value; it is spent once for all. In the inverted form, M—C—M, on the contrary, the buyer lays out money in order that, as a seller, he may recover money. By the purchase of his commodity he throws money into circulation, in order to withdraw it again by the sale of the same commodity. He lets the money go, but only with the sly intention of getting it back again. The money, therefore, is not spent, it is merely advanced.

In the circuit C—M—C, the same piece of money changes its place twice. The seller gets it from the buyer and pays it away to another seller. The complete circulation, which begins with the receipt, concludes with the payment, of money for commodities. It is the very contrary in the circuit M—C—M. Here it is not the piece of money that changes its place twice, but the commodity. The buyer takes it from the hands of the seller and passes it into the hands of another buyer. Just as in the simple circulation of commodities the double change of place of the same piece of money effects its passage from one hand into another, so here the double change of place of the same commodity brings about the reflux of the money to its point of departure.

Such reflux is not dependent on the commodity being sold for more than was paid for it. This circumstance influences only the amount of the money that comes back. The reflux itself takes place, so soon as the purchased commodity is resold, in other words, so soon as the circuit M—C—M is completed. We have here, therefore, a palpable difference between the circulation of money as capital, and its circulation as mere money.

The circuit C—M—C comes completely to an end, so soon as the money brought in by the sale of one commodity is abstracted again by the purchase of another.

If, nevertheless, there follows a reflux of money to its starting-point, this can only happen through a renewal or repetition of the operation. If I sell a quarter of corn for £3, and with this £3 buy clothes, the money, so far as I am concerned, is spent and done with. It belongs to the clothes merchant. If I now sell a second quarter of corn, money indeed flows back to me, not however as a sequel to the first transaction, but in consequence of its repetition. The money again leaves me, so soon as I complete this second transaction by a fresh purchase. Therefore, in the circuit C—M—C, the expenditure of money has nothing to do with its reflux. On the other hand, in M—C—M, the reflux of the money is conditioned by the very mode of its expenditure. Without this reflux, the operation fails, or the process is interrupted and incomplete, owing to the absence of its complementary and final phase, the sale.

The circuit C—M—C starts with one commodity and finishes with another, which falls out of circulation and into consumption. Consumption, the satisfaction of wants, in one word, use-value, is its end and aim. The circuit M—C—M, on the contrary, commences with money and ends with money. Its leading motive, and the goal that attracts it, is therefore mere exchange-value.

In the simple circulation of commodities, the two extremes of the circuit have the same economic form. They are both commodities, and commodities of equal value. But they are also use-values differing in their qualities, as, for example, corn and clothes. The exchange of products, of the different materials in which the labour of society is embodied, forms here the basis of the movement. It is otherwise in the circulation M—C—M, which at first sight appears purposeless, because tautological. Both extremes have the same economic form. They are both money, and therefore are not qualitatively different use-values; for money is but the converted form of commodities, in which their particular use-values vanish. To exchange £100 for cotton, and then this same cotton again for £100, is merely a roundabout way of exchanging money for money, the same for the same, and appears to be an operation just as purposeless as it is absurd. One sum of money is distinguishable from another only by its amount. The character and tendency of the process M—C—M is therefore not due to any qualitative difference between its extremes, both being money, but solely to their quantitative difference. More money is withdrawn from circulation at the finish than was thrown into it at the start. The cotton that was bought for £100 is perhaps resold for £100 + £10 or £110. The exact form of this process is therefore M—C—M′, where M′ = M + $\triangle$M = the original sum advanced, plus an increment. This increment or excess over the original value I call 'surplus value'. The value originally advanced, therefore, not only remains intact while in circulation, but adds to itself a surplus value or expands itself. It is this movement that converts it into capital.

Of course, it is also possible that in C—M—C the two extremes C—C, say corn and clothes, may represent different quantities of value. The farmer may sell his corn above its value, or may buy the clothes at less than their value. He may, on the other hand, 'be done' by the clothes merchant. Yet, in the form of circulation now under consideration, such differences in value are purely accidental. The fact that the corn and the clothes are equivalents does not deprive the process of all meaning, as it does in M—C—M. The equivalence of their values is rather a necessary condition to its normal course.

The repetition or renewal of the act of selling in order to buy, is kept within bounds by the very object it aims at, namely, consumption or the satisfaction of definite wants, an aim that lies altogether outside the sphere of circulation. But when we buy in order to sell, we, on the contrary, begin and end with the same thing, money, exchange-value; and thereby the movement becomes interminable. No doubt, M becomes M + $\triangle$M, £100 become

£110. But when viewed in their qualitative aspect alone, £110 are the same as £100, namely money; and considered quantitatively, £110 is, like £100, a sum of definite and limited value. If now the £110 be spent as money, they cease to play their part. They are no longer capital. Withdrawn from circulation, they become petrified into a hoard, and though they remained in that state till doomsday, not a single farthing would accrue to them. If, then, the expansion of value is once aimed at, there is just the same inducement to augment the value of the £110 as that of the £100; for both are but limited expressions for exchange-value, and therefore both have the same vocation to approach, by quantitative increase, as near as possible to absolute wealth. Momentarily, indeed, the value originally advanced, the £100 is distinguishable from the surplus value of £10 that is annexed to it during circulation; but the distinction vanishes immediately. At the end of the process, we do not receive with one hand the original £100, and with the other, the surplus value of £10. We simply get a value of £110, which is in exactly the same condition and fitness for commencing the expanding process, as the original £100 was. Money ends the movement only to begin it again. Therefore, the final result of every separate circuit, in which a purchase and consequent sale are completed, forms of itself the starting-point of a new circuit. The simple circulation of commodities—selling in order to buy—is a means of carrying out a purpose unconnected with circulation, namely, the appropriation of use-values, the satisfaction of wants. The circulation of money as capital is, on the contrary, an end in itself, for the expansion of value takes place only within this constantly renewed movement. The circulation of capital has therefore no limits.

As the conscious representative of this movement, the possessor of money becomes a capitalist. His person, or rather his pocket, is the point from which the money starts and to which it returns. The expansion of value, which is the objective basis or mainspring of the circulation M—C—M, becomes his subjective aim, and it is only in so far as the appropriation of ever more and more wealth in the abstract becomes the sole motive of his operations, that he functions as a capitalist, that is, as capital personified and endowed with consciousness and a will. Use-values must therefore never be looked upon as the real aim of the capitalist; neither must the profit on any single transaction. The restless never-ending process of profit-making alone is what he aims at. This boundless greed after riches, this passionate chase after exchange-value, is common to the capitalist and the miser; but while the miser is merely a capitalist gone mad, the capitalist is a rational miser. The never-ending augmentation of exchange-value, which the miser strives after, by seeking to save his money from circulation, is attained by the more acute capitalist by constantly throwing it afresh into circulation.

The independent form, i.e., the money-form, which the value of commodities assumes in the case of simple circulation, serves only one purpose, namely, their exchange, and vanishes in the final result of the movement. On the other hand, in the circulation M—C—M, both the money and the commodity represent only different modes of existence of value itself, the money its general mode, and the commodity its particular, or, so to say, disguised mode. It is constantly changing from one form to the other without thereby becoming lost, and thus assumes an automatically active character. If now we take in turn each of the two different forms which self-expanding value successively assumes in the course of its life, we then arrive at these two propositions: capital is money; capital is commodities. In truth, however, value is here the active factor in a process, in which, while constantly assuming the form in turn of money and commodities, is at the same time changes in magnitude, differentiates itself by throwing off surplus value from itself; the original value, in other words, expands spontaneously. For the movement, in the course of which it adds surplus value, is its own movement; its expansion, therefore, is automatic expansion. Because it is value, it has acquired the occult quality of being able to add value to itself. It brings forth living offspring, or, at the least, lays golden eggs.

Value, therefore, being the active factor in such a process, and assuming at one time the form of money, at another that of commodities, but through all these changes preserving itself and expanding, requires some independent form, by means of which its identity may at any time be established. And this form it possesses only in the shape of money. It is under the form of money that value begins and ends, and begins again, every act of its own spontaneous generation. It began by being £100, it is now £110, and so on. But the money itself is only one of the two forms of value. Unless it takes the form of some commodity, it does not become capital. There is here no antagonism, as in the case of hoarding, between the money and commodities. The capitalist knows that all commodities, however scurvy they may look, or however badly they may smell, are in faith and in truth money, inwardly circumcized Jews, and what is more, a wonderful means whereby out of money to make more money.

In simple circulation, C—M—C, the value of commodities attained at the most a form independent of their use-values, i.e., the form of money; but that same value now in the circulation M—C—M, or the circulation of capital, suddenly presents itself as an independent substance, endowed with a motion of its own, passing through a life process of its own, in which money and commodities are mere forms which it assumes and casts off in turn. Nay, more: instead of simply representing the relations of commodities, it enters now, so to say, into private relations with itself. It differentiates itself as original value from itself as surplus value; as the father differentiates himself from himself qua the son, yet both are one and of one age: for only by the surplus value of £10 does the £100 originally advanced become capital, and so soo as this takes place, so soon as the son, and by the son, the father, is begotten, so soon does their difference vanish, and they again become one, £110.

Value therefore now becomes value in process, money in process, and, as such, capital. It comes out of circulation, enters into it again, preserves and multiplies itself within its circuit, comes back out of it with expanded bulk, and begins the same round ever afresh. M—M', money which begets money, such is the description of capital from the mouths of its first interpreters, the mercantilists.

Buying in order to sell, or, more accurately, buying in order to sell dearer, M—C—M', appears certainly to be a form peculiar to one kind of capital alone, namely, merchants' capital. But industrial capital too is money, that is changed into commodities, and by the sale of these commodities, is re-converted into more money. The events that take place outside the sphere of circulation, in the interval between the buying and selling, do not affect the form of this movement. Lastly, in the case of interest-bearing capital, the circulation M—C—M' appears abridged. We have its result without the intermediate stage, in the form M—M', *en style lapidaire* [in lapidary style] so to say, money that is worth more money, value that is greater than itself.

M—C—M' is therefore in reality the general formula of capital as it appears *prima facie* within the sphere of circulation. . . .

The Sale of Labour Power

The change of value that occurs in the case of money intended to be converted into capital cannot take place in the money itself, since in its function of means of purchase and of payment, it does no more than realize the price of the commodity it buys or pays for; and, as hard cash, it is value petrified, never varying. Just as little can it originate in the second act of circulation, the re-sale of the commodity, which does no more than transform the article from its bodily form back again into its money-form. The change must, therefore, take place in the commodity bought by the first act, M—C, but not in its value, for equivalents are exchanged, and the commodity is paid for at its full value. We are, therefore, forced to the conclusion that the change originates in the use-value, as such, of the commodity, i.e., in its consumption. In order to be able to extract value from the consumption of a commodity, our friend Moneybags must be so lucky as to find, within the sphere of circulation, in the market, a commodity whose use-value possesses the peculiar property of being a source of value, whose actual consumption, therefore, is itself an embodiment of labour, and, consequently, a creation of value. The possessor of money does find on the market such a special commodity in capacity for labour or labour power.

By labour power or capacity for labour is to be understood the aggregate of those mental and physical capabilities existing in a human being, which he exercises whenever he produces a use-value of any description.

But in order that our owner of money may be able to find labour power offered for sale as a commodity, various conditions must first be fulfilled. The exchange of commodities of itself implies no other relations of dependence than those which result from its own nature. On this assumption, labour power can appear upon the market as a commodity, only if, and so far as, its possessor, the individual whose labour power it is, offers it for sale, or sells it, as a commodity. In order that he may be able to do this, he must have it at his disposal, must be the untrammelled owner of his capacity for labour, i.e., of his person. He and the owner of money meet in the market, and deal with each other as on the basis of equal rights, with this difference alone, that one is buyer, the other seller; both, therefore, equal in the eyes of the law. The continuance of this relation demands that the owner of the labour power should sell it only for a definite period, for if he were to sell it rump and stump, once for all, he would be selling himself, converting himself from a free man into a slave, from an owner of a commodity into a commodity. He must constantly look upon his labour power as his own property, his own commodity, and this he can only do by placing it at the disposal of the buyer temporarily, for a definite period of time. By this means alone can he avoid renouncing his rights of ownership over it.

The second essential condition to the owner of money finding labour power in the market as a commodity is this—that the labourer instead of being in the position to sell commodities in which his labour is incorporated, must be obliged to offer for sale as a commodity that very labour power, which exists only in his living self. . . .

The question why this free labourer confronts him in the market has no interest for the owner of money, who regards the labour-market as a branch of the general market for commodities. And for the present it interests us just as little. We cling to the fact theoretically, as he does practically. One thing, however, is clear—Nature does not produce on the one side owners of money or commodities, and on the other men possessing nothing but their own labour power. This relation has no natural basis, neither is its social basis one that is common to all historical periods. It is clearly the result of a past historical development, the product of many economical revolutions, of the extinction of a whole series of older forms of social production.

So, too, the economical categories already discussed by us bear the stamp of history. Definite historical conditions are necessary that a product may become a commodity. It must not be produced as the immediate means of subsistence of the producer himself. Had we gone further, and inquired under what circumstances all, or even the majority of products take the form of commodities, we should have found that this can only happen with production of a very specific kind, capitalist production. Such an inquiry, however, would have been foreign to the analysis of commodities. Production and circulation of commodities can take place, although the great mass of the objects produced are intended for the immediate requirements of their producers, are not turned into commodities, and consequently social production is not yet by a long way dominated in its length and breadth by exchange-value. The appearance of products as commodities presupposes such a development of the social division of labour, that the separation of use-value from exchange-value, a separation which first begins with barter, must already have been completed. But such a degree of development is common to many forms of society, which in other respects present the most varying historical features. On the other hand, if we consider money, its existence implies a definite stage in the exchange of commodities. The particular functions of money which it performs, either as the mere equivalent of commodities, or as means of circulation, or means of payment, as hoard or as universal money, point, according to the extent and relative preponderance of the one function or the other, to very different stages in the process of social production. Yet we know by experience that a circulation of commodities relatively primitive suffices for the production of all these forms. Otherwise with capital. The historical conditions of its existence are by no means given with the mere circulation of money and commodities. It can spring into life only when the owner of the means of production and subsistence meets in the market with the free labourer selling his labour power. And this one historical condition comprises a world's history. Capital, therefore, announces from its first appearance a new epoch in the process of social production.

We must now examine more closely this peculiar commodity, labour power. Like all others it has a value. How is that value determined?

The value of labour power is determined, as in the case of every other commodity, by the labour time necessary for the production, and consequently also the reproduction, of this special article. So far as it has value, it represents no more than a definite quantity of the average labour of society incorporated in it. Labour power exists only as a capacity, or power of the living individual. Its production consequently presupposes his existence. Given the individual, the production of labour power consists in his reproduction of himself or his maintenance. For his maintenance he requires a given quantity of the means of subsistence. Therefore the labour time requisite for the production of labour power reduces itself to that necessary for the production of those means of subsistence; in other words, the value of labour power is the value of the means of subsistence necessary for the maintenance of the labourer. Labour power, however, becomes a reality only by its exercise; it sets itself in action only by working. But thereby a definite quantity of human muscle, nerve, brain, etc., is wasted, and these require to be restored. This increased expenditure demands a larger income. If the owner of labour power works today, tomorrow he must again be able to repeat the same process in the same conditions as regards health and strength. His means of subsistence must therefore be sufficient to maintain him in his normal state as a labouring individual. His natural wants, such as food, clothing, fuel, and housing, vary according to the climatic and other physical conditions of his country. On the other hand, the number and extent of his so-called necessary wants, as also the modes of satisfying them, are themselves the product of historical development, and depend therefore to a great extent on the degree of civilization of a country, more particularly on the conditions under which, and consequently on the habits and degree of comfort in which, the class of free labourers has been formed. In contradistinction therefore to the case of other commodities, there enters into the determination of the value of labour power a historical and moral element. Nevertheless, in a given country, at a given period, the average quantity of the means of subsistence necessary for the labourer is practically known.

The owner of labour power is mortal. If then his appearance in the market is to be continuous, and the continuous conversion of money into capital assumes this, the seller of labour power must perpetuate himself, 'in the way that every living individual perpetuates himself, by procreation'. The labour power withdrawn from the market by wear and tear and death must be continually replaced by, at the very least, an equal amount of fresh labour power. Hence the sum of the means of subsistence necessary for the production of labour power must include the means necessary for the labourer's substitutes, i.e., his children, in order that this race of peculiar commodity-owners may perpetuate its appearance in the market.

In order to modify the human organism, so that it may acquire skill and handiness in a given branch of industry, and become labour power of a special kind, a special education or training is requisite, and this, on its part, costs an equivalent in commodities of a greater or less amount. This amount varies according to the more or less complicated character of the labour power. The expenses of this education (excessively small in the case of ordinary labour power), enter *pro tanto* into the total value spent in its production. . . .

We now know how the value paid by the purchaser to the possessor of this peculiar commodity, labour power, is determined. The use-value which the former gets in exchange manifests itself only in the actual usufruct, in the consumption of the labour power. The money-owner buys everything necessary for this purpose, such as raw material, in the market, and pays for it at its full value. The consumption of labour power is at one and the same time the production of commodities and of surplus value. The consumption of labour power is completed, as in the case of every other commodity, outside the limits of the market or of the sphere of circulation. Accompanied by Mr. Moneybags and by the possessor of labour power, we therefore take leave for a time of this noisy sphere, where everything takes place on the surface and in view of all men, and follow them both into the hidden abode of production, on whose threshold there stares us in the face 'No admittance except on business.' Here we shall see, not only how capital produces, but how capital is produced. We shall at last force the secret of profit-making.

This sphere that we are deserting, within whose boundaries the sale and purchase of labour power goes on, is in fact a very Eden of the innate rights of man. There alone rule Freedom, Equality, Property, and Bentham. Freedom, because both buyer and seller of a commodity, say of labour power, are constrained only by their own free will. They contract as free agents, and the agreement they come to is but the form in which they give legal expression to their common will. Equality, because each enters into relation with the other, as with a simple owner of commodities, and they exchange equivalent for equivalent. Property, because each disposes only of what is his own. And Bentham, because each looks only to himself. The only force that brings them together and puts them in relation with each other is the selfishness, the gain, and the private interests of each. Each looks to himself only, and no one troubles himself about the rest, and just because they do so, do they all, in accordance with the pre-established harmony of things, or under the auspices of an all-shrewd providence, work together to their mutual advantage, for the common weal and in the interest of all. . . .

The Production of Surplus Value

The capitalist buys labour power in order to use it; and labour power in use is labour itself. The purchaser of labour power consumes it by setting the seller of it to work. By working, the latter becomes actually what before he only was potentially, labour power in action, a labourer. In order that his labour may reappear in a commodity, he must, before all things, expend it on something useful, on something capable of satisfying a want of some sort. Hence, what the capitalist sets the labourer to produce, is a particular use-value, a specified article. The fact that the production of use-values, or goods, is carried on under the control of a capitalist and on his behalf, does not alter the general character of that production. We shall, therefore, in the first place, have to consider the labour process independently of the par-ticular form it assumes under given social conditions.

Labour is, in the first place, a process in which both man and Nature participate, and in which man of his own accord starts, regulates, and con-trols the material reactions between himself and Nature. He opposes himself to Nature as one of her own forces, setting in motion arms and legs, head and hands, the natural forces of his body, in order to appropriate Nature's productions in a form adapted to his own wants. By thus acting on the external world and changing it, he at the same time changes his own nature. He develops his slumbering powers and compels them to act in obedience to his sway. We are not now dealing with those primitive instinctive forms of labour that remind us of the mere animal. An immeasurable interval of time separates the state of things in which a man brings his labour power to market for sale as a commodity, from that state in which human labour was still in its first instinctive stage. We presuppose labour in a form that stamps it as exclusively human. A spider conducts operations that resemble those of a weaver, and a bee puts to shame many an architect in the construction of her cells. But what distinguishes the worst architect from the best of bees is this, that the architect raises his structure in imagination before he erects it in reality. At the end of every labour process, we get a result that already existed in the imagination of the labourer at its commencement. He not only effects a change of form in the material on which he works, but he also realizes a purpose of his own that gives the law to his *modus operandi*, and to which he must subordinate his will. And this subordination is no mere momentary act. Besides the exertion of the bodily organs, the process demands that, during the whole operation, the workman's will be steadily in consonance with his purpose. This means close attention. The less he is attracted by the nature of the work, and the mode in which it is carried on, and the less, therefore, he enjoys it as something which gives play to his bodily and mental powers, the more close his attention is forced to be.

The elementary factors of the labour process are 1, the personal activity of man, i.e., work itself, 2, the subject of that work, and 3, its instruments.

The soil (and this, economically speaking, includes water) in the virgin state in which it supplies man with necessaries or the means of subsistence ready to hand, exists independently of him, and is the universal subject of human labour. All those things which labour merely separates from immediate connection with their environment, are subjects of labour spontaneously provided by Nature. Such are fish which we catch and take from their element, water, timber which we fell in the virgin forest, and ores which we extract from their veins. If, on the other hand, the subject of labour has, so to say, been filtered through previous labour, we call it raw material; such is ore already extracted and ready for washing. All raw material is the subject of labour, but not every subject of labour is raw material; it can only become so, after it has undergone some alteration by means of labour.

An instrument of labour is a thing, or a complex of things, which the labourer interposes between himself and the subject of his labour, and which serves as the conductor of his activity. He makes use of the mechanical, physical, and chemical properties of some substances in order to make other substances subservient to his aims. Leaving out of consideration such ready-made means of subsistence as fruits, in gathering which a man's own limbs serve as the instruments of his labour, the first thing of which the labourer possesses himself is not the subject of labour but its instrument. Thus Nature becomes one of the organs of his activity, one that he annexes to his own bodily organs, adding stature to himself in spite of the Bible. As the earth is his original larder, so too it is his original tool house. It supplies him, for instance, with stones for throwing, grinding, pressing, cutting, etc. The earth itself is an instrument of labour, but when used as such in agriculture implies a whole series of other instruments and a comparatively high development of labour. No sooner does labour undergo the least development, than it requires specially prepared instruments. Thus in the oldest caves we find stone implements and weapons. In the earliest period of human history domesticated animals, i.e., animals which have been bred for the purpose, and have undergone modifications by means of labour, play the chief part as instruments of labour along with specially prepared stones, wood, bones, and shells. The use and fabrication of instruments of labour, although existing in the germ among certain species of animals, is specifically characteristic of the human labour process, and Franklin therefore defines man as a tool-making animal. Relics of bygone instruments of labour possess the same importance for the investigation of extinct economical forms of society, as do fossil bones for the determination of extinct species of animals. It is not the articles made, but how they are made, and by what instruments, that enables us to distinguish different economical epochs. Instruments of labour not only supply a standard of the degree of development to which human labour has attained, but they are also indicators of the social conditions under which that labour is carried on. Among the instruments of labour, those of a mechanical nature, which, taken as a whole, we may call the bone and muscles of production, offer much more decided characteristics of a given epoch of production than those which, like pipes, tubs, baskets, jars, etc., serve only to hold the materials for labour, which latter class we may in a general way call the vascular system of production. The latter first begins to play an important part in the chemical industries.

In a wider sense we may include among the instruments of labour, in addition to those things that are used for directly transferring labour to its subject, and which therefore, in one way or another, serve as conductors of activity, all such objects as are necessary for carrying on the labour process. These do not enter directly into the process, but without them it is either impossible for it to take place at all, or possible only to a partial extent. Once more we find the earth to be a universal instrument of this sort, for it furnishes a *locus standi* [established position] to the labourer and a field of employment for his activity. Among instruments that are the result of previous labour and also belong to this class, we find workshops, canals, roads, and so forth.

In the labour process, therefore, man's activity, with the help of the instruments of labour, effects an alteration, designed from the commencement, in the material worked upon. The process disappears in the product; the latter is a use-value, Nature's material adapted by a change of form to the wants of man. Labour has incorporated itself with its subject: the former is materialized, the latter transformed. That which in the labourer appeared as movement, now appears in the product as a fixed quality without motion. The blacksmith forges and the product is a forging.

If we examine the whole process from the point of view of its result, the product, it is plain that both the instruments and the subject of labour are means of production, and that the labour itself is productive labour.

Though a use-value, in the form of a product, issues from the labour process, yet other use-values, products of previous labour, enter into it as means of production. The same use-value is both the product of a previous process, and a means of production in a later process. Products are therefore not only results, but also essential conditions of labour.

With the exception of the extractive industries, in which the material for labour is provided immediately by Nature, such as mining, hunting, fishing, and agriculture (so far as the latter is confined to breaking up virgin soil), all branches of industry manipulate raw material, objects already filtered through labour, already products of labour. Such is seed in agriculture. Animals and plants, which we are accustomed to consider as products of Nature, are in their present form not only products of, say, last year's labour, but the result of a gradual transformation, continued through many generations, under man's superintendence, and by means of his labour. But in the great majority of cases, instruments of labour show, even to the most superficial observer, traces of the labour of past ages.

Raw material may either form the principal substance of a product, or it may enter into its formation only as an accessory. An accessory may be consumed by the instruments of labour, as coal under a boiler, oil by a wheel, hay by draft-horses, or it may be mixed with the raw material in order to produce some modification thereof, as chlorine into unbleached linen, coal with iron, dye-stuff with wool, or again, it may help to carry on the work itself, as in the case of the materials used for heating and lighting workshops. The distinction between principal substance and accessory vanishes in the true chemical industries, because there none of the raw material reappears, in its original composition, in the substance of the product.

Every object possesses various properties, and is thus capable of being applied to different uses. One and the same product may therefore serve as raw material in very different processes. Corn, for example, is a raw material for millers, starch-manufacturers, distillers, and cattle-breeders. It also enters as raw material into its own production in the shape of seed; coal, too, is at the same time the product of, and a means of production in, coal-mining.

Again, a particular product may be used in one and the same process, both as an instrument of labour and as raw material. Take, for instance, the fattening of cattle, where the animal is the raw material, and at the same time an instrument for the production of manure.

A product, though ready for immediate consumption, may yet serve as raw material for a further product, as grapes when they become the raw material for wine. On the other hand, labour may give us its product in such a form that we can use it only as raw material, as is the case with cotton, thread, and yarn. Such a raw material, though itself a product, may have to go through a whole series of different processes; in each of these in turn, it serves, with constantly varying form, as raw material, until the last process of the series leaves it a perfect product, ready for individual consumption, or for use as an instrument of labour.

Hence we see, that whether a use-value is to be regarded as raw material, as instrument of labour, or as product, this is determined entirely by its function in the labour process, by the position it there occupies; as this varies, so does its character.

Whenever therefore a product enters as a means of production into a new labour process, it thereby loses its character of product, and becomes a mere factor in the process. A spinner treats spindles only as implements for spinning, and flax only as the material that he spins. Of course it is impossible to spin without material and spindles; and therefore the existence of these things as products, at the commencement of the spinning operation, must be presumed: but in the process itself, the fact that they are products of previous labour is a matter of utter indifference; just as in the digestive process it is of no importance whatever that bread is the produce of the previous labour of the farmer, the miller, and the baker. On the contrary, it is generally by their imperfections as products, that the means of production in any process assert themselves in their character of products. A blunt knife or weak thread forcibly remind us of Mr. A, the cutler, or Mr. B, the spinner. In the finished product the labour by means of which it has acquired its useful qualities is not palpable, has apparently vanished.

A machine which does not serve the purposes of labour is useless. In addition, it falls a prey to the destructive influence of natural forces. Iron rusts and wood rots. Yarn with which we neither weave nor knit is cotton wasted. Living labour must seize upon these things and rouse them from their death-sleep, change them from mere possible use-values into real and effective ones. Bathed in the fire of labour, appropriated as part and parcel of labour's organism, and, as it were, made alive for the performance of their functions in the process, they are in truth consumed, but consumed with a purpose, as elementary constituents of new use-values, of new products, ever ready as means of subsistence for individual consumption, or as means of production for some new labour process.

If then, on the one hand, finished products are not only results, but also necessary conditions, of the labour process, on the other hand, their assumption into that process, their contact with living labour, is the sole means by which they can be made to retain their character of use-values, and be utilized.

Labour uses up its material factors, its subject and its instruments, consumes them, and is therefore a process of consumption. Such productive consumption is distinguished from individual consumption by this, that the latter uses up products, as means of subsistence for the living individual; the former, as means whereby alone, labour, the labour power of the living individual, is enabled to act. The product, therefore, of individual consumption is the consumer himself; the result of productive consumption is a product distinct from the consumer.

In so far, then, as its instruments and subjects are themselves products, labour consumes products in order to create products, or in other words, consumes one set of products by turning them into means of production for another set. But, just as in the beginning, the only participators in the labour process were man and the earth, which latter exists independently of man, so even now we still employ in the process many means of production, provided directly by Nature, that do not represent any combination of natural substances with human labour.

The labour process, resolved as above into its simple elementary factors, is human action with a view to the production of use-values, appropriation of natural substances to human requirements; it is the necessary condition for effecting exchange of matter between man and Nature; it is the everlasting Nature-imposed condition of human existence, and therefore is independent of every social phase of that existence, or rather, is common to every such phase. It was, therefore, not necessary to represent our labourer in connection with other labourers; man and his labour on one side, Nature and its materials on the other, sufficed. As the taste of the porridge does not tell you who grew the oats, no more does this simple process tell you of itself what are the social conditions under which it is taking place, whether under the slave-owner's brutal lash, or the anxious eye of the capitalist, whether Cincinnatus carries it on in tilling his modest farm or a savage in killing wild animals with stones.

Let us now return to our would-be capitalist. We left him just after he had purchased, in the open market, all the necessary factors of the labour process; its objective factors, the means of production, as well as its subjective factor, labour power. With the keen eye of an expert, he has selected the means of production and the kind of labour power best adapted to his particular trade, be it spinning, bootmaking, or any other kind. He then proceeds to consume the commodity, the labour power that he has just bought, by causing the labourer, the impersonation of that labour power, to consume the means of production by his labour. The general character of the labour process is evidently not changed by the fact that the labourer works for the capitalist instead of for himself; moreover, the particular methods and operations employed in bootmaking or spinning are not immediately changed by the intervention of the capitalist. He must begin by taking the labour power as he finds it in the market, and consequently be satisfied with labour of such a kind as would be found in the period immediately preceding the rise of capitalists. Changes in the methods of production by the subordination of labour to capital can take place only at a later period, and therefore will have to be treated of in a later chapter.

The labour process, turned into the process by which the capitalist consumed labour power, exhibits two characteristic phenomena. First, the labourer works under the control of the capitalist to whom his labour belongs; the capitalist taking good care that the work is done in a proper manner, and that the means of production are used with intelligence, so that there is no unnecessary waste of raw material, and no wear and tear of the implements beyond what is necessarily caused by the work.

Secondly, the product is the property of the capitalist and not that of the labourer, its immediate producer. Suppose that a capitalist pays for a day's labour power at its value; then the right to use that power for a day belongs to him, just as much as the right to use any other commodity, such as a horse that he has hired for the day. To the purchaser of a commodity belongs its use, and the seller of labour power, by giving his labour, does no more, in reality, than part with the use-value that he has sold. From the instant he steps into the workshop, the use-value of his labour power, and therefore also its use, which is labour, belongs to the capitalist. By the purchase of labour power, the capitalist incorporates labour, as a living ferment, with the lifeless constituents of the product. From his point of view, the labour process is nothing more than the consumption of the commodity purchased, i.e., of labour power; but this consumption cannot be effected except by supplying the labour power with the means of production. The labour process is a process between things that the capitalist has purchased, things that have become his property. The product of this process belongs, therefore, to him, just as much as does the wine which is the product of a process of fermentation completed in his cellar.

The product appropriated by the capitalist is a use-value, as yarn, for example, or boots. But, although boots are, in one sense, the basis of all social progress, and our capitalist is a decided 'progressist', yet he does not manufacture boots for their own sake. Use-value is by no means the thing 'qu'on aime pour lui-même' in the production of commodities. Use-values are only produced by capitalists, because, and in so far as, they are the material substratum, the depositories of exchange-value. Our capitalist has two objects in view: in the first place, he wants to produce a use-value that has a value in exchange, that is to say, an article destined to be sold, a commodity; and secondly, he desires to produce a commodity whose value shall be greater than the sum of the values of the commodities used in its production, that is, of the means of production and the labour power, that he purchased with his good money in the open market. His aim is to produce not only a use-value, but a commodity also; not only use-value, but value; not only value, but at the same time surplus value.

It must be borne in mind, that we are now dealing with the production of commodities, and that, up to this point, we have only considered one aspect of the process. Just as commodities are, at the same time, use-values and values, so the process of producing them must be a labour process, and at the same time, a process of creating value.

Let us now examine production as a creation of value.

We know that the value of each commodity is determined by the quantity of labour expended on and materialized in it, by the working-time necessary, under given social conditions, for its production. This rule also holds good in the case of the product that accrued to our capitalist, as the result of the labour process carried on for him. Assuming this product to be 10 lb of yarn, our first step is to calculate the quantity of labour realized in it.

For spinning the yarn, raw material is required; suppose in this case 10 lb of cotton. We have no need at present to investigate the value of this cotton, for our capitalist has, we will assume, bought it at its full value, say of ten shillings. In this price the labour required for the production of the cotton is already expressed in terms of the average labour of society. We will further assume that the wear and tear of the spindle, which, for our present purpose, may represent all other instruments of labour employed, amounts to the value of 2s. If, then, twenty-four hours' labour, or two working-days, are required to produce the quantity of gold represented by twelve shillings, we have here, to begin with, two days' labour already incorporated in the yarn.

We must not let ourselves be misled by the circumstance that the cotton has taken a new shape while the substance of the spindle has to a certain extent been used up. By the general law of value, if the value of 40 lb of yarn = the value of 40 lb of cotton + the value of a whole spindle, i.e., if the same working-time is required to produce the commodities on either side of this equation, then 10 lb of yarn are an equivalent for 10 lb of cotton, together with one-fourth of a spindle. In the case we are considering the same working-time is materialized in the 10 lb of yarn on the one hand, and in the 10 lb of cotton and the fraction of a spindle on the other. Therefore, whether value appears in cotton, in a spindle, or in yarn, makes no difference in the amount of that value. The spindle and cotton, instead of resting quietly side by side, join together in the process, their forms are altered, and they are turned into yarn; but their value is no more affected by this fact than it would be if they had been simply exchanged for their equivalent in yarn.

The labour required for the production of the cotton, the raw material of the yarn, is part of the labour necessary to produce the yarn, and is therefore contained in the yarn. The same applies to the labour embodied in the spindle, without whose wear and tear the cotton could not be spun.

Hence, in determining the value of the yarn, or the labour time required for its production, all the special processes carried on at various times and in different places, which were necessary, first to produce the cotton and the wasted portion of the spindle, and then with the cotton and spindle to spin the yarn, may together be looked on as different and successive phases of one and the same process. The whole of the labour in the yarn is past labour; and it is a matter of no importance that the operations necessary for the production of its constituent elements were carried on at times which, referred to the present, are more remote than the final operation of spinning. If a definite quantity of labour, say thirty days, is requisite to build a house, the total amount of labour incorporated in it is not altered by the fact that the work of the last day is done twenty-nine days later than that of the first. Therefore the labour contained in the raw material and the instruments of labour can be treated just as if it were labour expended in an earlier stage of the spinning process, before the labour of actual spinning commenced.

The values of the means of production, i.e., the cotton and the spindle, which values are expressed in the price of twelve shillings, are therefore constituent parts of the value of the yarn, or, in other words, of the value of the product.

Two conditions must nevertheless be fulfilled. First, the cotton and spindle must concur in the production of a use-value; they must in the present case become yarn. Value is independent of the particular use-value by which it is borne, but it must be embodied in a use-value of some kind. Secondly, the time occupied in the labour of production must not exceed the time really necessary under the given social conditions of the case. Therefore, if no more than 1 lb of cotton be requisite to spin 1 lb of yarn, care must be taken that no more than this weight of cotton is consumed in the production of 1 lb of yarn; and similarly with regard to the spindle. Though the capitalist have a hobby, and use a gold instead of a steel spindle, yet the only labour that counts for anything in the value of the yarn is that which would be required to produce a steel spindle, because no more is necessary under the given social conditions.

We now know what portion of the value of the yarn is owing to the cotton and the spindle. It amounts to twelve shillings or the value of two days' work. The next point for our consideration is, what portion of the value of the yarn is added to the cotton by the labour of the spinner.

We have now to consider this labour under a very different aspect from that which it had during the labour-process; there, we viewed it solely as that particular kind of human activity which changes cotton into yarn; there, the more the labour was suited to the work, the better the yarn, other circumstances remaining the same. The labour of the spinner was then viewed as specifically different from other kinds of productive labour, different on the one hand in its special aim, viz., spinning, different, on the other hand, in the special character of its operations, in the special nature of its means of production and in the special use-value of its product. For the operation of spinning, cotton and spindles are a necessity, but for making rifled cannon they would be of no use whatever. Here, on the contrary, where we consider the labour of the spinner only so far as it is value-creating, i.e., a source of value, his labour differs in no respect from the labour of the man who bores cannon, or (what here more nearly concerns us), from the labour of the cotton-planter and spindle-maker incorporated in the means of production.

It is solely by reason of this identity, that cotton-planting, spindle-making, and spinning are capable of forming the component parts, differing only quantitatively from each other, of one whole, namely, the value of the yarn. Here, we have nothing more to do with the quality, the nature, and the specific character of the labour, but merely with its quantity. And this simply requires to be calculated. We proceed upon the assumption that spinning is simple, unskilled labour, the average labour of a given state of society. Hereafter we shall see that the contrary assumption would make no difference.

While the labourer is at work, his labour constantly undergoes a transformation: from being motion, it becomes an object without motion; from being the labourer working, it becomes the thing produced. At the end of one hour's spinning, that act is represented by a definite quantity of yarn; in other words, a definite quantity of labour, namely that of one hour, has become embodied in the cotton. We say labour, i.e., the expenditure of his vital force by the spinner, and not spinning labour, because the special work of spinning counts here only so far as it is the expenditure of labour power in general, and not in so far as it is the specific work of the spinner.

In the process we are now considering, it is of extreme importance that no more time be consumed in the work of transforming the cotton into yarn than is necessary under the given social conditions. If under normal, i.e., average social conditions of production, a pounds of cotton ought to be made into b pounds of yarn by one hour's labour, then a day's labour does not count as 12 hours' labour unless 12 a pounds of cotton have been made into 12 b pounds of yarn; for in the creation of value, the time that is socially necessary alone counts.

Not only the labour, but also the raw material and the product now appear in quite a new light, very different from that in which we viewed them in the labour process pure and simple. The raw material serves now merely as an absorbent of a definite quantity of labour. By this absorption it is in fact changed into yarn, because it is spun, because labour power in the form of spinning is added to it; but the product, the yarn, is now nothing more than a measure of the labour absorbed by the cotton. If in one hour $1\frac{2}{3}$ lb of cotton can be spun into $1\frac{2}{3}$ lb of yarn, then 10 lb of yarn indicate the absorption of 6 hours' labour. Definite quantities of product, these quantities being determined by experience, now represent nothing but definite quantities of labour, definite masses of crystallized labour time. They are nothing more than the materialization of so many hours or so many days of social labour.

We are here no more concerned about the facts, that the labour is the specific work of spinning, that its subject is cotton and its product yarn, than we are about the fact that the subject itself is already a product and therefore raw material. If the spinner, instead of spinning, were working in a coal-mine, the subject of his labour, the coal, would be supplied by Nature;

nevertheless, a definite quantity of extracted coal, a hundredweight for example, would represent a definite quantity of absorbed labour.

We assumed, on the occasion of its sale, that the value of a day's labour power is three shillings, and that six hours' labour is incorporated in that sum; and consequently that this amount of labour is requisite to produce the necessaries of life daily required on an average by the labourer. If now our spinner by working for one hour, can convert $1\frac{2}{3}$ lb of cotton into $1\frac{2}{3}$ of yarn, it follows that in six hours he will convert 10 lb of cotton into 10 lb of yarn. Hence, during the spinning process, the cotton absorbs six hours' labour. The same quantity of labour is also embodied in a piece of gold of the value of three shillings. Consequently, by the mere labour of spinning, a value of three shillings is added to the cotton.

Let us now consider the total value of the product, the 10 lb of yarn. Two and a half days' labour has been embodied in it, of which two days were contained in the cotton and in the substance of the spindle worn away, and half a day was absorbed during the process of spinning. This two and a half days' labour is also represented by a piece of gold of the value of fifteen shillings. Hence, fifteen shillings is an adequate price for the 10 lb of yarn, or the price of one pound is eighteen pence.

Our capitalist stares in astonishment. The value of the product is exactly equal to the value of the capital advanced. The value so advanced has not expanded, no surplus value has been created, and consequently money has not been converted into capital. The price of the yarn is fifteen shillings, and fifteen shillings were spent in the open market upon the constituent elements of the product, or, what amounts to the same thing, upon the factors of the labour process; ten shillings were paid for the cotton, two shillings for the substance of the spindle worn away, and three shillings for the labour power. The swollen value of the yarn is of no avail, for it is merely the sum of the values formerly existing in the cotton, the spindle, and the labour power: out of such a simple addition of existing values, no surplus value can possibly arise. These separate values are now all concentrated in one thing; but so they were also in the sum of fifteen shillings, before it was split up into three parts, by the purchase of the commodities.

There is in reality nothing very strange in this result. The value of one pound of yarn being eighteen pence, if our capitalist buys 10 lb of yarn in the market, he must pay fifteen shillings for them. It is clear that, whether a man buys his house ready built, or gets it built for him, in neither case will the mode of acquisition increase the amount of money laid out on the house.

Our capitalist, who is at home in his vulgar economy, exclaims: 'Oh! but I advanced my money for the express purpose of making more money.' The way to Hell is paved with good intentions, and he might just as easily have intended to make money, without producing at all. He threatens all sorts of things. He won't be caught napping again. In future he will buy the commodities in the market, instead of manufacturing them himself. But if all his brother capitalists were to do the same, where would he find his commodities in the market? And his money he cannot eat. He tries persuasion. 'Consider my abstinence; I might have played ducks and drakes with the 15 shillings; but instead of that I consumed it productively, and made yarn with it.' Very well, and by way of reward he is now in possession of good yarn instead of a bad conscience; and as for playing the part of a miser, it would never do for him to relapse into such bad ways as that; we have seen before to what results such asceticism leads. Besides, where nothing is, the king has lost his rights; whatever may be the merit of his abstinence, there is nothing wherewith specially to remunerate it, because the value of the product is merely the sum of the values of the commodities that were thrown into the process of production. Let him therefore console himself with the reflection that virtue is its own reward. But no, he becomes importunate. He says: 'The yarn is of no use to me: I produced it for sale.' In that case let him sell it, or, still better, let him for the future produce only things for satisfying his personal wants, a remedy that his physician MacCulloch has already prescribed as infallible against an epidemic of overproduction. He now gets obstinate. 'Can the labourer,' he asks, 'merely with his arms and legs, produce commodities out of nothing? Did I not supply him with the materials, by means of which, and in which alone, his labour could be embodied? And as the greater part of society consists of such ne'er-do-wells, have I not rendered society incalculable service by my instruments of production, my cotton and my spindle, and not only society, but the labourer also, whom in addition I have provided with the necessaries of life? And am I to be allowed nothing in return for all this service?' Well, but has not the labourer rendered him the equivalent service of changing his cotton and spindle into yarn? Moreover, there is here no question of service. A service is nothing more than the useful effect of a use-value, be it of a commodity, or be it of labour. But here we are dealing with exchange-value. The capitalist paid to the labourer a value of three shillings, and the labourer gave him back an exact equivalent in the value of three shillings, added by him to the cotton: he gave him value for value. Our friend, up to this time so purse-proud, suddenly assumes the modest demeanour of his own workman, and exclaims: 'Have I myself not worked? Have I not performed the labour of superintendence and of overlooking the spinner? And does not this labour, too, create value?' His overlooker and his manager try to hide their smiles. Meanwhile, after a hearty laugh, he reassumes his usual mien. Though he chanted to us the whole creed of the economists, in reality, he says, he would not give a brass farthing for it. He leaves this and all such like subterfuges and juggling tricks to the professors of Political Economy, who are paid for it. He himself is a practical man; and though he does not always consider what he says outside his business, yet in his business he knows what he is about.

Let us examine the matter more closely. The value of a day's labour power amounts to three shillings, because on our assumption half a day's labour is embodied in that quantity of labour power, i.e., because the means of subsistence that are daily required for the production of labour power, cost half a day's labour. But the past labour that is embodied in the labour power, and the living labour that it can call into action; the daily cost of maintaining it, and its daily expenditure in work, are two totally different things. The former determines the exchange-value of the labour power, the latter is its use-value. The fact that half a day's labour is necessary to keep the labourer alive during twenty-four hours, does not in any way prevent him from working a whole day. Therefore, the value of labour power, and the value which that labour power creates in the labour process, are two entirely different magnitudes; and this difference of the two values was what the capitalist had in view, when he was purchasing the labour power. The useful qualities that labour power possesses, and by virtue of which it makes yarn or boots, were to him nothing more than a *conditio sine qua non*; for in order to create value, labour must be expended in a useful manner. What really influenced him was the specific use-value which this commodity possesses of being a source not only of value, but of more value than it has itself. This is the special service that the capitalist expects from labour power, and in this transaction he acts in accordance with the 'eternal laws' of the exchange of commodities. The seller of labour power, like the seller of any other commodity, realizes its exchange-value, and parts with its use-value. He cannot take the one without giving the other. The use-value of labour power, or in other words, labour, belongs just as little to its seller, as the use-value of oil after it has been sold belongs to the dealer who has sold it. The owner of the money has paid the value of a day's labour power; his, therefore, is the use of it for a day; a day's labour belongs to him. The circumstance that on the one hand the daily sustenance of labour power costs only half a day's labour, while on the other hand the very same labour power can work during a whole day, that consequently the value which its use during one day creates, is double what he pays for that use, this circumstance is, without doubt, a piece of good luck for the buyer, but by no means an injury to the seller.

Our capitalist foresaw this state of things, and that was the cause of his laughter. The labourer therefore finds, in the workshop, the means of production necessary for working, not only during six, but during twelve hours. Just as during the six hours' process our 10 lb of cotton absorbed six hours' labour, and became 10 lb of yarn, so now, 20 lb of cotton will absorb 12 hours' labour and be changed into 20 lb of yarn. Let us now examine the product of this prolonged process. There is now materialized in this 20 lb of yarn the labour of five days, of which four days are due to the cotton and the lost steel of the spindle, the remaining day having been absorbed by the cotton during the spinning process. Expressed in gold, the labour of five days is thirty shillings. This is therefore the price of the 20 lb of yarn, giving, as before, eighteen pence as the price of a pound. But the sum of the values of the commodities that entered into the process amounts to twenty-seven shillings. The value of the yarn is thirty shillings. Therefore the value of the product is $\frac{1}{9}$ greater than the value advanced for its production; twenty-seven shillings have been transformed into thirty shillings; a surplus value of three shillings has been created. The trick has at last succeeded; money has been converted into capital.

Every condition of the problem is satisfied, while the laws that regulate the exchange of commodities have been in no way violated. Equivalent has been exchanged for equivalent. For the capitalist as buyer paid for each commodity, for the cotton, the spindle, and the labour power, its full value. He then did what is done by every purchaser of commodities; he consumed their use-value. The consumption of the labour power, which was also the process of producing commodities, resulted in 20 lb of yarn, having a value of thirty shillings. The capitalist, formerly a buyer, now returns to market as a seller, of commodities. He sells his yarn at eighteen pence a pound, which is its exact value. Yet for all that he withdraws three shillings more from circulation than he originally threw into it. This metamorphosis, this conversion of money into capital, takes place both within the sphere of circulation and also outside it; within the circulation, because conditioned by the purchase of the labour power in the market; outside the circulation, because what is done within it is only a stepping-stone to the production of surplus value, a process which is entirely confined to the sphere of production. Thus 'tout est pour le mieux dans le meilleur des mondes possibles'. [Everything is for the best in the best of all possible worlds.]

By turning his money into commodities that serve as the material elements of a new product, and as factors in the labour process, by incorporating living labour with their dead substance, the capitalist at the same time converts value, i.e. past, materialized, and dead labour, into capital, into value big with value, a live monster that is fruitful and multiplies.

If we now compare the two processes of producing value and of creating surplus value, we see that the latter is nothing but the continuation of the former beyond a definite point. If on the one hand the process be not carried beyond the point where the value paid by the capitalist for the labour power is replaced by an exact equivalent, it is simply a process of producing value; if, on the other hand, it be continued beyond that point, it becomes a process of creating surplus value.

If we proceed further, and compare the process of producing value with the labour process, pure and simple, we find that the latter consists of the useful labour, the work, that produces use-values. Here we contemplate the labour as producing a particular article; we view it under its qualitative aspect alone, with regard to its end and aim. But viewed as a value-creating process, the same labour process presents itself under its quantitative aspect alone. Here it is a question merely of the time occupied by the labourer in doing the work; of the period during which the labour power is usefully expended. Here, the commodities that take part in the process do not count any longer as necessary adjuncts of labour power in the production of a definite, useful object. They count merely as depositories of so much absorbed or materialized labour; that labour, whether previously embodied in the means of production, or incorporated in them for the first time during the process by the action of labour power, counts in either case only according to its duration; it amounts to so many hours or days as the case may be.

Moreover, only so much of the time spent in the production of any article is counted as, under the given social conditions, is necessary. The consequences of this are various. In the first place, it becomes necessary that the labour should be carried on under normal conditions. If a self-acting mule is the implement in general use for spinning, it would be absurd to supply the spinner with a distaff and spinning-wheel. The cotton too must not be such rubbish as to cause extra waste in being worked, but must be of suitable quality. Otherwise the spinner would be found to spend more time in producing a pound of yarn than is socially necessary, in which case the excess of time would create neither value nor money. But whether the material factors of the process are of normal quality or not depends not upon the labourer, but entirely upon the capitalist. Then again, the labour power itself must be of average efficacy. In the trade in which it is being employed, it must possess the average skill, handiness, and quickness prevalent in that trade, and our capitalist took good care to buy labour power of such normal goodness. This power must be applied with the average amount of exertion and with the usual degree of intensity; and the capitalist is as careful to see that this is done as that his workmen are not idle for a single moment. He has bought the use of the labour power for a definite period, and he insists upon his rights. He has no intention of being robbed. Lastly, and for this purpose our friend has a penal code of his own, all wasteful consumption of raw material or instruments of labour is strictly forbidden, because what is so wasted represents labour superfluously expended, labour that does not count in the product or enter into its value.

We now see that the difference between labour, considered on the one hand as producing utilities, and on the other hand as creating value, a difference which we discovered by our analysis of a commodity, resolves itself into a distinction between two aspects of the process of production.

The process of production, considered on the one hand as the unity of the labour process and the process of creating value, is production of commodities; considered on the other hand as the unity of the labour process and the process of producing surplus value, it is the capitalist process of production, or capitalist production of commodities.

We stated, on a previous page, that in the creation of surplus value it does not in the least matter whether the labour appropriated by the capitalist be simple unskilled labour of average quality or more complicated skilled labour. All labour of a higher or more complicated character than average labour is expenditure of labour power of a more costly kind, labour power whose production has cost more time and labour, and which therefore has a higher value, than unskilled or simple labour power. This power being of higher value, its consumption is labour of a higher class, labour that creates in equal times proportionally higher values than unskilled labour does. Whatever difference in skill there may be between the labour of a spinner and that of a jeweller, the portion of his labour by which the jeweller merely replaces the value of his own labour power, does not in any way differ in quality from the additional portion by which he creates surplus value. In the making of jewellery, just as in spinning, the surplus value results only from a quantitative excess of labour, from a lengthening-out of one and the same labour process, in the one case of the process of making jewels, in the other of the process of making yarn.

But on the other hand, in every process of creating value, the reduction of skilled labour to average social labour, e.g., one day of skilled to six days of unskilled labour, is unavoidable. We therefore save ourselves a superfluous operation, and simplify our analysis, by the assumption, that the labour of the workman employed by the capitalist is unskilled average labour. . . .

Constant and Variable Capital

The various factors of the labour process play different parts in forming the value of the product.

The labourer adds fresh value to the subject of his labour by expending upon it a given amount of additional labour, no matter what the specific character and utility of that labour may be. On the other hand, the values of the means of production used up in the process are preserved, and present themselves afresh as constituent parts of the value of the product; the values of the cotton and the spindle, for instance, reappear again in the value of the yarn. The value of the means of production is therefore preserved, by being transferred to the product. This transfer takes place during the conversion of those means into a product, or in other words, during the labour process. It is brought about by labour; but how?

The labourer does not perform two operations at once, one in order to add value to the cotton, the other in order to preserve the value of the means of production, or, what amounts to the same thing, to transfer to the yarn, to the product, the value of the cotton on which he works, and part of the value of the spindle with which he works. But, by the very act of adding new value, he preserves their former values. Since, however, the addition of new value to the subject of his labour, and the preservation of its former value, are two entirely distinct results, produced simultaneously by the labourer, during one operation, it is plain that this twofold nature of the result can be explained only by the twofold nature of his labour; at one and the same time, it must in one character create value, and in another character preserve or transfer value.

Now, in what manner does every labourer add new labour and consequently new value? Evidently, only by labouring productively in a particular way; the spinner by spinning, the weaver by weaving, the smith by forging. But, while thus incorporating labour generally, that is value, it is by the particular form alone of the labour, by the spinning, the weaving, and the forging respectively, that the means of production, the cotton and spindle, the yarn and loom, and the iron and anvil become constituent elements of the product, of a new use-value. Each use-value disappears, but only to reappear under a new form in a new use-value. Now, we saw, when we were considering the process of creating value, that if a use-value be effectively consumed in the production of a new use-value, the quantity of labour expended in the production of the consumed article forms a portion of the quantity of labour necessary to produce the new use-value; this portion is therefore labour transferred from the means of production to the new product. Hence, the labourer preserves the values of the consumed means of production, or transfers them as portions of its value to the product, not by virtue of his additional labour, abstractedly considered, but by virtue of the particular useful character of that labour, by virtue of its special productive form. In so far then as labour is such specific productive activity, in so far as it is spinning, weaving, or forging, it raises, by mere contact, the means of production from the dead, makes them living factors of the labour process, and combines with them to form the new products. . . .

We have seen that the means of production transfer value to the new product, so far only as during the labour process they lose value in the shape of their old use-value. The maximum loss of value that they can suffer in the process is plainly limited by the amount of the original value with which they came into the process, or, in other words, by the labour time necessary for their production. Therefore, the means of production can never add more value to the product than they themselves possess independently of the process in which they assist. However useful a given kind of raw material, or a machine, or other means of production may be, though it may cost £150, or, say, 500 days' labour, yet it cannot, under any circumstances, add to the value of the product more than £150. Its value is determined not by the labour process into which it enters as a means of production, but by that out of which it has issued as a product. In the labour process it only serves as a mere use-value, a thing with useful properties, and could not, therefore, transfer any value to the product, unless it possessed such value previously.

While productive labour is changing the means of production into constituent elements of a new product, their value undergoes a metempsychosis. It deserts the consumed body, to occupy the newly created one. But this transmigration takes place, as it were, behind the back of the labourer. He is unable to add new labour, to create new value, without at the same time preserving old values, and this, because the labour he adds must be of a specific useful kind; and he cannot do work of a useful kind, without employing products as the means of production of a new product, and thereby transferring their value to the new product. The property therefore which labour power in action, living labour, possesses of preserving value, at the same time that it adds it, is a gift of Nature which costs the labourer nothing, but which is very advantageous to the capitalist in as much as it preserves the existing value of his capital. So long as trade is good, the capitalist is too much absorbed in money-grubbing to take notice of this gratuitous gift of labour. A violent interruption of the labour process by a crisis makes him sensitively aware of it.

As regards the means of production, what is really consumed is their use-value, and the consumption of this use-value by labour results in the product. There is no consumption of their value, and it would therefore be inaccurate to say that it is reproduced. It is rather preserved; not by reason of any operation it undergoes itself in the process; but because the articles in which it originally exists vanishes, it is true, but vanishes into some other article. Hence, in the value of the product, there is a reappearance of the value of the means of production, but there is, strictly speaking, no reproduction of that value. That which is produced is a new use-value in which the old exchange-value reappears. . . .

That part of capital then, which is represented by the means of production, by the raw material, auxiliary material and the instruments of labour, does not, in the process of production, undergo any quantitative alteration of value. I therefore call it the constant part of capital, or, more shortly, constant capital.

On the other hand, that part of capital represented by labour power does, in the process of production, undergo an alteration of value. It both reproduces the equivalent of its own value, and also produces an excess, a surplus value, which may itself vary, may be more or less according to circumstances. This part of capital is continually being transformed from a constant into a variable magnitude. I therefore call it the variable part of capital, or, shortly, variable capital. The same elements of capital which, from the point of view of the labour process, present themselves respectively as the objective and subjective factors, as means of production and labour power, present themselves, from the point of view of the process of creating surplus value, as constant and variable capital. . . .

The Rate of Surplus Value

. . . If we look at the means of production, in their relation to the creation of value, and to the variation in the quantity of value, apart from anything else, they appear simply as the material in which labour power, the value-creator, incorporates itself. Neither the nature, nor the value of this material is of any importance. The only requisite is that there be a sufficient supply to absorb the labour expended in the process of production. That supply once given, the material may rise or fall in value, or even be, as land and the sea, without any value in itself; but this will have no influence on the creation of value or on the variation in the quantity of value.

In the first place then we equate the constant capital to zero. The capital advanced is consequently reduced from c + v to v, and instead of the value of the product (c + v) + s we have now the value produced (v + s). Given the new value produced = £180, which sum consequently represents the whole labour expended during the process, then subtracting from it £90, the value of the variable capital, we have remaining £90, the amount of the surplus value. This sum of £90 or s expresses the absolute quantity of surplus value produced. The relative quantity produced, or the increase per cent of the variable capital, is determined, it is plain, by the ratio of the surplus value to the variable capital, or is expressed by s/v. In our example this ratio is $\frac{90}{90}$, which gives an increase of 140 per cent. This relative increase in the value of the variable capital, or the relative magnitude of the surplus value, I call, 'The rate of surplus value'.

We have seen that the labourer, during one portion of the labour process, produces only the value of his labour power, that is, the value of his means of subsistence. Now since his work forms part of a system, based on the social division of labour, he does not directly produce the actual necessaries which he himself consumes; he produces instead a particular commodity, yarn for example, whose value is equal to the value of those necessaries or of the money with which they can be bought. The portion of his day's labour devoted to this purpose will be greater or less, in proportion to the value of the necessaries that he daily requires on an average, or, what amounts to the same thing, in proportion to the labour time required on an average to produce them. If the value of those necessaries represent on an average the expenditure of six hours' labour, the workman must on an average work for six hours to produce that value. If instead of working for the capitalist, he worked independently on his own account, he would, other things being equal, still be obliged to labour for the same number of hours, in order to produce the value of his labour power, and thereby to gain the means of subsistence necessary for his conservation or continued reproduction. But as we have seen, during that portion of his day's labour in which he produces the value of his labour power, say three shillings, he produces only an equivalent for the value of his labour power already advanced by the capitalist; the new value created only replaces the variable capital advanced. It is owing to this fact, that the production of the new value of three shillings takes the semblance of a mere reproduction. That portion of the working-day, then, during which this reproduction takes place, I call 'necessary' labour time, and the labour expended during that time I call 'necessary' labour. Necessary, as regards the labourer, because independent of the particular social form of his labour; necessary, as regards capital, and the world of capitalists, because on the continued existence of the labourer depends their existence also.

During the second period of the labour process, that in which his labour is no longer necessary labour, the workman, it is true, labours, expends labour power; but his labour, being no longer necessary labour, he creates no value for himself. He creates surplus value which, for the capitalist, has all the charms of a creation out of nothing. This portion of the working-day, I name surplus labour time, and to the labour expended during that time, I give the name of surplus labour. It is every bit as important, for a correct understanding of surplus value, to conceive it as a mere congelation of surplus labour time, as nothing but materialized surplus labour, as it is, for a proper comprehension of value, to conceive it as a mere congelation of so many hours of labour, as nothing but materialized labour. The essential difference between the various economic forms of society, between, for instance, a society based on slave-labour, and one based on wage-labour, lies only in the mode in which this surplus labour is in each case extracted from the actual producer, the labourer.

Since, on the one hand, the values of the variable capital and of the labour power purchased by that capital are equal, and the value of this labour power determines the necessary portion of the working-day; and since, on the other hand, the surplus value is determined by the surplus portion of the working-day, it follows that surplus value bears the same ratio to variable capital, that surplus labour does to necessary labour, or in other words, the rate of surplus value

$$\frac{s}{v} = \frac{\text{surplus labour}}{\text{necessary labour}}.$$

Both ratios,

$$\frac{s}{v} \text{ and } \frac{\text{surplus labour}}{\text{necessary labour}},$$

express the same thing in different ways; in the one case by reference to materialized, incorporated labour, in the other by reference to living, fluent labour.

The rate of surplus value is therefore an exact expression for the degree of exploitation of labour power by capital, or of the labourer by the capitalist. . . .

The Working Day

What experience shows to the capitalist generally is a constant excess of population, i.e., an excess in relation to the momentary requirements of surplus-labour-absorbing capital, although this excess is made up of generations of human beings stunted, short-lived, swiftly replacing each other, plucked, so to say, before maturity. And, indeed, experience shows to the intelligent observer with what swiftness and grip the capitalist mode of production, dating historically speaking only from yesterday, has seized the vital power of the people by the very root—shows how the degeneration of the industrial population is only retarded by the constant absorption of primitive and physically uncorrupted elements from the country—shows how even the country labourers, in spite of fresh air and the principle of natural selection, that works so powerfully among them and only permits the survival of the strongest, are already beginning to die off. Capital that has such good reasons for denying the sufferings of the legions of workers that surround it, is in practice moved as much and as little by the sight of the coming degradation and final depopulation of the human race, as by the probable fall of the earth into the sun. In every stock-jobbing swindle everyone knows that some time or other the crash must come, but everyone hopes that it may fall on the head of his neighbour, after he himself has caught the shower of gold and placed it in safety. *Après moi le déluge!* [After me the flood!] is the watchword of every capitalist and of every capitalist nation. Hence Capital is reckless of the health or length of life of the labourer, unless under compulsion from society. To the outcry as to the physical and mental degradation, the premature death, the torture of overwork, it answers: Ought these to trouble us since they increase our profits? But looking at things as a whole, all this does not, indeed, depend on the good or ill will of the individual capitalist. Free competition brings out the inherent laws of capitalist production, in the shape of external coercive laws having power over every individual capitalist. . . .

It must be acknowledged that our labourer comes out of the process of production other than he entered. In the market he stood as owner of the commodity 'labour power' face to face with other owners of commodities, dealer against dealer. The contract by which he sold to the capitalist his labour power proved, so to say, in black and white that he disposed of himself freely. The bargain concluded, it is discovered that he was no 'free agent', that the time for which he is free to sell his labour power is the time for which he is forced to sell it, that in fact the vampire will not lose its hold on him 'so long as there is a muscle, a nerve, a drop of blood to be exploited'. For 'protection' against 'the serpent of their agonies', the labourers must put their heads together and, as a class, compel the passing of a law, an all-powerful social barrier that shall prevent the very workers from selling, by voluntary contract with capital, themselves and their families into slavery and death. In place of the pompous catalogue of the 'inalienable rights of man' comes the modest Magna Carta of a legally limited working-day, which shall make clear 'when the time which the worker sells is ended, and when his own begins'. *Quantum mutatus ab illo!* [How changed from what it was before!]

The Division of Labour

Division of labour in a society, and the corresponding tying-down of individuals to a particular calling, develops itself, just as does the division of labour in manufacture, from opposite starting-points. Within a family, and after further development within a tribe, there springs up naturally a division of labour, caused by differences of sex and age, a division that is consequently based on a purely physiological foundation, which division enlarges its materials by the expansion of the community, by the increase of population, and more especially, by the conflicts between different tribes, and the subjugation of one tribe by another. On the other hand, as I have before remarked, the exchange of products springs up at the points where different families, tribes, communities, come in contact; for, in the beginning of civilization, it is not private individuals but families, tribes, etc., that meet on an independent footing. Different communities find different means of production, and different means of subsistence in their natural environment. Hence, their modes of production, and of living, and their products are different. It is this spontaneously developed difference which, when different communities come in contact, calls forth the mutual exchange of products, and the consequent gradual conversion of those products into commodities. Exchange does not create the differences between the spheres of production, but brings what are already different into relation, and thus converts them into more or less interdependent branches of the collective production of an enlarged society. In the latter case, the social division of labour arises from the exchange between spheres of production, that are originally distinct and independent of one another. In the former, where the physiological division of labour is the starting-point, the particular organs of a compact whole grow loose, and break off, principally owing to the exchange of commodities with foreign communities, and then isolate themselves so far, that the sole bond still connecting the various kinds of work, is the exchange of the products as commodities. In the one case, it is the making dependent what was before independent; in the other case, the making independent what was before dependent.

The foundation of every division of labour that is well developed, and brought about by the exchange of commodities, is the separation between town and country. It may be said that the whole economical history of society is summed up in the movement of this antithesis. We pass it over, however, for the present. . . .

In manufacture, as well as in simple co-operation, the collective working organism is a form of existence of capital. The mechanism that is made up of numerous individual detail labourers belongs to the capitalist. Hence, the productive power resulting from a combination of labours appears to be the productive power of capital. Manufacture proper not only subjects the previously independent workman to the discipline and command of capital, but, in addition, creates a hierarchic gradation of the workmen themselves. While simple co-operation leaves the mode of working by the individual for the most part unchanged, manufacture thoroughly revolutionizes it, and seizes labour power by its very roots. It converts the labourer into a crippled monstrosity, by forcing his detail dexterity at the expense of a world of productive capabilities and instincts; just as in the States of La Plata they butcher a whole beast for the sake of his hide or his tallow. Not only is the detail work distributed to the different individuals, but the individual himself is made the automatic motor of a fractional operation, and the absurd fable of Menenius Agrippa, which makes man a mere fragment of his own body, becomes realized. If, at first, the workman sells his labour power to capital, because the material means of producing a commodity fail him, now his very labour power refuses its services unless it has been sold to capital. Its functions can be exercised only in an environment that exists in the workshop of the capitalist after the sale. By nature unfitted to make anything independently, the manufacturing labourer develops productive activity as a mere appendage of the capitalist's workshop. As the chosen people bore in their features the sign manual of Jehovah, so division of labour brands the manufacturing workman as the property of capital. . . .

The General Law of Capitalist Accumulation

In this chapter we consider the influence of the growth of capital on the lot of the labouring class. The most important factor in this inquiry is the composition of capital and the changes it undergoes in the course of the process of accumulation.

The composition of capital is to be understood in a twofold sense. On the side of value, it is determined by the proportion in which it is divided into constant capital or value of the means of production, and variable capital or value of labour power, the sum total of wages. On the side of material, as it functions in the process of production, all capital is divided into means of production and living labour power. This latter composition is determined by the relation between the mass of the means of production employed, on the one hand, and the mass of labour necessary for their employment on the other. I call the former the value-composition, the latter the technical composition of capital. Between the two there is a strict correlation. To express this, I call the value-composition of capital, in so far as it is determined by its technical composition and mirrors the changes of the latter, the organic composition of capital. Wherever I refer to the composition of capital, without further qualification, its organic composition is always understood.

The many individual capitals invested in a particular branch of production have, one with another, more or less different compositions. The average of their individual compositions gives us the composition of the total capital in this branch of production. Lastly, the average of these averages, in all branches of production, gives us the composition of the total social capital of a country, and with this alone are we, in the last resort, concerned in the following investigation.

Growth of capital involves growth of its variable constituent or of the part invested in labour power. A part of the surplus value turned into additional capital must always be re-transformed into variable capital, or additional labour-fund. If we suppose that, all other circumstances remaining the same, the composition of capital also remains constant (i.e., that a definite mass of means of production constantly needs the same mass of labour power to set it in motion), then the demand for labour and the subsistence-fund of the labourers clearly increase in the same proportion as the capital, and the more rapidly, the more rapidly the capital increases. Since the capital produces yearly a surplus value, of which one part is yearly added to the original capital; since this increment itself grows yearly along with the augmentation of the capital already functioning; since lastly, under special stimulus to enrichment, such as the opening of new markets, or of new spheres for the outlay of capital in consequence of newly developed social wants, etc., the scale of accumulation may be suddenly extended, merely by a change in the division of the surplus value or surplus product into capital and revenue, the requirements of accumulating capital may exceed the increase of labour power or of the number of labourers; the demand for labourers may exceed the supply, and, therefore, wages may rise. This must, indeed, ultimately be the case if the conditions supposed above continue. For since in each year more labourers are employed than in its predecessor, sooner or later a point must be reached, at which the requirements of accumulation begin to surpass the customary supply of labour, and, therefore, a rise of wages takes place. A lamentation on this score was heard in England during the whole of the fifteenth, and the first half of the eighteenth centuries. The more or less favourable circumstances in which the wage-working class supports and multiplies itself, in no way alter the fundamental character of capitalist production. As simple reproduction constantly reproduces the capital-relation itself, i.e., the relation of capitalists on the one hand, and wage-workers on the other, so reproduction on a progressive scale, i.e., accumulation, reproduces the capital-relation on a progressive scale, more capitalists or larger capitalists at this pole, more wage-workers at that. The reproduction of a mass of labour power, which must incessantly reincorporate itself with capital for that capital's self-expansion; which cannot get free from capital, and whose enslavement to capital is only concealed by the variety of individual capitalists to whom it sells itself, this reproduction of labour power forms, in fact, an essential of the reproduction of capital itself. Accumulation of capital is, therefore, increase of the proletariat.

The law of capitalist production, that is at the bottom of the pretended 'natural law of population', reduces itself simply to this: The correlation between accumulation of capital and rate of wages is nothing else than the correlation between the unpaid labour transformed into capital, and the additional paid labour necessary for the setting in motion of this additional capital. It is therefore in no way a relation between two magnitudes, independent one of the other: on the one hand, the magnitude of the capital; on the other, the number of the labouring population; it is rather, at bottom, only the relation between the unpaid and the paid labour of the same labouring population. If the quantity of unpaid labour supplied by the working-class, and accumulated by the capitalist class, increases so rapidly that its conversion into capital requires an extraordinary addition of paid labour, then wages rise, and, all other circumstances remaining equal, the unpaid labour diminishes in proportion. But as soon as this diminution touches the point at which the surplus labour that nourishes capital is no longer supplied in normal quantity, a reaction sets in: a smaller part of revenue is capitalized, accumulation lags, and the movement of rise in wages receives a check. The rise of wages therefore is confined within limits that not only leave intact the foundations of the capitalistic system, but also secure its reproduction on a progressive scale. The law of capitalistic accumulation, metamorphosed by economists into a pretended law of Nature, in reality merely states that the very nature of accumulation excludes every diminution in the degree of exploitation of labour, and every rise in the price of labour, which could seriously imperil the continual reproduction, on an ever-enlarging scale, of the capitalistic relation. It cannot be otherwise in a mode of production in which the labourer exists to satisfy the needs of self-expansion of existing values, instead of, on the contrary, material wealth existing to satisfy the needs of development on the part of the labourer. As in religion man is governed by the products of his own brain, so in capitalistic production, he is governed by the products of his own hand. . . .

But if a surplus labouring population is a necessary product of accumulation or of the development of wealth on a capitalist basis, this surplus population becomes, conversely, the lever of capitalistic accumulation, nay, a condition of existence of the capitalist mode of production. It forms a disposable industrial reserve army, that belongs to capital quite as absolutely as if the latter had bred it at its own cost. Independently of the limits of the actual increase of population, it creates, for the changing needs of the self-expansion of capital, a mass of human material always ready for exploitation. With accumulation, and the development of the productiveness of labour that accompanies it, the power of sudden expansion of capital grows also; it grows, not merely because the elasticity of the capital already functioning increases, not merely because the absolute wealth of society expands, of which capital only forms an elastic part, not merely because credit, under every special stimulus, at once places an unusual part of this wealth at the disposal of production in the form of additional capital; it grows, also, because the technical conditions of the process of production themselves—machinery, means of transport, etc.—now admit of the rapidest transformation of masses of surplus product into additional means of production. The mass of social wealth, overflowing with the advance of accumulation, and transformable into additional capital, thrusts itself frantically into old branches of production, whose market suddenly expands, or into newly formed branches, such as railways, etc., the need for which grows out of the development of the old ones. In all such cases, there must be the possibility of throwing great masses of men suddenly on the decisive points without injury to the scale of production in other spheres. Overpopulation supplies these masses. The course characteristic of modern industry, viz., a decennial cycle (interrupted by smaller oscillations) of periods of average activity, production at high pressure, crisis and stagnation, depends on the constant formation, the greater or less absorption, and the re-formation of the industrial reserve army or surplus population. In their turn, the varying phases of the industrial cycle recruit the surplus population, and become one of the most energetic agents of its reproduction. This peculiar course of modern industry, which occurs in no earlier period of human history, was also impossible in the childhood of capitalist production. The composition of capital changed but very slowly. With its accumulation, therefore, there kept pace, on the whole, a corresponding growth in the demand for labour. Slow as was the advance of accumulation compared with that of more modern times, it found a check in the natural limits of the exploitable labouring population, limits which could only be got rid of by forcible means to be mentioned later. The expansion by fits and starts of the scale of production is the preliminary to its equally sudden contraction; the latter again evokes the former, but the former is impossible without disposable human material, without an increase in the number of labourers independently of the absolute growth of the population. This increase is effected by the simple process that constantly 'sets free' a part of the labourers; by methods which lessen the number of labourers employed in proportion to the increased production. The whole form of the movement of modern industry depends, therefore, upon the constant transformation of a part of the labouring population into unemployed or half-employed hands. The superficiality of Political Economy shows itself in the fact that it looks upon the expansion and contraction of credit, which is a mere symptom of the periodic changes of the industrial cycle, as their cause. As the heavenly bodies, once thrown into a certain definite motion, always repeat this, so is it with social production as soon as it is once thrown into this movement of alternate expansion and contraction. Effects, in their turn, become causes, and the varying accidents of the whole process, which always reproduces its own conditions, take on the form of periodicity. When this periodicity is once consolidated, even Political Economy then sees that the production of a relative surplus population—i.e., surplus with regard to the average needs of the self-expansion of capital—is a necessary condition of modern industry. . . .

The industrial reserve army, during the periods of stagnation and average prosperity, weighs down the active labour-army; during the periods of over-production and paroxysm it holds its pretensions in check. Relative surplus population is therefore the pivot upon which the law of demand and supply of labour works. It confines the field of action of this law within the limits absolutely convenient to the activity of exploitation and to the domination of capital. . . .

The relative surplus population exists in every possible form. Every labourer belongs to it during the time when he is only partially employed or wholly unemployed. Not taking into account the great periodically recurring forms that the changing phases of the industrial cycle impress on it, now an acute form during the crisis, then again a chronic form during dull times—it has always three forms, the floating, the latent, the stagnant. . . .

The lowest sediment of the relative surplus population finally dwells in the sphere of pauperism. Exclusive of vagabonds, criminals, prostitutes, in a word, the 'dangerous' classes, this layer of society consists of three categories. First, those able to work. One need only glance superficially at the statistics of English pauperism to find that the quantity of paupers increases with every crisis, and diminishes with every revival of trade. Second, orphans and pauper children. These are candidates for the industrial reserve army, and are, in times of great prosperity, as 1860, e.g., speedily and in large numbers enrolled in the active army of labourers. Third, the demoralized and ragged, and those unable to work, chiefly people who succumb to their incapacity for adaptation, due to the division of labour; people who have passed the normal age of the labourer; the victims of industry, whose number increases with the increase of dangerous machinery, of mines, chemical works, etc., the mutilated, the sickly, the widows, etc. Pauperism is the hospital of the active labour-army and the dead weight of the industrial reserve army. Its production is included in that of the relative surplus population, its necessity in theirs; along with the surplus population, pauperism forms a condition of capitalist production, and of the capitalist development of wealth. It enters into the *faux frais* [unnecessary expenditure] of capitalist production; but capital knows how to throw these, for the most part, from its own shoulders on to those of the working-class and the lower middle class.

The greater the social wealth, the functioning capital, the extent and energy of its growth, and, therefore, also the absolute mass of the proletariat and the productiveness of its labour, the greater is the industrial reserve army. The same causes which develop the expansive power of capital develop also the labour power at its disposal. The relative mass of the industrial reserve army increases therefore with the potential energy of wealth. But the greater this reserve army in proportion to the active labour-army, the greater is the mass of a consolidated surplus population, whose misery is in inverse ratio to its torment of labour. The more extensive, finally, the lazurus-layers of the working-class, and the industrial reserve army, the greater is official pauperism. *This is the absolute general law of capitalist accumulation.* Like all other laws it is modified in its working by many circumstances, the analysis of which does not concern us here.

The folly of the economic wisdom that preaches to the labourers the accommodation of their number to the requirements of capital is now patent. The mechanism of capitalist production and accumulation constantly effects this adjustment. The first word of this adaptation is the creation of a relative surplus population, or industrial reserve army. Its last word is the misery of constantly extending strata of the active army of labour, and the dead weight of pauperism.

The law by which a constantly increasing quantity of means of produc-tion, thanks to the advance in the productiveness of social labour, may be set in movement by a progressively diminishing expenditure of human power, this law, in a capitalist society—where the labourer does not employ the means of production, but the means of production employ the labourer —undergoes a complete inversion and is expressed thus: the higher the productiveness of labour, the greater is the pressure of the labourers on the means of employment, the more precarious, therefore, becomes their condi-tion of existence, viz., the sale of their own labour power for the increasing of another's wealth, or for the self-expansion of capital. The fact that the means of production, and the productiveness of labour, increase more rapidly than the productive population, expresses itself, therefore, capitalis-tically in the inverse form that the labouring population always increases more rapidly than the conditions under which capital can employ this increase for its own self-expansion. . . .

Within the capitalist system all methods for raising the social produc-tiveness of labour are brought about at the cost of the individual labourer; all means for the development of production transform themselves into means of domination over, and exploitation of, the producers; they mutilate the labourer into a fragment of a man, degrade him to the level of an appendage of a machine, destroy every remnant of charm in his work and turn it into a hated toil; they estrange from him the intellectual potentialities of the labour process in the same proportion as science is incorporated in it as an indepen-dent power; they distort the conditions under which he works, subject him during the labour process to a despotism the more hateful for its meanness;

they transform his lifetime into working-time, and drag his wife and child beneath the wheels of the Juggernaut of capital. But all methods for the production of surplus value are at the same time methods of accumulation; and every extension of accumulation becomes again a means for the development of those methods. It follows therefore that in proportion as capital accumulates, the lot of the labourer, be his payment high or low, must grow worse. The law, finally, that always equilibrates the relative surplus population, or industrial reserve army, to the extent and energy of accumulation, this law rivets the labourer to capital more firmly than the wedges of Vulcan did Prometheus to the rock. It establishes an accumulation of misery, corresponding with accumulation of capital. Accumulation of wealth at one pole is, therefore, at the same time accumulation of misery, agony of toil, slavery, ignorance, brutality, mental degradation, at the opposite pole, i.e., on the side of the class that produces its own product in the form of capital. . . .

Primitive Accumulation

We have seen how money is changed into capital; how through capital surplus value is made, and from surplus value more capital. But the accumulation of capital presupposes surplus value; surplus value presupposes capitalistic production; capitalistic production presupposes the pre-existence of considerable masses of capital and of labour power in the hands of producers of commodities. The whole movement, therefore, seems to turn in a vicious circle, out of which we can only get by supposing a primitive accumulation (previous accumulation of Adam Smith) preceding capitalistic accumulation; an accumulation not the result of the capitalist mode of production, but its starting-point.

This primitive accumulation plays in Political Economy about the same part as original sin in theology. Adam bit the apple, and thereupon sin fell on the human race. Its origin is supposed to be explained when it is told as an anecdote of the past. In times long gone by there were two sorts of people; one, the diligent, intelligent, and, above all, frugal élite; the other, lazy rascals, spending their substance, and more, in riotous living. The legend of theological original sin tells us certainly how man came to be condemned to eat his bread in the sweat of his brow; but the history of economic original sin reveals to us that there are people to whom this is by no means essential. Never mind! Thus it came to pass that the former sort accumulated wealth, and the latter sort had at last nothing to sell except their own skins. And from this original sin dates the poverty of the great majority that, despite all its labour, has up to now nothing to sell but itself, and the wealth of the few that increases constantly although they have long ceased to work. Such insipid childishness is every day preached to us in the defence of property. M. Thiers, e.g., had the assurance to repeat it with all the solemnity of a statesman, to the French people, once so *spirituel*. But as soon as the question of property crops up, it becomes a sacred duty to proclaim the intellectual food of the infant as the one thing fit for all ages and for all stages of development. In actual history it is notorious that conquest, enslavement, robbery, murder, briefly force, play the great part. In the tender annals of Political Economy, the idyllic reigns from time immemorial. Right and 'labour' were from all time the sole means of enrichment, the present year of course always excepted. As a matter of fact, the methods of primitive accumulation are anything but idyllic.

In themselves money and commodities are no more capital than are the means of production and of subsistence. They want transforming into capital. But this transformation itself can only take place under certain circumstances that centre in this, viz., that two very different kinds of commodity-possessors must come face to face and into contact; on the one hand, the owners of money, means of production, means of subsistence, who are eager to increase the sum of values they possess, by buying other people's labour power; on the other hand, free labourers, the sellers of their own labour power, and therefore the sellers of labour. Free labourers, in the double sense that neither they themselves form part and parcel of the means of production, as in the case of slaves, bondsmen, etc., nor do the means of production belong to them, as in the case of peasant-proprietors; they are, therefore, free from, unencumbered by, any means of production of their own. With this polarization of the market for commodities, the fundamental conditions of capitalist production are given. The capitalist system presupposes the complete separation of the labourers from all property in the means by which they can realize their labour. As soon as capitalist production is once on its own legs, it not only maintains this separation, but reproduces it on a continually extending scale. The process, therefore, that clears the way for the capitalist system, can be none other than the process which takes away from the labourer the possession of his means of production; a process that transforms, on the one hand, the social means of subsistence and of production into capital, on the other, the immediate producers into wage-labourers. The so-called primitive accumulation, therefore, is nothing else than the historical process of divorcing the producer from the means of production. It appears as primitive, because it forms the pre-historic stage of capital and of the mode of production corresponding with it.

The economic structure of capitalistic society has grown out of the economic structure of feudal society. The dissolution of the latter set free the elements of the former.

The immediate producer, the labourer, could only dispose of his own person after he had ceased to be attached to the soil and ceased to be the slave, serf, or bondman of another. To become a free seller of labour power, who carries his commodity wherever he finds a market, he must further have escaped from the regime of the guilds, their rules for apprentices and journeymen, and the impediments of their labour regulations. Hence, the historical movement which changes the producers into wage-workers, appears, on the one hand, as their emancipation from serfdom and from the fetters of the guilds, and this side alone exists for our bourgeois historians. But, on the other hand, these new freedmen became sellers of themselves only after they had been robbed of all their own means of production, and of all the guarantees of existence afforded by the old feudal arrangements. And the history of this, their expropriation, is written in the annals of mankind in letters of blood and fire.

The industrial capitalists, these new potentates, had on their part not only to displace the guild masters of handicrafts, but also the feudal lords, the possessors of the sources of wealth. In this respect their conquest of social power appears as the fruit of a victorious struggle both against feudal lordship and its revolting prerogatives, and against the guilds and the fetters they laid on the free development of production and the free exploitation of man by man. The chevaliers d'industrie, however, only succeeded in supplanting the chevaliers of the sword by making use of events of which they themselves were wholly innocent. They have risen by means as vile as those by which the Roman freedman once on a time made himself the master of his *patronus*.

The starting-point of the development that gave rise to the wage-labourer as well as to the capitalist was the servitude of the labourer. The advance consisted in a change of form of this servitude, in the transformation of feudal exploitation into capitalist exploitation. To understand its march, we need not go back very far. Although we come across the first beginnings of capitalist production as early as the fourteenth or fifteenth century, sporadically, in certain towns of the Mediterranean, the capitalistic era dates from the sixteenth century. Wherever it appears, the abolition of serfdom has been long effected, and the highest development of the Middle Ages, the existence of sovereign towns, has been long on the wane.

In the history of primitive accumulation, all revolutions are epoch-making that act as levers for the capitalist class in course of formation; but, above all, those moments when great masses of men are suddenly and forcibly torn from their means of subsistence, and hurled as free and 'unattached' proletarians on the labour-market. The expropriation of the agricultural producer, of the peasant, from the soil, is the basis of the whole process. The history of this expropriation, in different countries, assumes different aspects, and runs through its various phases in different orders of succession, and at different periods. In England alone, which we take as our example, has it the classic form. . . .

The Historical Tendency of Capitalist Accumulation

What does the primitive accumulation of capital, i.e., its historical genesis, resolve itself into? In so far as it is not immediate transformation of slaves and serfs into wage-labourers, and therefore a mere change of form, it only means the expropriation of the immediate producers, i.e., the dissolution of private property based on the labour of its owner. Private property, as the antithesis to social, collective property, exists only where the means of labour and the external conditions of labour belong to private individuals. But according as these private individuals are labourers or not labourers, private property has a different character. The numberless shades, that it at first sight presents, correspond to the intermediate stages lying between these two extremes. The private property of the labourer in his means of production is the foundation of petty industry, whether agricultural, manufacturing, or both; petty industry, again, is an essential condition for the development of social production and of the free individuality of the labourer himself. Of course, this petty mode of production exists also under slavery, serfdom, and other states of dependence. But it flourishes, it lets loose its whole energy, it attains its adequate classical form, only where the labourer is the private owner of his own means of labour set in action by himself: the peasant of the land which he cultivates, the artisan of the tool which he handles as a virtuoso. This mode of production presupposes parcelling of the soil, and scattering of the other means of production. As it excludes the concentration of these means of production, so also it excludes co-operation, division of labour within each separate process of production, the control over, and the productive application of the forces of Nature by society, and the free development of the social productive powers. It is compatible only with a system of production, and a society, moving within narrow and more or less primitive bounds. To perpetuate it would be, as Pecqueur rightly says, 'to decree universal mediocrity'. At a certain stage of development it brings forth the material agencies for its own dissolution. From that moment new forces and new passions spring up in the bosom of society; but the old social organization fetters them and keeps them down. It must be annihilated; it is annihilated. Its annihilation, the transformation of the individualized and scattered means of production into socially concentrated ones, of the pigmy property of the many into the huge property of the few, the expropriation of the great mass of the people from the soil, from the means of subsistence, and from the means of labour, this fearful and painful expropriation of the mass of the people forms the prelude to the history of capital. It comprises a series of forcible methods, of which we have passed in review only those that have been epoch-making as methods of the primitive accumulation of capital. The expropriation of the immediate producers was accomplished with merciless Vandalism, and under the stimulus of passions the most infamous, the most sordid, the pettiest, the most meanly odious. Self-earned private property, that is based, so to say, on the fusing together of the isolated, independent labouring-individual with the conditions of his labour, is supplanted by capitalistic private property, which rests on exploitation of the nominally free labour of others, i.e., on wages-labour.

As soon as this process of transformation has sufficiently decomposed the old society from top to bottom, as soon as the labourers are turned into proletarians, their means of labour into capital, as soon as the capitalist mode of production stands on its own feet, then the further socialization of labour and further transformation of the land and other means of production into socially exploited and, therefore, common means of production, as well as the further expropriation of private proprietors, takes a new form. That which is now to be expropriated is no longer the labourer working for himself, but the capitalist exploiting many labourers. This expropriation is accomplished by the action of the immanent laws of capitalistic production itself, by the centralization of capital. One capitalist always kills many. Hand in hand with this centralization, or this expropriation of many capitalists by few, develop, on an ever-extending scale, the co-operative form of the labour process, the conscious technical application of science, the methodical cultivation of the soil, the transformation of the instruments of labour into instruments of labour only usable in common, the economizing of all means of production by their use as the means of production of combined, socialized labour, the entanglement of all peoples in the net of the world market, and with this, the international character of the capitalistic regime. Along with the constantly diminishing number of the magnates of capital, who usurp and monopolize all advantages of this process of transformation, grows the mass of misery, oppression, slavery, degradation, exploitation; but with this too grows the revolt of the working-class, a class always increasing in numbers, and disciplined, united, organized by the very mechanism of the process of capitalist production itself. The monopoly of capital becomes a fetter upon the mode of production, which has sprung up and flourished along with, and under it. Centralization of the means of production and socialization of labour at last reach a point where they become incompatible with their capitalist integument. This integument is burst asunder. The knell of capitalist private property sounds. The expropriators are expropriated.

The capitalist mode of appropriation, the result of the capitalist mode of production, produces capitalist private property. This is the first negation of individual private property, as founded on the labour of the proprietor. But capitalist production begets, with the inexorability of a law of Nature, its own negation. It is the negation of negation. This does not re-establish private property for the producer, but gives him individual property based on the acquisitions of the capitalist era: i.e., on co-operation and the possession in common of the land and of the means of production.

The transformation of scattered private property, arising from individual labour, into capitalist private property is, naturally, a process, incomparably more protracted, violent, and difficult, than the transformation of capitalistic private property, already practically resting on socialized production, into socialized property. In the former case, we had the expropriation of the mass of the people by a few usurpers; in the latter, we have the expropriation of a few usurpers by the mass of the people. . . .

From Volume Three

The Tendency of the Rate of Profit to Fall

. . . Since the development of the productivity of labour proceeds very disproportionately in the various lines of industry, and not only disproportionately in degree but frequently also in opposite directions, it follows that the mass of average profit (= surplus value) must be substantially below the level one would naturally expect after the development of the productiveness in the most advanced branches of industry. The fact that the development of the productivity in different lines of industry proceeds at substantially different rates and frequently even in opposite directions, is not due merely to the anarchy of competition and the peculiarity of the bourgeois mode of production. Productivity of labour is also bound up with natural conditions, which frequently become less productive as productivity grows—inasmuch as the latter depends on social conditions. Hence the opposite movements in these different spheres— progress here, and retrogression there. Consider the mere influence of the seasons, for instance, on which the bulk of raw materials depends for its mass, the exhaustion of forest lands, coal and iron mines, etc.

While the circulating part of constant capital, such as raw materials, etc., continually increases its mass in proportion to the productivity of labour, this is not the case with fixed capital, such as buildings, machinery, and lighting and heating facilities, etc. Although in absolute terms a machine becomes dearer with the growth of its bodily mass, it becomes relatively cheaper. If five labourers produce ten times as much of a commodity as before, this does not increase the outlay for fixed capital tenfold; although the value of this part of constant capital increases with the development of the productiveness, it does not by any means increase in the same proportion. We have frequently pointed out the difference in the ratio of constant to variable capital as expressed in the fall of the rate of profit, and the difference in the same ratio as expressed in relation to the individual commodity and its price with the development of the productivity of labour. . . .

Under competition, the increasing minimum of capital required with the increase in productivity for the successful operation of an independent industrial establishment, assumes the following aspect: as soon as the new, more expensive equipment has become universally established, smaller capitals are henceforth excluded from this industry. Smaller capitals can carry on independently in the various spheres of industry only in the infancy of mechanical inventions. Very large undertakings, such as railways, on the other hand, which have an unusually high proportion of constant capital, do not yield the average rate of profit, but only a portion of it, only an interest. Otherwise the general rate of profit would have fallen still lower. But this offers direct employment to large concentrations of capital in the form of stocks.

Growth of capital, hence accumulation of capital, does not imply a fall in the rate of profit, unless it is accompanied by the aforementioned changes in the proportion of the organic constituents of capital. Now it so happens that in spite of the constant daily revolutions in the mode of production, now this and now that larger or smaller portion of the total capital continues to accumulate for certain periods on the basis of a given average proportion of those constituents, so that there is no organic change with its growth, and consequently no cause for a fall in the rate of profit. This constant expansion of capital, hence also an expansion of production, on the basis of the old method of production which goes quietly on while new methods are already being introduced at its side, is another reason, why the rate of profit does not decline as much as the total capital of society grows.

The increase in the absolute number of labourers does not occur in all branches of production, and not uniformly in all, in spite of the relative decrease of variable capital laid out in wages. In agriculture, the decrease of the element of living labour may be absolute.

At any rate, it is but a requirement of the capitalist mode of production that the number of wage-workers should increase absolutely, in spite of its relative decrease. Labour power becomes redundant for it as soon as it is no longer necessary to employ it for twelve to fifteen hours daily. A development of productive forces which would diminish the absolute number of labourers, i.e. enable the entire nation to accomplish its total production in a shorter time span, would cause a revolution, because it would put the bulk of the population out of the running. This is another manifestation of the specific barrier of capitalist production, showing also that capitalist production is by no means an absolute form for the development of the productive forces and for the creation of wealth, but rather that at a certain point it comes into collision with this development. This collision appears partly in periodical crises, which arise from the circumstance that now this and now that portion of the labouring population becomes redundant under its old mode of employment. The limit of capitalist production is the excess time of the labourers. The absolute spare time gained by society does not concern it. The development of productivity concerns it only in so far as it increases the surplus labour time of the working-class, not because it decreases the labour time for material production in general. It moves thus in contradiction.

We have seen that the growing accumulation of capital implies its growing concentration. Thus grows the power of capital, the alienation of the conditions of social production personified in the capitalist from the real producers. Capital comes more and more to the fore as a social power, whose agent is the capitalist. This social power no longer stands in any possible relation to that which the labour of a single individual can create. It becomes an alienated, independent, social power, which stands opposed to society as an object, and as an object that is the capitalist's source of power. The contradiction between the general social power into which capital develops, on the one hand, and the private power of the individual capitalists over these social conditions of production, on the other, becomes ever more irreconcilable, and yet contains the solution of the problem, because it implies at the same time the transformation of the conditions of production into general, common, social, conditions. This transformation stems from the development of the productive forces under capitalist production, and from the ways and means by which this development takes place.

No capitalist ever voluntarily introduces a new method of production, no matter how much more productive it may be, and how much it may increase the rate of surplus value, so long as it reduces the rate of profit. Yet every such new method of production cheapens the commodities. Hence, the capitalist sells them originally above their prices of production, or, perhaps, above their value. He pockets the difference between their costs of production and the market prices of the same commodities produced at higher costs of production. He can do this, because the average labour time required socially for the production of these latter commodities is higher than the labour time required for the new methods of production. His method of production stands above the social average. But competition makes it general and subject to the general law. There follows a fall in the rate of profit—perhaps first in this sphere of production, and eventually it achieves a balance with the rest—which is, therefore, wholly independent of the will of the capitalist.

It is still to be added to this point, that this same law also governs those spheres of production, whose product passes neither directly nor indirectly into the consumption of the labourers, or into the conditions under which their necessities are produced; it applies, therefore, also to those spheres of production, in which there is no cheapening of commodities to increase the relative surplus value or cheapen labour power. (At any rate, a cheapening of constant capital in all these lines may increase the rate of profit, with the exploitation of labour remaining the same.) As soon as the new production method begins to spread, and thereby to furnish tangible proof that these commodities can actually be produced more cheaply, the capitalists working with the old methods of production must sell their product below its full price of production, because the value of this commodity has fallen, and because the labour time required by them to produce it is greater than the social average. In one word—and this appears as an effect of competition—these capitalists must also introduce the new method of production, in which the proportion of variable to constant capital has been reduced.

All the circumstances which lead to the use of machinery cheapening the price of a commodity produced by it come down in the last analysis to a reduction of the quantity of labour absorbed by a single piece of the commodity; and secondly, to a reduction in the wear-and-tear portion of the machinery, whose value goes into a single piece of the commodity. The less rapid the wear of machinery, the more the commodities over which it is distributed, and the more living labour it replaces before its term of reproduction arrives. In both cases quantity and value of the fixed constant capital increase in relation to the variable.

'All other things being equal, the power of a nation to save from its profits varies with the rate of profits: is great when they are high, less, when low; but as the rate of profits declines, all other things do not remain equal. . . . A low rate of profits is ordinarily accompanied by a rapid rate of accumulation, relatively to the numbers of the people, as in England . . . a high rate of profit by a slower rate of accumulation, relatively to the numbers of the people. Examples: Poland, Russia, India, etc.' (Richard Jones, *An Introductory Lecture on Political Economy*, London, 1833, pp. 50 ff.) Jones emphasizes correctly that in spite of the falling rate of profit the inducements and faculties to accumulate are augmented; first, on account of the growing relative overpopulation; second, because the growing productivity of labour is accompanied by an increase in the mass of use-values represented by the same exchange-value, hence in the material elements of capital; third, because the branches of production become more varied; fourth, due to the development of the credit system, the stock companies, etc., and the resultant case of converting money into capital without becoming an industrial capitalist; fifth, because the wants and the greed for wealth increase; and, sixth, because the mass of investments in fixed capital grows, etc.

Three cardinal facts of capitalist production:

(1) Concentration of means of production in few hands, whereby they cease to appear as the property of of the immediate labourers and turn into social production capacities. Even if initially they are the private property of capitalists. These are the trustees of bourgeois society, but they pocket all the proceeds of this trusteeship.

(2) Organization of labour itself into social labour: through co-operation, division of labour, and the uniting of labour with the natural sciences.

In these two senses, the capitalist mode of production abolishes private property and private labour, even though in contradictory forms.

(3) Creation of the world-market.

The stupendous productivity developing under the capitalist mode of production relative to population, and the increase, if not in the same proportion, of capital-values (not just of their material substance), which grow much more rapidly than the population, contradict the basis, which constantly narrows in relation to the expanding wealth, and for which all this immense productiveness works. They also contradict the conditions under which this swelling capital augments its value. Hence the crises. . . .

The Trinity Formula

I

Capital–profit (profit of enterprise plus interest), land–ground-rent, labour–wages, this is the trinity formula which comprises all the secrets of the social production process.

Furthermore, since as previously demonstrated interest appears as the specific characteristic product of capital and profit of enterprise on the contrary appears as wages independent of capital, the above trinity formula reduces itself more specifically to the following:

Capital–interest, land–ground-rent, labour–wages, where profit, the specific characteristic form of surplus value belonging to the capitalist mode of production, is fortunately eliminated.

On closer examination of this economic trinity, we find the following:

First, the alleged sources of the annually available wealth belong to widely dissimilar spheres and are not at all analogous with one another. They have about the same relation to each other as lawyer's fees, beetroot, and music.

Capital, land, labour! However, capital is not a thing, but rather a definite social production relation, belonging to a definite historical formation of society, which is manifested in a thing and lends this thing a specific social character. Capital is not the sum of the material and produced means of production. Capital is rather the means of production transformed into capital, which in themselves are no more capital than gold or silver in itself is money. It is the means of production monopolized by a certain section of society, confronting living labour power as products and working conditions rendered independent of this very labour power, which are personified through this antithesis in capital. It is not merely the products of labourers turned into independent powers, products as rulers and buyers of their producers, but rather also the social forces and the future . . . form of this labour, which confront the labourers as properties of their products. Here, then, we have a definite and, at first glance, very mystical, social form of one of the factors in a historically produced social production process.

And now alongside of this we have the land, inorganic nature as such, *rudis indigestaque moles* [crude and undigested mass], in all its primeval wildness. Value is labour. Therefore surplus value cannot be earth. Absolute fertility of the soil effects nothing more than the following: a certain quantity of labour produces a certain product—in accordance with the natural fertility of the soil. The difference in soil fertility causes the same quantities of labour and capital, hence the same value, to be manifested in different quantities of agricultural products; that is, causes these products to have different individual values. The equalization of these individual values into market-values is responsible for the fact that the 'advantages of fertile over inferior soil ... are transferred from the cultivator or consumer to the landlord'. (Ricardo, *Principles,* London, 1821, p. 62.)

And finally, as third party in this union, a mere ghost—'the' Labour, which is no more than an abstraction and taken by itself does not exist at all, or, if we take ... the productive activity of human beings in general, by which they promote the interchange with Nature, divested not only of every social form and well-defined character, but even in its bare natural existence, independent of society, removed from all societies, and as an expression and confirmation of life which the still non-social man in general has in common with the one who is in any way social.

II

Capital–interest; landed property, private ownership of the Earth, and, to be sure, modern and corresponding to the capitalist mode of production—rent; wage-labour–wages. The connection between the sources of revenue is supposed to be represented in this form. Wage-labour and landed property, like capital, are historically determined social forms; one of labour, the other of monopolized terrestrial globe, and indeed both forms corresponding to capital and belonging to the same economic formation of society.

The first striking thing about this formula is that side by side with capital, with this form of an element of production belonging to a definite mode of production, to a definite historical form of social process of production, side by side with an element of production amalgamated with and represented by a definite social form are indiscriminately placed: the land on the one hand and labour on the other, two elements of the real labour process, which in this material form are common to all modes of production, which are the material elements of every process of production and have nothing to do with its social form.

Secondly. In the formula: capital–interest, land–ground-rent, labour–wages, capital, land, and labour appear respectively as sources of interest (instead of profit), ground-rent and wages, as their products, or fruits; the former are the basis, the latter the consequence, the former are the cause, the latter the effect; and indeed, in such a manner that each individual source is related to its product as to that which is ejected and produced by it. All the proceeds, interest (instead of profit), rent, and wages, are three components of the value of the products, i.e., generally speaking, components of value or expressed in money, certain money components, price components. The formula: capital–interest is now indeed the most meaningless formula of capital, but still one of its formulas. But how should land create value, i.e., a socially defined quantity of labour, and moreover that particular portion of the value of its own products which forms the rent? Land, e.g., takes part as an agent of production in creating a use-value, a material product, wheat. But it has nothing to do with the production of the value of wheat. In so far as value is represented by wheat, the latter is merely considered as a definite quantity of materialized social labour, regardless of the particular substance in which this labour is manifested or of the particular use-value of this substance. This nowise contradicts that (1) other circumstances being equal, the cheapness or dearness of wheat depends upon the productivity of the soil. The productivity of agricultural labour is dependent on natural conditions, and the same quantity of labour is represented by more or fewer products, use-values, in accordance with such productivity. How large the quantity of labour represented in one bushel of wheat depends upon the number of bushels yielded by the same quantity of labour. It depends, in this case, upon the soil productivity in what quantities of product the value shall be manifested. But this value is given, independent of this distribution. Value is represented in use-value; and use-value is a prerequisite for the creation of value; but it is folly to create an antithesis by placing a use-value, like land, on one side and on the other side value, and a particular portion of value at that.

III

Vulgar economy actually does no more than interpret, systematize, and defend in doctrinaire fashion the conceptions of the agents of bourgeois production who are entrapped in bourgeois production relations. It should not astonish us, then, that vulgar economy feels particularly at home in the estranged outward appearances of economic relations in which these *prima facie* absurd and perfect contradictions appear and that these relations seem the more self-evident the more their internal relationships are concealed from it, although they are understandable to the popular mind. But all science would be superfluous if the outward appearance and the essence of things directly coincide. Thus, vulgar economy has not the slightest suspicion that the trinity which it takes as its point of departure, namely, land–rent, capital–interest, labour–wages or the price of labour, are *prima facie* three impossible combinations. First we have the use-value land, which has no value, and the exchange-value rent: so that a social relation conceived as a thing is made proportional to Nature, i.e., two incommensurable magnitudes are supposed to stand in a given ratio to one another. Then capital –interest. If capital is conceived as a certain sum of values represented independently by money, then it is *prima facie* nonsense to say that a certain value should be worth more than it is worth. It is precisely in the form: capital–interest that all intermediate links are eliminated, and capital is reduced to its most general formula, which therefore in itself is also inexplicable and absurd. The vulgar economist prefers the formula capital–interest with its occult quality of making a value unequal to itself, to the formula capital–profit, precisely for the reason that this already more nearly approaches actual capitalist relations. Then again, driven by the disturbing thought that 4 is not 5 and that 100 taler cannot possibly be 110 taler, he flees from capital as value to the material substance of capital; to its use-value as a condition of production of labour, to machinery, raw materials, etc. Thus, he is able once more to substitute in place of the first incomprehensible relation, whereby 4 = 5, a wholly incommensurable one between a use-value, a thing on one side, and a definite social production relation, surplus-value, on the other, as in the case of landed property. As soon as the vulgar economist arrives at this incommensurable relation, everything becomes clear to him, and he no longer feels the need for further thought. For he has arrived precisely at the 'rational' in bourgeois conception. Finally, labour–wages, or price of labour, is an expression, as shown in Book I, which *prima facie* contradicts the conception of value as well as of price—the latter generally being but a definite expression of value. And 'price of labour' is just as irrational as a yellow logarithm. But here the vulgar economist is all the more satisfied, because he has gained the profound insight of the bourgeois, namely, that he pays money for labour, and since precisely the contradiction between the formula and the conception of value relieves him from all obligation to understand the latter.

We have seen that the capitalist process of production is a historically determined form of the social process of production in general. The latter is as much a production process of material conditions of human life as a process taking place under specific historical and economic production relations, producing and reproducing these production relations themselves, and thereby also the bearers of this process, their material conditions of existence and their mutual relations, i.e., their particular socio-economic form. For the aggregate of these relations, in which the agents of this production stand with respect to Nature and to one another, and in which they produce, is precisely society, considered from the standpoint of its economic structure. Like all its predecessors, the capitalist process of production proceeds under definite material conditions, which are, however, simultaneously the bearers of definite social relations entered into by individuals in the process of reproducing their life. Those conditions, like these relations, are on the one hand prerequisites, on the other hand results and creations of the capitalist process of production; they are produced and reproduced by it. We saw also that capital—and the capitalist is merely capital personified and functions in the process of production solely as the agent of capital—in its corresponding social process of production, pumps a definite quantity of surplus labour out of the direct producers, or labourers; capital obtains this surplus labour without an equivalent, and in essence it always remains forced labour—no matter how much it may seem to result from free contractual agreement. This surplus labour appears as surplus value, and this surplus value exists as a surplus product. Surplus labour in general, as labour performed over and above the given requirements, must always remain. In the capitalist as well as in the slave system, etc., it merely assumes an antagonistic form and is supplemented by complete idleness of a stratum of society. A definite quantity of surplus labour is required as insurance against accidents, and by the necessary and progressive expansion of the process of reproduction in keeping with the development of the needs and the growth of population, which is called accumulation from the viewpoint of the capitalist. It is one of the civilizing aspects of capital that it enforces this surplus labour in a manner and under conditions which are more advantageous to the development of the productive forces, social relations, and the creation of the elements for a new and higher form than under the preceding forms of slavery, serfdom, etc. Thus it gives rise to a stage, on the one hand, in which coercion and monopolization of social development (including its material and intellectual advantages) by one portion of society at the expense of the other are eliminated; on the other hand, it creates the material means and embryonic conditions, making it possible in a higher form of society to combine this surplus labour with a greater reduction of time devoted to material labour in general. For, depending on the development of labour productivity, surplus labour may be large in a small total working-day, and relatively small in a large total working-day. If the necessary labour time $= 3$ and the surplus labour $= 3$, then the total working-day $= 6$ and the rate of surplus labour $= 100$ per cent. If the necessary labour $= 9$ and the surplus labour $= 3$, then the total working-day $= 12$ and the rate of surplus labour only $= 33\frac{1}{3}$ per cent. In that case, it depends upon the labour productivity how much use-value shall be produced in a definite time, hence also in a definite surplus labour time. The actual wealth of society, and the possibility of constantly expanding its reproduction process, therefore, do not depend upon the duration of surplus labour, but upon its productivity and the more or less copious conditions of production under which it is performed. In fact, the realm of freedom actually begins only where labour which is determined by necessity and mundane considerations ceases; thus in the very nature of things it lies beyond the sphere of actual material production. Just as the savage must wrestle with Nature to satisfy his wants, to maintain and reproduce life, so must civilized man, and he must do so in all social formations and under all possible modes of production. With his development this realm of physical necessity expands as a result of his wants; but, at the same time, the forces of production which satisfy these wants also increase. Freedom in this field can only consist in socialized man, the associated producers, rationally regulating their interchange with Nature, bringing it under their common control, instead of being ruled by it as by the blind forces of Nature; and achieving this with the least expenditure of energy and under conditions most favourable to, and worthy of, their human nature. But it none the less still remains a realm of necessity. Beyond it begins that development of human energy which is an end in itself, the true realm of freedom, which, however, can blossom forth only with this realm of necessity as its basis. The shortening of the working-day is its basic prerequisite.

In a capitalist society, this surplus value, or this surplus product (leaving aside chance fluctuations in its distribution and considering only its regulating law, its standardizing limits), is divided among capitalists as dividends proportionate to the share of the social capital each holds. In this form surplus value appears as average profit which falls to the share of capital, an average profit which in turn divides into profit of enterprise and interest, and which under these two categories may fall into the laps of different kinds of capitalists. This appropriation and distribution of surplus value, or surplus product, on the part of capital, however, has its barrier in landed property. Just as the operating capitalist pumps surplus labour, and thereby surplus value and surplus product in the form of profit, out of the labourer, so the landlord in turn pumps a portion of this surplus value, or surplus product, out of the capitalist in the form of rent in accordance with the laws already elaborated.

Hence, when speaking here of profit as that portion of surplus value falling to the share of capital, we mean average profit (equal to profit of enterprise plus interest) which is already limited by the deduction of rent from the aggregate profit (identical in mass with aggregate surplus value); the deduction of rent is assumed. Profit of capital (profit of enterprise plus interest) and ground-rent are thus no more than particular components of surplus value, categories by which surplus value is differentiated depending on whether it falls to the share of capital or landed property, headings which in no whit however alter its nature. Added together, these form the sum of social surplus value. Capital pumps the surplus labour, which is represented by surplus value and surplus product, directly out of the labourers. Thus, in this sense, it may be regarded as the producer of surplus value. Landed property has mothing to do with the actual process of production. Its role is confined to transferring a portion of the produced surplus value from the pockets of capital to its own. However, the landlord plays a role in the capitalist process of production not merely through the pressure he exerts upon capital, nor merely because large landed property is a prerequisite and condition of capitalist production since it is a prerequisite and condition of the expropriation of the labourer from the means of production, but particularly because he appears as the personification of one of the most essential conditions of production.

Finally, the labourer in the capacity of owner and seller of his individual labour power receives a portion of the product under the label of wages, in which that portion of his labour appears which we call necessary labour, i.e., that required for the maintenance and reproduction of this labour power, be the conditions of this maintenance and reproduction scanty or bountiful, favourable or unfavourable.

Whatever may be the disparity of these relations in other respects, they all have this in common: capital yields a profit year after year to the capitalist, land a ground-rent to the landlord, and labour power, under normal conditions and so long as it remains useful labour power, a wage to the labourer. These three portions of total value annually produced, and the corresponding portions of the annually created total product (leaving aside for the present any consideration of accumulation), may be annually consumed by their respective owners, without exhausting the source of their reproduction. They are like the annually consumable fruits of a perennial tree, or rather three trees; they form the annual incomes of three classes, capitalist, land-owner and labourer, revenues distributed by the functioning capitalist in his capacity as direct extorter of surplus labour and employer of labour in general. Thus, capital appears to the capitalist, land to the landlord, and labour power, or rather labour itself, to the labourer (since he actually sells labour power only as it is manifested, and since the price of labour power, as previously shown, inevitably appears as the price of labour under the capitalist mode of production), as three different sources of their specific revenues, namely, profit, ground-rent, and wages. They are really so in the sense that capital is a perennial pumping-machine of surplus labour for the capitalist, land a perennial magnet for the landlord, attracting a portion of the surplus value pumped out by capital, and finally, labour the constantly self-renewing condition and ever self-renewing means of acquiring under the title of wages a portion of the value created by the labourer and thus a part of the social product measured by this portion of value, i.e., the necessities of life. They are so, furthermore, in the sense that capital fixes a portion of the value and thereby of the product of the annual labour in the form of profit; landed property fixes another portion in the form of rent; and wage-labour fixes a third portion in the form of wages, and precisely by this transformation converts them into revenues of the capitalist, landowner, and labourer, without, however, creating the substance itself which is transformed into these various categories. The distribution rather presupposes the existence of this substance, namely, the total value of the annual product, which is nothing but materialized social labour. Nevertheless, it is not in this form that the matter appears to the agents of production, the bearers of the various functions in the production process, but rather in a distorted form. Why this takes place will be developed in the further course of our analysis. Capital, landed property, and labour appear to those agents of production as three different, independent sources, from which as such there arise three different components of the annually produced value—and thereby the product in which it exists; thus, from which there arise not merely the different forms of this value as revenues falling to the share of particular factors in the social process of production, but from which this value itself arises, and thereby the substance of these forms of revenue. . . .

Differential rent is bound up with the relative soil fertility, in other words, with properties arising from the soil as such. But, in the first place, in so far as it is based upon the different individual values of the products of different soil types, it is but the determination just mentioned; secondly, in so far as it is based upon the regulating general market-value, which differs from these individual values, it is a social law carried through by means of competition, which has to do neither with the soil nor the different degrees of its fertility.

It might seem as if a rational relation were expressed at least in 'labour–wages'. But this is no more the case than with 'land–ground-rent'. In so far as labour is value-creating, and is manifested in the value of commodities, it has nothing to do with the distribution of this value among various categories. In so far as it has the specifically social character of wage-labour, it is not value-creating. It has already been shown in general that wages of labour, or price of labour, is but an irrational expression for the value, or price of labour power; and the specific social conditions, under which this labour power is sold, have nothing to do with labour as a general agent in production. Labour is also materialized in that value component of a commodity which as wages forms the price of labour power; it creates this portion just as much as the other portions of the product; but it is materialized in this portion no more and no differently than in the portions forming rent or profit. And, in general, when we establish labour as value-creating, we do not consider it in its concrete form as a condition of production, but in its social delimitation which differs from that of wage-labour.

Even the expression 'capital–profit' is incorrect here. If capital is viewed in the only relation in which it produces surplus value, namely, its relation to the labourer whereby it extorts surplus labour by compulsion exerted upon labour power, i.e., the wage-labourer, then this surplus value comprises, outside of profit (profit of enterprise plus interest), also rent, in short, the entire undivided surplus value. Here, on the other hand, as a source of revenue, it is placed only in relation to that portion falling to the share of the capitalist. This is not the surplus value which it extracts generally but only that portion which it extracts for the capitalist. Still more does all connection vanish no sooner the formula is transformed into 'capital–interest'.

If we at first considered the disparity of the above three sources, we now note that their products, their offshoots, or revenues, on the other hand, all belong to the same sphere, that of value. However, this is compensated for (this relation not only between incommensurable magnitudes, but also between wholly unlike, mutually unrelated, and non-comparable things) in that capital, like land and labour, is simply considered as a material substance, that is, simply as a produced means of production, and thus is abstracted both as a relation to the labourer and as value.

Thirdly, if understood in this way, the formula, capital–interest (profit), land–rent, labour–wages, presents a uniform and symmetrical incongruity. In fact, since wage-labour does not appear as a socially determined form of labour, but rather all labour appears by its nature as wage-labour (thus appearing to those in the grip of capitalist production relations), the definite specific social forms assumed by the material conditions of labour—the produced means of production and the land—with respect to wage-labour (just as they, in turn, conversely presuppose wage-labour), directly coincide with the material existence of these conditions of labour or with the form possessed by them generally in the actual labour process, independent of its concrete historically determined social form, or indeed independent of any social form. The changed form of the conditions of labour i.e., alienated from labour and confronting it independently, whereby the produced means of production are thus transformed into capital, and the land into monopolized land, or landed property—this form belonging to a definite historical period thereby coincides with the existence and function of the produced means of production and of the land in the process of production in general. These means of production are in themselves capital by nature; capital is merely an 'economic appellation' for these means of production; and so, in itself land is by nature the earth monopolized by a certain number of landowners. Just as products confront the producer as an independent force in capital and capitalists—who actually are but the personification of capital—so land becomes personified in the landlord and likewise gets on its hind legs to demand, as an independent force, its share of the product created with its help. Thus, not the land receives its due portion of the product for the restoration and improvement of its productivity, but instead the landlord takes a share of this product to chaffer away or squander. It is clear that capital presupposes labour as wage-labour. But it is just as clear that if labour as wage-labour is taken as the point of departure, so that the identity of labour in general with wage-labour appears to be self-evident, then capital and monopolized land must also appear as the natural form of the conditions of labour in relation to labour in general. To be capital, then, appears as the natural form of the means of labour and thereby as the purely real character arising from their function in the labour process in general. Capital and produced means of production thus become identical terms. Similarly, land and land monopolized through private ownership become identical. The means of labour as such, which are by nature capital, thus become the source of profit, much as the land as such becomes the source of rent.

Labour as such, in its simple capacity as purposive productive activity, relates to the means of production, not in their social determinate form, but rather in their concrete substance, as material and means of labour; the latter likewise are distinguished from one another merely materially, as use-values, i.e., the land as unproduced, the others as produced, means of labour. If, then, labour coincides with wage-labour, so does the particular social form in which the conditions of labour confront labour coincide with their material existence. The means of labour as such are then capital, and the land as such is landed property. The formal independence of these conditions of labour in relation to labour, the unique form of this independence with respect to wage-labour, is then a property inseparable from them as things, as material conditions of production, an inherent, immanent, intrinsic character of them as elements of production. Their definite social character in the process of capitalist production bearing the stamp of a definite historical epoch is a natural and intrinsic substantive character belonging to them, as it were, from time immemorial, as elements of the production process. Therefore, the respective part played by the earth as the original field of activity of labour, as the realm of forces of Nature, as the pre-existing arsenal of all objects of labour, and the other respective part played by the produced means of production (instruments, raw materials, etc.) in the general process of production, must seem to be expressed in the respective shares claimed by them as capital and landed property, i.e., which fall to the share of their social representatives in the form of profit (interest) and rent, like to the labourer—the part his labour plays in the process of production is expressed in wages. Rent, profit, and wages thus seem to grow out of the role played by the land, produced means of production, and labour in the simple labour process, even when we consider this labour process as one carried on merely between man and Nature, leaving aside any historical determination. It is merely the same thing again, in another form, when it is argued: the product in which a wage-labourer's labour for himself is manifested, his proceeds or revenue, is simply wages, the portion of value (and thereby the social product measured by this value) which his wages represent. Thus, if wage-labour coincides with labour generally, then so do wages with the produce of labour, and the value portion representing wages with the value created by labour generally. But in this way the other portions of value, profit and rent also appear independent with respect to wages, and must arise from sources of their own, which are specifically different and independent of labour; they must arise from the participating elements of production, to the share of whose owners they fall; i.e., profit arises from the means of production, the material elements of capital, and rent arises from the land, or Nature, as represented by the landlord (Roscher) [contemporary German economist].

Landed property, capital, and wage-labour are thus transformed from sources of revenue—in the sense that capital attracts to the capitalist, in the form of profit, a portion of the surplus value extracted by him from labour, that monopoly in land attracts for the landlord another portion in the form of rent; and that labour grants the labourer the remaining portion of value in the form of wages—from sources by means of which one portion of value is transformed into the form of profit, another into the form of rent, and a third into the form of wages—into actual sources from which these value portions and respective portions of the product in which they exist, or for which they are exchangeable, arise themselves, and from which, therefore, in the final analysis, the value of the product itself arises.

In the case of the simplest categories of the capitalist mode of production, and even of commodity-production, in the case of commodities and money, we have already pointed out the mystifying character that transforms the social relations, for which the material elements of wealth serve as bearers in production, into properties of these things themselves (commodities) and still more pronouncedly transforms the production relation itself into a thing (money). All forms of society, in so far as they reach the stage of commodity-production and money circulation, take part in this perversion. But under the capitalist mode of production and in the case of capital, which forms its dominant category, its determining production relation, this enchanted and perverted world develops still more. If one considers capital, to begin with, in the actual process of production as a means of extracting surplus labour, then this relationship is still very simple, and the actual connection impresses itself upon the bearers of this process, the capitalists themselves, and remains in their consciousness. The violent struggle over the limits of the working-day demonstrates this strikingly. But even within this non-mediated sphere, the sphere of direct action between labour and capital, matters do not rest in this simplicity. With the development of relative surplus value in the actual specifically capitalist mode of production, whereby the productive powers of social labour are developed, these productive powers and the social interrelations of labour in the direct labour process seem transferred from labour to capital. Capital thus becomes a very mystic being since all of labour's social productive forces appear to be due to capital, rather than labour as such, and seem to issue from the womb of capital itself. Then the process of circulation intervenes, with its changes of substance and form, on which all parts of capital, even agricultural capital, devolve to the same degree that the specifically capitalist mode of production develops. This is a sphere where the relations under which value is originally produced are pushed completely into the background. In the direct process of production the capitalist already acts simultaneously as producer of commodities and manager of commodity-production. Hence this process of production appears to him by no means simply as a process of producing surplus value. But whatever may be the surplus value extorted by capital in the actual production process and appearing in commodities, the value and surplus value contained in the commodities must first be realized in the circulation process. And both the restitution of the values advanced in production and, particularly, the surplus-value contained in the commodities seem not merely to be realized in the circulation, but actually to arise from it; an appearance which is especially reinforced by two circumstances: first, the profit made in selling depends on cheating, deceit, inside knowledge, skill, and a thousand favourable market opportunities; and then by the circumstance that added here to labour time is a second determining element—time of circulation. This acts, in fact, only as a negative barrier against the formation of value and surplus value, but it has the appearance of being as definite a basis as labour itself and of introducing a determining element that is independent of labour and resulting from the nature of capital. In Book II we naturally had to present this sphere of circulation merely with reference to the form determinations which it created and to demonstrate the further development of the structure of capital taking place in this sphere. But in reality this sphere is the sphere of competition, which, considered in each individual case, is dominated by chance; where, thcn, the inner law, which prevails in these accidents and regulates them, is only visible when these accidents are grouped together in large numbers, where it remains, therefore, invisible and unintelligible to the individual agents in production. But furthermore: the actual process of production, as a unity of the direct production process and the circulation process, gives rise to new formations, in which the vein of internal connections is increasingly lost, the production relations are rendered independent of one another, and the component values become ossified into forms independent of one another.

The conversion of surplus value into profit, as we have seen, is determined as much by the process of circulation as by the process of production. Surplus value, in the form of profit, is no longer related back to that portion of capital invested in labour from which it arises, but to the total capital. The rate of profit is regulated by laws of its own, which permit, or even require, it to change while the rate of surplus value remains unaltered. All this obscures more and more the true nature of surplus value and thus the actual mechanism of capital. Still more is this achieved through the transformation of profit into average profit and of values into prices of production, into the regulating averages of market prices. A complicated social process intervenes here, the equalization process of capitals, which divorces the relative average prices of the commodities from their values, as well as the average profits in the various spheres of production (quite aside from the individual investments of capital in each particular sphere of production) from the actual exploitation of labour by the particular capitals. Not only does it appear so, but it is true in fact that the average price of commodities differs from their value, thus from the labour realized in them, and the average profit of a particular capital differs from the surplus value which this capital has extracted from the labourers employed by it. The value of commodities appears, directly, solely in the influence of fluctuating productivity of labour upon the rise and fall of the prices of production, upon their movement and not upon their ultimate limits. Profit seems to be determined only secondarily by direct exploitation of labour, in so far as the latter permits the capitalist to realize a profit deviating from the average profit at the regulating market prices, which apparently prevail independent of such exploitation. Normal average profits themselves seem immanent in capital and independent of exploitation; abnormal exploitation, or even average exploitation under favourable, exceptional conditions, seems to determine only the deviations from average profit, not this profit itself. The division of profit into profit of enterprise and interest (not to mention the intervention of commercial profit and profit from money-dealing, which are founded upon circulation and appear to arise completely from it, and not from the process of production itself) consummates the individualization of the form of surplus value, the ossification of its form as opposed to its substance, its essence. One portion of profit, as opposed to the other, separates itself entirely from the relationship of capital as such and appears as arising not out of the function of exploiting wage-labour, but out of the wage-labour of the capitalist himself. In contrast thereto, interest then seems to be independent both of the labourer's wage-labour and the capitalist's own labour, and to arise from capital as its own independent source. If capital originally appeared on the surface of circulation as a fetishism of capital, as a value-creating value, so it now appears again in the form of interest-bearing capital, as in its most estranged and characteristic form. Wherefore also the formula capital–interest, as the third to land–rent and labour–wages, is much more consistent than capital–profit, since in profit there still remains a recollection of its origin, which is not only extinguished in interest, but is also placed in a form thoroughly antithetical to this origin.

Finally, capital as an independent source of surplus value is joined by landed property, which acts as a barrier to average profit and transfers a portion of surplus value to a class that neither works itself, nor directly exploits labour, nor can find morally edifying rationalizations, as in the case of interest-bearing capital, e.g., risk and sacrifice of lending capital to others. Since here a part of the surplus value seems to be bound up directly with a natural element, the land, rather than with social relations, the form of mutual estrangement and ossification of the various parts of surplus value is completed, the inner connection completely disrupted, and its source entirely buried, precisely because the relations of production, which are bound to the various material elements of the production process, have been rendered mutually independent.

In capital–profit, or still better capital–interest, land–rent, labour–wages, in this economic trinity represented as the connection between the component parts of value and wealth in general and its sources, we have the complete mystification of the capitalist mode of production, the conversion of social relations into things, the direct coalescence of the material production relations with their historical and social determination. It is an enchanted, perverted, topsy-turvy world, in which Monsieur le Capital and Madame la Terre do their ghost-walking as social characters and at the same time directly as mere things. It is the great merit of classical economy to have destroyed this false appearance and illusion, this mutual independence and ossification of the various social elements of wealth, this personification of things and conversion of production relations into entities, this religion of everyday life. It did so by reducing interest to a portion of profit, and rent to the surplus above average profit, so that both of them converge in surplus value; and by representing the process of circulation as a mere metamorphosis of forms, and finally reducing value and surplus value of commodities to labour in the direct production process. Nevertheless, even the best spokesmen of classical economy remain more or less in the grip of the world of illusion which their criticism had dissolved, as cannot be otherwise from a bourgeois standpoint, and thus they all fall more or less into inconsistencies, half-truths, and unsolved contradictions. On the other hand, it is just as natural for the actual agents of production to feel completely at home in these estranged and irrational forms of capital–interest, land–rent, labour–wages, since these are precisely the forms of illusion in which they move about and find their daily occupation. It is therefore just as natural that vulgar economy, which is no more than a didactic, more or less dogmatic, translation of everyday conceptions of the actual agents of production, and which arranges them in a certain rational order, should see precisely in this trinity, which is devoid of all inner connection, the natural and indubitable lofty basis for its shallow pompousness. This formula simultaneously corresponds to the interests of the ruling classes by proclaiming the physical necessity and eternal justification of their sources of revenue and elevating them to a dogma.

In our description of how production relations are converted into entities and rendered independent in relation to the agents of production, we leave aside the manner in which the interrelations, due to the world-market, its conjunctures, movements of market prices, periods of credit, industrial and commercial cycles, alternations of prosperity and crisis, appear to them as overwhelming natural laws that irresistibly enforce their will over them, and confront them as blind necessity. We leave this aside because the actual movement of competition belongs beyond our scope, and we need present only the inner organization of the capitalist mode of production, in its ideal average, as it were.

In preceding forms of society this economic mystification arose principally with respect to money and interest-bearing capital. In the nature of things it is excluded, in the first place, where production for the use-value, for immediate personal requirements, predominates; and, secondly, where slavery or serfdom form the broad foundation of social production, as in antiquity and during the Middle Ages. Here, the domination of the producers by the conditions of production is concealed by the relations of dominion and servitude, which appear and are evident as the direct motive power of the process of production. In early communal societies in which primitive communism prevailed, and even in the ancient communal towns, it was this communal society itself with its conditions which appeared as the basis of production, and its reproduction appeared as its ultimate purpose. Even in the medieval guild system neither capital nor labour appear untrammelled, but their relations are rather defined by the corporate rules, and by the same associated relations, and corresponding conceptions of professional duty, craftsmanship, etc. Only when the capitalist mode of production . . . [The manuscript breaks off here.—*Ed*.]

Classes

. . . The owners merely of labour power, owners of capital, and landowners, whose respective sources of income are wages, profit, and ground-rent, in other words, wage-labourers, capitalists, and landowners, constitute then three big classes of modern society based upon the capitalist mode of production.

In England, modern society is indisputably most highly and classically developed in economic structure. Nevertheless, even here the stratification of classes does not appear in its pure form. Middle and intermediate strata even here obliterate lines of demarcation everywhere (although incomparably less in rural districts than in the cities). However, this is immaterial for our analysis. We have seen that the continual tendency and law of development of the capitalist mode of production is more and more to divorce the means of production from labour, and more and more to concentrate the scattered means of production into large groups, thereby transforming labour into wage-labour and the means of production into capital. And to this tendency, on the other hand, corresponds the independent separation of landed property from capital and labour, or the transformation of all landed property into the form of landed property corresponding to the capitalist mode of production.

The first question to be answered is this: What constitutes a class?—and the reply to this follows naturally from the reply to another question, namely: What makes wage-labourers, capitalists, and landlords constitute the three great social classes?

At first glance—the identity of revenues and sources of revenue. There are three great social groups whose members, the individuals forming them, live on wages, profit, and ground-rent respectively, on the realization of their labour power, their capital, and their landed property.

However, from this standpoint, physicians and officials, e.g., would also constitute two classes, for they belong to two distinct social groups, the members of each of these groups receiving their revenue from one and the same source. The same would also be true of the infinite fragmentation of interest and rank into which the division of social labour splits labourers as well as capitalists and landlords—the latter, e.g., into owners of vineyards, farm owners, owners of forests, mine owners, and owners of fisheries. . . .

BIBLIOGRAPHY

Original: MEW, Vols. 23–5.

Present translation: K. Marx, *Capital*, Vol. 1, trans. S. Moore and E. Aveling, Moscow, 1954, pp. 7 ff., 35 ff., 71 ff., 85 f., 109 ff., 146 ff., 167 ff., 177 ff., 199 ff., 215 ff., 269 f., 351 f., 360 f., 612 ff., 713 ff., 761 ff.; Vol. 3, trans. E. Untermann, Moscow, 1971, pp. 814 ff., pp. 885 ff. Reproduced by kind permission of Lawrence and Wishart Ltd.

Other translations: Vol. 1, trans. E. Untermann, Chicago, 1906, Vol. 1, trans. E. and C. Paul, London, 1926. Vol. 1, trans. B. Fowkes, London and New York, 1976.

Commentaries: A. Bose, *Marxian and Post-Marxian Political Economy*, London, 1975, Part 1.
G. D. H. Cole, *A History of Socialist Thought*, London, 1953, Vol. 2, Ch. 11.
M. Dobb, *Marx as an Economist*, London, 1943.
B. Fine, *Marx's 'Capital'*, London, 1975.
R. Garaudy, *Karl Marx: The Evolution of his Thought*, London, 1967, Part 3.
N. Geras, 'Marx and the Critique of Political Economy', *Ideology in Social Science*, ed. R. Blackburn, London, 1972.
M. Godelier, 'Structure and Contradiction in *Capital*', *Ideology in Social Science*, ed. R. Blackburn, London, 1972.
D. Horowitz, ed., *Marx and Modern Economics*, London, 1968.
M. Howard and J. King, *The Political Economy of Marx*, London, 1975, particularly sections 5 and 6.
E. Mandel, *An Introduction to Marxist Economic Theory*, New York, 1969.
I. Rubin, *Essays on Marx's Theory of Value*, Detroit, 1972.
J. Schumpeter, *Capitalism, Socialism and Democracy*, London, 1943, Ch. 3.
I. Zeitlin, *Marxism: A Re-examination*, New York, 1967.

33. Results of the Immediate Process of Production

The following selections come from the 'sixth chapter' of *Capital* (published for the first time in 1933), which was to have followed the section on capitalist accumulation which forms the final section of *Capital* Volume One in its present form. It is not known why Marx did not include this chapter in the published version. It contains interesting sections on alienation and the growth of service industries in the context of a general discussion of the transformation undergone by the simple commodity in a developed capitalist system.

Alienation in the Productive Process

... But everything is changed when we examine the process of valorization. Here it is not the worker who uses the means of production, but the means of production which use the worker. It is not living labour which realizes itself in material labour as its objective organ, but it is material labour which conserves itself and grows by absorbing living labour to such an extent that it becomes value creating value, capital in movement. ...

In circulation, the capitalist and the worker only face each other as sellers of commodities. But, owing to the specifically bipolar nature of the commodities that they sell each other, the worker necessarily enters into the productive process as an integral part of the use-value or real mode of existence or value-existence of capital: thus their relationship is only realized *inside* the productive process and the capitalist potential (which has bought labour) only really becomes capitalist when the labourer (potential wage-earner through the sale of his labour-power) really passes under the direction of capital in the productive process.

The functions exercised by the capitalist are no more than the consciously and wilfully executed functions of capital-value which valorizes itself by absorbing living labour. The capitalist only functions as the personification of capital, capital in person, just as the worker is only labour personified, labour which belongs to the worker as far as its hardship and effort goes and to the capitalist as far as it is a substance creating ever greater riches; in short, the worker appears as an element incorporated into capital within the productive process, as its living and variable factor.

The domination of the capitalist over the worker is thus the domination of the thing over man, of dead labour over living labour, of the product over the producer; for the commodities which become the means of domination (in fact only over the worker) are themselves merely the results of the productive process, *its* products.

At the level of material production, the true process of social life—which is nothing but the productive process—we find the same relationship as obtains at the level of ideology, in religion: the subject is transformed into object and vice versa.

From the historical point of view, this inversion represents a transitional phase which is necessary in order to force the majority of humanity to produce wealth for itself by inexorably developing the productive forces of social labour which alone can constitute the material basis for a free human society. It is necessary to go through this antagonistic form, just as it is necessary at first to give man's spiritual forces a religious form by erecting them into autonomous power over against him.

This is the 'process of alienation' of man's own labour. From the start, the worker is superior to the capitalist in that the capitalist is rooted in his 'process of alienation' and is completely content therein, whereas the worker who is its victim finds himself from the beginning in a state of rebellion against it and experiences the process as one of enslavement. . . . The self-valorization of capital—the creation of surplus value—is the determining, supreme, and dominant aim of the capitalist, the complete motive and content of his actions, the rationalized instinct and aim of the miser—a poor content which demonstrates that the capitalist is in the same slavish relation to capital as the worker, although at the opposite pole.

Man can only live if he produces means of subsistence, but he can only produce these means if he holds the means of production, the material conditions for labour. It is easy to understand that, if the worker is deprived of the means of production, he is also deprived of the means of subsistence, just as, inversely, if he is deprived of the means of subsistence, he cannot create his means of production.

What first of all—even before the real transformation of money or the commodity into capital—gives the character of *capital* to the conditions of labour is not the nature of money, commodities, or material use-values as means of subsistence and means of production; it is the fact that this money and these commodities, these means of subsistence, arise as autonomous powers, personified by their owners, over against labour power which is deprived of all material wealth; it is the fact that the material conditions, indispensable to the realization of labour, are alienated from the worker and, what is more, appear as fetishes endowed with their own will and soul; it is finally the fact that commodities figure as buyers of people. . . .

Money cannot become capital without first of all being exchanged for the labour power that the worker sells as a commodity; on the other hand, labour can only earn wages from the moment when the worker's own objective conditions arise over against him as autonomous forces, the property of someone else, value existing for itself and bringing everything back to itself

—in short, capital. Therefore, if from the point of view of its matter—that is, its use-value—capital is reduced to the objective conditions of labour, from the formal point of view these conditions must oppose labour as autonomous and alien powers, as value (objectified labour) which treats living labour simply as a means of conserving and increasing itself. Wage-labour is thus a form of labour that is socially necessary for capitalist production, just as capital—value concentrated into a power—in the socially necessary form that the objective conditions of labour must assume in order for it to be wage-labour.

It follows that wage-labour is the necessary condition for the formation of capital and always remains the necessary premiss of capitalist production. This is why, even if the first process—that of the exchange of money for labour power or purchase of labour power—does not as such form part of the immediate process of production, it does on the other hand form part of the general production of the relationship.

In the eyes of the capitalist, the worker, and the economist (who cannot conceive of the work process outside the process of capitalist appropriation) the material elements of the labour process appear, in virtue of their material properties, as capital. This is why the economist is incapable of distinguishing between these simple factors in the labour process and the social property that is amalgamated with them—that of capital. He is incapable because, in reality, it is one and the same labour process (in which the means of production, by their material properties, simply serve as aids to labour) which transforms these means of production into simple means for the absorption of labour.

In the labour process considered in itself, the worker was the means of production; in the labour process which is at the same time the capitalist process of production, the means of production employ the worker in such a way that the labour is no more than a means by which a given sum of values, or a determinate mass of objectified labour, absorbs living labour in order to conserve and increase itself. The labour process is thus the process of the self-valorization of objectified labour thanks to living labour.

Thus capital uses the worker, the worker does not use capital: and capital is only composed of the objects which employ the worker and thus have an existence, a will, and a consciousness personified in the capitalist. . . .

Classical political economy had the great merit of conceiving the whole process of production as taking place between objectified labour and living labour, living labour being opposed to capital, simple objectified labour, that is, value which valorizes itself thanks to living labour.

Its only faults are: (1) It has not been able to show how this exchange of a greater quantity of living labour against a lesser quantity of objectified labour does not contradict the law of the exchange of commodities, in other words, the determination of the value of commodities by labour time. (2) As a result, it identifies the exchange of a determinate quantity of objectified labour for labour power in the circulation process purely and simply with the assumption of living labour in the production process by objectified labour under the form of the means of production; whence the confusion between the exchange process of variable capital for labour power and the assumption process of living labour by constant capital in the production process.

These errors are explicable in terms of the hold that capital exercises on the economists. In fact, the exchange of a lesser quantity of objectified labour against a larger quantity of living labour appears as a single and unique process without, in the capitalist's eyes, any intermediary: does he not pay for the labour only after it has been valorized? . . .

The axiom of classical economics is the mobility of the labour force and the fluidity of capital. This is accurate in as much as the capitalist mode of production tends in that direction remorselessly, despite all the obstacles that it creates—for the most part of its own accord. In any case, in order to lay bare the laws of economics in their purity, we must make an abstraction of these obstacles, just as in pure mechanics we ignore secondary frictions which, in each particular case, must be removed so that the law can be applied. . . .

Capitalism as a Stage towards Socialism

In the course of their development, society's forces of production or productive forces of labour are socialized, and become directly social (collective) thanks to co-operation, the division of labour on the shopfloor, the use of mechanization and, in general, the changes which the process of production undergoes thanks to the conscious use of the natural sciences, mechanics, chemistry, etc. applied to definite technological ends, and thanks to all that is related to labour performed on a large scale, etc. It is only this socialized labour that is able to apply the general products of human development—for example mathematics—to the process of immediate production, as the development of these sciences is in its turn determined by the level reached by the material process of production.

All this development of the productive force of socialized labour—just like the application of science, that general product of social development—is opposed to the more or less remote and scattered labour of the particular individual. This is all the more true inasmuch as everything is presented directly a productive force of capital, and not as a productive force of labour, whether it is the force of the isolated worker, or of workers associated in the process of production, or even of a productive force of labour which could be identified with capital.

This mystification, proper to the capitalist relationship in general, will develop from now on much more than could be the case in the simple formal submission of labour to capital. Furthermore, it is only at this level that the historical significance of capitalist production appears in a striking (specific) way, precisely through the changes undergone by the immediate process of production and through the development of the productive social forces of labour.

Factory Worker and Artisan

It is evident that the worker works with more continuity for the capitalist than does the artisan for his chance customers; his labour is not limited by the fortuitous needs of particular buyers but only by the exploitation needs of the capital which employs him. Compared to the labour of the slave, that of the free worker is more productive, because it is more intense. The slave only works swayed by fear, and it is not his existence itself which is at stake, since it is guaranteed to him even if it does not belong to him. The free worker, on the other hand, is impelled by his needs. The consciousness (or rather the idea) of being solely determined by himself, of being free, as well as the feeling (sense) of the responsibility which is connected with this, make him a much better worker, because like every seller of goods, he is responsible for the goods which he supplies and obliged to supply them at a certain quality, at the risk of being ousted by the other sellers of the same goods.

The continuity of the relationship between the slave and the supporter of slavery was assured by the constraint experienced directly by the slave. On the other hand, the free worker himself is obliged to assure the continuity of his relationship, for his existence and that of his family depend on the continued renewal of the sale of his labour force to the capitalist.

Productive and Unproductive Labour

In capitalist production, the absolute rule becomes, on the one hand, the production of articles in the form of commodities and, on the other hand, wage-earning labour. A great number of functions and activities which, now adorned with a halo and considered as ends in themselves, were formerly carried on gratuitously or remunerated in an indirect way (in Britain, for example, the learned professions, doctors, barristers, etc., could not and still cannot bring an action in law to be paid), change directly into wage-earning labour (however diverse their content) or fall within the scope of laws regulating wage value, as far as concerns the estimation of their value and the cost of the different services, from that of the whore to that of the king. . . .

With the development of capitalist production, all services change into wage-earning labour and all those who carry them on into wage-earning workers, so that they acquire this character in common with the productive workers. This is what leads some people to confuse these two categories, inasmuch as wages are a phenomenon and a creation characterizing capitalist production. Moreover, this gives capital's apologists the opportunity to change the productive worker, under the pretence that he is a wage-earner, with a worker who simply exchanges his services (that is his labour as use-value) for money. This is to overlook a little conveniently what characterize in a fundamental way the productive worker and capitalist production: the production of surplus value and the self-valorization process of capital which incorporates living labour as simple money. The soldier is a wage-earner, if he is a mercenary, but he is not for all that a productive worker.

The process of capitalist production does not simply create commodities, it absorbs unpaid labour and changes the means of production into means of absorbing unpaid labour.

From the above it follows that productive labour in no way implies that it has an exact content, a particular usefulness, a definite use-value in which it is materialized. This is what explains why labour of the same content can be either productive or unproductive.

For example, Milton, the author of *Paradise Lost*, is an unproductive worker, whereas a writer who supplies his publisher with a manufactured work is a productive worker. Milton has produced his poem as a silkworm produces silk, by expressing his nature through his activity. By later selling his product for the sum of £5, he was, to this extent, a merchant. On the other hand, the proletarian man of letters in Leipzig who, at his publisher's command, produces books—for example, economics texts—is similar to the productive worker to the extent that his production is subject to capital and only exists with a view to its valorization.

A singer who sings like a bird is an unproductive worker, to the extent that she sells her song for money she is a wage-earner and a merchant. But this very singer becomes a productive worker, when she is engaged by a contractor to sing and make money, since she directly produces capital. The teacher in the class-room is not a productive worker; but he becomes productive if he is engaged with others as a wage-earner in order to valorize, with his labour, the money of the contractor of an establishment which exploits learning. Indeed, most of these labours are scarcely subject formally to capital: they are transitory forms.

On the whole, labour which can only be utilized as service, because its products are inseparable from their author so that they cannot become autonomous commodities (which does not prevent them, nevertheless, from being exploited in a directly capitalist way), represents a derisory amount in relation to that of capitalist production.

The same labour (for example, that of a gardener, a tailor) can be performed by the same worker on behalf of a capitalist or on immediate uses. In the two cases, he is wage-earning or hired by the day, but, if he works for the capitalist, he is a productive worker, since he produces capital, whereas if he works for a direct user he is unproductive. Indeed, in the first case, his labour represents an element of the process of self-valorization of capital, but not in the second.

A great part of the annual product which is consumed as income and no longer returns to production as means of production consists of the most nefarious products (use-values), satisfying the most unhealthy envies and caprices. However that may be, their content is completely indifferent as regards determining productive labour. (It is obvious, however, that if a disproportionate part was consumed in this way, at the expense of the means of production and subsistence which enter into the reproduction whether of goods or of the labour force, the development of wealth would suffer a stoppage.) This sort of productive labour creates use-values, and is crystallized in products destined solely for unproductive consumption and themselves deprived of any use-value for the process of reproduction.

They would only be able to acquire usefulness by exchanging themselves for use-values destined for reproduction. However, that would be a simple displacement, as they must necessarily be consumed somewhere in a non-reproductive way.

Here are some comments in advance about this subject: current economics is incapable of saying anything sensible whatever—even from the capitalist point of view—on the limits of the production of luxury products. None the less, the question becomes very simple, if we analyse the elements of the process of reproduction correctly. From the capitalist point of view, luxury becomes condemnable from the time when the process of reproduction—or its progress accomplished by the simple natural progression of population— meets a check in the disproportionate application of productive labour to the creation of articles which are not used for reproduction, so that there is insufficient reproduction of the necessary means of subsistence and means of production. Furthermore, luxury is an absolute necessity for a mode of production which, creating wealth for non-producers, must give it forms which only permit its appropriation by those who are sybarites.

For the worker, this productive labour is, like any other, the sole means of which he disposes to reproduce his necessary means of subsistence. For his capitalist, who is indifferent to the nature of use-value and to the character of utilized concrete labour, it is only a means of coining money and of producing surplus value.

The supreme ideal of capitalist production is—at the same time as it increases net produce in a relative way—to reduce as much as possible the number of those who live on wages, and to increase as much as possible the number of those who live on net produce.

Alienated Labour

Even in the purely formal relationship—valid in general for any capitalist production, since the latter conserves, even in its full development, the characteristics of its underdeveloped mode—the means of production, the material conditions of labour, are not subject to the worker, but it is he who is subject to them: it is capital which employs labour. In this simple manner, this relationship enhances the personification of objects and the reification of people.

But the relationship becomes more complex and apparently more mysterious when, with the development of the specifically capitalist mode of production, it is no longer only the objects—those products of labour, use-values and exchange-values—which rear themselves as capital over against the worker, but also the social forms of labour which present themselves as forms of development of capital, so that the productive forces, thus developed, of social labour appear as productive forces of capital: as such, they are 'capitalized', over against labour. Indeed, collective unity is found in co-operation, the association and division of labour, the utilization of natural forces, sciences and products of labour in the form of machines. All that is opposed to the individual worker as something which is alien to him and exists at the outset in a material form; what is more, it seems to him that he has not contributed anything, or even that all this exists despite what he does.

In short, all the things become independent of him, simple modes of existence of the means of labour which dominate him as objects. The intelligence and will of the collective workshop seem embodied in its representatives—the capitalist or his lieutenants—to the extent that the workers are associated in the workshop and the functions of capital embodied in the capitalist are opposed to them.

The social forms of labour of the individual workers—subjectively as well as objectively—or, in other words, the form of their own social labour, are relationships established according to a mode quite independent of them: by being subject to capital, the workers become elements of these social formations, which rear up over against them like forms of capital itself, as if they belonged to it—unlike the labour power of the workers—and as if they were derived from capital and were immediately incorporated in it.

All this adopts forms that are all the more real in that, on the one hand, the labour power itself is modified by these forms to the point where it becomes powerless when it is separated from them. In other words, its autonomous productive force is shattered when it is no longer in the capitalist relationship. On the other hand, machinery develops so that the conditions of labour manage, even from the technological point of view, to dominate labour at the same time as they replace it, suppress it, and make it superfluous in the forms where it is autonomous.

In this process, the social characteristics of labour appear to the workers as if they were capitalized over against them: in machinery, for example, the visible products of labour appear to dominate labour. The same of course goes for the forces of nature and science (that product of general historical development in its abstract quintessence), which oppose the worker as powers of capital, by actually being separate from the art and knowledge of the individual worker. Although they are, in their origin, the product of labour, they appear as if they are incorporated in capital, and the worker scarcely enters into the process of labour. The capitalist who uses a machine does not need to understand it (cf. Ure); yet science realized in the machine appears as capital over against the workers. In fact, all these applications—based on associated labour—of science, of the forces of nature, and of the products of labour in bulk appear solely as means of exploiting labour and appropriating surplus labour, and therefore as forces themselves belonging to capital. Of course, capital utilizes all these means with the sole aim of exploiting labour, but, to do this, it has to apply them to production. It is in this way that the development of productive social forces of labour and the conditions of this development appear as the work of capital, and the worker finds himself, confronted with all this, in a relationship which is not only passive but also antagonistic.

Capital thus becomes a quite mysterious being.

The conditions of labour are gathered as social forces over against the worker, and it is in this form that they are capitalized.

Capital appears therefore as productive:

(1) Because it compels the worker to perform surplus labour. Now if labour is productive, it is precisely due to the fact that it performs surplus labour and that a difference is effected between the value of labour power and that of its valorization;

(2) Because it personifies and represents, in an objectified form, the 'forces of the social production of labour' or productive forces of social labour.

We have already seen that the law of capitalist production—the creation of surplus value, etc.—imposes itself as a compulsion which the capitalists exert on one another as well as on the workers; in short it is a law of capital which works against both.

The force, of a social nature, of labour does not develop in the process of valorization as such, but in the process of real labour. This is why it appears as a characteristic quality inherent in capital as a thing, like its use-value. Productive (of value) labour continues to oppose capital as the labour of individual workers, whatever the social combination in which these workers enter the process of production. While capital is opposed to workers as a social force of labour, productive labour itself always appears over against capital as the labour of individual workers.

By analysing the process of accumulation, we have seen that it is as the immanent force of capital that there appears the element thanks to which past labour—in the form of productive forces and of conditions of production already produced—increases reproduction. This happens from the point of view of use-value as well as of exchange-value, whose value aggregate is conserved by a definite quantity of living labour, just as the aggregate of use-values is produced afresh. In fact, objectified labour always operates by capitalizing itself over against the worker.

The Reproduction of the Capitalist Relationship

The product of capitalist production is not only surplus value—it is capital.

Even after its change into factors of the labour process—into means of production, constant capital, and into the labour power into which variable capital changes—the sum of money or of value advanced is only capital in itself and potentially, but it was this still more before this transformation. It is only in the heat of the production process, when in reality living labour is incorporated into the material elements of capital and when the additional labour is in fact absorbed, that not only this labour, but also the sum of value advanced, becomes real and active capital. Instead of just possible capital, capital by destination.

What is happening during the total process? The worker sells the disposal of his capacity to labour in exchange for a certain value, determined by the value of his labour power, so as to make sure of the necessary means of subsistence. What is the result of it for him? Simply and purely, the reproduction of his labour power. What does he give in exchange? The activity which conserves, creates, and increases value: his labour. Setting aside the wastage of his labour power, it comes out of the process as it went in, a simple subjective labour power which has to run through the same cycle to conserve itself.

Capital, on the other hand, does not come out of the process as it went into it: it is only in this process that it changes into true capital, into value which valorizes itself. The global product which is born of it is now the form of existence of realized capital: as such, it again meets the worker as the property of the capitalist, as an autonomous power, although it has been created by labour.

Thus this process not only reproduces capital but also produces it. In the beginning, the production conditions oppose the worker as capital to the extent that he finds them objectified beforehand over against him; now it is the very product of his labour that he finds before him, as production conditions changed into capital. What was a premiss has now become the result of the production process.

To say that the production process creates capital is only another way of saying that it creates surplus value. But that is not all.

Surplus value is reconverted into additional capital and therefore appears as the creation of new capital or of enlarged capital. Capital not only realizes itself but also creates more capital. Thus the accumulation process is immanent to the capitalist production process: it implies the creation of new wage-earning workers and new means of realizing and increasing the existing capital, whether capital subjects to itself strata of the population which up till then still escaped it, for example women and children, or whether it acquires an increased number of workers by the natural rise in population.

On closer inspection we notice that capital regulates, according to the demands of its exploitation, the production of the labour forces and of the exploited human masses. Thus capital produces not only capital but also a growing mass of workers, thanks to which alone it can operate as additional capital. Labour therefore does not only produce—in opposition to itself and on a continually increased scale—labour conditions in the form of capital; capital produces, on an always enlarged scale, the productive wage-earning workers whom it needs. Labour produces its own production conditions like capital, and capital produces labour in a wage-earning form as a means of realizing it as capital.

Capitalist production is not only reproduction of the relationship, it is its reproduction on an ever wider scale. To the very extent that the force of the social production of labour develops, with the capitalist mode of production, the wealth accumulated over against the worker increases and dominates him as capital: the world of wealth swells before the worker like a world which is alien to him and which dominates him, in proportion as poverty, want, and dependence increase for him. His destitution accompanies this plethora, while there increases yet further the mass of this living means of production of capital which is the working proletariat.

The growth of capital there goes hand in hand with the increase in the proletariat: they are two products burgeoning at opposite poles of one and the same process.

The relationship is not only reproduced, but continues to produce on a still more immense scale, by always creating more workers, by continually subjecting to itself new branches of production. While it is reproduced—as we have emphasized, when describing the specifically capitalist mode of production—in conditions ever more favourable to one of these poles, that of the capitalists, it is reproduced in conditions ever more unfavourable to the other, that of the wage-earning workers.

From the point of view of the continuity of the production process, wages are only the part of the product which, after being created by the worker, changes into means of subsistence, in other words into means of conserving and increasing the labour power necessary to capital for its self-valorization and its vital process. This conservation and this increase in labour force —the result of the process—therefore appear simply as the reproduction and enlargement of the conditions of reproduction and enlargement of the conditions of reproduction and accumulation which belong to capital.

Thus there disappears even the appearance which still continued to exist on the surface, namely that capitalist and worker free each other as owners of commodities, equal in law, in circulation, and in the market and that like all the other owners of commodities the only thing that distinguishes them is the material content of their goods, the particular use-value of the goods which they sell to each other. In other words, this original form of relationship only continues to exist as the pure reflection of the underlying capitalist relationship.

Here we must distinguish two stages: the reproduction of the relationship itself on an ever wider scale, such as it results from the capitalist production process, and the form that it assumes historically at the time of its birth and that it assumes again continually on the surface of developed capitalist society.

1. The introductory process, which develops inside the sphere of circulation, is the sale and purchase of labour power. The capitalist process of production is not only transformation into capital of the value of the goods which the capitalist throws on to the market or puts back into the labour process: these changed products are not his products, but those of the worker. The capitalist constantly sells, for labour, a part of this product—the means of subsistence necessary for the conservation and growth of labour power, and he constantly lends to it another of them—the objective conditions of labour—as means of self-valorization of capital, as capital. In this way, whereas the worker reproduces his products as capital, the capitalist reproduces the worker as wage-earner, that is as a seller of his own labour. The relationship between simple sellers of goods would imply that they exchanged their own labour incorporated in different use-values. The purchase/sale of labour force, as a constant result of capitalist production implies, on the contrary, that the worker is constantly buying again a fraction of his own product, in exchange for his living labour.

It is in this way that there disappears the appearance of the simple relationship between owners of goods: the constant act of purchase/sale of labour power and the ceaseless confrontation of the goods produced by the worker himself as purchaser of his labour power and as variable capital are only forms which mediate his subjection to capital, living labour being only a simple means of conservation and increase of objectified labour, become autonomous over against it.

The form of mediation inherent in the capitalist mode of production is therefore used to perpetuate the relationship between the capital which purchases the labour, and the worker who sells it; however, it is only different in form from the other more or less direct modes of subjection and appropriation of labour by the owners of the conditions of production. It conceals under the simple monetary relationship the true transaction and the dependence perpetuated thanks to the mediation of the sale/purchase act which is renewed constantly. This relationship continually reproduces not only the conditions of this trade but also its results, namely what one purchases and what the other sells. The perpetual renewal of this relationship of purchase/sale only serves to mediate the continuity of the specific relationship of dependence, by giving it the mystifying appearance of a transaction, a contract between owners of good endowed with equal rights and similarly equally free one in front of the other. Thus, the initial relationship becomes itself an immanent stage in the domination of living labour by objectified labour which has established itself with capitalist production.

Both the following groups are therefore mistaken:

Those who consider wage-earning labour, the sale of labour to capital, in short the status of the wage-earner, as external to capitalist production. In fact, wage-earning labour is a form of mediation that is essential and continually reproduced by the relationship of capitalist production.

Those who see the substance itself in the superficial relationship of purchase/sale, in this essential formality and in this reflection of the capitalist relationship, and who as a result claim to subordinate the relationship between workers and capitalists to the general relationship between owners of goods, in order to vindicate it and efface its specific differences.

2. The premises of the formation of the capitalist relationship in general arise at a definite historical level of social production. It is necessary that in the womb of the previous mode of production, the means of production and circulation, and even needs must be developed to the point when they tend to surpass the former relations of production and change them into capitalist relationships. In the long term, it is sufficient for them to permit a formal surrender of labour to capital. On the basis of this new relationship, there develops a specifically different mode of production which, on the one hand, creates new material productive forces and, on the other hand, develops on this basis to create new actual conditions. It is a question of a complete economic revolution: on the one hand capital begins by producing the actual conditions of the domination of capital over labour—then it perfects them and gives them an adequate form; on the other hand, as regards the productive forces of labour, the conditions of production are relationships of circulation developed by it in opposition to the workers, it creates the actual conditions of a new mode of production which, by abolishing the antagonistic form of capitalism, lays the material foundations of a new social life and of a new form of society.

Our conception differs fundamentally from that of the economists who, imprisoned in the capitalist system, see clearly how one produces in the capitalist relationship, but not how this relationship itself is produced and at the same time creates the material conditions of its dissolution, doing away at the same stroke with its historical justification as a necessary form of the economic development and the production of social wealth.

On the contrary, we have seen not only how capital produces but also how it is itself produced, and how it comes out of the production process essentially different from what it was on going in to it. In fact, on the one hand, it changes the former mode of production; on the other hand, this change, together with a given level of the development of the material productive forces, form the basis and the preliminary condition of its own revolution.

Man is distinct from all other animal species in that his needs have no limits and are completely elastic; no other animal can repress his needs in as extraordinary a way, and limit his conditions of life to such a minimum. In short, there is no animal with more tendency to *irelandization*. In the value of labour power, there is no room for the consideration of this psychological minimum of existence.

The price of labour power—like that of any other commodity—can rise above or fall below its value, in other words deviate, in one direction or the other, from the price which is the monetary expression of value. The level of vital needs, whose sum total represents the value of labour power, can increase or diminish. . . .

Later on in our study, we shall see that, for the analysis of capital, it is quite immaterial whether we presuppose a low or high level of the needs of the workers. Moreover, in theory as in practice, we start from the value of labour power as a given quantity. For example, if a fortunate individual wishes to convert his money into capital—into, let us say, capital to exploit a cotton factory—he will first of all inquire about the average level of wages in the area where he intends to establish himself. He knows that wages—like the price of cotton—continually deviate from the average, but that these fluctuations in the end counterbalance each other. In the settlement of his accounts, he therefore takes wages as a given quantity of value.

Moreover, the value of labour power constitutes the rational and declared basis of the Trades Unions, whose importance for the working class it is vital not to underestimate. The aim of the Trade Unions is to prevent the level of wages from falling below the amount traditionally paid in the various branches of industry, and to prevent the price of the labour power from falling below its value. Of course they know that if the relationship between offer and demand changes, the market price also changes. However, on the one hand, such a change is far from being the simple unilateral act of the purchaser, in our case of the capitalist; on the other hand, there is a great difference between, on the one hand, the amount of wages as it is determined by supply and demand (that is the amount resulting from the 'honest' operation of the exchange of goods, when purchaser and seller deal on an equal footing) and, on the other hand, the amount of wages which the seller—the worker—is forced to accept, when the capitalist deals with each worker taken in isolation and imposes a low wage on him, exploiting the exceptional adversity of the solitary worker, independently of the general relationship of supply and demand.

BIBLIOGRAPHY

Original: *Arkhiv Marksa i Engelsa*, Vol. 2, Part 7, Moscow, 1933, pp. 4 ff.

Present translation: By the editor.

Other translations: Appendix to *Capital*, Vol. 1, trans. B. Fowkes, London and New York, 1976.

Commentaries: see under sections 31 and 32.

34. Letters 1858–1868

Marx on his 'Economics'

Marx to Lassalle, 22 Feb. 1858

... The first work in question is critique of the economic categories, or, if you like, the system of bourgeois economy critically presented. It is a presentation of the system and simultaneously, through this presentation, a criticism of it. I am by no means sure how many printer's sheets the whole thing will add up to. If I had the time, leisure, and means to give the whole thing the necessary finish before I hand it over to the public I would greatly condense it, as I have always liked the method of condensation. This way, however, printed in successive booklets, it may perhaps be easier for the public to understand, but it will surely work to the detriment of its form and the thing will necessarily be somewhat drawn out. Remember, as soon as you find out whether you can get this job done in Berlin or not please write to me, because if it will not work out there I shall try Hamburg. Another point is that I must get paid by the publisher who undertakes the job, a necessity which may shipwreck the whole business in Berlin.

The presentation, that is, the manner of treatment, is wholly scientific, hence not in violation of any police regulations in the ordinary sense. The whole thing is divided into six books. (1) Capital (contains some introductory chapters). (2) Landed Property. (3) Wage Labour. (4) The State. (5) International Trade. (6) World Market. Naturally, I cannot refrain from criticizing other economists now and then, and particularly not from polemizing against Ricardo, in so far as he himself, as a bourgeois, cannot help making blunders even from the strictly economic point of view. However, the critique and history of political economy and of socialism as a whole was to form the subject of another work. Finally, the brief historical sketch of the development of the economic categories, or relationships, was to be a third work. After all, I have a presentiment that now, when after fifteen years of study I have got so far as to be able to get down to the thing, turbulent movements without will most likely interfere. But never mind. If I finish too late to find the world still interested in that sort of thing, the fault will obviously be my own. ...

Marx to Engels, 8 Jan. 1868

... It is strange that the fellow [a reviewer of *Capital*] does not sense the three fundamentally new elements of the book:

(1) That in contrast to all former political economy, which from the very

outset treats the particular fragments of surplus value with their fixed forms of rent, profit, and interest as already given, I first deal with the general form of surplus value, in which all these fragments are still undifferentiated—in solution, as it were.

(2) That the economists, without exception, have missed the simple point that if the commodity has a double character—use-value and exchange-value—then the labour represented by the commodity must also have a twofold character, while the bare analysis of labour without more, as in Smith, Ricardo, etc., is bound to come up everywhere against the inexplicable. This is, in fact, the whole secret of the critical conception.

(3) That for the first time wages are shown to be the irrational form in which a relation hidden behind them appears, and that this is exactly represented in the two forms of wages—time wages and piece wages. (It was a help to me that similar formulas are often found in higher mathematics.)
. . .

Marx to Kugelmann, 11 July 1868

. . . The unfortunate fellow does not see that, even if there were no chapter on 'value' in my book, the analysis of the real relations which I give would contain the proof and demonstration of the real value relation. All that palaver about the necessity of proving the concept of value comes from complete ignorance both of the subject dealt with and of scientific method. Every child knows that a nation which ceased to work, I will not say for a year, but even for a few weeks, would perish. Every child knows, too, that the masses of products corresponding to the different needs require different and quantitatively determined masses of products corresponding to the different needs require different and quantitatively determined masses of the total labour of society. That this necessity of the distribution of social labour in definite proportions cannot possibly be done away with by a particular form of social production but can only change the mode of its appearance, is self-evident. No natural laws can be done away with. What can change in historically different circumstances is only the form in which these laws assert themselves. And the form in which this proportional distribution of labour asserts itself, in a state of society where the interconnection of social labour is manifested in the private exchange of the individual products of labour, is precisely the exchange value of these products.

Science consists precisely in demonstrating how the law of value asserts itself. So that if one wanted at the very beginning to 'explain' all the phenomena which seemingly contradict that law, one would have to present the science before science. It is precisely Ricardo's mistake that in his first chapter on value he takes as given all possible and still to be developed categories in order to prove their conformity with the law of value.

On the other hand, as you correctly assumed, the history of the theory certainly shows that the concept of the value relation has always been the same—more or less clear, hedged more or less with illusions or scientifically more or less definite. Since the thought process itself grows out of conditions, is itself a natural process, thinking that really comprehends must always be the same, and can vary only gradually, according to maturity of development, including the development of the organ by which the thinking is done. Everything else is drivel.

The vulgar economist has not the faintest idea that the actual everyday exchange relations cannot be directly identical with the magnitudes of value. The essence of bourgeois society consists precisely in this, that *a priori* there is no conscious social regulation of production. The rational and naturally necessary asserts itself only as a blindly working average. And then the vulgar economist thinks he has made a great discovery when, as against the revelation of the inner interconnection, he proudly claims that in appearance things look different. In fact, he boasts that he holds fast to appearance, and takes it for the ultimate. Why, then, have any science at all?

But the matter has also another background. Once the interconnection is grasped, all theoretical belief in the permanent necessity of existing conditions collapses before their collapse in practice. Here, therefore, it is absolutely in the interest of the ruling classes to perpetuate this senseless confusion. And for what other purpose are the sycophantic babblers paid, who have no other scientific trump to play save that in political economy one should not think at all? . . .

Marx to Engels, 24 Aug. 1867

. . . The best points in my book are: (1) the twofold character of labour, according to whether it is expressed in use-value or exchange-value. (All understanding of the facts depends upon this.) It is emphasized immediately, in the first chapter; (2) the treatment of surplus value independently of its particular forms as profit, interest, ground rent, etc. This will come out especially in the second volume. The treatment of the particular forms by classical economy, which always mixes them up with the general form, is a regular hash. . . .

On Darwin

Marx to Lassalle, 16 Jan. 1862

. . . Darwin's book is very important and serves me as a natural-scientific basis for the class struggle in history. One has to put up with the crude English method of development, of course. Despite all deficiencies, not only is the death-blow dealt here for the first time to 'teleology' in the natural sciences but its rational meaning is empirically explained. . . .

Marx to Engels, 18 June 1862

... Darwin, whom I have looked up again, amuses me when he says he is applying the 'Malthusian' theory also to plants and animals, as if with Mr. Malthus the whole point were not that he does not apply the theory to plants and animals but only to human beings—and with geometrical progression—as opposed to plants and animals. It is remarkable how Darwin recognizes among beasts and plants his English society with its division of labour, competition, opening-up of new markets, 'inventions', and the Malthusian 'struggle for existence'. It is Hobbes's *bellum omnium contra omnes*, and one is reminded of Hegel's *Phenomenology*, where civil society is described as a 'spiritual animal kingdom', while in Darwin the animal kingdom figures as civil society. ...

On Machinery

Marx to Engels, 28 Jan. 1863

... You may or may not know, for in itself the question does not matter, that there is a great dispute as to what distinguishes a machine from a tool. The English (mathematical) mechanists, in their crude way, call a tool a simple machine and a machine a complex tool. The English technologists, however, who pay rather more attention to economics, base the distinction between the two on the fact (and in this they are followed by many, by most, of the English economists) that in one case the motive power is derived from human beings, in the other from a natural force. The German asses, who are great at these small things, have therefore concluded that, for instance, a plough is a machine, while the most complex jenny, etc., in so far as it is worked by hand, is not. But now if we look at the machine in its elementary form there is no question at all that the industrial revolution starts not from the motive power but from that section of the machinery which the English call the working machine; hence not, for instance, from the substitution of water or steam for the foot which turns the spinning-wheel, but from the transformation of the immediate process of spinning itself and from the displacement of that portion of human labour which is not merely 'exertion of power' (as in treading a wheel) but which concerns the processing, the direct action on the material to be worked up. On the other hand it is likewise not open to question that as soon as the point at issue is no longer the historical development of machinery but machinery on the basis of the present mode of production, the working machine (for instance, in the case of the sewing-machine) is the only determining factor; for once this process has been mechanized everyone nowadays knows that the thing can be moved by hand, water-power, or a steam-engine, depending on its size.

To pure mathematicians these questions are immaterial, but they become very important when it is a question of proving the connection between the social relations of men and the development of these material modes of production.

The re-reading of my excerpts bearing on the history of technology has led me to the opinion that, apart from the discovery of gunpowder, the compass, and printing—those necessary prerequisites of bourgeois development—the two material bases on which the preparations for machine-operated industry proceeded within manufacture during the period from the sixteenth to the middle of the eighteenth century (the period in which manufacture was developing from handicraft into large-scale industry proper) were the clock and the mill (at first the corn mill, specifically, the water-mill). Both were inherited from the ancients. (The water-mill was introduced into Rome from Asia Minor at the time of Julius Caesar.) The clock was the first automatic machine applied to practical purposes; the whole theory of the production of regular motion was developed through it. Its nature is such that it is based on a combination of half-artistic handicraft and direct theory. Cardanus, for instance, wrote about (and gave practical formulas for) the construction of clocks. German authors of the sixteenth century called clockmaking 'learned (non-guild) handicraft', and it would be possible to show from the development of the clock how entirely different the relation between erudition and practice was on the basis of handicraft from what it is, for instance, in modern large-scale industry. There is also no doubt that in the eighteenth century the idea of applying automatic devices (moved by springs) to production was first suggested by the clock. It can be proved historically that Vaucanson's experiments on these lines had a tremendous influence on the imagination of the English inventors.

The mill, on the other hand, from the very beginning, as soon as the water-mill came into existence, possessed the essential elements of the organism of a machine. The mechanical motive power. Firstly, the motor, which depends on it; the transmitting mechanism; and, finally, the working machine, which deals with the material—each with an existence independent of the others. The theory of friction, and also the investigations into the mathematical forms of gear-wheels, cogs, etc., were all developed in connection with the mill; ditto as to the theory of measurement of the degree of motive power, of the best way of employing it, etc. Almost all the great mathematicians after the middle of the seventeenth century, so far as they occupied themselves with practical mechanics and its theoretical side, started from the simple corn-grinding water-mill. And indeed this was why the name Mühle and mill, which arose during the manufacturing period, came to be applied to all mechanical forms of motive power adapted to practical purposes.

But with the mill, as with the press, the forge, the plough, etc., the work proper, that of beating, crushing, grinding, pulverizing, etc., has been performed from the very first without human labour, even though the moving force was human or animal. This kind of machinery is therefore very ancient, at least in its origins, and mechanical propulsion proper was first applied to it. Hence it is practically the only machinery found in the manufacturing period. The industrial revolution begins as soon as mechanisms are employed where from ancient times the final result has required human labour; hence not where, as with the tools mentioned above, the material actually to be worked up has never been dealt with by the human hand, but where, in the nature of things, man has not from the very first acted merely as power. If one is to follow the German asses in calling the use of animal power (which is just as much voluntary movement as human power) machinery, then remember, the use of this kind of locomotive is at any rate much older than the simplest handicraft tool. . . .

NOTE: all extracts taken from *MESC*. Reproduced by kind permission of Lawrence and Wishart Ltd.

V. Later Political Writings
1864–1882

35. Inaugural Address to the First International

In September 1864 was founded the International Workingmen's Association, commonly known as the First International. It brought together European Trades Union leaders in a loose federation to exchange ideas and concert action for political and social reform. Marx sat on the General Council in London and soon became the dominant personality there. One of his first tasks was to draw up an Inaugural Address on behalf of the General Council. Marx described it as 'a sort of review of the fortunes of the working classes' since 1845. It is a skilful piece of writing, well adapted to his audience (mainly British Trades Unionists), and includes material that later appeared in *Capital*. It begins with one of Marx's clearest statements of relative immiserization and goes on to comment on the weakness of working-class movements since 1848.

Working Men,

It is a great fact that the misery of the working masses has not diminished from 1848 to 1864, and yet this period is unrivalled for the development of its industry and the growth of its commerce. In 1850, a moderate organ of the British middle class, of more than average information, predicted that if the exports and imports of England were to rise 50 per cent, English pauperism would sink to zero. Alas! on 7 April 1864, the Chancellor of the Exchequer delighted his parliamentary audience by the statement that the total import and export trade of England had grown in 1863 'to £443,955,000! that astonishing sum about three times the trade of the comparatively recent epoch of 1843!' With all that, he was eloquent upon 'poverty'. 'Think', he exclaimed, 'of those who are on the border of that region', upon 'wages . . . not increased'; upon 'human life . . . in nine cases out of ten but a struggle of existence!' He did not speak of the people of Ireland, gradually replaced by machinery in the north, and by sheep-walks in the south, though even the sheep in that unhappy country are decreasing, it is true, not at so rapid a rate as the men. He did not repeat what then had been just betrayed by the highest representatives of the upper ten thousand in a sudden fit of terror. When the garotte panic had reached a certain height, the House of Lords caused an inquiry to be made into, and a report to be published upon, transportation and penal servitude. Out came the murder in the bulky Blue Book of 1863, and proved it was, by official facts and figures, that the worst of the convicted criminals, the penal serfs of England and Scotland, toiled much less and fared far better than the agricultural labourers of England and Scotland. But this was not all. When, consequent upon the Civil War in America, the operatives of Lancashire and Cheshire were thrown upon the streets, the same House of Lords sent to the manufacturing districts a physician commissioned to investigate into the smallest possible amount of carbon and nitrogen, to be administered in the cheapest and plainest form, which on an average might just suffice to 'avert starvation diseases'. Dr. Smith, the medical deputy, ascertained that 28,000 grains of carbon, and 1330 grains of nitrogen were the weekly allowance that would keep an average adult . . . just over the level of starvation diseases, and he found furthermore that quantity pretty nearly to agree with the scanty nourishment to which the pressure of extreme distress had actually reduced the cotton operatives. But now mark! The same learned Doctor was later on again deputed by the medical officer of the Privy Council to inquire into the nourishment of the poorer labouring classes. The results of his researches are embodied in the 'Sixth Report on Public Health', published by order of Parliament in the course of the present year. What did the Doctor discover? That the silk weavers, the needle women, the kid glovers, the stocking weavers, and so forth, received, on an average, not even the distress pittance of the cotton operatives, not even the amount of carbon and nitrogen 'just sufficient to avert starvation diseases'.

'Moreover,' we quote from the report, 'as regards the examined families of the agricultural population, it appeared that more than a fifth were with less than the estimated sufficiency of carbonaceous food, that more than one-third were with less than the estimated sufficiency of nitrogenous food, and that in three counties (Berkshire, Oxfordshire, and Somersetshire) insufficiency of nitrogenous food was the average local diet.' 'It must be remembered', adds the official report, 'that privation of food is very reluctantly borne, and that, as a rule, great poorness of diet will only come when other privations have preceded it. . . . Even cleanliness will have been found costly or difficult, and if there still be self-respectful endeavours to maintain it, every such endeavour will represent additional pangs of hunger.' 'These are painful reflections, especially when it is remembered that the poverty to which they advert is not the deserved poverty of idleness; in all cases it is the poverty of working populations. Indeed, the work which obtains the scanty pittance of food is for the most part excessively prolonged.' The report brings out the strange, and rather unexpected fact. 'That of the divisions of the United Kingdom', England, Wales, Scotland, and Ireland, 'the agricultural population of England', the richest division, 'is considerably the worst fed'; but that even the agricultural labourers of Berkshire, Oxfordshire, and Somersetshire fare better than great numbers of skilled indoor operatives of the East of London.

Such are the official statements published by order of Parliament in 1864, during the millennium of free trade, at a time when the Chancellor of the Exchequer told the House of Commons that 'the average condition of the British labourer has improved in a degree we know to be extraordinary and unexampled in the history of any country or any age'. Upon these official congratulations jars the dry remark of the official Public Health Report: 'The public health of a country means the health of its masses, and the masses will scarcely be healthy unless, to their very base, they be at least moderately prosperous.'

Dazzled by the 'Progress of the Nation' statistics dancing before his eyes, the Chancellor of the Exchequer exclaims in wild ecstasy: 'From 1842 to 1852 the taxable income of the country increased by 6 per cent; in the eight years from 1853 to 1861, it has increased from the basis taken in 1853 20 per cent! the fact is so astonishing as to be almost incredible! . . . This intoxicating augmentation of wealth and power', adds Mr. Gladstone, 'is entirely confined to classes of property!'

If you want to know under what conditions of broken health, tainted morals, and mental ruin that 'intoxicating augmentation of wealth and power entirely confined to classes of property' was and is being produced by the classes of labour, look to the picture hung up in the last 'Public Health Report' of the workshops of tailors, printers, and dressmakers! Compare the 'Report of the Children's Employment Commission' of 1863, where it is stated, for instance, that: 'The potters as a class, both men and women, represent a much degenerated population, both physically and mentally', that 'the unhealthy child is an unhealthy parent in his turn', that 'a progressive deterioration of the race must go on', and that 'the degenerescence of the population of Staffordshire would be even greater were it not for the constant recruiting from the adjacent country, and the intermarriages with more healthy races'. Glance at Mr. Tremenheere's Blue Book on the 'Grievances complained of by the Journeymen Bakers'! And who has not shuddered at the paradoxical statement made by the inspectors of factories, all illustrated by the Registrar-General, that the Lancashire operatives, while put upon the distress pittance of food, were actually improving in health, because of their temporary exclusion by the cotton famine from the cotton factory, and that the mortality of the children was decreasing, because their mothers were now at last allowed to give them, instead of Godfrey's cordial, their own breasts.

Again reverse the medal! The Income and Property Tax Returns laid before the House of Commons on 20 July 1864 teach us that the persons with yearly incomes valued by the tax-gatherer at £50,000 and upwards, had, from 5 April 1862 to 5 April 1863, been joined by a dozen and one, their number having increased in that single year from 67 to 80. The same returns disclose the fact that about 3000 persons divide among themselves a yearly income of about £25,000,000 sterling, rather more than the total revenue doled out annually to the whole mass of the agricultural labourers of England and Wales. Open the census of 1861, and you will find that the number of the male landed proprietors of England and Wales had decreased from 16,934 in 1851 to 15,066 in 1861, so that the concentration of land had grown in ten years 11 per cent. If the concentration of the soil of the country in a few hands proceed at the same rate, the land question will become singularly simplified, as it had become in the Roman empire, when Nero grinned at the discovery that half the Province of Africa was owned by six gentlemen.

We have dwelt so long upon these 'facts so astonishing to be almost incredible' because England heads the Europe of commerce and industry. It will be remembered that some months ago one of the refugee sons of Louis Philippe publicly congratulated the English agricultural labourer on the superiority of his lot over that of his less florid comrade on the other side of the Channel. Indeed, with local colours changed, and on a scale somewhat contracted, the English facts reproduce themselves in all the industrious and progressive countries of the Continent. In all of them there has taken place, since 1848, an unheard-of development of industry, and an undreamed-of expansion of imports and exports. In all of them 'the augmentation of wealth and power entirely confined to classes of property' was truly 'intoxicating'. In all of them, as in England, a minority of the working classes got their real wages somewhat advanced; while in most cases the monetary rise of wages denoted no more a real access of comforts than the inmate of the metropolitan poor-house or orphan asylum, for instance, was in the least benefited by his first necessaries costing £9 15s. 8d. in 1861 against £7 7s. 4d. in 1852. Everywhere the great mass of the working classes were sinking down to a lower depth, at the same rate at least that those above them were rising in the social scale. In all countries of Europe it has now become a truth demonstrable to every unprejudiced mind, and only denied by those whose interest it is to hedge other people in a fool's paradise, that no improvement of machinery, no appliance of science to production, no contrivances of communication, no new colonies, no emigration, no opening of markets, no free trade, nor all these things put together, will do away with the miseries of the industrious masses; but that, on the present false base, every fresh development of the productive powers of labour must tend to deepen social contrasts and point social antagonisms. Death of starvation rose almost to the rank of an institution, during this intoxicating epoch of economical progress, in the metropolis of the British Empire. That epoch is marked in the annals of the world by the quickened return, the widening compass, and the deadlier effects of the social pest called a commercial and industrial crisis.

After the failure of the Revolutions of 1848, all party organizations and party journals of the working classes were, on the Continent, crushed by the iron hand of force, the most advanced sons of labour fled in despair to the Transatlantic Republic, and the short-lived dreams of emancipation vanished before an epoch of industrial fever, moral marasmus, and political reaction. The defeat of the Continental working classes, partly owed to the diplomacy of the English Government, acting then as now in fraternal solidarity with the Cabinet of St. Petersburg, soon spread its contagious effects to this side of the Channel. While the rout of their Continental brethren unmanned the English working classes, and broke their faith in their own cause, it restored to the landlord and the money-lord their somewhat shaken confidence. They insolently withdrew concessions already advertised. The discoveries of new goldlands led to an immense exodus, leaving an irreparable void in the ranks of the British proletariat. Others of its formerly active members were caught by the temporary bribe of greater work and wages, and turned into 'political blacks'. All the efforts made at keeping up, or remodelling, the Chartist Movement failed signally; the press organs of the working class died one by one of the apathy of the masses, and, in point of fact, never before seemed the English working class so thoroughly reconciled to a state of political nullity. If, then, there had been no solidarity of action between the British and the Continental working classes, there was, at all events, a solidarity of defeat.

And yet the period passed since the Revolutions of 1848 has not been without its compensating features. We shall here only point to two great facts.

After a thirty years' struggle, fought with most admirable perseverance, the English working classes, improving a momentary split between the landlords and money-lords, succeeded in carrying the Ten Hours' Bill. The immense physical, moral, and intellectual benefits hence accruing to the factory operatives, half-yearly chronicled in the reports of the inspectors of factories, are now acknowledged on all sides. Most of the Continental governments had to accept the English Factory Act in more or less modified forms, and the English Parliament itself is every year compelled to enlarge its sphere of action. But besides its practical import, there was something else to exalt the marvellous success of this working men's measure. Through their most notorious organs of science, such as Dr. Ure, Professor Senior, and other sages of that stamp, the middle class had predicted, and to their heart's content proved, that any legal restriction of the hours of labour must sound the death knell of British industry, which, vampire-like, could but live by sucking blood, and children's blood, too. In olden times, child murder was a mysterious rite of the religion of Moloch, but it was practised on some very solemn occasions only, once a year perhaps, and then Moloch had no exclusive bias for the children of the poor. This struggle about the legal restriction of the hours of labour raged the more fiercely since, apart from frightened avarice, it told indeed upon the great contest between the blind rule of the supply and demand laws which form the political economy of the middle class, and social production controlled by social foresight, which forms the political economy of the working class. Hence the Ten Hours' Bill was not only a great practical success; it was the victory of a principle; it was the first time that in broad daylight the political economy of the middle class succumbed to the political economy of the working class.

But there was in store a still greater victory of the political economy of labour over the political economy of property. We speak of the co-operative movement, especially the co-operative factories raised by the unassisted efforts of a few bold 'hands'. The value of these great social experiments cannot be overrated. By deed, instead of by argument, they have shown that production on a large scale, and in accord with the behests of modern science, may be carried on without the existence of a class of masters employing a class of hands; that to bear fruit, the means of labour need not be monopolized as a means of dominion over, and of extortion against, the labouring man himself; and that, like slave labour, like serf labour, hired labour is but a transitory and inferior form, destined to disappear before associated labour plying its toil with a willing hand, a ready mind, and a joyous heart. In England, the seeds of the co-operative system were sown by Robert Owen; the working men's experiments, tried on the Continent, were, in fact, the practical upshot of the theories, not invented, but loudly proclaimed, in 1848.

At the same time, the experience of the period from 1848 to 1864 has proved beyond doubt that, however excellent in principle, and however useful in practice, co-operative labour, if kept within the narrow circle of the casual efforts of private workmen, will never be able to arrest the growth in geometrical progression of monopoly, to free the masses, nor even to lighten perceptibly the burden of their miseries. It is perhaps for this very reason that plausible noblemen, philanthropic middle-class spouters, and even keen political economists have all at once turned nauseously complimentary to the very co-operative labour system they had vainly tried to nip in the bud by deriding it as the Utopia of the dreamer, or stigmatizing it as the sacrilege of the Socialist. To save the industrious masses, co-operative labour ought to be developed to national dimensions, and consequently, to be fostered by national means. Yet, the lords of land and the lords of capital will always use their political privileges for the defence and perpetuation of their economical monopolies. So far from promoting, they will continue to lay every possible impediment in the way of the emancipation of labour. Remember the sneer with which, last session, Lord Palmerston put down the advocates of the Irish Tenants' Right Bill. The House of Commons, cried he, is a house of landed proprietors. To conquer political power has therefore become the great duty of the working classes. They seem to have comprehended this, for in England, Germany, Italy, and France there have taken place simultaneous revivals, and simultaneous efforts are being made at the political reorganization of the working men's party.

One element of success they possess—numbers; but numbers weigh only in the balance, if united by combination and led by knowledge. Past experience has shown how disregard of that bond of brotherhood which ought to exist between the workmen of different countries, and incite them to stand firmly by each other in all their struggles for emancipation, will be chastized by the common discomfiture of their incoherent efforts. This thought prompted the working men of different countries assembled on 28 September 1864, in public meeting at St. Martin's Hall, to found the International Association.

Another conviction swayed that meeting.

If the emancipation of the working classes requires their fraternal concurrence, how are they to fulfil that great mission with a foreign policy in pursuit of criminal designs, playing upon national prejudices, and squandering in piratical wars the people's blood and treasure? It was not the wisdom of the ruling classes, but the heroic resistance to their criminal folly by the working classes of England that saved the West of Europe from plunging headlong into an infamous crusade for the perpetuation and propagation of slavery on the other side of the Atlantic. The shameless approval, mock sympathy, or idiotic indifference with which the upper classes of Europe have witnessed the mountain fortress of the Caucasus falling a prey to, and heroic Poland being assassinated by, Russia; the immense and unresisted encroachments of that barbarous power, whose head is at St. Petersburg, and whose hands are in every cabinet of Europe, have taught the working classes the duty to master themselves the mysteries of international politics; to watch the diplomatic acts of their respective Governments; to counteract them, if necessary, by all means in their power; when unable to prevent, to combine in simultaneous denunciations, and to vindicate the simple laws of morals and justice, which ought to govern the relations of private individuals, as the rules paramount of the intercourse of nations.

The fight for such a foreign policy forms part of the general struggle for the emancipation of the working classes.

Proletarians of all countries, Unite!

BIBLIOGRAPHY

Original (in English): *MESW*, Vol. 1, pp. 377 ff. Reproduced by kind permission of Lawrence and Wishart Ltd.

Commentaries: J. Braunthal, *History of the International*, London, 1967, Vol. 1, pp. 75 ff.
H. Collins and C. Abramsky, *Karl Marx and the British Labour Movement: Years of the First International*, London, 1965.
R. Harrison, *Before the Socialists*, London, 1965.
D. McLellan, *Karl Marx, His Life and Thought*, London and New York, 1973, pp. 360 ff.

36. On Trade Unions

The following is from a speech delivered to a German Trade Union delegation in Hanover in 1869. Although its strictures on political parties were to some extent influenced by the situation prevailing at the time in Germany, it does demonstrate the confidence that Marx placed in trade union potential.

... If they wish to accomplish their task, trade unions ought never to be attached to a political association or place themselves under its tutelage; to do so would be to deal themselves a mortal blow. Trade unions are the schools of socialism. It is in trade unions that workers educate themselves and become socialists, because under their very eyes and every day the struggle with capital is taking place. Any political party, whatever its nature and without exception, can only hold the enthusiasm of the masses for a short time, momentarily; unions, on the other hand, lay hold on the masses in a more enduring way; they alone are capable of representing a true working-class party and opposing a bulwark to the power of capital. The great mass of workers, whatever party they belong to, have at last understood that their material situation must become better. But once the worker's material situation has become better, he can consecrate himself to the education of his children; his wife and children do not need to go to the factory, he himself can cultivate his mind more, look after his body better, and he becomes socialist without noticing it. ...

BIBLIOGRAPHY

Original: *Volksstaat*, 27 Nov. 1869.
Present translation: by the editor.

37. *The Civil War in France*

The defeat of France in the Franco–Prussian war of 1870 meant the fall of Louis Napoleon and the creation of a Provisional Government under Thiers. The Provisional Government began to negotiate peace terms, but the workers and lower middle class in Paris rose in revolt and held the city for two months against both the Thiers government and the Prussians.

Despite popular belief, the original revolt and the subsequent Paris Commune had only the most tenuous links with the International. But the General Council was urged to issue a statement and Marx was asked to draft it. Marx produced two drafts (passages of which are excerpted below) before the final version. He delayed until the Commune had been bloodily suppressed, and the statement became an obituary.

The Civil War in France begins with a brilliantly vicious series of sketches of the members of the Provisional Government. The most important section is the third one, where Marx describes the political organization of the Commune—both actual and potential. The model here appears to be much more decentralized than the model in parallel passages in, for example, the *Communist Manifesto*. There is, however, some controversy about how free Marx felt to express his views in what amounted to an obituary.

From the Published Version

... On the dawn of the eighteenth of March, Paris arose to the thunderburst of 'Vive la Commune!' What is the Commune, that sphinx so tantalizing to the bourgeois mind?

'The proletarians of Paris,' said the Central Committee in its manifesto of the 18th March, 'amidst the failures and treasons of the ruling classes, have understood that the hour has struck for them to save the situation by taking into their own hands the direction of public affairs. ... They have understood that it is their imperious duty and their absolute right to render themselves masters of their own destinies, by seizing upon the governmental power.'

But the working class cannot simply lay hold of the ready-made state machinery, and wield it for its own purposes.

The centralized State power, with its ubiquitous organs of standing army, police, bureaucracy, clergy, and judicature—organs wrought after the plan of a systematic and hierarchic division of labour—originates from the days of absolute monarchy, serving nascent middle-class society as a mighty weapon in its struggles against feudalism. Still, its development remained clogged by all manner of medieval rubbish, seigniorial rights, local privileges, municipal and guild monopolies, and provincial constitutions. The gigantic broom of the French Revolutions of the eighteenth century swept away all these relics of bygone times, thus clearing simultaneously the social soil of its last hindrances to the superstructure of the modern State edifice raised under the First Empire, itself the offspring of the coalition wars of old semi-feudal Europe against modern France. During the subsequent regimes the Government, placed under parliamentary control—that is, under the direct control of the propertied classes—became not only a hotbed of huge national debts and crushing taxes; with its irresistible allurements of place, pelf, and patronage, it became not only the bone of contention between the rival factions and adventurers of the ruling classes; but its political character changed simultaneously with the economic changes of society. At the same pace at which the progress of modern industry developed, widened, intensified the class antagonism between capital and labour, the State power assumed more and more the character of the national power of capital over labour, of a public force organized for social enslavement, of an engine of class despotism. After every revolution marking a progressive phase in the class struggle, the purely repressive character of the State power stands out in bolder and bolder relief. The Revolution of 1830, resulting in the transfer of Government from the landlords to the capitalists, transferred it from the more remote to the more direct antagonists of the working men. The bourgeois Republicans who, in the name of the Revolution of February, took the State power, used it for the June massacres in order to convince the working class that 'social' republic meant the Republic ensuring their social subjection, and in order to convince the royalist bulk of the bourgeois and landlord class that they might safely leave the cares and emoluments of Government to the bourgeois 'Republicans'. However, after their one heroic exploit of June, the bourgeois Republicans had, from the front, to fall back to the rear of the 'Party of Order'—a combination formed by all the rival fractions and factions of the appropriating class in their now openly declared antagonism to the producing classes. The proper form of their joint-stock Government was the Parliamentary Republic, with Louis Bonaparte for its President. Theirs was a regime of avowed class terrorism and deliberate insult towards the 'vile multitude'. If the Parliamentary Republic, as M. Thiers said, 'divided them (the different fractions of the ruling class) least', it opened an abyss between that class and the whole body of society outside their spare ranks. The restraints by which their own divisions had under former regimes still checked the State power were removed by their union; and in view of the threatening upheaval of the proletariat, they now used that State power mercilessly and ostentatiously as the national war-engine of capital against labour. In their uninterrupted crusade against the producing masses they were, however, bound not only to invest the executive with continually increased powers of repression, but at the same time to divest their own parliamentary stronghold—the National Assembly—one by one, of all its own means of defence against the Executive. The Executive, in the person of

Louis Bonaparte, turned them out. The natural offspring of the 'Party-of-Order' Republic was the Second Empire.

The empire, with the *coup d'état* for its certificate of birth, universal suffrage for its sanction, and the sword for its sceptre, professed to rest upon the peasantry, the large mass of producers not directly involved in the struggle of capital and labour. It professed to save the working class by breaking down Parliamentarism, and, with it, the undisguised subserviency of Government to the propertied classes. It professed to save the propertied classes by upholding their economic supremacy over the working class; and, finally, it professed to unite all classes by reviving for all the chimera of national glory. In reality, it was the only form of government possible at a time when the bourgeoisie had already lost, and the working class had not yet acquired, the faculty of ruling the nation. It was acclaimed throughout the world as the saviour of society. Under its sway, bourgeois society, freed from political cares, attained a development unexpected even by itself. Its industry and commerce expanded to colossal dimensions; financial swindling celebrated cosmopolitan orgies; the misery of the masses was set off by a shameless display of gorgeous, meretricious, and debased luxury. The State power, apparently soaring high above society, was at the same time itself the greatest scandal of that society and the very hotbed of all its corruptions. Its own rottenness, and the rottenness of the society it had saved, were laid bare by the bayonet of Prussia, herself eagerly bent upon transferring the supreme seat of that regime from Paris to Berlin. Imperialism is, at the same time, the most prostituted and the ultimate form of the State power which nascent middle-class society had commenced to elaborate as a means of its own emancipation from feudalism, and which full-grown bourgeois society had finally transformed into a means for the enslavement of labour by capital.

The direct antithesis to the empire was the Commune. The cry of 'social republic', with which the revolution of February was ushered in by the Paris proletariat, did but express a vague aspiration after a Republic that was not only to supersede the monarchical form of class-rule, but class-rule itself. The Commune was the positive form of that Republic.

Paris, the central seat of the old governmental power, and, at the same time, the social stronghold of the French working class, had risen in arms against the attempt of Thiers and the Rurals to restore and perpetuate that old governmental power bequeathed to them by the empire. Paris could resist only because, in consequence of the siege, it had got rid of the army, and replaced it by a National Guard, the bulk of which consisted of working men. This fact was now to be transformed into an institution. The first decree of the Commune, therefore, was the suppression of the standing army, and the substitution for it of the armed people.

The Commune was formed of the municipal councillors, chosen by universal suffrage in the various wards of the town, responsible and revocable at short terms. The majority of its members were naturally working men, or acknowledged representatives of the working class. The Commune was to be a working, not a parliamentary, body, executive and legislative at the same time. Instead of continuing to be the agent of the Central Government, the police was at once stripped of its political attributes, and turned into the responsible and at all times revocable agent of the Commune. So were the officials of all other branches of the Administration. From the members of the Commune downwards, the public service had to be done at workmen's wages. The vested interests and the representation allowances of the high dignitaries of State disappeared along with the high dignitaries themselves. Public functions ceased to be the private property of the tools of the Central Government. Not only municipal administration, but the whole initiative hitherto exercised by the State was laid into the hands of the Commune.

Having once got rid of the standing army and the police, the physical force elements of the old Government, the Commune was anxious to break the spiritual force of repression, the 'parson-power', by the disestablishment and disendowment of all churches as proprietary bodies. The priests were sent back to the recesses of private life, there to feed upon the alms of the faithful in imitation of their predecessors, the Apostles. The whole of the educational institutions were opened to the people gratuitously, and at the same time cleared of all interference of Church and State. Thus, not only was education made accessible to all, but science itself freed from the fetters which class prejudice and governmental force had imposed upon it.

The judicial functionaries were to be divested of that sham independence which had but served to mask their abject subserviency to all succeeding governments to which, in turn, they had taken, and broken, the oaths of allegiance. Like the rest of public servants, magistrates and judges were to be elective, responsible, and revocable.

The Paris Commune was, of course, to serve as a model to all the great industrial centres of France. The communal regime once established in Paris and the secondary centres, the old centralized Government would in the provinces, too, have to give way to the self-government of the producers. In a rough sketch of national organization which the Commune had no time to develop, it states clearly that the Commune was to be the political form of even the smallest country hamlet, and that in the rural districts the standing army was to be replaced by a national militia, with an extremely short term of service. The rural communes of every district were to administer their common affairs by an assembly of delegates in the central town, and these district assemblies were again to send deputies to the National Delegation in Paris, each delegate to be at any time revocable and bound by the *mandat impératif* [formal instructions] of his constituents. The few but important functions which still would remain for a central government were not to be suppressed, as has been intentionally misstated, but were to be discharged by Communal, and therefore strictly responsible, agents. The unity of the nation was not to be broken but, on the contrary, to be organized by the Communal Constitution and to become a reality by the destruction of the State power which claimed to be the embodiment of that unity independent of, and superior to, the nation itself, from which it was but a parasitic excrescence. While the merely repressive organs of the old governmental power were to be amputated, its legitimate functions were to be wrested from an authority usurping pre-eminence over society itself, and restored to the responsible agents of society. Instead of deciding once in three or six years which member of the ruling class was to misrepresent the people in Parliament, universal suffrage was to serve the people, constituted in Communes, as individual suffrage serves every other employer in the search for the workmen and managers in his business. And it is well known that companies, like individuals, in matters of real business generally know how to put the right man in the right place, and, if they for once make a mistake, to redress it promptly. On the other hand, nothing could be more foreign to the spirit of the Commune than to supersede universal suffrage by hierarchic investiture.

It is generally the fate of completely new historical creations to be mistaken for the counterpart of older and even defunct forms of social life, to which they may bear a certain likeness. Thus, this new Commune, which breaks the modern State power, has been mistaken for a reproduction of the medieval Communes, which first preceded, and afterwards became the substratum of, that very State power. The Communal Constitution has been mistaken for an attempt to break up into a federation of small States, as dreamt of by Montesquieu and the Girondins, that unity of great nations which, if originally brought about by political force, has now become a powerful coefficient of social production. The antagonism of the Commune against the State power has been mistaken for an exaggerated form of the ancient struggle against over-centralization. Peculiar historical circumstances may have prevented the classical development, as in France, of the bourgeois form of government, and may have allowed, as in England, to complete the great central State organs by corrupt vestries, jobbing councillors, and ferocious poor-law guardians in the towns, and virtually hereditary magistrates in the counties. The Communal Constitution would have restored to the social body all the forces hitherto absorbed by the State parasite feeding upon, and clogging the free movement of, society. By this one act it would have initiated the regeneration of France. The provincial French middle class saw in the Commune an attempt to restore the sway their order had held over the country under Louis Philippe, and which, under Louis Napoleon, was supplanted by the pretended rule of the country over the towns. In reality, the Communal Constitution brought the rural producers under the intellectual lead of the central towns of their districts, and these secured to them, in the working men, the natural trustees of their interests.

The very existence of the Commune involved, as a matter of course, local municipal liberty, but no longer as a check upon the, now superseded, State power. It could only enter into the head of a Bismarck, who, when not engaged on his intrigues of blood and iron, always likes to resume his old trade, so befitting his mental calibre, of contributor to *Kladderadatsch* [the Berlin *Punch*], it could only enter into such a head, to ascribe to the Paris Commune aspirations after that caricature of the old French municipal organization of 1791, the Prussian municipal constitution which degrades the town governments to mere secondary wheels in the police-machinery of the Prussian State. The Commune made that catchword of bourgeois revolutions, cheap government, a reality, by destroying the two greatest sources of expenditure—the standing army and State functionarism. Its very existence presupposed the non-existence of monarchy, which, in Europe at least, is the normal encumbrance and indispensable cloak of class-rule. It supplied the Republic with the basis of really democratic institutions. But neither cheap Government nor the 'true Republic' was its ultimate aim; they were its mere concomitants.

The multiplicity of interpretations to which the Commune has been subjected, and the multiplicity of interests which construed it in their favour, show that it was a thoroughly expansive political form, while all previous forms of government had been emphatically repressive. Its true secret was this. It was essentially a working-class government, the produce of the struggle of the producing against the appropriating class, the political form at last discovered under which to work out the economic emancipation of labour.

Except on this last condition, the Communal Constitution would have been an impossibility and a delusion. The political rule of the producer cannot coexist with the perpetuation of his social slavery. The Commune was therefore to serve as a lever for uprooting the economical foundations upon which rests the existence of classes, and therefore of class-rule. With labour emancipated, every man becomes a working man, and productive labour ceases to be a class attribute.

It is a strange fact. In spite of all the tall talk and all the immense literature, for the last sixty years, about Emancipation of Labour, no sooner do the working men anywhere take the subject into their own hands with a will, than uprises at once all the apologetic phraseology of the mouthpieces of present society with its two poles of Capital and Wages Slavery (the landlord now is but the sleeping partner of the capitalist), as if capitalist society was still in its purest state of virgin innocence, with its antagonisms still undeveloped, with its delusions still unexploded, with its prostitute realities not yet laid bare. The Commune, they exclaim, intends to abolish property, the basis of all civilization! Yes, gentlemen, the Commune intended to abolish that class-property which makes the labour of the many the wealth of the few. It aimed at the expropriation of the expropriators. It wanted to make individual property a truth by transforming the means of production, land and capital, now chiefly the means of enslaving and exploiting labour, into mere instruments of free and associated labour. But this is Communism, 'impossible' Communism! Why, those members of the ruling classes who are intelligent enough to perceive the impossibility of continuing the present system—and they are many—have become the obtrusive and full-mouthed apostles of co-operative production. If co-operative production is not to remain a sham and a snare; if it is to supersede the Capitalist system; if united co-operative societies are to regulate national production upon a common plan, thus taking it under their own control, and putting an end to the constant anarchy and periodical convulsions which are the fatality of Capitalist production—what else, gentlemen, would it be but Communism, 'possible' Communism?

The working class did not expect miracles from the Commune. They have no ready-made Utopias to introduce *par décret du peuple* [by decree of the people]. They know that in order to work out their own emancipation, and along with it that higher form to which present society is irresistibly tending by its own economical agencies, they will have to pass through long struggles, through a series of historic processes, transforming circumstances and men. They have no ideals to realize, but to set free the elements of the new society with which old collapsing bourgeois society itself is pregnant. In the full consciousness of their historic mission, and with the heroic resolve to act up to it, the working class can afford to smile at the coarse invective of the gentlemen's gentlemen with the pen and inkhorn, and at the didactic patronage of well-wishing bourgeois-doctrinaires, pouring forth their ignorant platitudes and sectarian crotchets in the oracular tone of scientific infallibility.

When the Paris Commune took the management of the revolution in its own hands; when plain working men for the first time dared to infringe upon the Governmental privilege of their 'natural superiors', and, under circumstances of unexampled difficulty, performed their work modestly, conscientiously, and efficiently—performed it at salaries the highest of which barely amounted to one-fifth of what, according to high scientific authority, is the minimum required for a secretary to a certain metropolitan school board—the old world writhed in convulsions of rage at the sight of the Red Flag, the symbol of the Republic of Labour, floating over the Hôtel de Ville.

And yet this was the first revolution in which the working class was openly acknowledged as the only class capable of social initiative, even by the great bulk of the Paris middle class—shopkeepers, tradesmen, merchants—the wealthy capitalists alone excepted. The Commune had saved them by a sagacious settlement of that ever-recurring cause of dispute among the middle classes themselves—the debtor and creditor accounts. The same portion of the middle class, after they had assisted in putting down the working men's insurrection of June 1848, had been at once unceremoniously sacrificed to their creditors by the then Constituent Assembly. But this was not their only motive for now rallying round the working class. They felt that there was but one alternative—the Commune or the Empire—under whatever name it might reappear. The Empire had ruined them economically by the havoc it made of public wealth, by the wholesale financial swindling it fostered, by the props it lent to the artificially accelerated centralization of capital, and the concomitant expropriation of their own ranks. It had suppressed them politically, it had shocked them morally by its orgies, it had insulted their Voltairianism by handing over the education of their children to the *frères Ignorantins* [ignorantine Brothers], it had revolted their national feeling as Frenchmen by precipitating them headlong into a war which left only one equivalent for the ruins it made—the disappearance of the Empire. In fact, after the exodus from Paris of the high Bonapartist and capitalist bohème, the true middle-class Party of Order came out in the shape of the 'Union Républicaine', enrolling themselves under the colours of the Commune and defending it against the wilful misconstruction of Thiers. Whether the gratitude of this great body of the middle class will stand the present severe trial, time must show.

The Commune was perfectly right in telling the peasants that 'its victory was their only hope'. Of all the lies hatched at Versailles and re-echoed by the glorious European penny-a-liner, one of the most tremendous was that the Rurals represented the French peasantry. Think only of the love of the French peasant for the men to whom, after 1815, he had to pay the milliard of indemnity. In the eyes of the French peasant, the very existence of a great landed proprietor is in itself an encroachment on his conquests of 1789. The bourgeois, in 1848, had burdened his plot of land with the additional tax of forty-five cents in the franc; but then he did so in the name of the revolution; while now he had fomented a civil war against the revolution; to shift on to the peasant's shoulders the chief load of the five milliards of indemnity to be paid to the Prussian. The Commune, on the other hand, in one of its first proclamations, declared that the true originators of the war would be made to pay its cost. The Commune would have delivered the peasant of the blood tax—would have given him a cheap government—transformed his present blood-suckers, the notary, advocate, executor, and other judicial vampires, into salaried communal agents, elected by, and responsible to, himself. It would have freed him of the tyranny of the *garde champêtre* [gamekeeper], the gendarme, and the prefect; would have put enlightenment by the schoolmaster in the place of stultification by the priest. And the French peasant is, above all, a man of reckoning. He would find it extremely reasonable that the pay of the priest, instead of being extorted by the taxgatherer, should only depend upon the spontaneous action of the parishioners' religious instincts. Such were the great immediate boons which the rule of the Commune—and that rule alone—held out to the French peasantry. It is, therefore, quite superfluous here to expatiate upon the more complicated but vital problems which the Commune alone was able, and at the same time compelled, to solve in favour of the peasant, viz., the hypothecary debt, lying like an incubus upon his parcel of soil, the *prolétariat foncier* [the rural proletariat], daily growing upon it, and his expropriation from it enforced, at a more and more rapid rate, by the very development of modern agriculture and the competition of capitalist farming.

The French peasant had elected Louis Bonaparte president of the Republic; but the Party of Order created the Empire. What the French peasant really wants he commenced to show in 1849 and 1850, by opposing his *maire* to the Government's prefect, his schoolmaster to the Government's priest, and himself to the Government's gendarme. All the laws made by the Party of Order in January and February 1850 were avowed measures of repression against the peasant. The peasant was a Bonapartist, because the great Revolution, with all its benefits to him, was, in his eyes, personified in Napoleon. This delusion, rapidly breaking down under the Second Empire (and in its very nature hostile to the Rurals), this prejudice of the past, how could it have withstood the appeal of the Commune to the living interests and urgent wants of the peasantry?

The Rurals—this was, in fact, their chief apprehension—knew that three months' free communication of Communal Paris with the provinces would bring about a general rising of the peasants, and hence their anxiety to establish a police blockade around Paris, so as to stop the spread of the rinderpest.

If the Commune was thus the true representative of all the healthy elements of French society, and therefore the truly national Government, it was, at the same time, as a working men's Government, as the bold champion of the emancipation of labour, emphatically international. Within sight of the Prussian army, that had annexed to Germany two French provinces, the Commune annexed to France the working people all over the world.

The Second Empire had been the jubilee of cosmopolitan blacklegism, the rakes of all countries rushing in at its call for a share in its orgies and in the plunder of the French people. Even at this moment the right hand of Thiers is Ganesco, the foul Wallachian, and his left hand is Markovsky, the Russian spy. The Commune admitted all foreigners to the honour of dying for an immortal cause. Between the foreign war lost by their treason, and the civil war fomented by their conspiracy with the foreign invader, the bourgeoisie had found the time to display their patriotism by organizing police-hunts upon the Germans in France. The Commune made a German working man its Minister of Labour. Thiers, the bourgeoisie, the Second Empire, had continually deluded Poland by loud professions of sympathy, while in reality betraying her to, and doing the dirty work of, Russia. The Commune honoured the heroic sons of Poland by placing them at the head of the defenders of Paris. And, to broadly mark the new era of history it was conscious of initiating, under the eyes of the conquering Prussians, on the one side, and of the Bonapartist army, led by Bonapartist generals, on the other, the Commune pulled down that colossal symbol of martial glory, the Vendôme column.

The great social measure of the Commune was its own working existence. Its special measures could but betoken the tendency of a government of the people. Such were the abolition of the nightwork of journeymen bakers; the prohibition, under penalty, of the employers' practice to reduce wages by levying upon their work-people fines under manifold pretexts—a process in which the employer combines in his own person the parts of legislator, judge, and executor, and filches the money to boot. Another measure of this class was the surrender, to associations of workmen, under reserve of compensation, of all closed workshops and factories, no matter whether the respective capitalists had absconded or preferred to strike work.

The financial measures of the Commune, remarkable for their sagacity and moderation, could only be such as were compatible with the state of a besieged town. Considering the colossal robberies committed upon the city of Paris by the great financial companies and contractors, under the protection of Haussmann, the commune would have had an incomparably better title to confiscate their property than Louis Napoleon had against the Orleans family. The Hohenzollern and the English oligarchs, who both have derived a good deal of their estates from Church plunder, were, of course, greatly shocked at the Commune clearing but 8000 *f.* out of secularization.

While the Versailles Government, as soon as it had recovered some spirit and strength, used the most violent means against the Commune; while it put down the free expression of opinion all over France, even to the forbidding of meetings of delegates from the large towns; while it subjected Versailles and the rest of France to an espionage far surpassing that of the Second Empire; while it burned by its gendarme inquisitors all papers printed at Paris, and sifted all correspondence from and to Paris; while in the National Assembly the most timid attempts to put in a word for Paris were howled down in a manner unknown even to the *Chambre introuvable* [Chamber of Deputies, notorious for its reaction] of 1816; with the savage warfare of Versailles outside, and its attempts at corruption and conspiracy inside Paris—would the Commune not have shamefully betrayed its trust by affecting to keep up all the decencies and appearances of liberalism as in a time of profound peace? Had the Government of the Commune been akin to that of M. Thiers, there would have been no more occasion to suppress Party-of-Order papers at Paris than there was to suppress Communal papers at Versailles.

It was irritating indeed to the Rurals that at the very same time they declared the return to the church to be the only means of salvation for France, the infidel Commune unearthed the peculiar mysteries of the Picpus nunnery, and of the Church of Saint Laurent. It was a satire upon M. Thiers that, while he showered grand crosses upon the Bonapartist generals in acknowledgement of their mastery in losing battles, signing capitulations, and turning cigarettes at Wilhelmshöhe, the Commune dismissed and arrested its generals whenever they were suspected of neglecting their duties. The expulsion from, and arrest by, the Commune of one of its members who had slipped in under a false name, and had undergone at Lyons six days' imprisonment for simple bankruptcy, was it not a deliberate insult hurled at the forger, Jules Favre, then still the foreign minister of France, still selling France to Bismarck, and still dictating his orders to that paragon Government of Belgium? But indeed the Commune did not pretend to infallibility, the invariable attribute of all governments of the old stamp. It published its doings and sayings, it initiated the public into all its shortcomings.

In every revolution there intrude, at the side of its true agents, men of a different stamp; some of them survivors of and devotees to past revolutions, without insight into the present movement, but preserving popular influence by their known honesty and courage, or by the sheer force of tradition; others mere bawlers, who, by dint of repeating year after year the same set of stereotyped declamations against the Government of the day, have sneaked into the reputation of revolutionists of the first water. After the eighteenth of March, some such men did also turn up, and in some cases contrived to play pre-eminent parts. As far as their power went, they hampered the real action of the working class, exactly as men of that sort have hampered the full development of every previous revolution. They are an unavoidable evil: with time they are shaken off; but time was not allowered to the Commune.

Wonderful, indeed, was the change the Commune had wrought in Paris! No longer any trace of the meretricious Paris of the Second Empire. No longer was Paris the rendezvous of British landlords, Irish absentees, American ex-slaveholders, and shoddy men, Russian ex-serfowners, and Wallachian boyards. No more corpses at the morgue, no nocturnal burglaries, scarcely any robberies; in fact, for the first time since the days of February 1848, the streets of Paris were safe, and that without any police of any kind. 'We,' said a member of the Commune, 'hear no longer of assassination, theft, and personal assault; it seems indeed as if the police had dragged along with it to Versailles all its Conservative friends.'

The *cocottes* had refound the scent of their protectors—the absconding men of family, religion, and, above all, of property. In their stead, the real women of Paris showed again at the surface—heroic, noble, and devoted, like the women of antiquity. Working, thinking, fighting, bleeding Paris—almost forgetful, in its incubation of a new society, of the cannibals at its gates—radiant in the enthusiasm of its historic initiative!

Opposed to this new world at Paris, behold the old world at Versailles—that assembly of the ghouls of all defunct regimes, Legitimists and Orleanists, eager to feed upon the carcass of the nation—with a tail of antediluvian Republicans, sanctioning, by their presence in the Assembly, the slaveholders' rebellion, relying for the maintenance of their Parliamentary Republic upon the vanity of the senile mountebank at its head, and caricaturing 1789 by holding their ghastly meetings in the *Jeu de Paume*. There it was, this Assembly, the representative of everything dead in France, propped up to the semblance of life by nothing but the swords of the generals of Louis Bonaparte. Paris all truth, Versailles all lie; and that lie vented through the mouth of Thiers.

Thiers tells a deputation of the mayors of the Seine-et-Oise—'You may rely upon my word, which I have never broken!'

He tells the Assembly itself that 'it was the most freely elected and most Liberal Assembly France ever possessed'; he tells his motley soldiery that it was 'the admiration of the world, and the finest army France ever possessed'; he tells the provinces that the bombardment of Paris by him was a myth: 'If some cannon-shots have been fired, it is not the deed of the army of Versailles, but of some insurgents trying to make believe that they are fighting, while they dare not show their faces.'

He again tells the provinces that 'the artillery of Versailles does not bombard Paris, but only cannonades it'.

He tells the Archbishop of Paris that the pretended executions and reprisals (!) attributed to the Versailles troops were all moonshine. He tells Paris that he was only anxious 'to free it from the hideous tyrants who oppress it', and that, in fact, the Paris of the Commune was 'but a handful of criminals'.

The Paris of M. Thiers was not the real Paris of the 'vile multitude', but a phantom Paris, the Paris of the *francs-fileurs* [runaways], the Paris of the Boulevards, male and female—the rich, the capitalist, the gilded, the idle Paris, now thronging with its lackeys, its black-legs, its literary *bohême*, and its *cocottes* at Versailles, Saint-Denis, Rueil, and Saint-Germain; considering the civil war but an agreeable diversion, eyeing the battle going on through telescopes, counting the rounds of cannon, and swearing by their own honour and that of their prostitutes, that the performance was far better got up then it used to be at the Porte St. Martin. The men who fell were really dead; the cries of the wounded were cries in good earnest; and, besides, the whole thing was so intensely historical. . . .

. . . After Whit-Sunday 1871, there can be neither peace nor truce possible between the working men of France and the appropriators of their produce. The iron hand of a mercenary soldiery may keep for a time both classes tied down in common oppression. But the battle must break out again and again in evergrowing dimensions, and there can be no doubt as to who will be the victor in the end—the appropriating few, or the immense working majority. And the French working class is only the advanced guard of the modern proletariat.

While the European governments thus testify, before Paris, to the international character of class-rule, they cry down the International Working

Men's Association—the international counter-organization of labour against the cosmopolitan conspiracy of capital—as the head fountain of all these disasters. Thiers denounced it as the despot of labour, pretending to be its liberator. Picard ordered that all communications between the French Internationals and those abroad should be cut off; Count Jaubert, Thiers's mummified accomplice of 1835, declares it the great problem of all civilized governments to weed it out. The Rurals roar against it, and the whole European press joins the chorus. An honourable French writer, completely foreign to our Association, speaks as follows:

> The members of the Central Committee of the National Guard, as well as the greater part of the members of the Commune, are the most active, intelligent, and energetic minds of the International Working Men's Association; . . . men who are thoroughly honest, sincere, intelligent, devoted, pure, and fanatical in the good sense of the word.

The police-tinged bourgeois mind naturally figures to itself the International Working Men's Association as acting in the manner of a secret conspiracy, its central body ordering, from time to time, explosions in different countries. Our Association is, in fact, nothing but the international bond between the most advanced working men in the various countries of the civilized world. Wherever, in whatever shape, and under whatever conditions the class struggle obtains any consistency, it is but natural that members of our Association should stand in the foreground. The soil out of which it grows is modern society itself. It cannot be stamped out by any amount of carnage. To stamp it out, the Governments would have to stamp out the despotism of capital over labour—the condition of their own parasitical existence.

Working men's Paris, with its Commune, will be for ever celebrated as the glorious harbinger of a new society. Its martyrs are enshrined in the great heart of the working class. Its exterminators history has already nailed to that eternal pillory from which all the prayers of their priests will not avail to redeem them.

From the Drafts

. . . The centralized state machinery which, with its ubiquitous and complicated military, bureaucratic, clerical, and judiciary organs, entoils (enmeshes) the living civil society like a boa constrictor, was first forged in the days of absolute monarchy as a weapon of nascent modern society in its struggle of emancipation from feudalism. The seigniorial privileges of the medieval lords and cities and clergy were transformed into the attributes of a unitary state power, displacing the feudal dignitaries by salaried state functionaries, transferring the arms from medieval retainers of the landlords and the corporations of townish citizens to a standing army, substituting for the checkered (party-coloured) anarchy of conflicting medieval powers the regulated plan of a state power, with a systematic and hierarchic division of labour. The first French Revolution with its task to found national unity (to create a nation) had to break down all local, territorial, townish, and provincial independence. It was, therefore, forced to develop what absolute monarchy had commenced, the centralization and organization of state power, and to expand the circumference and the attributes of the state power, the number of its tools, its independence, and its supernaturalist sway over real society, which in fact took the place of the medieval supernaturalist heaven with its saints. Every minor solitary interest engendered by the relations of social groups was separated from society itself, fixed and made independent of it and opposed to it in the form of state interest, administered by state priests with exactly determined hierarchical functions.

This parasitical excrescence upon civil society, pretending to be its ideal counterpart, grew to its full development under the sway of the first Bonaparte. The Restoration and the Monarchy of July added nothing to it but a greater division of labour, growing at the same measure in which the division of labour within civil society created new groups of interest, and, therefore, new material for state action. In their struggle against the Revolution of 1848, the parliamentary Republic of France and the governments of all continental Europe were forced to strengthen, with their measures of repression against the popular movement, the means of action and the centralization of that governmental power. All revolutions thus only perfected the state machinery instead of throwing off this deadening incubus. The factions and parties of the ruling classes, which alternately struggled for supremacy, considered the occupancy (control) (seizure) and the direction of this immense machinery of government as the main booty of the victor. It centred in the creation of immense standing armies, a host of state vermin, and huge national debts. During the time of the absolute monarchy it was a means of the struggle of modern society against feudalism, crowned by the French Revolution, and under the first Bonaparte it served not only to subjugate the Revolution and annihilate all popular liberties; it was an instrument of the French Revolution to strike abroad, to create for France on the Continent, instead of feudal monarchies, more or less states after the image of France. Under the Restoration and the Monarchy of July it became not only a means of the forcible class domination of the middle class, and a means of adding to the direct economic exploitation a second exploitation of the people by assuring to their families all the rich places of the state household. During the time of the revolutionary struggle of 1848 at last it served as a means of annihilating that Revolution and all aspirations for the emancipation of the popular masses. But the state parasite received only its last development during the Second Empire. The governmental power with its standing army, its all-directing bureaucracy, its stultifying clergy, and its servile tribunal hierarchy had grown so independent of society itself that a grotesquely mediocre adventurer with a hungry band of desperadoes behind him sufficed to wield it. It did not any longer want the pretext of an armed coalition of old Europe against the modern world founded by the Revolution of 1789. It appeared no longer as a means of class domination, subordinate to its parliamentary ministry or legislature. Humbling under its sway even the interests of the ruling classes, whose parliamentary show work it supplanted by self-elected *Corps Législatifs* and self-paid senates, sanctioned in its absolute sway by universal suffrage, the acknowledged necessity for keeping up 'order', that is, the rule of the landowner and the capitalist over the producer, cloaking under the tatters of a masquerade of the past the orgies of the corruption of the present and the victory of the most parasite faction, the financial swindler, the debauchery of all the reactionary influences of the past let loose—a pandemonium of infamies—the state power had received its last and supreme expression in the Second Empire. Apparently the final victory of this governmental power over society, it was in fact the orgy of all the corrupt elements of that society. To the eye of the uninitiated it appeared only as the victory of the Executive over the Legislative, of the final defeat of the form of class rule pretending to be the autocracy of society [by] its form pretending to be a superior power to society. But in fact it was only the last degraded and the only possible form of that class rule, as humiliating to those classes themselves as to the working classes which they kept fettered by it.

The 4th of September was only the revindication of the Republic against the grotesque adventurer that had assassinated it. The true antithesis to the Empire itself—that is, the state power, the centralized Executive, of which the Second Empire was only the exhausting formula—was the Commune. This state power forms in fact the creation of the middle class, first a means to break down feudalism, then a means to crush the emancipatory aspirations of the producers, of the working class. All reactions and all revolutions had only served to transfer that organized power—that organized force of the slavery of labour—from one hand to the other, from one faction of the ruling classes to the other. It had served the ruling classes as a means of subjugation and of pelf. It had sucked new forces from every new change. It had served as the instrument of breaking down every popular rise and served it to crush the working classes after they had fought and been ordered to secure its transfer from one part of its oppressors to the other. This was, therefore, a Revolution not against this or that Legitimate, Constitutional, Republican, or Imperialist form of state power. It was a Revolution against the State itself, of this supernaturalist abortion of society, a resumption by the people for the people of its own social life. It was not a Revolution to transfer it from one faction of the ruling classes to the other, but a Revolution to break down this horrid machinery of class domination itself. It was not one of those dwarfish struggles between the executive and the parliamentary forms of class domination, but a revolt against both these forms, integrating each other, and of which the parliamentary form was only the deceitful bywork of the executive. The Second Empire was the final form of this state usurpation. The Commune was its definite negation and, therefore, the initiation of the social Revolution of the nineteenth century. Whatever, therefore, its fate at Paris, it will make *le tour du monde* [world tour]. It was at once acclaimed by the working class of Europe and the United States as the magic word of delivery. The glories and the antediluvian deeds of the Prussian conqueror seemed only hallucinations of a bygone past.

It was only the working class that could formulate by the word 'Commune' and initiate by the fighting Commune of Paris this new aspiration. Even the last expression of that state power in the Second Empire, although humbling for the pride of the ruling classes and casting to the winds their parliamentary pretensions of self-government, had been only the last possible form of their class rule. While politically dispossessing them, it was the orgy under which all the economic and social infamies of their regime got full sway. The middling bourgeoisie and the petty middle class were by their economical conditions of life excluded from initiating a new revolution and induced to follow in the track of the ruling classes or [be] the followers of the working class. The peasants were the passive economical basis of the Second Empire, of that last triumph of a State separate from and independent of society. Only the proletarians, fired by a new social task to accomplish by them for all society, to do away with all classes and class rule, were the men to break the instrument of that class rule—the State, the centralized and organized governmental power usurping to be the master instead of the servant of society. In the active struggle against them by the ruling classes, supported by the passive adherence of the peasantry, the Second Empire—the last crowning and at the same time the most signal prostitution of the State, which had taken the place of the medieval church— had been engendered. It had sprung into life against them. By them it was broken, not as a peculiar form of governmental (centralized) power, but as its most powerful, elaborated into seeming independence from society, expression, and, therefore, also its most prostitute reality, covered with infamy from top to bottom, having centred in absolute corruption at home and absolute powerlessness abroad.

But this one form of class rule had only broken down to make the Executive, the governmental state machinery, the great and single object of attack to the Revolution.

Parliamentarism in France had come to an end. Its last term and fullest sway was the parliamentary Republic from May 1848 to the *coup d'état*. The Empire that killed it, was its own creation. Under the Empire with its *Corps Législatif* and its Senate—and in this form it has been reproduced in the military monarchies of Prussia and Austria—it had been a mere farce, a mere by-work of despotism in its crudest form. Parliamentarism then was dead in France and the workmen's Revolution certainly was not to awaken it from the death.

The Commune—the reabsorption of the state power by society as its own living forces instead of as forces controlling and subduing it, by the popular masses themselves, forming their own force instead of the organized force of their suppression—the political form of their social emancipation, instead of the artificial force appropriated by their oppressors (their own force opposed to and organized against them) of society wielded for their oppression by their enemies. This form was simple like all great things. The reaction of former revolutions—the time wanted for all historical developments, and in the past always lost in all revolutions in the very days of popular triumph, whenever it had rendered its victorious arms to be turned against itself—[the Commune] first displaced the army by the National Guard. 'For the first time since the 4th September the Republic is liberated from the government of its enemies. . . . In the city [is] a national militia that defends the citizens against the power (the government) instead of a permanent army that defends the government against the citizens.' (Proclamation of Central Committee of 22 March.)

(The people had only to organize this militia on a national scale, to have done away with the standing armies; the first economical *conditio sine qua non* [essential condition] for all social improvements, discarding at once this source of taxes and state debt, and this constant danger of government usurpation of class rule—of the regular class rule or an adventurer pretending to save all classes); at the same time the safest guarantee against foreign aggression and making in fact the costly military apparatus impossible in all other states; the emancipation of the peasant from the blood-tax and [from being] the most fertile source of all state taxation and state debts. Here already [is] the point in which the Commune is a bait for the peasant, the first word of his emancipation. With the 'independent police' abolished, and its ruffians supplanted by servants of the Commune. The general suffrage, till now abused either for the parliamentary sanction of the Holy State Power, or a play in the hands of the ruling classes, only employed by the people to sanction (choose the instruments of) parliamentary class rule once in many years, adapted to its real purposes, to choose by the Communes their own functionaries of administration and initiation. [Gone is] the delusion as if administration and political governing were mysteries, transcendent functions only to be trusted to the hands of a trained caste—state parasites, richly paid sycophants and sinecurists, in the higher posts, absorbing the intelligence of the masses and turning them against themselves in the lower places of the hierarchy. Doing away with the state hierarchy altogether and replacing the haughteous masters of the people by its always removable servants, a mock responsibility by a real responsibility, as they act continuously under public supervision. Paid like skilled workmen, £12 a month, the highest salary not exceeding £240 a year, a salary somewhat more than a fifth, according to a great scientific authority, Professor Huxley, [needed] to satisfy a clerk for the Metropolitan School Board. The whole sham of state mysteries and state pretensions was done away [with] by a Commune, mostly consisting of simple working men, organizing the defence of Paris, carrying on war against the pretorians of Bonaparte, securing the supplies for that immense town, filling all the posts hitherto divided between government, police, and prefecture, doing their work publicly, simply, under the most difficult and complicated circumstances, and doing it, as Milton did his *Paradise Lost*, for a few pounds, acting in bright daylight, with no pretensions to infallibility, not hiding itself behind circumlocution offices, not ashamed to confess blunders by correcting them. Making in one order the public functions—military, administrative, political—real workmen's functions, instead of the hidden attributes of a trained caste; (keeping order in the turbulence of civil war and revolution) (initiating measures of general regeneration). Whatever the merits of the single measures of the Commune, its greatest measure was its own organization, extemporized with the foreign enemy at one door, and the class enemy at the other, proving by its life its vitality, confirming its theories by its action. Its appearance was a victory over the victors of France. Captive Paris resumed by one bold spring the leadership of Europe, not depending on brute force, but by taking the lead of the social movement, by giving body to the aspirations of the working class of all countries.

With all the great towns organized into Communes after the model of Paris, no government could have repressed the movement by the surprise of sudden reaction. Even by this preparatory step the time of incubation, the guarantee of the movement, won. All France would have been organized into self-working and self-governing communes, the standing army replaced by the popular militias, the army of state parasites removed, the clerical hierarchy displaced by the schoolmasters, the state judge transformed into Communal organs, the suffrage for national representation not a matter of sleight of hand for an all-powerful government, but the deliberate expression of the organized communes, the state functions reduced to a few functions for general national purposes.

Such is the Commune—the political form of the social emancipation, of the liberation of labour from the usurpation (slaveholding) of the monopolists of the means of labour, created by the labourers themselves or forming the gift of nature. As the state machinery and parliamentarism are not the real life of the ruling classes, but only the organized general organs of their dominion, the political guarantees and forms the expressions of the old order of things, so the Commune is not the social movement of the working class and, therefore, of a general regeneration of mankind, but the organized means of action. The Commune does not [do] away with the class struggles, through which the working classes strive for the abolition of all classes, and, therefore, of all [class rule] (because it does not represent a peculiar interest. It represents the liberation of 'labour', that is, the fundamental and natural condition of individual and social life which only by usurpation, fraud, and artificial contrivances can be shifted from the few upon the many), but it affords the rational medium in which that class struggle can run through its different phases in the most rational and humane way. It could start violent reactions and as violent revolutions. It begins the emancipation of labour —its great goal—by doing away with the unproductive and mischievous work of the state parasites, by cutting away the springs which sacrifice an immense portion of the national produce to the feeding of the state monster, on the one side, by doing, on the other, the real work of administration, local and national, for working men's wages. It begins therefore with an immense saving, with economical reform as well as political transformation.

The Communal organization once firmly established on a national scale, the catastrophes it might still have to undergo would be sporadic slaveholders' insurrections, which, while for a moment interrupting the work of peaceful progress, would only accelerate the movement, by putting the sword into the hand of the Social Revolution.

The working classes know that they have to pass through different phases of class struggle. They know that the superseding of the economical conditions of the slavery of labour by the conditions of free and associated labour can only be the progressive work of time (that economical transformation), that they require not only a change of distribution, but a new organization of production, or rather the delivery (setting free) of the social forms of production in present organized labour (engendered by present industry) of the trammels of slavery, of their present class character, and their harmonious national and international co-ordination. They know that this work of regeneration will be again and again relented and impeded by the resistance of vested interests and class egotisms. They know that the present 'spontaneous action of the natural laws of capital and landed property' can only be superseded by 'the spontaneous action of the laws of the social economy of free and associated labour' in a long process of development of new conditions, as was the 'spontaneous action of the economic laws of slavery' and the 'spontaneous action of the economical laws of serfdom'. But they know at the same time that great strides may be [made] at once through the Communal form of political organization and that the time has come to begin that movement for themselves and mankind. . . .

BIBLIOGRAPHY

Original (in English): *MESW*, Vol. 1, pp. 499 ff. Drafts: K. Marx and F. Engels, *On the Paris Commune*, Moscow, 1971. Reproduced by kind permission of Lawrence and Wishart Ltd.

Commentaries: S. Avineri, *The Social and Political Thought of Karl Marx*, Cambridge, 1968, pp. 239 ff.

H. Collins and C. Abramsky, *Karl Marx and the British Labour Movement*, London, 1965, pp. 57 ff.

H. Draper, Introduction to K. Marx and F. Engels, *Writings on the Paris Commune*, New York, 1971.

C. Hitchens, Introduction to *Marx on the Paris Commune*, London, 1971.

M. Johnstone, 'Marx and the Dictatorship of the Proletariat', *Massachussetts Review*, 1972.

B. Wolfe, *Marxism, 100 Years in the Life of a Doctrine*, London, 1967, Part 3.

38. Preface to the Second German Edition of the *Communist Manifesto*

This preface was hastily written in 1872 when the need for a second edition became obvious. Its importance lies chiefly in the modifications that the experience of the Paris Commune led Marx to make in his previous ideas.

... However much the state of things may have altered during the last twenty-five years, the general principles laid down in this Manifesto are, on the whole, as correct today as ever. Here and there some detail might be improved. The practical application of the principles will depend, as the Manifesto itself states, everywhere and at all times, on the historical conditions for the time being existing, and, for that reason, no special stress is laid on the revolutionary measures proposed at the end of Section II. That passage would, in many respects, be very differently worded today. In view of the gigantic strides of Modern Industry in the last twenty-five years, and of the accompanying improved and extended party organization of the working class, in view of the practical experience gained, first in the February Revolution, and then, still more, in the Paris Commune, where the proletariat for the first time held political power for two whole months, this programme has in some details become antiquated. One thing especially was proved by the Commune, viz., that 'the working class cannot simply lay hold of the ready-made State machinery, and wield it for its own purposes'. (See *The Civil War in France*; *Address of the General Council of the International Working Men's Association*, London, Truelove, 1871, p. 15, where this point is further developed.) Further, it is self-evident that the criticism of socialist literature is deficient in relation to the present time, because it comes down only to 1847; also, that the remarks on the relation of the Communists to the various opposition parties (Section IV), although in principle still correct, yet in practice are antiquated, because the political situation has been entirely changed, and the progress of history has swept from off the earth the greater portion of the political parties there enumerated.

But, then, the Manifesto has become a historical document which we have no longer any right to alter. A subsequent edition may perhaps appear with an introduction bridging the gap from 1847 to the present day; this reprint was too unexpected to leave us time for that. ...

BIBLIOGRAPHY

Original: *MEW*, Vol. 4, pp. 573 f.

Present translation: *MESW*, Vol. 1, pp. 21 f. Reproduced by kind permission of Lawrence and Wishart Ltd.

39. On Bakunin's *Statism and Anarchy*

The following excerpts are taken from marginal jottings that Marx made in late 1874 while reading Bakunin's book *Statism and Anarchy*. They are particularly interesting for Marx's apparent optimism about the prospects for revolution in countries where the majority of the population still consisted of peasants and for his remarks on government in communist society.

Bakunin: Once the proletariat is the ruling class, over whom will it rule?

Marx: This means, that as long as the other classes, and in particular the capitalist class, still exist, as long as the proletariat is still struggling with it (because, with the proletariat's conquest of governmental power its enemies and the old organization of society have not yet disappeared), it must use coercive means, hence governmental means; it is still a class and the economic conditions on which the class struggle and the existence of classes depend, have not yet disappeared and must be removed by force, or transformed and their process of transformation speeded up by force. . . .

Where the mass of the peasants are still owners of private property, where they even form a more or less important majority of the population, as they do in the states of the Western European continent, where they have not yet disappeared and been replaced by agricultural wage labourers, as in England; in these cases the following situation arises: either the peasantry hinders every workers' revolution and causes it to fail, as it has done in France up till now; or the proletariat (for the landowning peasant does not belong to the proletariat and even when his own position causes him to belong to it, he does not think he belongs to it) must as a government inaugurate measures which directly improve the situation of the peasant and which thus win him for the revolution; measures which in essence facilitate the transition from private to collective property in land so that the peasant himself is converted for economic reasons; the proletariat must not, however, come into open collision with peasantry by, for example, proclaiming the abolition of inheritance or the abolition of property; this latter is only possible where the capitalist landlord has expropriated the peasant and the real worker of the land is just as much a proletarian wage labourer as the city worker, and thus has directly the same interests. . . .

Schoolboy's asininity! A radical social revolution is tied to certain historical conditions of economic development; these are its prerequisites. It is therefore only possible where, with capitalist production, the industrial proletariat occupies at least a significant position among the mass of the people. And so in order to have any chance whatever of victory, it must at least be able to do as much immediately for the peasants, *mutatis mutandis*, as the French bourgeoisie did in its revolution for the then existing French peasants. A fine idea, that the rule of labour includes the suppression of rural labour!

But there the innermost thought of Mr. Bakunin comes to light. He does not understand a thing about social revolution, only the political phrases about it; its economic conditions do not exist for him. Now since all hitherto existing economic forms, developed or undeveloped, include the servitude of the worker (be it in the form of the wage-worker, peasant, etc.) he believes that in all of them a radical revolution is equally possible. But even more! He wants the European social revolution, founded on the economic basis of capitalist production, to take place at the level of the Russian or Slav agricultural and pastoral people. Will, not economic conditions, is the foundation of his social revolution. . . .

Bakunin: What does it mean to say that the proletariat is organized as a ruling class?

Marx: It means that the proletariat, instead of fighting piecemeal against the economically privileged classes, has obtained enough strength and organization to use general means of forcibly expressing itself in this struggle; but it can only use economic means which abolish its own character as wage-labourers, that is as a class; with its complete victory, therefore, its domination is at an end because its character as a class has disappeared.

Bakunin: Will, perhaps, the whole of the proletariat be at the head of the government?

Marx: In a trade union, for example, is the executive committee composed of the whole of the union? Will all division of labour and the different functions that it entails disappear? And in Bakunin's construction from the bottom to the top will everyone be at the top? Then there will be no bottom. Will all members of the Commune manage the common interests of the enterprise at the same time? Then there is no distinction between enterprise and commune.

Bakunin: There are about 40 million Germans. Will, for example, all the forty million be members of the government?

Marx: Certainly! For the thing begins with the self-government of the Commune.

Bakunin: The whole people will govern and there will be no one to be governed.

Marx: According to this principle, when a man rules himself, he does not rule himself; since he is only himself and no one else.

Bakunin: Then there will be no government, no State, but if there is a State in existence there will also be governors and slaves.

Marx: This merely means: when class rule has disappeared, there will no longer be any state in the present political sense of the word. . . .

Marx: Asinine! This is democratic verbiage, political drivel! An election is a political form, both in the smallest Russian commune and in the Artel. The character of the election does not depend on this description, but on the economic basis, the economic interrelations of the electors, and as soon as the functions have ceased to be political, then there exists (1) no governmental function; (2) the distribution of general functions has become a business matter which does not afford any room for domination; (3) the election has none of its present political character.

Bakunin: Universal suffrage by the whole people of representatives and rulers of the State—this is the last word of the Marxists as well as of the democratic school. They are lies behind which lurks the despotism of a governing minority, lies all the more dangerous in that this minority appears as the expression of the so-called people's will.

Marx: Under collective property, the so-called will of the people disappears in order to make way for the real will of the co-operative.

Bakunin: Result: rule of the great majority of the people by a privileged minority. But, the Marxists say, this minority will consist of workers. Yes, indeed, but of ex-workers, who, once they become only representatives or rulers of the people, cease to be workers.

Marx: No more than a manufacturer today ceases to be a capitalist when he becomes a member of the municipal council.

Bakunin: And from the heights of the State they begin to look down upon the whole common world of the workers. From that time on they represent not the people but themselves and their own claims to govern the people. Those who can doubt this know nothing at all about human nature.

Marx: If Mr. Bakunin were in the know, if only with the position of a manager in a workers' co-operative, he would send all his nightmares about authority to the devil. He should have asked himself: what form can administrative functions assume on the basis of that workers' state, if it pleases him to call it thus? . . .

BIBLIOGRAPHY

Original: *MEW*, Vol. 18, pp. 634 ff.

Present translation: by the editor.

Other translations: In H. Mayer, 'Marx on Bakunin: A Neglected Text', *Cahiers de l'ISEA*, 1959.

Commentaries: Article by Mayer cited above.

40. Critique of the Gotha Programme

In May 1875 the two wings of the German socialists—the one led by Marx's disciple Liebknecht and the other consisting of the followers of Lassalle—met at Gotha, a small town in central Germany, and decided to unite on a common programme. Marx disapproved of the programme, which he thought made too many concessions to the Lassalleans, and sent the following comments to Liebknecht for private circulation among the members of his party. They contain two main points: the first was a criticism of the programme's proposals for distributing the national product on the basis of vague phrases about 'equality' and 'the proceeds of labour'. The second point was an attack on the notion of a 'free state', which Marx found to be a contradiction in terms. Together they add up to Marx's most important statement on organization in the future communist society.

. . . Free state—what is this?

It is by no means the aim of the workers, who have got rid of the narrow mentality of humble subjects, to set the state free. In the German Empire the 'state' is almost as 'free' as in Russia. Freedom consists in converting the state from an organ superimposed upon society into one completely subordinate to it, and today, too, the forms of state are more free or less free to the extent that they restrict the 'freedom of the state'.

The German workers' party—at least if it adopts the programme—shows that its socialist ideas are not even skin-deep; in that, instead of treating existing society (and this holds good for any future one) as the basis of the existing state (or of the future state in the case of future society), it treats the state rather as an independent entity that possesses its own intellectual, ethical, and libertarian bases.

And what of the riotous misuse which the programme makes of the words 'present-day state', 'present-day society', and of the still more riotous misconception it creates in regard to the state to which it addresses its demands?

'Present-day society' is capitalist society, which exists in all civilized countries, more or less free from medieval admixture, more or less modified by the particular historical development of each country, more or less developed. On the other hand, the 'present-day state' changes with a country's frontier. It is different in the Prusso-German Empire from what it is in Switzerland, and different in England from what it is in the United States. 'The present-day state' is, therefore, a fiction.

Nevertheless, the different states of the different civilized countries, in spite of their motley diversity of form, all have this in common, that they are based on modern bourgeois society, only one more or less capitalistically developed. They have, therefore, also certain essential characteristics in common. In this sense it is possible to speak of the 'present-day states', in contrast with the future, in which its present root, bourgeois society, will have died off.

The question then arises: what transformation will the state undergo in communist society? In other words, what social functions will remain in existence there that are analogous to present state functions? This question can only be answered scientifically, and one does not get a flea-hop nearer to the problem by a thousandfold combination of the word people with the word state.

Between capitalist and communist society lies the period of the revolutionary transformation of the one into the other. Corresponding to this is also a political transition period in which the state can be nothing but the revolutionary dictatorship of the proletariat.

Now the programme does not deal with this nor with the future state of communist society.

Its political demands contain nothing beyond the old democratic litany familiar to all: universal suffrage, direct legislation, popular rights, a people's militia, etc. They are a mere echo of the bourgeois People's Party, of the League of Peace and Freedom. They are all demands which, in so far as they are not exaggerated in fantastic presentation, have already been realized. Only the state to which they belong does not lie within the borders of the German Empire, but in Switzerland, the United States, etc. This sort of 'state of the future' is a present-day state, although existing outside the 'framework' of the German Empire.

But one thing has been forgotten. Since the German workers' party expressly declares that it acts within 'the present-day national state', hence within its own state, the Prusso-German Empire—its demands would indeed otherwise be largely meaningless, since one only demands what one has not got—it should not have forgotten the chief thing, namely, that all those pretty little gewgaws rest on the recognition of the so-called sovereignty of the people and hence are appropriate only in a democratic republic.

Since one has not the courage—and wisely so, for the circumstances demand caution—to demand the democratic republic, as the French workers' programmes under Louis Philippe and under Louis Napoleon did, one should not have resorted, either, to the subterfuge, neither 'honest' nor decent, of demanding things which have meaning only in a democratic republic from a state which is nothing but a police-guarded military despotism, embellished with parliamentary forms, alloyed with a feudal admixture, already influenced by the bourgeoisie and bureaucratically carpentered, and then to assure this state into the bargain that one imagines one will be able to force such things upon it 'by legal means'.

Even vulgar democracy, which sees the millennium in the democratic republic and has no suspicion that it is precisely in this last form of state of bourgeois society that the class struggle has to be fought out to a confusion —even it towers mountains above this kind of democratism which keeps within the limits of what is permitted by the police and not permitted by logic.

That, in fact, by the word 'state' is meant the government machine, or the state in so far as it forms a special organism separated from society through division of labour, is shown by the words 'the German workers' party demands as the economic basis of the state: a single progressive income tax', etc. Taxes are the economic basis of the government machinery and of nothing else. In the state of the future, existing in Switzerland, this demand has been pretty well fulfilled. Income tax presupposes various sources of income of the various social classes, and hence capitalist society. It is, there-fore, nothing remarkable that the Liverpool financial reformers, bourgeois headed by Gladstone's brother, are putting forward the same demand as the programme. . . .

'The emancipation of labour demands the promotion of the instruments of labour to the common property of society and the co-operative regulation of the total labour with a fair distribution of the proceeds of labour.'

'Promotion of the instruments of labour to the common property' ought obviously to read their 'conversion into the common property'; but this only in passing.

What are 'proceeds of labour'? The product of labour or its value? And in the latter case, is it the total value of the product or only that part of the value which labour has newly added to the value of the means of production consumed?

'Proceeds of labour' is a loose notion which Lassalle has put in the place of definite economic conceptions.

What is 'a fair distribution'?

Do not the bourgeois assert that the present-day distribution is 'fair'? And is it not, in fact, the only 'fair' distribution on the basis of the present-day mode of production? Are economic relations regulated by legal conceptions or do not, on the contrary, legal relations arise from economic ones? Have not also the socialist sectarians the most varied notions about 'fair' distribu-tion?

To understand what is implied in this connection by the phrase 'fair distribution', we must take the first paragraph and this one together. The latter presupposes a society wherein 'the instruments of labour are common property and the total labour is co-operatively regulated', and from the first paragraph we learn that 'the proceeds of labour belong undiminished with equal right to all members of society'.

'To all members of society'? To those who do not work as well? What remains then of the 'undiminished proceeds of labour'? Only to those members of society who work? What remains then of the 'equal right' of all members of society?

But 'all members of society' and 'equal right' are obviously mere phrases. The kernel consists in this, that in this communist society every worker must receive the 'undiminished' Lassallean 'proceeds of labour'.

Let us take first of all the words 'proceeds of labour' in the sense of the product of labour; then the co-operative proceeds of labour are the total social product.

From this must now be deducted:

First, cover for replacement of the means of production used up.

Secondly, additional portion for expansion of production.

Thirdly, reserve or insurance funds to provide against accidents, dislocations caused by natural calamities, etc.

These deductions from the 'undiminished proceeds of labour' are an economic necessity and their magnitude is to be determined according to available means and forces, and partly by computation of probabilities, but they are in no way calculable by equity.

There remains the other part of the total product, intended to serve as means of consumption.

Before this is divided among the individuals, there has to be deducted again from it:

First, the general costs of administration not belonging to production.

This part will, from the outset, be very considerably restricted in comparison with present-day society and it diminishes in proportion as the new society develops.

Secondly, that which is intended for the common satisfaction of needs, such as schools, health services, etc.

From the outset this part grows considerably in comparison with present-day society and it grows in proportion as the new society develops.

Thirdly, funds for those unable to work, etc., in short, for what is included under so-called official poor relief today.

Only now do we come to the 'distribution' which the programme, under Lassallean influence, alone has in view in its narrow fashion, namely, to that part of the means of consumption which is divided among the individual producers of the co-operative society.

The 'undiminished proceeds of labour' have already unnoticeably become converted into the 'diminished' proceeds, although what the producer is deprived of in his capacity as a private individual benefits him directly or indirectly in his capacity as a member of society.

Just as the phrase of the 'undiminished proceeds of labour' has disappeared, so now does the phrase of the 'proceeds of labour' disappear altogether.

Within the co-operative society based on common ownership of the means of production, the producers do not exchange their products; just as little does the labour employed on the products appear here as the value of these products, as a material quality possessed by them, since now, in contrast to capitalist society, individual labour no longer exists in an indirect fashion but directly as a component part of the total labour. The phrase 'proceeds of labour', objectionable also today on account of its ambiguity, thus loses all meaning.

What we have to deal with here is a communist society, not as it has developed on its own foundations, but, on the contrary, just as it emerges from capitalist society; which is thus in every respect, economically, morally, and intellectually, still stamped with the birth marks of the old society from whose womb it emerges. Accordingly, the individual producer receives back from society—after the deductions have been made—exactly what he gives to it. What he has given to it is his individual quantum of labour. For example, the social working day consists of the sum of the individual hours of work; the individual labour time of the individual producer is the part of the social working day contributed by him, his share in it. He receives a certificate from society that he has furnished such and such an amount of labour (after deducting his labour for the common funds), and with this certificate he draws from the social stock of means of consumption as much as costs the same amount of labour. The same amount of labour which he has given to society in one form he receives back in another.

Here obviously the same principle prevails as that which regulates the exchange of commodities, as far as this is exchange of equal values. Content and form are changed, because under the altered circumstances no one can give anything except his labour, and because, on the other hand, nothing can pass to the ownership of individuals except individual means of consumption. But, as far as the distribution of the latter among the individual producers is concerned, the same principle prevails as in the exchange of commodity-equivalents: a given amount of labour in one form is exchanged for an equal amount of labour in another form.

Hence, equal right here is still in principle—bourgeois right, although principle and practice are no longer at loggerheads, while the exchange of equivalents in commodity exchange only exists on the average and not in the individual case.

In spite of this advance, this equal right is still constantly stigmatized by a bourgeois limitation. The right of the producers is proportional to the labour they supply; the equality consists in the fact that measurement is made with an equal standard, labour.

But one man is superior to another physically or mentally and so supplies more labour in the same time, or can labour for a longer time; and labour, to serve as a measure, must be defined by its duration or intensity, otherwise it ceases to be a standard of measurement. This equal right is an unequal right for unequal labour. It recognizes no class differences, because everyone is only a worker like everyone else; but it tacitly recognizes unequal individual endowment and thus productive capacity as natural privileges. It is, there-fore, a right of inequality, in its content, like every right. Right by its very nature can consist only in the application of an equal standard; but unequal individuals (and they would not be different individuals if they were not unequal) are measurable only by an equal standard in so far as they are brought under an equal point of view, are taken from one definite side only, for instance, in the present case, are regarded only as workers and nothing more is seen in them, everything else being ignored. Further, one worker is married, another not; one has more children than another, and so on and so forth. Thus, with an equal performance of labour, and hence an equal share in the social consumption fund, one will in fact receive more than another, one will be richer than another, and so on. To avoid all these defects, right instead of being equal would have to be unequal.

But these defects are inevitable in the first phase of communist society as it is when it has just emerged after prolonged birth pangs from capitalist society. Right can never be higher than the economic structure of society and its cultural development conditioned thereby.

In a higher phase of communist society, after the enslaving subordination of the individual to the division of labour, and therewith also the antithesis between mental and physical labour, has vanished; after labour has become not only a means of life but life's prime want; after the productive forces have also increased with the all-round development of the individual, and all the springs of co-operative wealth flow more abundantly—only then can the narrow horizon of bourgeois right be crossed in its entirety and society inscribe on its banners: from each according to his ability, to each according to his needs!

I have dealt more at length with the 'undiminished proceeds of labour', on the one hand, and with 'equal right' and 'fair distribution', on the other, in order to show what a crime it is to attempt, on the one hand, to force on our Party again, as dogmas, ideas which in a certain period had some meaning but have now become obsolete verbal rubbish, while again perverting, on the other, the realistic outlook, which it cost so much effort to instil into the Party but which has now taken root in it, by means of ideological nonsense about right and other trash so common among the democrats and French Socialists.

Quite apart from the analysis so far given, it was in general a mistake to make a fuss about so-called distribution and put the principal stress on it.

Any distribution whatever of the means of consumption is only a con-sequence of the distribution of the conditions of production themselves. The latter distribution, however, is a feature of the mode of production itself. The capitalist mode of production, for example, rests on the fact that the material conditions of production are in the hands of non-workers in the form of property in capital and land, while the masses are only owners of the personal condition of production, of labour power. If the elements of production are so distributed, then the present-day distribution of the means of consumption results automatically. If the material conditions of production are the co-operative property of the workers themselves, then there likewise results a distribution of the means of consumption different from the present one. Vulgar socialism (and from it in turn a section of the democracy) has taken over from the bourgeois economists the consideration and treatment of distribution as independent of the mode of production and hence the presentation of socialism as turning principally on distribution. After the real relation has long been made clear, why retrogress again? . . .

BIBLIOGRAPHY

Original: *MEW*, Vol. 19, pp. 18 ff.

Present translation: *MESW*, Vol. 2, pp. 29 ff. Reproduced by kind permission of Lawrence and Wishart Ltd.

Other translations: None.

Commentaries: G. D. H. Cole, *A History of Socialist Thought*, London, 1961, Vol. 2, pp. 244 ff.

41. Letter to Mikhailovsky

In 1877 one of the leading Populist theoreticians in Russia, Mikhailovsky, criticized *Capital* on the grounds that it condemned the efforts of Russians who worked for a development in their country which would bypass the capitalist stage. Marx defended himself in the following letter which claimed that the question was an open one, and that his materialist conception of history was not some sort of 'formula' that could be applied irrespective of particular circumstances.

... The chapter on primitive accumulation does not pretend to do more than trace the path by which, in Western Europe, the capitalist order of economy emerged from the womb of the feudal order of economy. It therefore describes the historical movement which by divorcing the producers from their means of production converts them into wage workers (proletarians in the modern sense of the word) while it converts those who possess the means of production into capitalists. In that history 'all revolutions are epoch-making that act as levers for the advancement of the capitalist class in course of formation; above all, those which, by stripping great masses of men of their traditional means of production and subsistence, suddenly hurl them on the labour market. But the basis of this whole development is the expropriation of the agricultural producer. This has been accomplished in radical fashion only in England ... but all the countries of Western Europe are going through the same movement', etc. (*Capital*, French edition, page 315.) At the end of the chapter the historical tendency of production is summed up thus: That it 'itself begets its own negation with the inexorability which governs the metamorphoses of nature'; that it has itself created the elements of a new economic order, by giving the greatest impulse at once to the productive forces of social labour and to the integral development of every individual producer; that capitalist property, resting already, as it actually does, on a collective mode of production, cannot but transform itself into social property. At this point I have not furnished any proof, for the good reason that this statement is itself nothing else but a general summary of long expositions previously given in the chapters on capitalist production.

Now what application to Russia could my critic make of this historical sketch? Only this: If Russia is tending to become a capitalist nation after the example of the West European countries—and during the last few years she has been taking a lot of trouble in this direction—she will not succeed without having first transformed a good part of her peasants into proletarians; and after that, once taken to the bosom of the capitalist regime, she will experience its pitiless laws like other profane peoples. That is all. But that is too little for my critic. He feels he absolutely must metamorphose my historical sketch of the genesis of capitalism in Western Europe into a historico-philosophic theory of the general path every people is fated to tread, whatever the historical circumstances in which it finds itself, in order that it may ultimately arrive at the form of economy which ensures, together with the greatest expansion of the productive powers of social labour, the most complete development of man. But I beg his pardon. (He is both honouring and shaming me too much.) Let us take an example.

In several parts of *Capital* I allude to the fate which overtook the plebeians of ancient Rome. They were originally free peasants, each cultivating his own piece of land on his own account. In the course of Roman history they were expropriated. The same movement which divorced them from their means of production and subsistence involved the formation not only of big landed property but also of big money capital. And so one fine morning there were to be found on the one hand free men, stripped of everything except their labour power, and on the other, in order to exploit this labour, those who held all the acquired wealth in their possession. What happened? The Roman proletarians became not wage-labourers but a mob of do-nothings more abject than the former 'poor whites' in the South of the United States, and alongside of them there developed a mode of production which was not capitalist but based on slavery. Thus events strikingly analogous but taking place in different historical surroundings led to totally different results. By studying each of these forms of evolution separately and then comparing them one can easily find the clue to this phenomenon, but one will never arrive there by using as one's master key a general historico-philosophical theory, the supreme virtue of which consists in being super-historical. . . .

BIBLIOGRAPHY

Original: *MEW*, Vol. 19, p. 107.

Present translation: *MESC*, pp. 311 ff. [Reproduced by kind permission of Lawrence and Wishart Ltd.]

Other translations: None.

Commentaries: J. Billington, *Mikhailovsky and Russian Populism*, Oxford, 1958.

E. Hobsbawm, Introduction to *Pre-Capitalist Economic Formations*, New York and London, 1964.

J. Sanderson, *An Interpretation of the Political Ideas of Marx and Engels*, London, 1969, Appendix.

F. Venturi, *The Roots of Revolution*, New York and London, 1960.

A. Walicki, *The Controversy over Capitalism*, Oxford, 1969.

42. Circular Letter

The following excerpts are from a long letter sent by Marx and Engels in 1879 to Bebel, Liebknecht, and other leaders of the German Social Democratic Party. Marx and Engels oppose tendencies in the Party that they see as too quietist and not enough concerned with class struggle and revolution.

... It is the representatives of the petty bourgeoisie who are here making themselves heard, full of anxiety that the proletariat, under the pressure of its revolutionary position, may 'go too far'. Instead of determined political opposition, general mediation; instead of struggle against government and bourgeoisie, an attempt to win over and persuade them; instead of defiant resistance to ill-treatment from above, humble submission and confession that the punishment was deserved. Historically necessary conflicts are all interpreted as misunderstandings, and all discussion ends with the assurance that after all we are all agreed on the main point. The people who came out as bourgeois democrats in 1848 could just as well call themselves Social-Democrats now. To the former the democratic republic was as unattainably remote as the overthrow of the capitalist system is to the latter, and therefore is of absolutely no importance in present-day practical politics; one can mediate, compromise, and philanthropize to one's heart's content. It is just the same with the class struggle between proletariat and bourgeoisie. It is recognized on paper because its existence can no longer be denied, but in practice it is hushed up, diluted, attenuated. The Social-Democratic Party is not to be a workers' party, is not to incur the odium of the bourgeoisie or of anyone else; it should above all conduct energetic propaganda among the bourgeoisie, instead of laying stress on far-reaching aims which frighten away the bourgeoisie and after all are not attainable in our generation, it should rather devote its whole strength and energy to those petty-bourgeois patchwork reforms which, by providing the old order of society with new props, may perhaps transform the ultimate catastrophe into a gradual, piecemeal, and as far as possible peaceful process of dissolution. These are the same people who, ostensibly engaged in indefatigable activity, not only do nothing themselves but try to prevent anything happening at all except—chatter; the same people whose fear of every form of action in 1848 and 1849 obstructed the movement at every step and finally brought about its downfall, the same people who never see reaction and are then quite astonished to find themselves in the end in a blind alley where neither resistance nor flight is possible, the same people who want to confine history within their narrow philistine horizon and over whose heads history invariably proceeds to the order of the day.

As to their socialist worth, this has been adequately criticized already in the *Manifesto*, the chapter on 'German, or "True", Socialism'. Where the class struggle is pushed aside as a disagreeable 'coarse' phenomenon, nothing remains as a basis for socialism but 'true love of humanity' and empty phraseology about 'justice'.

It is an inevitable phenomenon, rooted in the course of development, that people from what have hitherto been the ruling classes should also join the militant proletariat and supply it with educative elements. We clearly stated this in the *Manifesto*. But here two points are to be noted:

First, in order to be of use to the proletarian movement these people must bring real educative elements into it. But with the great majority of the German bourgeois converts that is not the case. Neither the *Zukunft* nor the *Neue Gesellschaft* have contributed anything which could advance the movement one step further. Here there is an absolute lack of real educational material, whether factual or theoretical. In its place there are attempts to bring superficially mastered socialist ideas into harmony with the exceedingly varied theoretical standpoints which these gentlemen have brought with them from the universities or elsewhere and of which, owing to the process of decomposition which the remnants of German philosophy are at present undergoing, one was more confused than the other. Instead of thoroughly studying the new science themselves to begin with, each of them preferred to trim it to fit the point of view he had brought along, made himself forthwith a private science of his own, and at once came forward with the pretension of wanting to teach it. Hence, there are about as many points of view among these gentry as there are heads; instead of producing clarity in a single case they have only produced desperate confusion—fortunately almost exclusively among themselves. Educative elements whose first principle is to teach what they have not learnt can very well be dispensed with by the Party.

Secondly. If people of this kind from other classes join the proletarian movement, the first condition must be that they should not bring any remnants of bourgeois, petty-bourgeois, etc., prejudices with them but should wholeheartedly adopt the proletarian outlook. But these gentlemen, as has been proved, are chock-full of bourgeois and petty-bourgeois ideas. In such a petty-bourgeois country as Germany these ideas certainly have their justification. But only outside the Social-Democratic Workers' Party. If these gentlemen constitute themselves into a Social-Democratic petty-bourgeois party they have a perfect right to do so; one could then negotiate with them, form a bloc according to circumstances, etc. But in a workers' party they are an adulterating element. If reasons exist for tolerating them there for the moment it is our duty only to tolerate them, to allow them no influence in the Party leadership, and to remain aware that a break with them is only a matter of time. That time, moreover, seems to have come. How the Party can tolerate the authors of this article in its midst any longer is incomprehensible to us. But if even the leadership of the Party should fall more or less into the hands of such people, the Party would simply be castrated and there would be an end of proletarian incisiveness.

As for ourselves, in view of our whole past there is only one road open to us. For almost forty years we have stressed the class struggle as the immediate driving power of history, and in particular the class struggle between bourgeoisie and proletariat as the great lever of the modern social revolution; it is, therefore, impossible for us to co-operate with people who wish to expunge this class struggle from the movement. When the International was formed we expressly formulated the battle-cry: The emancipation of the working classes must be conquered by the working classes themselves. We cannot therefore co-operate with people who openly state that the workers are too uneducated to emancipate themselves and must be freed from above by philanthropic big bourgeois and petty bourgeois. If the new Party organ adopts a line that corresponds to the views of these gentlemen, that is bourgeois and not proletarian, then nothing remains for us, much though we should regret it, but publicly to declare our opposition to it, and to dissolve the bonds of the solidarity with which we have hitherto represented the German Party abroad. But it is to be hoped that things will not come to such a pass. . . .

BIBLIOGRAPHY

Original: *MEW*, Vol. 34, pp. 405 ff.

Present translation: *MESC*, pp. 325 ff. Reproduced by kind permission of Lawrence and Wishart Ltd.

Other translations: None.

43. Letter to Vera Sassoulitch

In February 1881 a Russian Populist exile in Geneva, Vera Sassoulitch, wrote to Marx saying that his authority was being claimed in Russia for the view that the traditional peasant commune was doomed to disappear, and asking him to clarify his position. Marx drafted three separate lengthy answers, excerpted below, before finally sending a terse and ambivalent reply.

Letter

Dear Citizeness,

A nervous disease that I have been suffering from periodically for the last ten years has prevented me from replying earlier to your letter of 16 February. To my regret I am unable to give you a succinct answer, prepared for publication, to the question which you have graciously submitted to me. Months have passed since I promised to write something on the same subject to the St. Petersburg Committee. However I hope a few lines will suffice to remove all doubt in your mind about the misunderstanding concerning my so-called theory.

In analysing the genesis of capitalist production I say:

'The foundation of the capitalist system is therefore the utmost separation of the producer from the means of production. . . . The basis of this whole development is the expropriation of the agricultural producer. This has been accomplished in radical fashion only in England. . . . But all the countries of Western Europe are going through the same movement.' (*Capital*, French ed., p. 315.)

Hence the 'historical inevitability' of this movement is expressly limited to the countries of Western Europe. The reason for this limitation is indicated in the following passage of Chapter XXXII:

'Self-earned private property . . . will be supplanted by capitalistic private property, which rests on the exploitation of the labour of others, on wages-labour.' (Ibid., p. 341.)

In this western movement the point in question therefore is the transformation of one form of private property into another form of private property. With the Russian peasants one would on the contrary have to transform their common property into private property.

Thus the analysis given in *Capital* assigns no reasons for or against the vitality of the rural community, but the special research into this subject which I conducted, the materials for which I obtained from original sources, has convinced me that this community is the mainspring of Russia's social regeneration, but in order that it might function as such one would first have to eliminate the deleterious influences which assail it from every quarter and then to ensure the conditions normal for spontaneous development.

I have the honour, dear citizeness, to be wholly devoted.

Yours,

Karl Marx

From the Drafts

. . . If capitalist production is to establish its reign in Russia, the large majority of the peasants, that is of the Russian people, must become wage-earners and thus have been expropriated by the previous abolition of its communist property. But in every instance, the western precedent would prove nothing at all about the 'historical inevitability' of this process.

In this western movement, what is at issue, therefore, is the transformation of one form of private property into another form of private property. With the Russian peasants, on the contrary, their common property would have to be transformed into private property. Whether one affirms or denies the inevitability of that transformation, the reasons for and the reasons against have nothing to do with my analysis of the genesis of the capitalist regime. The most that one could infer would be that, given the actual state of the large majority of Russian peasants, the act of their conversion into small-scale owners would only be the prologue to their rapid expropriation.

If, at the moment of emancipation the rural communes had first of all been placed in conditions of normal prosperity; if, then, the huge public debt raised for the most part at the expense of the peasantry—together with the other enormous sums furnished by the intermediary of the state (and always at the expense of the peasantry) to the 'new pillars of society' transformed into capitalists—if all these expenditures had been used for the further development of the rural commune; then no one today would dream of the 'historical inevitability' of the annihilation of the commune, and everyone would recognize in it the element of the regeneration of Russian society and an element of superiority over the countries still enslaved by the capitalist regime.

The Russian 'Marxists' of whom you speak are quite unknown to me. The Russians with whom I have personal contact entertain, as far as I know, views that are quite the contrary.

The most serious argument that has been put against the Russian commune amounts to this: return to the origins of western societies and you will find everywhere common ownership of land; with social progress this disappeared everywhere to give rise to private property; therefore it cannot escape the same destiny in Russia alone. . . .

Theoretically speaking the Russian 'rural commune' can keep its land—by developing its basis which is the common ownership of land and by eliminating the principle of private property which is also implicit in it. It can become a direct starting-point for the economic system towards which modern society is tending. It can acquire a new skin without beginning by its suicide. It can obtain the fruits with which capitalist production has enriched humanity without passing through the capitalist regime, a regime which, considered exclusively from the point of view of its possible duration, scarcely counts in the life of society. But we must come down from pure theory to Russian reality.

Russia is the only European country where the 'agricultural commune' has been maintained on a national scale until today. It is not the prey of a foreign conqueror, as in India. Nor is its life isolated from the modern world. From one aspect, its common property in land enables it directly and gradually to transform the system of individual plots into a collective agriculture that the Russian peasants already practise on the undivided prairies.

If the spokesmen of the 'new pillars of society' denied the theoretical possibility of the evolution of the modern rural commune that I have indicated, they would have to be asked whether Russia has been forced, like the west, to pass through a long period of the incubation of mechanical industry to arrive at machines, steam engines, railways, etc? They would have to be asked further how they have managed, at the drop of a hat, to introduce into their country all the mechanisms of exchange such as banks and joint stock companies whose elaboration cost centuries of elaboration to the West? ...

From the historical point of view, a circumstance very favourable to the preservation of the 'agricultural commune' by continuing its further development is that it is not only the contemporary of western capitalist production and can thus obtain its fruits without enslaving itself to its *modus operandi*; the commune has also survived beyond the period when the capitalist system was still intact and, on the contrary, it now finds it both in western Europe and the United States struggling both with the mass of workers and with science and with the very productive forces it has engendered—in a word, in a crisis which will end with its elimination and a return by modern societies to a superior form of an 'archaic' type of property and collective production.

It is understood that the evolution of the commune would be gradual and the first step would be to put it in normal conditions on its actual basis, for the peasant is everywhere the enemy of any sudden change.

To expropriate the cultivators, it is not necessary to expel them from their lands as was done in England and elsewhere; nor is it necessary to abolish common property by an *ukase*. Snatch from the peasants the product of their labour beyond a certain point and, despite your police force and your army, you will not succeed in chaining them to their fields. In the last stages of the

Roman Empire, provincial decurions—not peasants but landed proprietors—fled their houses, abandoned their lands, and even sold themselves into slavery, and all only to get rid of property which was merely an official pretext for bringing pressure to bear on them without mercy or pity.

Since the so-called emancipation of the serfs, the Russian commune was placed by the state in abnormal economic conditions, and since that time the state has not ceased to heap on it all the social forces concentrated in its hands. Weakened by fiscal exactions, it became inert matter easily exploited by commerce, landed property, and usury. This external oppression let loose inside the commune itself the conflict of interests that was already present and rapidly developed the seeds of its decomposition. But that is not all. At the expense of the peasantry, the state has cultivated, in a hot-house, branches of the western capitalist system which, without in any way developing the productive bases of its agriculture, are precisely calculated to facilitate and precipitate the theft of its fruits by unproductive intermediaries. It has thus co-operated in the production of a new capitalist vermin sucking the blood of the 'rural commune' that was already so impoverished.

In a word, the state has given its assistance in precociously developing the technical and economic means most calculated to facilitate and precipitate the exploitation of the cultivator, that is, of the largest productive force in Russia, and to enrich the 'new pillars of society'.

This concourse of destructive influences, unless broken by a powerful reaction, must naturally result in the death of the rural commune.

But the question arises: Why have all these interests (I include large industries placed under governmental tutelage) found advantages in the present state of the rural commune, why do they deliberately conspire to kill the goose that lays the golden eggs? Precisely because they consider 'this present state' to be no longer tenable and that therefore the present method of exploiting it is no longer fitting. Already the poverty of the cultivator has spread to the earth which is becoming sterile. Good harvests are balanced by famines. The average of the ten last years revealed an agricultural production not only stagnant but also retrogressive. For the first time Russia must import cereals instead of exporting them. There is thus no more time to be lost. The issue must be decided.

Since so many diverse interests, and above all those of the 'new social pillars' erected under the benign empire of Alexander II, have made gains from the present state of the 'rural commune', why would they deliberately move to conspire at its death? Why do their spokesmen denounce the wounds inflicted on the commune as so many irrefutable proofs of its natural decline? Simply because the economic facts, whose analysis would take me too far, have unveiled the mystery that the present state of the commune is no longer tenable and that through the necessity of things alone the present state of exploiting the popular masses will no longer be suitable. So something new is necessary and that something, insinuated under the most diverse forms, always amounts to this: abolish common property, allow the more or less well off minority of peasants to form a rural middle class, and transform the large majority into proletarians pure and simple.

On the one hand, the 'rural commune' is almost reduced to its last extremity, and on the other there is a powerful conspiracy to give it the *coup de grâce*. At the same time as the commune is being bled and tortured and its earth sterilized and pauperized, the literary lackeys of the 'new pillars of society' ironically describe the wounds inflicted on it as so many signs of its spontaneous and incontestable decrepitude, that it is dying a natural death, and that they are doing it a favour by shortening its agony. Here, there is no problem to be solved; there is quite simply an enemy to be beaten. It is thus no longer a theoretical problem. To save the Russian commune, a Russian revolution is necessary. Moreover, the Russian government and the 'new pillars of society' are doing their best to prepare the masses for such a catastrophe. If the revolution comes at an opportune moment, if it concentrates all its forces to ensure the free development of the rural commune, this commune will soon develop into an element that regenerates Russian society and guarantees superiority over countries enslaved by the capitalist regime.
. . .

BIBLIOGRAPHY

Original: K. Marx, *Œuvres*, ed. M. Rubel, Paris, 1968, Vol. 2, pp. 1557 ff.

Present translation: *MESC*, pp. 339 f. The Drafts are translated by the present editor. Reproduced by kind permission of Lawrence and Wishart Ltd.

Other translations: Selections in K. Marx and F. Engels, *The Russian Menace to Europe*, ed. P. Blackstock and B. Hoselitz, London, 1953.

Commentaries: P. Blackstock and B. Hoselitz, edition cited above.
A. Walicki, *The Controversy over Capitalism*, Oxford, 1969.
See also Commentaries cited under Section 41 above.

44. Comments on Adolph Wagner

This short selection is from Marx's last extended comment on economics. It was written in the second half of 1880 and consisted in a criticism of Adolph Wagner's *Lehrbuch der politischen Oekonomie*. Wagner was a Berlin professor who taught a sort of state socialism. Marx here explains his conception of man as a being conditioned by his practical activities.

. . . Man? If it is man as a category who is being referred to here, the he has absolutely no needs; if it is man as a being who individually confronts nature, he should be considered as a non-gregarious individual; whereas if it is man as already living in some form or other of society (and this is what Mr. Wagner means, for his Man possesses, if not a university education, then at least a language), then it is necessary to begin by describing the specific character of this social man, that is, the character of the community in which he lives because its production—the process that enables him to earn his living—already possesses a certain social character.

But for a pedantic schoolmaster the relationships of man to nature are not *practical* relationships based on action, but theoretical relationships. . . .

Men do not in any way begin by 'finding themselves in a theoretical relationship to the things of the external world'. Like every animal, they begin by *eating, drinking*, etc. that is, not by 'finding themselves' in a relationship, but by behaving actively, gaining possession of certain things in the external world by their actions, thus satisfying their needs. (They thus begin by production.) By repetition of this process, the property that those things have of 'satisfying their needs' is impressed on their brain; men, like animals, also learn to distinguish 'theoretically' the external things which, above all others, serve to satisfy their needs. At a certain point in their evolution, after the multiplication and development of their needs and of the activities to assuage them, men will baptize with the aid of words the whole category of these things that experience has enabled them to distinguish from the rest of the external world. This is an inevitable result; for, during the process of production (that is, the process of acquiring these things), men continuously create active relationships with each other and with these things, and soon they will have to struggle with each other for their possession. But this linguistic denomination only expresses, in the form of a representation, what has become an acquired experience by constant repetition: that certain external things serve to satisfy the needs of men who live in given social relationships (which results necessarily from the existence of language). Men only give a generic name to these things because they already know that they serve to satisfy their needs and that they attempt to acquire them by frequently repeated acts and thus keep them in their possession; they perhaps call them 'goods' or something else, which means that they use these things practically, that they are useful to them and they bestow on this thing the character of utility as something properly belonging to it—although a sheep could only with difficulty imagine that the fact of being eatable by men belonged to its 'useful' properties.

Thus, men begin effectively by appropriating certain things in the external world as means of satisfying their own needs, etc.; later they came also to give them a verbal designation according to the function they seem to fulfil in their practical experience, that is, as means of satisfying these needs. . . .

BIBLIOGRAPHY

Original: *MEW*, Vol. 19, pp. 355 ff.

Present translation: by the editor.

Other translations: *Karl Marx on Method*, ed. T. Carver, Oxford, 1974.

Commentaries: Introduction by T. Carver to the edition cited above.

45. Preface to the Russian Edition of the *Communist Manifesto*

This is Marx's last published writing. In it he reiterates his view of the various possibilities for revolution in Russia—a view he had already elaborated in his reply to Vera Sassoulitch.

What a limited field the proletarian movement still occupied at that time (December 1847) is most clearly shown by the last section of the Manifesto: the position of the Communists in relation to the various opposition parties in the various countries. Precisely Russia and the United States are missing here. It was the time when Russia constituted the last great reserve of all European reaction, when the United States absorbed the surplus proletarian forces of Europe through immigration. Both countries provided Europe with raw materials and were at the same time markets for the sale of its industrial products. At that time both were, therefore, in one way or another, pillars of the existing European order.

How very different today! Precisely European immigration fitted North America for a gigantic agricultural production, whose competition is shaking the very foundations of European landed property—large and small. In addition it enabled the United States to exploit its tremendous industrial resources with an energy and on a scale that must shortly break the industrial monopoly of Western Europe, and especially of England, existing up to now. Both circumstances react in revolutionary manner upon America itself. Step by step the small and middle landownership of the farmers, the basis of the whole political constitution, is succumbing to the competition of giant farms; simultaneously, a mass proletariat and a fabulous concentration of capitals are developing for the first time in the industrial regions.

And now Russia! During the Revolution of 1848–9 not only the European princes, but the European bourgeois as well, found their only salvation from the proletariat, just beginning to awaken, in Russian intervention. The tsar was proclaimed the chief of European reaction. Today he is a prisoner of war of the revolution, in Gatchina, and Russia forms the vanguard of revolutionary action in Europe.

The Communist Manifesto had as its object the proclamation of the inevitably impending dissolution of modern bourgeois property. But in Russia we find, face to face with the rapidly developing capitalist swindle and bourgeois landed property, just beginning to develop, more than half the land owned in common by the peasants. Now the question is: Can the

Russian *obshchina*, though greatly undermined, yet a form of the primeval common ownership of land, pass directly to the higher form of communist common ownership? Or, on the contrary, must it first pass through the same process of dissolution as constitutes the historical evolution of the West?

The only answer to that possible today is this: If the Russian Revolution becomes the signal for a proletarian revolution in the West, so that both complement each other, the present Russian common ownership of land may serve as the starting-point for a communist development.

BIBLIOGRAPHY

Original: *MEW*, Vol. 4, p. 575 f.

Present translation: *MESW*, Vol. 1, pp. 23 f. Reproduced by kind permission of Lawrence and Wishart Ltd.

Other translations: None.

Commentaries: See under Section 41 above.

46. Letters 1863–1881

On Working-class Consciousness

Marx to Engels, 6 Apr. 1863

... How soon the English workers will free themselves from their apparent bourgeois infection one must wait and see. For the rest, as far as the main points in your book [*Outlines of a Critique of Political Economy*, 1844] are concerned, they have been confirmed down to the smallest detail by developments since 1844. You see, I have myself compared the book again with my notes on the later period. Only the small German petty-bourgeois, who measure world history by the yard and the latest 'interesting news in the papers', would imagine that in developments of such magnitude twenty years are more than a day—though later on days may come again in which twenty years are embodied.

Re-reading your book has made me regretfully aware of our increasing age. How freshly and passionately, with what bold anticipations and no learned and scientific doubts, the thing is still dealt with here! And the very illusion that the result will leap into the daylight of history tomorrow or the day after gives the whole thing a warmth and vivacious humour—compared with which the later 'gray in gray' makes a damned unpleasant contrast. ...

Marx to Schweitzer, 13 Feb. 1865

... Combinations and the trade unions growing out of them are of the utmost importance not only as a means of organizing the working class for struggle against the bourgeoisie. This importance appears, for instance, in the fact that even workers of the United States, despite their franchise and their republic, cannot do without them. The right of combination in Prussia and in Germany at large means furthermore a breach in the rule of the police and bureaucracy; it tears to bits the Rules Governing Servants and the control of the nobility in the rural districts. In short it is a measure for the conversion of 'subjects' into full-fledged citizens, which the Progressive Party, i.e., any bourgeois opposition party in Prussia which is not crazy, could allow a hundred times sooner than the Prussian Government, and above all the government of a Bismarck! On the other hand support for co-operative societies from the Royal Prussian Government—and anyone who knows Prussian conditions knows beforehand its necessarily minute dimensions—is of no value whatever as an economic measure, while at the same time it extends the system of tutelage, corrupts a section of the workers, and castrates the movement. The bourgeois party in Prussia discredited itself and brought on its present misery chiefly because it seriously believed that with the 'new era' power, by the grace of the Prince Regent, had fallen into its lap. But the workers' party will discredit itself far more if it imagines that in the Bismarck era or any other Prussian era the golden apples will drop into its mouth by the grace of the king. That disappointment will follow Lassalle's hapless illusion that a Prussian Government would carry out a socialist intervention is beyond all doubt. The logic of things will tell. But the honour of the workers' party demands that it should reject such hallucinations even before their hollowness is exposed by experience. The working class is revolutionary or it is nothing. . . .

Marx to Kugelmann, 9 Oct. 1866

. . . I had great fears for the first Congress at Geneva. On the whole however it turned out better than I expected. The effect in France, England, and America was unhoped for. I could not, and did not want to go there, but wrote the programme for the London delegates. I deliberately restricted it to those points which allow of immediate agreement and concerted action by the workers, and give direct nourishment and impetus to the requirements of the class struggle and the organization of the workers into a class. The Parisian gentlemen had their heads full of the emptiest Proudhonist phrases. They babble about science and know nothing. They scorn all revolutionary action, that is, action arising out of the class struggle itself, all concentrated, social movements, and therefore also those which can be carried through by political means (for instance the legal shortening of the working day). . . .

Marx to Schweitzer, 13 Oct. 1868

. . . To begin with, as far as the Lassallean Association is concerned, it was founded in a period of reaction. Lassalle—and this remains his immortal service—re-awakened the workers' movement in Germany after its fifteen years of slumber. But he committed great mistakes. He allowed himself to be governed too much by the immediate circumstances of the time. He made a minor starting-point—his opposition to a dwarf like Schulze-Delitzsch—into the central point of his agitation—state aid versus self-help. In so doing he merely took up again the watchword which Buchez, the leader of French Catholic socialism, had given out in 1843 seqq. against the genuine workers' movement in France. Much too intelligent to regard this watchword as anything but a temporary makeshift, Lassalle could only justify it on the ground of its immediate (as he alleges!) practicability. For this purpose he had to maintain that it could be carried out in the near future. Hence the 'State' transformed itself into the Prussian State. Thus he was driven into making concessions to the Prussian monarchy, the Prussian reaction (feudal party), and even the clericals. With Buchez's state aid for associations he combined the Chartist cry of universal suffrage. He overlooked the fact that conditions in Germany and England were different. He overlooked the lessons of the Second Empire with regard to universal suffrage in France. Moreover, like everyone who maintains that he has a panacea for the sufferings of the masses in his pocket, he gave his agitation from the outset a religious and sectarian character. Every sect is in fact religious. Furthermore, just because he was the founder of a sect, he denied all natural connection with the earlier working-class movement both inside and outside of Germany. He fell into the same mistake as Proudhon: instead of looking among the genuine elements of the class movement for the real basis of his agitation, he wanted to prescribe the course to be followed by this movement according to a certain doctrinaire recipe.

Most of what I am now saying, *post factum*, I had already told Lassalle in 1862, when he came to London and urged me to place myself with him at the head of the new movement.

You yourself have experienced in your own person the opposition between the movement of a sect and the movement of a class. The sect sees the justification for its existence and its point of honour not in what it has in common with the class movement but in the particular shibboleth which distinguishes it from the movement. Therefore when at Hamburg you proposed the congress for the formation of trade unions you were able to smash the opposition of the sect only by threatening to resign from the office of president. In addition you were obliged to assume a dual personality and to announce that in one case you were acting as the head of the sect and in the other as the organ of the class movement.

The dissolution of the General Association of German Workers gave you the opportunity to take a great step forward and to declare, to prove if necessary, that a new stage of development had now been reached, and that the moment was ripe for the sectarian movement to merge in the class movement and make an end of all sectarianism. As for the true content of the sect it would, as was the case with all previous working-class sects, be carried on into the general movement as an element enriching it. . . .

Marx to Bolte, 23 Nov. 1871

. . . The International was founded in order to replace the socialist or semi-socialist sects by a real organization of the working class for struggle. The original Rules and the Inaugural Address show this at a glance. On the other hand the International could not have maintained itself if the course of history had not already smashed sectarianism. The development of socialist sectarianism and that of the real working-class movement always stand in inverse ratio to each other. Sects are justified (historically) so long as the working class is not yet ripe for an independent historical movement. As soon as it has attained this maturity all sects are essentially reactionary.

Nevertheless, what history exhibits everywhere was repeated in the history of the International. What is antiquated tries to re-establish itself and maintain its position within the newly acquired form.

And the history of the International was a continual struggle of the General Council against the sects and amateur experiments, which sought to assert themselves within the International against the real movement of the working class. This struggle was conducted at the congresses, but far more in the private negotiations between the General Council and the individual sections.

In Paris, as the Proudhonists (Mutualists) were cofounders of the Association, they naturally held the reins there for the first few years. Later, of course, collectivist, positivist, etc., groups were formed there in opposition to them.

In Germany—the Lassalle clique. I myself corresponded with the notorious Schweitzer for two years and proved to him irrefutably that Lassalle's organization was a mere sectarian organization and, as such, hostile to the organization of the real workers' movement striven for by the International. He had his 'reasons' for not understanding.

At the end of 1868 the Russian Bakunin joined the International with the aim of forming inside it a second International under the name of 'Alliance de la Démocratie Socialiste' and with himself as leader. He—a man devoid of all theoretical knowledge—laid claim to representing in that separate body the scientific propaganda of the International, and wanted to make such propaganda the special function of that second International within the International.

His programme was a hash superficially scraped together from the Right and from the Left—equality of classes (!), abolition of the right of inheritance as the starting-point of the social movement (St. Simonist nonsense), atheism as a dogma dictated to the members, etc., and as the main dogma (Proudhonist): abstention from the political movement.

This children's primer found favour (and still has a certain hold) in Italy and Spain, where the real conditions for the workers' movement are as yet little developed, and among a few vain, ambitious, and empty doctrinaires in Latin Switzerland and in Belgium.

To Mr. Bakunin doctrine (the mess he has brewed from bits of Proudhon, St. Simon, and others) was and is a secondary matter—merely a means to his personal self-assertion. Though a nonentity as a theoretician he is in his element as an intriguer.

For years the General Council had to fight against this conspiracy (supported up to a certain point by the French Proudhonists, especially in the South of France). At last, by means of Conference Resolutions 1, 2 and 3, IX, XVI and XVII, it delivered its long-prepared blow.

It goes without saying that the General Council does not support in America what it combats in Europe. Resolutions 1, 2, 3 and IX now give the

New York Committee the legal weapons with which to put an end to all sectarianism and amateur groups, and, if necessary, to expel them. . . .

The political movement of the working class has as its ultimate object, of course, the conquest of political power for this class, and this naturally requires a previous organization of the working class developed up to a certain point and arising precisely from its economic struggles.

On the other hand, however, every movement in which the working class comes out as a class against the ruling classes and tries to coerce them by pressure from without is a political movement. For instance, the attempt in a particular factory or even in a particular trade to force a shorter working day out of individual capitalists by strikes, etc., is a purely economic movement. On the other hand the movement to force through an eight-hour, etc., law, is a political movement. And in this way, out of the separate economic movements of the workers there grows up everywhere a political movement, that is to say, a movement of the class, with the object of enforcing its interests in a general form, in a form possessing general, socially coercive force. While these movements presuppose a certain degree of previous organization, they are in turn equally a means of developing this organization.

Where the working class is not yet far enough advanced in its organization to undertake a decisive campaign against the collective power, i.e., the political power of the ruling classes, it must at any rate be trained for this by continual agitation against this power and by a hostile attitude towards the policies of the ruling classes. Otherwise it remains a plaything in their hands, as the September revolution in France showed, and as is also proved to a certain extent by the game that Messrs. Gladstone & Co. have been successfully engaged in in England up to the present time. . . .

Marx to Sorge, 19 Oct. 1877

. . . The workers themselves, when, like Herr Most & Co., they give up work and become professional literary men, always breed 'theoretical' mischief and are always ready to join muddleheads from the allegedly 'learned' caste. Utopian socialism especially which for decades we have been clearing out of the German workers' heads with so much effort and labour—their freedom from it having made them theoretically (and therefore also practically) superior to the French and English—utopian socialism, playing with fantastic pictures of the future structure of society, is again spreading like wildfire, and in a much more futile form, not only compared with the great French and English utopians, but even with—Weitling. It is natural that utopianism, which before the era of materialistically critical socialism concealed the latter within itself in embryo, can now, coming belatedly, only be silly, stale, and reactionary from the roots up. . . .

On Ireland

Marx to Engels, 30 Nov. 1867

... The question now is, what shall we advise the English workers? In my opinion they must make the repeal of the Union (in short, the affair of 1783, only democratized and adapted to the conditions of the time) an article of their *pronunziamento*. This is the only legal and therefore only possible form of Irish emancipation which can be admitted in the programme of an English party. Experience must show later whether a mere personal union can continue to subsist between the two countries. I half think it can if it takes place in time.

What the Irish need is:

(1) Self-government and independence from England.

(2) An agrarian revolution. With the best intentions in the world the English cannot accomplish this for them, but they can give them the legal means of accomplishing it for themselves.

(3) Protective tariffs against England. Between 1783 and 1801 every branch of Irish industry flourished. The Union, which overthrew the protective tariffs established by the Irish Parliament, destroyed all industrial life in Ireland. The bit of linen industry is no compensation whatever. The Union of 1801 had just the same effect on Irish industry as the measures for the suppression of the Irish woollen industry, etc., taken by the English Parliament under Anne, George II, and others. Once the Irish are independent, necessity will turn them into protectionists, as it did Canada, Australia, etc. Before I present my views in the Central Council (next Tuesday, this time fortunately without reporters), I would like you to give me your opinion in a few lines. ...

Marx to Kugelmann, 29 Nov. 1869

... I have become more and more convinced—and the only question is to drive this conviction home to the English working class—that it can never do anything decisive here in England until it separates its policy with regard to Ireland most definitely from the policy of the ruling classes, until it not only makes common cause with the Irish but actually takes the initiative in dissolving the Union established in 1801 and replacing it by a free federal relationship. And this must be done, not as a matter of sympathy with Ireland but as a demand made in the interests of the English proletariat. If not, the English people will remain tied to the leading-strings of the ruling classes, because it will have to join with them in a common front against Ireland. Every one of its movements in England itself is crippled by the strife with the Irish, who form a very important section of the working class in England. The prime condition of emancipation here—the overthrow of the English landed oligarchy—remains impossible because its position here cannot be stormed so long as it maintains its strongly entrenched outposts in Ireland. But there, once affairs are in the hands of the Irish people itself, once it is made its own legislator and ruler, once it becomes autonomous, the abolition of the landed aristocracy (to a large extent the same persons as the English landlords) will be infinitely easier than here, because in Ireland it is not merely a simple economic question but at the same time a national question, since the landlords there are not, like those in England, the traditional dignitaries and representatives of the nation, but its mortally hated oppressors. And not only does England's internal social development remain crippled by her present relations with Ireland; her foreign policy, and particularly her policy with regard to Russia and the United States of America, suffers the same fate.

But since the English working class undoubtedly throws the decisive weight into the scale of social emancipation generally, the lever has to be applied here. As a matter of fact, the English republic under Cromwell met shipwreck in—Ireland. *Non bis in idem!* [Not the same twice over!] But the Irish have played a capital joke on the English government by electing the 'convict felon' O'Donovan Rossa to Parliament. The government papers are already threatening a renewed suspension of the Habeas Corpus Act, a renewed system of terror. In fact, England never has and never can—so long as the present relations last—rule Ireland otherwise than by the most abominable reign of terror and the most reprehensible corruption. . . .

Marx to Meyer and Vogt, 9 Apr. 1870

. . . Every industrial and commercial centre in England now possesses a working class divided into two hostile camps, English proletarians and Irish proletarians. The ordinary English worker hates the Irish worker as a competitor who lowers his standard of life. In relation to the Irish worker he feels himself a member of the ruling nation and so turns himself into a tool of the aristocrats and capitalists of his country against Ireland, thus strengthening their domination over himself. He cherishes religious, social, and national prejudices against the Irish worker. His attitude towards him is much the same as that of the 'poor whites' to the 'niggers' in the former slave states of the U.S.A. The Irishman pays him back with interest in his own money. He sees in the English worker at once the accomplice and the stupid tool of the English rule in Ireland.

This antagonism is artificially kept alive and intensified by the press, the pulpit, the comic papers, in short, by all the means at the disposal of the ruling classes. This antagonism is the secret of the impotence of the English working class, despite its organization. It is the secret by which the capitalist class maintains its power. And that class is fully aware of it.

But the evil does not stop here. It continues across the ocean. The antagonism between English and Irish is the hidden basis of the conflict between the United States and England. It makes any honest and serious co-operation between the working classes of the two countries impossible. It enables the governments of both countries, whenever they think fit, to break the edge off the social conflict by their mutual bullying, and, in case of need, by war with one another.

England, being the metropolis of capital, the power which has hitherto ruled the world market, is for the present the most important country for the workers' revolution, and moreover the only country in which the material conditions for this revolution have developed up to a certain degree of maturity. Therefore to hasten the social revolution in England is the most important object of the International Workingmen's Association. The sole means of hastening it is to make Ireland independent.

Hence it is the task of the International everywhere to put the conflict between England and Ireland in the foreground, and everywhere to side openly with Ireland. And it is the special task of the Central Council in London to awaken a consciousness in the English workers that for them the national emancipation of Ireland is no question of abstract justice or human-itarian sentiment but the first condition of their own social emancipation. . . .

On the Commune
Marx to Liebknecht, 6 Apr. 1871

. . . It seems the Parisians are succumbing. It is their own fault, but a fault which really was due to their too great decency. The Central Committee and later the Commune gave that mischievous abortion Thiers time to centralize hostile forces, in the first place by their folly of not wanting to start a civil war—as if Thiers had not already started it by his attempt at the forcible disarming of Paris, as if the National Assembly, summoned to decide the question of war or peace with the Prussians, had not immediately declared war on the Republic! Secondly, in order that the appearance of having usurped power should not attach to them they lost precious moments (they should immediately have advanced on Versailles after the defeat (Place Vendôme) of the reaction in Paris) by the election of the Commune, the organization of which, etc., cost yet more time. . . .

Marx to Kugelmann, 12 Apr. 1871

. . . If you look at the last chapter of my Eighteenth Brumaire, you will find that I declare that the next attempt of the French Revolution will be no longer, as before, to transfer the bureaucratic-military machine from one hand to another, but to smash it, and this is the preliminary condition for every real people's revolution on the Continent. And this is what our heroic Party comrades in Paris are attempting. What elasticity, what historical initiative, what a capacity for sacrifice in these Parisians! After six months of hunger and ruin, caused by internal treachery more even than by the external enemy, they rise, beneath Prussian bayonets, as if there had never been a war between France and Germany and the enemy were not still at the gates of Paris! History has no like example of like greatness! If they are defeated only their 'good nature' will be to blame. They should have marched at once on Versailles after first Vinoy and then the reactionary section of the Paris National Guard had themselves retreated. They missed their opportunity because of conscientious scruples. They did not want to start a civil war, as if that mischievous abortion Thiers had not already started the civil war with his attempt to disarm Paris! Second mistake: The Central Committee surrendered its power too soon, to make way for the Commune. Again from a too 'honourable' scrupulousness! However that may be, the present rising in Paris—even if it be crushed by the wolves, swine, and vile curs of the old society—is the most glorious deed of our Party since the June insurrection in Paris. Compare these Parisians, storming heaven, with the slaves to heaven of the German-Prussian Holy Roman Empire, with its posthumous masquerades reeking of the barracks, the Church, cabbage-squirearchy and above all, of the philistine. . . .

Marx to Kugelmann, 17 Apr. 1871

. . . World history would indeed be very easy to make if the struggle were taken up only on condition of infallibly favourable chances. It would on the other hand be of a very mystical nature, if 'accidents' played no part. These accidents naturally form part of the general course of development and are compensated by other accidents. But acceleration and delay are very much dependent upon such 'accidents', including the 'accident' of the character of the people who first head the movement.

The decisively unfavourable 'accident' this time is by no means to be sought in the general conditions of French society, but in the presence of the Prussians in France and their position right before Paris. Of this the Parisians were well aware. But of this, the bourgeois canaille of Versailles were also well aware. Precisely for that reason they presented the Parisians with the alternative of either taking up the fight or succumbing without a struggle. The demoralization of the working class in the latter case would have been a far greater misfortune than in the succumbing of any number of 'leaders'. With the struggle in Paris the struggle of the working class against the capitalist class and its state has entered upon a new phase. Whatever the immediate outcome may be, a new point of departure of world-wide importance has been gained. . . .

Marx to Domela-Nieuwenhuis, 22 Feb. 1881

. . . a socialist government does not come into power in a country unless conditions are so developed that it can immediately take the necessary measures for intimidating the mass of the bourgeoisie sufficiently to gain time—the first *desideratum*—for permanent action.

Perhaps you will refer me to the Paris Commune; but apart from the fact that this was merely the rising of a city under exceptional conditions, the majority of the Commune was in no wise socialist, nor could it be. With a modicum of common sense, however, it could have reached a compromise with Versailles useful to the whole mass of the people—the only thing that could be reached at the time. The appropriation of the Bank of France alone would have been enough to put an end with terror to the vaunt of the Versailles people, etc., etc. . . .

On Violent Revolution

Marx to Hyndman, 8 Dec. 1880

. . . If you say that you do not share the views of my party for England I can only reply that the party considers an English revolution not necessary, but—according to historic precedents—possible. If the unavoidable evolution turn into a revolution, it would not only be the fault of the ruling classes, but also of the working class. Every pacific concession of the former has been wrung from them by 'pressure from without'. Their action kept pace with that pressure and if the latter has more and more weakened, it is only because the English working class know not how to wield their power and use their liberties, both of which they possess legally.

In Germany the working class were fully aware from the beginning of their movement that you cannot get rid of a military despotism but by a Revolution. At the same time they understood that such a Revolution, even if at first successful, would finally turn against them without previous organization, acquirement of knowledge, propaganda, and . . . [gap in manuscript] Hence they moved within strictly legal bounds. The illegality was all on the side of the government, which declared them *en dehors la loi*. Their crimes were not deeds, but opinions unpleasant to their rulers. Fortunately, the same government—the working class having been pushed to the background with the help of the bourgeoisie—becomes now more and more unbearable to the latter, whom it hits on their most tender point—the pocket. This state of things cannot last long. . . .

Speech in Amsterdam, 1872, *MEW*, Vol. 18, p. 160

. . . We are aware of the importance that must be accorded to the institutions, customs, and traditions of different countries; and we do not deny that there are countries like America, England (and, if I knew your institutions better, I would add Holland), where the workers can achieve their aims by peaceful means. However true that may be, we ought also to recognize that, in most of the countries on the Continent, it is force that must be the lever of our revolutions; it is to force that it will be necessary to appeal for a time in order to establish the reign of labour. . . .

NOTE: all extracts taken from *MESC*. Reproduced by kind permission of Lawrence and Wishart Ltd.

Chronological Table

Bibliography

A. Previous Collected Writings

K. Marx, *Selected Essays*, ed. H. Stenning (London and New York, 1926, reprinted 1968). A collection of seven essays from the early Marx, most of them minor.

K. Marx, F. Engels, *Selected Works* (Moscow, 1935, several reprints). The 'classical' anthology. None of the early writings are included and less than half the material is by Marx. Nevertheless provides complete and faithful translations of many of Marx's works.

K. Marx, *Capital, The Communist Manifesto and Other Writings*, ed. M. Eastman (New York, 1932). Concentrates on *Capital* to the complete exclusion of early writings.

K. Marx, F. Engels, *Basic Writings on Politics and Philosophy*, ed. L. Feuer (New York, 1959). Concentrates on Marx's historical writings, with a useful selection of letters and essays at the end.

K. Marx, *Selected Writings in Sociology and Social Philosophy*, ed. T. Bottomore and M. Rubel (London, 1956). In many ways the best anthology, drawing on all Marx's writings whether available in English or not.

K. Marx, *Early Writings*, ed. T. Bottomore (London, 1963). Contains the essays in the *Deutsch–Französische Jahrbücher* and the complete text of the 'Paris Manuscripts'.

Writings of the Young Marx on Philosophy and Society, ed. L. Easton and K. Guddat (New York, 1967). A comprehensive collection of Marx's writings from 1841 to 1847. Contains extracts from *The Holy Family* and *The German Ideology*.

The Essential Writings of Karl Marx, ed. D. Caute (London and New York, 1967). Small excerpts with emphasis on the philosophical and revolutionary aspects of Marx.

Marxist Social Thought, ed. R. Freedman (New York, 1968). Fairly comprehensive on the sociological aspects of Marx's later works. Little reference to economics or to Marx's early writings.

K. Marx, *The Early Texts*, ed. D. McLellan (Oxford, 1971). A comprehensive selection of writings up to and including 1844, with letters.

The Portable Marx, ed. E. Kamenka (New York, 1971). A selection containing longer extracts and some newly translated material.

Karl Marx on Economy, Class and Social Revolution, ed. Z. Jordan (London, 1971). A comprehensive collection, aimed at the sociologist.

Marx–Engels Reader, ed. R. Tucker (New York, 1971). A more balanced, but shorter, version of the Moscow edition above.

K. Marx, *The Essential Writings*, ed. R. Bender (New York, 1972). A large collection, well put together, with due emphasis on the economic writings.

There are also collections of texts, mostly newspaper articles, on the following specific themes:

On Britain (London, 1953).
On Ireland (London, 1970).
Marx on China (London, 1968).
First Indian War of Independence (Moscow, 1960).
Revolution in Spain (London, 1939).
On Colonialism (Moscow, 1960).
Karl Marx on Colonialism and Modernization, ed. S. Avineri (New York, 1968).
On Malthus (London, 1953).
On Literature and Art (Bombay, 1956).
On Religion (Moscow, 1957).
On Revolution, ed. S. Padover (New York, 1971).

There is no comprehensive English translation of Marx's works. Penguin Vintage are currently bringing out an eight-volume selection; and McGraw-Hill are producing a thirteen volume one edited by Saul Padover. The translation of the Collected Works of Marx and Engels to comprise fifty-one volumes is being undertaken by Lawrence and Wishart in London and International Publishers New York in conjunction with the Institute for Marxism-Leninism in Moscow. The first four volumes were published early in 1975 and the whole will be completed in the next ten or twelve years.

B. Commentaries

H. B. Acton, *The Illusion of the Epoch* (London, 1955). A critique of Marxism-Leninism as a philosophical creed.

H. B. Acton, *What Marx Really Said* (London, 1967). A short critical exposition, concentrating on Marx's ideas of historical materialism.

H. P. Adams, *Karl Marx in His Earlier Writings*, 2nd ed. (London, 1965). The first examination in English of Marx's early writings up to, and including, *The Holy Family*. Dated.

L. Althusser, *For Marx* (London, 1970). A controversial interpretation of Marx using structuralist and Freudian concepts. Supports the idea of a radical break between the young and the old Marx.

L. Althusser, *Reading Capital* (London, 1971). An attempt to analyse *Capital* in a scientific manner and give an account of the philosophy underlying it.

W. Ash, *Marxism and Moral Concepts* (New York, 1964). A good introduction.

S. Avineri, *The Social and Political Thought of Karl Marx* (Cambridge, 1968). An important and interesting book which emphasizes the continuity of Marx's thought from its earliest formulations and the influence of Hegel.

K. Axelos, *Alienation, Praxis and Techne in the Thought of Karl Marx* (London, 1977). Links Marx's concept of alienation to problems of modern technology.

J. Barzun, *Darwin, Marx and Wagner* (Boston, 1946). Good in placing Marx in an intellectual tradition.

I. Berlin, *Karl Marx. His Life and Environment* (Oxford, 1939). A very readable short biography.

S. F. Bloom, *The World of Nations. A Study of the National Implications in the Work of Marx* (New York, 1941). An exposition of Marx's views on the position of nation states in the development of communism.

W. Blumenberg, *Karl Marx* (London, 1971). An excellent short biography mainly using Marx's own words with a varied selection of photographs.

M. M. Bober, *Karl Marx's Interpretation of History*, 2nd ed. (New York, 1965; original ed., 1927). The oldest and fullest discussion of historical materialism in English.

A. Bose, *Marxian and Post-Marxian Political Economy* (London, 1975). A rigorous theoretical discussion.

T. Bottomore (ed.), *Karl Marx* (New York, 1971). A collection of commentaries on Marx, with an introduction, in the 'Makers of Modern Social Science' series.

T. Bottomore, *The Sociological Theory of Marxism* (London, 1973). Contains an analysis of Marx's theories on classes, the state, revolution, and so on.

L. B. Boudin, *The Theoretical System of Karl Marx in the Light of Recent Criticism* (Chicago 1907; reprinted New York, 1967). A defence of Marx's materialist conception of history and economic doctrine in face of the criticisms of Revisionists.

E. R. Browder, *Marx and America* (London, 1959). A useful brief overview of the position of America in Marx's thought.

S. de Brunhoff, *Marx on Money* (London, 1976). A detailed account of Marx's views on the subject.

R. N. Carew-Hunt, *The Theory and Practice of Communism* (London, 1963). Contains a rather over-schematized and unreliable section on Marx.

J. Carmichael, *Karl Marx. The Passionate Logician* (London, 1968). A shortish biography.

E. H. Carr, *Karl Marx. A Study in Fanaticism* (London, 1943). A well-written critical biography of medium length.

S. Chang, *The Marxian Theory of the State* (Philadelphia 1931; new ed. 1965). A good exposition, but one which conflates the ideas of Marx and Lenin.

G. D. H. Cole, *What Marx Really Meant* (London, 1934). A sympathetic and systematic exposition of Marx's ideas.

G. D. H. Cole, *A History of Socialist Thought* (London, 1953, vols 1 and 2). A measured and well-researched placing of Marx in the history of socialist thought.

H. Collins and C. Abramsky, *Karl Marx and the British Labour Movement. Years of the First International* (London, 1965). A very well-documented account of Marx's part in the First International with special reference to Britain.

M. Curtis (ed.), *Marxism* (New York, 1970). A wide-ranging reprint of articles on Marx's thought.

P. Demetz, *Marx, Engels and the Poets* (Chicago, 1967). An assessment of the views of Marx and Engels as literary critics.

M. Dobb, *Marx as an Economist* (London, 1943). One of the best short introductions to Marx as an economist.

H. Draper, *Karl Marx's Theory of Revolution* (New York, 1976), Vol. 1. A comprehensive survey of Marx on the State.

R. Dunayevskaya, *Marxism and Freedom* (New York, 1958). Contains sections on the philosophical aspects of the *1844 Manuscripts* and *Capital*.

L. Dupré, *The Philosophical Foundations of Marxism* (New York, 1966). A straightforward discussion of Marx's thought up to the *Communist Manifesto*, with some preliminary chapters on Hegel.

M. Evans, *Karl Marx* (New York and London, 1975). An excellent introduction, stressing the historical and political aspects.

I. Fetscher, *Marx and Marxism* (New York, 1971). Contains articles on the continuity in Marx's thought, bureaucracy, future communist society, and so on.

B. Fine, *Marx's 'Capital'* (London, 1975). A short, lucid commentary.

E. Fischer, *Marx in His Own Words* (London, 1970). A slight, but faithful, run-through of Marx's main ideas.

H. Fleischer, *Marxism and History* (London and New York, 1973). Good exposition of historical materialism.

E. Fromm, *Marx's Concept of Man* (New York, 1961). This introduction to selections from the '1844 Manuscripts' portrays Marx as a humanist and existentialist thinker.

A. Gamble and P. Walton, *From Alienation to Surplus Value* (London, 1972). Concentrates on labour and surplus value as unifying themes in Marx's works with special attention paid to the *Grundrisse* and *Theories of Surplus Value*.

R. Garaudy, *Karl Marx: The Evolution of His Thought* (London, 1967). A reliable and readable account by (at the time of writing) an orthodox communist.

H. Gemkow and others, *Karl Marx. A Biography* (Berlin, 1970). A well-documented, but quite uncritical, piece of hagiography.

J. M. Gillman, *The Falling Rate of Profit. Marx's Law and Its Significance to 20th Century Capitalism* (London, 1958). An examination of the limitations of Marx's law when applied to monopoly capitalism.

M. Godelier, *Rationality and Irrationality in Economics* (London, 1972). Examines the basic structures of Marx's economic views.

A. J. Gregor, *A Survey of Marxism* (New York, 1965). The first few chapters discuss the philosophical aspects of Marx.

O. Hammen, *The Red 48-ers* (New York, 1968). A comprehensive account of Marx's revolutionary activities during this period.

S. Hook, *Towards the Understanding of Karl Marx* (New York, 1933). Still a good introduction to the more systematic parts of Marx's thought.

S. Hook, *From Hegel to Marx*, 2nd ed. (Ann Arbor, 1962). A study of the relationships of Hegel and Marx and the young Hegelians.

D. Horowitz (ed.), *Marx and Modern Economics* (London, 1968). Contains essays examining the relevance today of particularly the more abstract of Marx's economic theories.

D. Howard, *The Development of the Marxian Dialectic* (Carbondale, 1972). A reliable treatment of Marx's early thought.

M. Howard and J. King, *The Political Economy of Karl Marx* (London, 1975). A good introduction, duly stressing debt to classical economists.

R. Hunt, *The Political Ideas of Marx and Engels* (Pittsburg and London, 1975). Vol. 1. A thorough examination of Marx's ideas and activities around 1848.

J. Hyppolite, *Studies on Marx and Hegel* (London, 1969). Contains profound assessments of Marx's critique of Hegel's *Philosophy of Right* and of *Capital*.

Z. Jordan, *The Evolution of Dialectical Materialism* (London, 1967). The early chapters contain a good account of naturalism and materialism in Marx.

E. Kamenka, *The Ethical Foundation of Marxism*, 2nd ed. (London, 1972). A description and critique of Marx's ethics from an analytical philosophical position.

A. C. Kettle, *Karl Marx, Founder of Modern Communism* (London, 1963). A good short biography by a communist.

L. Kolakowski, *Marxism and Beyond* (London, 1968). Contains essays highlighting the relationship between the individual and history in Marx's thought.

L. Kolakowski, *Main Currents in Marxism* (Oxford, 1976), Vol. 1. A balanced and lucid account of the more philosophical aspects of Marx's thought.

H. Koren, *Marx and the Authentic Man* (Duquesne, 1967). A short description of Marx's 'humanist' conception of man.

K. Korsch, *Karl Marx* (New York, 1936). An insightful biography by an ex-communist.

K. Korsch, *Marxism and Philosophy* (London, 1971). A brilliant reassessment of the Hegelian elements in Marx.

L. Krader, *The Asiatic Mode of Production. Sources, Development and Critique in the Work of Karl Marx* (Assen, 1975). A careful and detailed account.

H. Lefebvre, *The Sociology of Marx* (London, 1968). An excellent introduction.

G. Leff, *The Tyranny of Concepts* (London, 1961). An important critique of Marx's materialist conception of history.

N. Levine, *The Tragic Deception. Marx contra Engels* (Santa Barbara, 1975). A detailed attempt to counterpose the ideas of Marx to those of Engels.

J. Lewis, *The Life and Teaching of Karl Marx* (London, 1965). A good medium-length biography presenting Marx in a favourable light.

J. Lewis, *The Marxism of Marx* (London, 1972). A wise and humane commentary by a veteran communist.

G. Lichtheim, *Marxism, an Historical and Critical Study* (London, 1961). An excellent study of the development of Marxist doctrines from their origins up to 1917.

N. Lobkowicz, *Theory and Practice, The History of a Marxist Concept* (Notre Dame, 1967). An examination of Marx's concept of 'praxis' against a Young Hegelian background.

N. Lobkowicz (ed.), *Marx and the Western World* (Notre Dame, 1967). A large collection of articles on the relevance of Marx's thought today.

G. Lukacs, *History and Class Consciousness* (London, 1970). An extremely influential re-emphasis of Hegel's influence on Marx.

E. Mandel, *The Formation of Marx's Economic Thought* (London, 1971). An excellent analysis of the development of Marx's economic thought up to and including the *Grundrisse*.

H. Marcuse, *Reason and Revolution* (London, 1941). An account of Marx's notion of labour.

W. McBride, *The Philosophy of Marx* (London, 1977). A balanced philosophical discussion of Marx's methodology.

D. McLellan, *The Young Hegelians and Karl Marx* (London, 1969). An examination of the social and political thought of the Young Hegelians and its influence on the genesis of Marx's thought.

D. McLellan, *Marx Before Marxism* (London and New York, 1970). A detailed description of the development of Marx's thought up to and including the *1844 Manuscripts*.

D. McLellan, *The Thought of Karl Marx* (London and New York, 1971). A chronological and thematic introduction to Marx's thought.

D. McLellan, *Karl Marx: His Life and Thought* (London and New York, 1973). A comprehensive biography, dealing with Marx's personal, political, and intellectual activities.

J. Maguire, *Marx's Paris Writings* (Dublin, 1972). A well-informed and thorough commentary on the writings of 1844.

R. L. Meek, *Studies in the Labour Theory of Value* (London, 1956). Best treatment in English of this subject.

F. Mehring, *Karl Marx* (London, 1936). The classical biography of Marx; somewhat out of date and slightly hagiographical.

A. G. Meyer, *Marxism: The Unity of Theory and Practice. A Critical Essay* (Cambridge, Mass., 1954). Presents a functional interpretation of Marx's sociology.

R. Miliband, *Marxism and Politics* (Oxford, 1977). A good introduction.

D. Mitrany, *Marx against the Peasant* (London, 1951). An attack on the views of Marx and his followers on the peasants.

A. C. MacIntyre, *Marxism: an Interpretation* (London, 1953). A short and sharp philosophical assessment of Marx.

M. Morishima, *Marx's Economics* (Cambridge, 1973). A complex examination of Marx's theoretical economics.

B. Nicolaievsky and O. Maenchen-Helfen, *Karl Marx, Man and Fighter* (London, 1933; 3rd ed. 1973). An excellent biography emphasizing Marx's political activities.

B. Ollman, *Alienation: Marx's critique of Man in Capitalist Society* (Cambridge, 1971). An original and well-documented study of alienation in Marx, paying close attention to the way Marx uses his concepts.

F. Pappenheim, *The Alienation of Modern Man* (New York, 1959). Puts Marx's concept of alienation in a modern context.

B. Parekh, *Marx* (London and New York, 1977). An interesting account of the more philosophical aspects of Marx.

R. Payne, *Marx, A Biography* (London, 1968). A lot of information on Marx's private life, though the author's understanding of Marx's ideas is extremely deficient.

G. Petrovic, *Marx in the Mid-Twentieth Century* (Garden City, 1967). Emphasizes the humanist relevance of Marx today.

J. Plamenatz, *German Marxism and Russian Communism* (London, 1954). Contains one of the classical discussions of historical materialism as outlined in Marx's *Preface*.

J. Plamenatz, *Man and Society*, Vol. 2 (London, 1963). A clear, critical analysis of the main social and political themes in Marx.

J. Plamenatz, *Karl Marx's Philosophy of Man* (Oxford, 1975). A lengthy critical analysis.

K. R. Popper, *The Open Society and its Enemies*, Vol. 2 (London, 1952). An attack on Marx as a totalitarian thinker.

S. Prawer, *Marx and World Literature* (Oxford, 1976). An exhaustive account of Marx's literary sources.

Joan Robinson, *An Essay in Marxian Economics* (London, 1942). An impressive attempt to revitalize Marx's main economic doctrines.

604 Bibliography

C. L. Rossiter, *Marxism: The View from America* (New York, 1960). Contrasts Marx's ideas—more or less equated with those of his disciples—with the American way of life.

N. Rotenstreich, *Basic Problems of Marx's Philosophy* (New York, 1965). A philosophical commentary on Marx's Theses on Feuerbach.

M. Rubel and M. Manale, *Marx without Myth* (Oxford, 1976). Precise chronological account of Marx's life and work.

O. Rühle, *Karl Marx. His Life and Work* (New York, 1929). A full biography with emphasis in the psychological.

D. Ryazanov, *Karl Marx, Man, Thinker and Revolutionist* (New York, 1927). A well-informed series of lectures on Marx's life.

J. Sanderson, *An Interpretation of the Political Ideas of Marx and Engels* (London, 1969). A short book which seeks to put together the main texts of Marx and Engels on historical materialism, the state, revolution and future communist society.

R. Schlesinger, *Marx, His Time and Ours* (London, 1950). An important book investigating the continued relevance of Marx's ideas for the twentieth century.

A. Schmidt, *The Concept of Nature in Marx* (London, 1971). An important and well-documented consideration of the importance of Marx's materialism.

L. Schwartzchild, *Karl Marx: The Red Prussian* (New York, 1948). A strongly critical biography.

P. Sloan, *Marx and the Orthodox Economists* (Oxford, 1973). A defence of Marx against subsequent economic thinking.

J. Spargo, *Karl Marx, His Life and Works* (New York, 1910). The first biography of Marx in English.

C. J. S. Sprigge, *Karl Marx* (London, 1938; New York, 1962). A short biography.

I. Steedman, *Marx after Sraffa* (London, 1977). A reconstruction of Marx's theories in the light of neo-Ricardianism.

Elena A. Stepanova, *Karl Marx* (Moscow, 1962). A short piece of pure hagiography.

P. M. Sweezey, *The Theory of Capitalist Development* (New York, 1942). The best modern continuation of Marx's economic ideas.

R. Tucker, *Philosophy and Myth in Karl Marx* (Cambridge, 1961). A highly original—though in places also highly dubious—interpretation of Marx's thought as a continuity based on certain eschatological assumptions.

R. Tucker, *The Marxian Revolutionary Idea* (London, 1970). A series of essays dealing with the state and revolution in Marx.

V. Venables, *Human Nature, the Marxian View* (New York, 1945). One of the best statements of the Marxist view of man.

P. Walton and S. Hall, eds., *Situating Marx* (London, 1972). A series of essays centring on Marx's *Grundrisse*.

R. Williams, *Marxism and Literature* (Oxford, 1977). A good introduction.

E. Wilson, *To the Finland Station* (London, 1940; latest ed. 1970). A very readable (though occasionally inaccurate) account of the ideas of Marx as well as those of his predecessors and successors.

B. Wolfe, *Marxism: 100 years in the Life of a Doctrine* (London, 1967). A study of the evolution of Marxist doctrines with sections on Marx's political ideas in 1848 and 1871.

M. Wolfson, *Karl Marx* (New York, 1971). A short critique of Marx's main economic doctrines.

D. Wright, *The Trouble with Marx* (New Rochelle, 1967). A 'no holds barred' attack on Marx's ideas of history and economics.

C. Wright Mills, *The Marxists* (New York, 1962). Contains an acute account of Marx's sociological ideas.

I. Zeitlin, *Marxism: A Re-examination* (Princeton, 1967). A short and interesting book presenting in a favourable light the sociological elements in Marx's thought.

Indexes

Index of Names

Index of Subjects